CZECHOSLOVAK ACADEMY OF SCIENCES

General Topology and its Relations to Modern Analysis and Algebra III

Proceedings of the Third Prague Topological Symposium, 1971

PREVIOUS CONFERENCE PROCEEDINGS EDITED BY NOVÁK

I GENERAL TOPOLOGY AND ITS RELATIONS TO MODERN ANALYSIS AND ALGEBRA
Proceedings of the Symposium held in Prague in September, 1961
Published in 1962

II GENERAL TOPOLOGY AND ITS RELATIONS TO MODERN ANALYSIS AND ALGEBRA II
Proceedings of the Second Prague Topological Symposium, 1966
Published in 1967

III GENERAL TOPOLOGY AND ITS RELATIONS TO MODERN ANALYSIS AND ALGEBRA
Proceedings of the Kanpur Topological Conference, 1968
Published in 1971

IV GENERAL TOPOLOGY AND ITS RELATIONS TO MODERN ANALYSIS AND ALGEBRA III
Proceedings of the Third Prague Topological Symposium, 1971
Published in 1973

General Topology and its Relations to Modern Analysis and Algebra III

Proceedings of the
Third Prague Topological Symposium, 1971

ACADEMIA
Publishing House of the
Czechoslovak Academy
of Sciences
Prague

ACADEMIC PRESS
New York
and London

1972

Published in the U.S.A., Canada, Mexico, Central America, South America, Australia and New Zealand by
Academic Press, Inc., 111 Fifth Avenue, New York, New York 10003

Library of Congress Catalog Card No. 72-76014
ISBN 0-12-522566-0

Printed in Czechoslovakia

PREFACE

In August 1971, the Czechoslovak Academy of Sciences organized in Prague the Third Symposium on General Topology and its Relations to Modern Analysis and Algebra. This Symposium, the third in the series of the Prague Topological Symposia, formed a continuation of the previous two held in 1961 and 1966. The organizational work was done by an Organizing Committee consisting of J. Novák (chairman), Z. Frolík, J. Hejcman, M. Hušek, M. Katětov, V. Koutník, V. Pták, M. Sekanina, Š. Schwarz. The Czechoslovak Academy of Sciences, the Slovak Academy of Sciences, the Charles University and the Association of Czechoslovak Mathematicians and Physicists invited a number of prominent specialists from abroad. The International Mathematical Union granted financial support towards the travel expenses of some of the invited speakers. The Organizing Committee has the pleasant duty to express its sincere thanks to all these institutions who greatly contributed to the success of the Symposium.

The Third Topological Symposium was opened on August 30, 1971 by the Chairman of the Organizing Committee. The scientific program of the Symposium included 22 invited addresses and 88 scientific communications. The invited addresses had 30 and 40 minutes and the communications 20 minutes in length. The communications were mostly delivered in two parallel sections, with the exception of one day when three sections were held.

The Third Topological Symposium was attended by 158 mathematicians among whom 107 were from abroad and 51 from Czechoslovakia. A special program was arranged for the associate members (23 persons).

During the Symposium, a reception was given at the Ministry of Education for a number of foreign and Czechoslovak participants.

The Symposium was closed by a speech of K. Kuratowski at a plenary session on September 3. In the evening, a banquet was given for all participants by the Chairman of the Czechoslovak Academy of Sciences, J. Kožešník. The following day was devoted to a coach trip to the North-East of Bohemia.

Most of the communications presented at the Symposium as well as a few submitted in writing only will appear in this volume. The Organizing Committee heartily thanks all authors for their contributions. The editor wishes to express appreciation to Z. Frolík, M. Katětov, V. Koutník and V. Pták for editorial assistance.

J. Novák

LIST OF PARTICIPANTS

J. M. Aarts (Delft), O. T. Alas (São Paulo), J. Albrycht (Poznań), *R. A. Alò (Pittsburgh), R. D. Anderson (Baton Rouge), M. Ja. Antonovskii (Moskva), A. V. Arhangelskii (Moskva), S. P. Arya (New Delhi), C. E. Aull (Blacksburg), P. C. Baayen (Amsterdam), B. Banaschewski (Hamilton), W. Barit (Amsterdam), W. Bauer (Salzburg), A. Bellen (Trieste), N. Bezikašvili (Tbilisi), E. Binz (Mannheim), W. J. Blok (Amsterdam), J. Bočková (Praha), C. R. Borges (Davis), K. Borsuk (Warszawa), H. G. Bothe (Berlin), H. Breger (Heidelberg), J. Bruyning (Amsterdam), D. Bushaw (Pullman), J. J. Charatonik (Wrocław), J. Chvalina (Brno), S. Ciampa (Pisa), P. Čihák (Praha), P. Collins (Oxford), W. W. Comfort (Middletown), A. Császár (Budapest), K. Császár (Budapest), J. van Dalen (Amsterdam), J. Daneš (Praha), I. Dobrakov (Bratislava), M. Duchoň (Bratislava), R. Duda (Wrocław), R. Dyckhoff (Oxford), R. Engelking (Warszawa), M. Fiedler (Praha), J. Flachsmeyer (Greifswald), P. Fletcher (Blacksburg), R. Frič (Praha), Z. Frolík (Praha), S. Gähler (Berlin), K. M. Garg (Edmonton), J. Gerlits (Budapest), G. R. Gordh, Jr. (Lexington), L. Górniewicz (Gdańsk), G. Grimeisen (Stuttgart), J. de Groot (Amsterdam), S. Guazzone (Perugia), P. Gvozdková (Praha), A. W. Hager (Middletown), P. Hamburger (Budapest), J. Hanák (Brno), K. Hardy (Ottawa), Z. Hedrlín (Praha), J. Hejcman (Praha), M. Henriksen (Claremont), H. Herrlich (Bielefeld), E. Hewitt (Seattle), M. Hušek (Praha), R. Isler (Trieste), A. A. Ivanov (Leningrad), I. Ivanšić (Zagreb), J. E. Jayne (New York), K John (Praha), F. B. Jones (Riverside), I. Juhász (Budapest), M. Katětov (Praha), A. Kirkor (Warszawa), R. Kok (Amsterdam), J. Kolomý (Praha), V. Kořínek (Praha), P. Kostyrko (Bratislava), V. Koutník (Praha), P. Kratochvíl (Praha), P. B. Krikelis (Athens), E. H. Kronheimer (London), L. Kučera (Praha), K. Kuratowski (Warszawa), O. V. Lokutsievskii (Moskva), J. Lukeš (Praha), K. D. Magill, Jr. (Buffalo), E. Makai, Jr. (Budapest), A. A. Maltsev (Moskva), G. de Marco (Padova), S. Mardešić (Zagreb), P. R. Meyer (New York), E. Michael (Seattle), J. Mioduszewski (Katowice), L. Mišík (Bratislava), J. Mlček (Praha), B. Mrozowski (Wrocław), H. Musielak (Poznań), J. Musielak (Poznań), J. Nagata (Pittsburgh), S. A. Naimpally (Thunder Bay), S. Negrepontis (Montréal), L. D. Nel (Ottawa), T. Neubrunn (Bratislava), J. Novák (Praha), P. Novotný (Brno), P. Nyikos (Pittsburgh), J. Pachl (Praha),

A. R. Pears (London), J. Pechanec (Praha), A. Pietsch (Jena), N. Popa (Bucureşti), H. Poppe (Greifswald), D. Preiss (Praha), G. Preuss (Berlin), P. Pták (Praha), V. Pták (Praha), A. Pultr (Praha), H. C. Reichel (Wien), B. Riečan (Bratislava), T. Rishel (Halifax), J. Rosický (Brno), T. Šalát (Bratislava), H. M. Schaerf (Montreal), H. H. Schirmer (Ottawa), Š. Schwarz (Bratislava), J. Segal (Seattle), M. Sekanina (Brno), *L. I. Sennott (Fairfax), P. Simon (Praha), L. Skula (Brno), J. Smítal (Bratislava), J. C. Smith (Blacksburg), S. Spathis (Paris), A. K. Steiner (Edmonton), E. F. Steiner (Edmonton), G. E. Strecker (Pittsburgh), S. Švarc (Brno), S. Swaminathan (Halifax), F. D. Tall (Toronto), B. Taylor (Montreal), J. C. Taylor (Montreal), R. Telgársky (Bratislava),G. Tironi (Trieste), V. Trnková (Praha), L. Vašák (Praha), M. Vencelj (Ljubljana), A. Verbeek (Amsterdam), M. Vlach (Praha), A. Volčič (Trieste), P. Vopěnka (Praha), *R. Voreadou (Chicago), E. Wattel (Amsterdam), J. E. West (Ithaca), K. Wichterle (Praha), R. Y. T. Wong (Santa Barbara), A. V. Zarelua (Tbilisi), D. Zaremba (Wrocław), P. Zenor (Auburn), S. P. Zervos (Athens), A. B. Zižčenko (Moskva), V. Zizler (Praha), I. Zuzčák (Bratislava).

*) did not attend.

LIST OF INVITED ADDRESSES

Anderson, R. D.:	Some Open Questions in Infinite-Dimensional Topology.
Антоновский, М. Я.:	Несимметрические близости, равномерности и разрывные метрики.
Arhangelskii, A. V.:	On Cardinal Invariants.
Arya, S. P.:	Sum Theorems for Topological Spaces.
Banaschewski, B.:	On Profinite Universal Algebras.
Borsuk, K.:	Some Remarks Concerning the Theory of Shape in Arbitrary Metrizable Spaces.
Frolík, Z.:	Topological Methods in Measure Theory and the Theory of Measurable Spaces.
De Groot, J.:	On the Topological Characterization of Manifolds.
Herrlich, H.:	A Generalization of Perfect Maps.
Hewitt, E.:	Harmonic Analysis and Topology.
Jones, F. B.:	The Utility of Empty Inverse Limits.
Katětov, M.:	On Descriptive Classification of Functions.
Kuratowski, K.:	A General Approach to the Theory of Set-Valued Mappings.
Mardešić, S.:	A Survey of the Shape Theory of Compacta.
Michael, E.:	On Two Theorems of V. V. Filippov.
Nagata, J.:	A Survey of the Theory of Generalized Metric Spaces.
Pietsch, A.:	Ideals of Operators on Banach Spaces and Nuclear Locally Convex Spaces.
Pták, V.	Banach Algebras with Involution.
Steiner, A. K.:	On the Lattice of Topologies.
Taylor, J. C.:	The Martin Compactification in Axiomatic Potential Theory.
West, J. E.:	Identifying Hilbert Cubes: General Methods and their Application to Hyperspaces by Schori and West.
Zarelua, A. V.:	On Infinite-Dimensional Spaces.

LIST OF COMMUNICATIONS

Aarts, J. M.	Complementary Inductive Invariants and Dimension.
Alas, O. T.	Uniform Continuity in Paracompact Spaces.
Alò, R. A.	Some Tietze Type Extension Theorems (*Presented by P. Nyikos*).
Aull, C. E.	Point-Countable Bases and Quasi-Developments.
Barit, W.	Contraction of some Spaces of Homeomorphisms.
Bellen, A.: Bellen, A. and Volčič, A.	Non-Cyclic Transformations and Uniform Convergence of Picard Sequences (*Presented by A. Volčič*).
Binz, E.	Recent Results in the Functional Analytic Investigations of Convergence Spaces.
Borges, C. R.	Four Generalizations of Stratifiable Spaces.
Bothe, H. G.	About 1-Dimensional Compacta in E^3.
Charatonik, J. J.	On the Fixed Point Property for Set-Valued Mappings of Hereditarily Decomposable Continua.
Chvalina, J.: Chvalina, J. and Sekanina, M.	Realizations of Closure Spaces by Set Systems (*Presented by M. Sekanina*).
Ciampa, S.	On Refinement of a Metrization Theorem.
Comfort, W. W.: Comfort, W. W. and Negrepontis, S.	Continuous Functions on Products with Strong Topologies (*Presented by W. W. Comfort*).
Császár, Á.	Hyperextensions of Topological Spaces.
Császár, K.	*H*-Closed Extensions of Topological Spaces.
van Dalen, J.	The Large Inductive Dimension of a Product of *m* Connected Linearly Ordered Spaces.
Daneš, J.	On Norms and Subsets of Linear Spaces.
Duda, R.	One Result on Inverse Limits and Hyperspaces.
Efimov, B. A.	On the Imbedding of Extremally Disconnected Spaces into Bicompacta (*Presented in writing only*).
Flachsmeyer, J.	Normal and Category Measures on Topological Spaces.
Fletcher, P.: Fletcher, P. and Lindgren, W. F.	Topological Spaces which Admit a Compatible Complete Quasi-Uniformity (*Presented by P. Fletcher*).
Frič, R.	Sequential Envelope and Subspaces of the Čech-Stone Compactification.
Gagrat, M.: Gagrat, M. and Naimpally, S. A.	Proximity Approach to Topological Problems (*Presented by S. A. Naimpally*).

GÄHLER, S.	Über 2-Banach-Räume.
GARG, K. M.: GARG, K. M. and NAIMPALLY, S. A.	On some Pretopologies Associated with a Topology (*Presented by K. M. Garg*).
GERLITS, J.	On 𝔪-adic Spaces.
GORDH, G. R., JR.	On Monotone Decompositions of Smooth Continua.
GRIMEISEN, G.	On the Saturation of a Topological Partial Algebra with Respect to a Congruence Relation.
HAGER, A. W.	Three Classes of Uniform Spaces.
HAJEK, D. W.: HAJEK, D. W. and STRECKER, G. E.	Direct Limits of Hausdorff Spaces (*Presented by G. E. Strecker*).
HAMBURGER, P.	On Internal Characterizations of Complete Regularity and Wallman-Type Compactifications.
HANÁK, J.	Game-Theoretical Approach to some Modifications of Generalized Topologies.
HEDRLÍN, Z.	A Universal Topological Structure.
HEJCMAN, J.	Remarks on Dimensions of Mappings.
HENRIKSEN, M.	On Difficulties in Embedding Lattice-Ordered Integral Domains in Lattice-Ordered Fields.
HURSCH, J. L.: HURSCH, J. L. and VERBEEK, A.	A Class of Connected Spaces with many Ramifications (*Presented by A. Verbeek*).
HUŠEK, M.	Simple Categories of Topological Spaces.
ISLER, R.: TIRONI, G. and ISLER, R.	On some Problems of Local Approximability in Compact Spaces (*Presented by R. Isler*).
IVANOV, A. A.: IVANOVA, V. M. and IVANOV, A. A.	Continuous Mappings of Extensions of a Topological Space (*Presented by A. A. Ivanov*).
IVANOVA, V. M.: IVANOVA, V. M. and IVANOV, A. A.	Continuous Mappings of Extensions of a Topological Space (*Presented by A. A. Ivanov*).
IVANŠIĆ, I.	Disconnected Bounded *PL* Manifolds in Euclidean Spaces.
JAYNE, J. E.	Topological Representations of Measurable Spaces.
JUHÁSZ, I.	Cardinal Functions on Products.
KANNAN, V.: KANNAN, V. and RAJAGOPALAN, M.	On Rigidity and Groups of Homeomorphisms (*Presented in writing only*).
KIRKOR, A.	On Mild and Wicked Embeddings.
KOK, R.	A Connected T_1-Space with the Connected Intersection Property Is a Tree-Like Space.
KOLOMÝ, J.	Some Mapping and Fixed Point Theorems.
KOUTNÍK, V.	On some Convergence Closures Generated by Functions.
KRATOCHVÍL, P.	On a Convergence Property of Set Algebras.
KRIKELIS, P. B.	On Condensation Numbers.
KRONHEIMER, E. H.	Very Unlatticelike Ordered Spaces.
KUČERA, L.: KUČERA, L. and PULTR, A.	On a Mechanism of Choosing Morphisms in Concrete Categories (*Presented by L. Kučera*).

Kučera, L.: Kučera, L. and Pultr, A.	The Category of Compact Hausdorff Spaces Is not Algebraic if there Are too many Measurable Cardinals (*Presented by A. Pultr*).
Kurepa, D.	Factorials and the General Continuum Hypothesis (*Presented in writing only*).
Lindgren, W. F.: Fletcher, P. and Lindgren, W. F.	Topological Spaces which Admit a Compatible Complete Quasi-Uniformity (*Presented by P. Fletcher*).
Lokutsievskii, O. V.	A Note on Fixed Point Property.
Magill, K. D., Jr	Semigroups and Near-Rings of Continuous Functions.
Makai, E., Jr.	The Space of Bounded Maps into a Banach Space.
Мальцев, А. А.	Итеративные алгебры на топологическом пространстве.
Meyer, P. R.	On Total Orderings in Topology.
Mioduszewski, J.	On a Method which Leads to Extremally Disconnected Covers.
Mitiagin, B.	Homotopical Structure of Linear Groups of Banach Spaces (*Presented by M. Ja. Antonovskii*).
Mlček, J.	L'espace $\beta(X)$ (*Presented by P. Vopěnka*).
Musielak, J.: Musielak, J. and Waszak, A.	A Contribution to the Theory of Modular Spaces (*Presented by J. Musielak*).
Naimpally, S. A.: Gagrat, M. and Naimpally, S. A.	Proximity Approach to Topological Problems (*Presented by S. A. Naimpally*).
Naimpally, S. A.: Garg, K. M. and Naimpally, S. A.	On some Pretopologies Associated with a Topology (*Presented by K. M. Garg*).
Negrepontis, S.	Ramification Systems and Spaces of Ultrafilters.
Negrepontis, S.: Comfort, W. W. and Negrepontis, S.	Continuous Functions on Products with Strong Topologies (*Presented by W. W. Comfort*).
Novák, J.	On Completions of Convergence Commutative Groups.
Nyikos, P.	Strongly Zero-Dimensional Spaces.
Popa, N.	Un théorème du graphe Mackey-fermé.
Poppe, H.	A Compactness Criterion for Hausdorff Admissible (Jointly Continuous) Convergence Structures in Function Spaces.
Preiss, D.	Metric Spaces in which Prohorov's Theorem Is not Valid.
Preuss, G.	A Categorical Generalization of Completely Hausdorff Spaces.
Pultr, A.: Kučera, L. and Pultr, A.	On a Mechanism of Choosing Morphisms in Concrete Categories (*Presented by L. Kučera*).
Pultr, A.: Kučera, L. and Pultr, A.	The Category of Compact Hausdorff Spaces Is not Algebraic if there Are too many Measurable Cardinals (*Presented by A. Pultr*).

RAJAGOPALAN, M.: KANNAN, V. and RAJAGOPALAN, M.	On Rigidity and Groups of Homeomorphisms (*Presented in writing only*).
REITERMAN, J.: TRNKOVÁ, V. and REITERMAN, J.	When Categories of Presheaves Are Binding (*Presented by V. Trnková*).
RIEČAN, B.	On Topological Entropy.
RISHEL, T.	Nice Spaces; Nice Maps.
RUDIN, M. E.	Box Products (*Presented in writing only*).
SCHAERF, H. M.	Cardinalities of Bases.
SEGAL, J.	On the Shape Classification of Manifold-Like Continua.
SEKANINA, M.: CHVALINA, J. and SEKANINA, M.	Realizations of Closure Spaces by Set Systems (*Presented by M. Sekanina*).
SENNOTT, L. I.	Extending Point-Finite Covers (*Presented in writing only*).
SIMON, P.	A Note on Rudin's Example of Dowker Space.
SKULA, L.	Die Fortsetzung stetiger Homomorphismen von δ-Halbgruppen.
SMITH, J. C.	Properties of Expandable Spaces.
STRECKER, G. E.: HAJEK, D. W. and STRECKER, G. E.	Direct Limits of Hausdorff Spaces (*Presented by G. E. Strecker*).
SWAMINATHAN, S.	On a Closed Range Theorem for Nonlinear Operators.
TALL, F. D.	Some Set-Theoretic Consistency Results in Topology.
TAYLOR, B.	A Criterion for the Metrizability of a Compact Convex Set in Terms of the Set of Extreme Points.
TELGÁRSKY, R.	Covering Properties and Product Spaces.
Тихомиров, В. М.: Тихомиров, В. М. и Тумаркин, Л. А.:	О поперечниках Урысона n-мерной эвклидовой сферы (*Presented by O. V. Lokutsievskii*).
TIRONI, G.: TIRONI, G. and ISLER, R.	On some Problems of Local Approximability in Compact Spaces (*Presented by R. Isler*).
TRNKOVÁ, V.: TRNKOVÁ, V. and REITERMAN, J.	When Categories of Presheaves Are Binding (*Presented by V. Trnková*).
Тумаркин, Л. А.: Тихомиров, В. М. и Тумаркин, Л. А.:	О поперечниках Урысона n-мерной эвклидовой сферы (*Presented by O. V. Lokutsievskii*).
VERBEEK, A.: HURSCH, J. L. and VERBEEK, A.	A Class of Connected Spaces with many Ramifications (*Presented by A. Verbeek*).
VLACH, M.	Separation Properties of Families of Convex Sets in Linear Topological Spaces.
VOLČIČ, A.: BELLEN, A. and VOLČIČ, A.	Non-Cyclic Transformations and Uniform Convergence of Picard Sequences (*Presented by A. Volčič*).
VOPĚNKA, P.	Théorie des demiensembles.

WASZAK, A.: MUSIELAK, J. and WASZAK, A.	A Contribution to the Theory of Modular Spaces (*Presented by J. Musielak*).
WATTEL, E.	A General Fixed Point Theorem.
WICHTERLE, K.	Relations between $\mathfrak{B}$-Completeness and $\mathfrak{m}$-Paracompactness.
WONG, R. Y. T.	On Homeomorphisms of ∞-Dimensional Bundles.
ZAREMBA, D.	On Pseudo-Open Mappings.
ZENOR, P.	Spaces with Regular G_δ-Diagonals.
ZERVOS, S. P.	Lattices and Topology.
ZIZLER, V.	On Extremal Structure of Weakly Locally Compact Convex Sets in Banach Spaces.

Remark. Russian names are, as a rule, transliterated according to World Directory of Mathematicians 1970.

CONTRIBUTED PAPERS

COMPLEMENTARY INDUCTIVE INVARIANTS AND DIMENSION

J. M. AARTS

Delft

All spaces under discussion are assumed to be metrizable. Let $\mathscr{P}$ be a non-empty class of spaces which is closed for topological mappings. Then the following topological invariants can be defined.

(1) The strong (weak) inductive invariant $\mathscr{P}$-Ind X ($\mathscr{P}$-ind X) induced by the class $\mathscr{P}$ is inductively defined in a similar way as Ind X (ind X), but starting with the definition that $\mathscr{P}$-Ind X $(=\mathscr{P}\text{-ind } X) = -1$ iff $X \in \mathscr{P}$.

Of course, inductive dimension ($\mathscr{P} = \{\emptyset\}$) is the best explored inductive invariant. The concept of an inductive invariant has been introduced by Lelek [5].

(2) The deficiency of X with respect to $\mathscr{P}$ is defined as follows: $\mathscr{P}$-def $X \leqq n$ if there exists $Y \in \mathscr{P}$ such that $X \subset Y$ and $\dim Y \setminus X \leqq n$.

The case that $\mathscr{P}$ is the class of all compact spaces was first discussed by de Groot [2]. To these invariants we add

(3) The surplus of X with respect to $\mathscr{P}$ is defined by $\mathscr{P}$-sur $X \leqq n$ if there exists $Y \in \mathscr{P}$ such that $Y \subset X$ and $\dim X \setminus Y \leqq n$.

$\mathscr{P}$-def $X = n$, $\mathscr{P}$-def $X = \infty$ etc. are defined as usual. E.g. $\{\emptyset\}$-def $X = \infty$, whenever $X \neq \emptyset$.

It can be shown quite easily that $\mathscr{P}$-Ind $X \leqq \mathscr{P}$-sur X for every space X, if the class $\mathscr{P}$ is closed monotone (i.e. $Z \in \mathscr{P}$, whenever $Y \in \mathscr{P}$ and Z is a closed subset of Y). Furthermore, $\mathscr{P}$-Ind $X \leqq \mathscr{P}$-def X for every space X, if the class $\mathscr{P}$ is closed monotone and open monotone.

By $\mathscr{M}(\alpha)$ and $\mathscr{A}(\alpha)$ we denote the class of all sets of absolute multiplicative and additive Borel class α respectively. (See [4] for definitions. Recall that $\mathscr{A}(0) = \{\emptyset\}$, $\mathscr{M}(0)$ is the class of compact spaces, $\mathscr{A}(1)$ is the class of σ-locally compact spaces [7] and $\mathscr{M}(1)$ is the class of topologically complete spaces.)

Theorem 1. *Let* $\mathscr{P} = \mathscr{A}(\alpha)$ *or* $\mathscr{P} = \mathscr{M}(\alpha)$ *for* $\alpha \geqq 2$. *Then* $\mathscr{P}$-Ind $X \leqq n$ *if and only if there exist* $Y, Z \in \mathscr{P}$ *satisfying* $Y \subset X \subset Z$ *and* $\dim Z \setminus Y \leqq n$. *In particular* $\mathscr{P}$-Ind $X = \mathscr{P}$-def $X = \mathscr{P}$-sur X *for every space* X.

Theorem 2. (See [1].) $\mathscr{M}(1)$-Ind $X = \mathscr{M}(1)$-def X *for every space* X.

Theorem 3. $\mathscr{A}(1)$-Ind $X = \mathscr{A}(1)$-sur X *for every space* X.

Problems. Are the equalities $\mathscr{M}(1)$-Ind $X = \mathscr{M}(1)$-sur X and $\mathscr{A}(1)$-Ind $X = \mathscr{A}(1)$-def X valid for every space X ? To prove the second equality for separable spaces is a problem[1]) posed by Nagata [6]. As follows from the corollary below these equalities are closely related. It is a long unsolved problem whether or not $\mathscr{M}(0)$-ind $X = \mathscr{M}(0)$-def X for every separable space X ([2], [3]).

Definition. Let $\mathscr{P}$, $\mathscr{Q}$ and $\mathscr{R}$ be topologically closed classes of spaces. $\mathscr{P}$ and $\mathscr{Q}$ are *complementary with respect to* $\mathscr{R}$ if for every $Z \in \mathscr{R}$ and for all X and Y with $X \cup Y = Z$ and $X \cap Y = \emptyset$ the equality $\mathscr{P}$-Ind $X = \mathscr{Q}$-Ind Y holds.

Theorem 4. $\mathscr{A}(1)$ *and* $\mathscr{M}(1)$ *are complementary with respect to* $\mathscr{M}(0)$. $\mathscr{A}(\alpha)$ *and* $\mathscr{M}(\alpha)$ *are complementary with respect to* $\mathscr{M}(1)$ *for* $\alpha \geqq 2$.

Corollary. *If* $\mathscr{M}(1)$-Ind $X = \mathscr{M}(1)$-sur X *for every separable space X, then* $\mathscr{A}(1)$-Ind $X = \mathscr{A}(1)$-def X *for every separable space X.*

Example. It is known [1] that for the product X of the rationals and the n-dimensional cube I^n, we have $\mathscr{M}(1)$-Ind $X = n$.

By Theorem 4 it follows that for the product Y of the irrationals and I^n we have $\mathscr{A}(1)$-Ind $Y = n$.

The proofs of Theorems 1, 3, and 4 will be published in forthcoming papers.

References

[1] *J. M. Aarts:* Completeness degree. A generalization of dimension. Fund. Math. *63* (1968), 27–41.

[2] *J. de Groot:* Topologische studien. Thesis, Groningen, 1942.

[3] *J. de Groot and T. Nishiura:* Inductive compactness as a generalization of semi-compactness. Fund. Math. *58* (1966), 201–218.

[4] *C. Kuratowski:* Topologie I. Warszawa, 1958.

[5] *A. Lelek:* Dimension and mappings of spaces with finite deficiency. Colloq. Math. *12* (1964), 221–227.

[6] *J. Nagata:* Some aspects of extension theory in general topology. Contributions to Extension Theory of Topological Structures (Proc. Sympos., Berlin, 1967). Deutsch. Verlag Wissensch., Berlin, 1969, 157–161.

[7] *A. H. Stone:* Absolute F_σ spaces. Proc. Amer. Math. Soc. *13* (1962), 495–499.

[1]) Added in proof: This problem has been solved in the negative by J. M. Aarts and T. Nishiura.

UNIFORM CONTINUITY IN PARACOMPACT SPACES

O. T. ALAS

São Paulo

Let E_1 and E_2 be two nondiscrete paracompact Hausdorff spaces, $\mathfrak{U}_1$ and $\mathfrak{U}_2$ uniformities in E_1 and E_2, respectively, and $\mathfrak{U}_1 \otimes \mathfrak{U}_2$ the product uniformity in $E_1 \times E_2$. On $[0, 1]$ we consider the usual metric topology (thus there is a unique uniformity $\mathfrak{U}'$ in $[0, 1]$). We shall study the following question: Under what conditions every continuous function of $E_1 \times E_2$ (with the product topology) into $[0, 1]$ is a uniformly continuous map of $(E_1 \times E_2, \mathfrak{U}_1 \otimes \mathfrak{U}_2)$ into $([0, 1], \mathfrak{U}')$?

1. Preliminaries

Let E be a nondiscrete completely regular Hausdorff space.

Definition. The *index* of E is the least cardinal number for which there is a family (with this cardinality) of open subsets of E, whose intersection is not an open set. (Let us denote by m the index of E.)

Let E_1 and E_2 be two nondiscrete completely regular Hausdorff spaces. For $i = 1, 2$ let $\mathfrak{U}_i$ be a uniformity in E_i (i.e. compatible with the topology for E_i); $\mathfrak{U}_1 \otimes \mathfrak{U}_2$ denotes the product uniformity in $E_1 \times E_2$, i.e. the uniformity in $E_1 \times E_2$ for which the set $\{U_1 \otimes U_2 \mid U_1 \in \mathfrak{U}_1, U_2 \in \mathfrak{U}_2\}$, where $U_1 \otimes U_2 = \{((x_1, x_2), (y_1, y_2)) \mid (x_i, y_i) \in U_i,\ i = 1, 2\}$, is a basis.

Proposition. *Suppose $m > \aleph_0$ and every locally finite open covering of E has cardinality less than m. We have:*

1) *if E is a normal space, then every subset of E of cardinality m has an accumulation point (in E) and every open covering of E of cardinality m has a subcovering of cardinality less than m;*

2) *if E is a normal space, then every closed subset of E which is the intersection of at most m open subsets of E has a fundamental system $\mathfrak{B}$ of open neighborhoods, whose cardinality $|\mathfrak{B}|$ is less than or equal to m;*

3) *if m is the pseudoweight at some point $x \in E$, then m is the weight at the point x;*

4) *if E is a topological group and m is the pseudoweight at the neutral element of E, then E is paracompact.*

Proof. By the definition of index, m is a regular cardinal number. Since $m > \aleph_0$ and E is completely regular, every point of E has a fundamental system of neighborhoods, which are open-closed in E. (The closed G_δ-subsets of E are open.) The union of less than m closed subsets of E is closed in E.

Assertion 1) follows easily from the above consideration.

Assertions 2) and 3) are proved by using the same technique. We shall prove 3). Let us denote by μ the first ordinal number of cardinality m and by M the set of all ordinals less than μ. Since m is the pseudoweight at the point $x \in E$, there is a family $(V_i)_{i \in M}$ of neighborhoods of x, whose intersection is $\{x\}$. (We can and will suppose that the V_i are open-closed in E.) Put $W_0 = V_0$ and $W_i = \bigcap_{j<i} V_i$ for each $i \in M - \{0\}$. The family $(W_i)_{i \in M}$ is a fundamental system of neighborhoods of x. Indeed, let U be an open-closed neighborhood of x (x has a fundamental system of neighborhoods of this type). Consider the set $\{W_i - (W_{i'} \cup U) \mid i \in M - \{0\}\}$, where i' is the ordinal successor of i. It is a discrete collection of open-closed subsets of E. So there is $p \in M$ such that $W_i - (W_{i'} \cup U) = \emptyset$ for every $i \in M$, $i > p$. Thus $W_{p'} \subset U$. (If it were $t \in W_{p'}$ and $t \notin U$ there would be a minimal $k \in M$ such that $t \notin V_k$. By the construction of $W_{p'}$, k is greater than p; and t belongs to W_k and does not belong to $W_{k'}$; but W_k is contained in $W_{k'} \cup U$, which is a contradiction.)

Proof of 4). By virtue of assertion 3) the neutral element of E has a fundamental system $\mathfrak{B}$ of neighborhoods, with $|\mathfrak{B}| = m$. Since m is greater than $\aleph_0$ we can choose elements of $\mathfrak{B}$ such that $VV = V^{-1} = V$ for every $V \in \mathfrak{B}$. For each $V \in \mathfrak{B}$, $\{Vx \mid x \in E\}$ is a discrete open covering of E. The paracompactness of E follows from the fact that $\bigcup_{V \in \mathfrak{B}} \{Vx \mid x \in E\}$ is an open basis of the topology on E.

2. Main results

Let E_1 and E_2 be two nondiscrete completely regular Hausdorff spaces and m_1 and m_2 their indices.

Theorem 1. *Let $\mathfrak{U}_1$ and $\mathfrak{U}_2$ be uniformities in E_1 and E_2, respectively. If every continuous function of $E_1 \times E_2$ (with the product topology) into $[0, 1]$ is a uniformly continuous map of $(E_1 \times E_2, \mathfrak{U}_1 \otimes \mathfrak{U}_2)$ into $([0, 1], \mathfrak{U}')$, then every locally finite open covering of E_i has cardinality less than m_j $(i, j = 1, 2)$.*

Proof. We shall prove, for instance, that every locally finite open covering of E_1 has cardinality less than m_2. It is sufficient to prove that every discrete family of nonempty open subsets of E_1 has cardinality less than m_2.

On the contrary, let us suppose that there exists a discrete family of nonempty open subsets of E_1, $(W_t)_{t \in T}$, whose cardinality $|T|$ is equal to m_2. There is a point $d \in E_2$ and a family of open neighborhoods of d, $(V_t)_{t \in T}$, such that $\bigcap_{t \in T} V_t$ is not a neighborhood of d. For each $t \in T$ we fix a point $a_t \in W_t$ and two continuous functions $f_t : E_1 \to [0, 1]$ and $g_t : E_2 \to [0, 1]$ satisfying the conditions:

1) $f_t(a_t) = 1$, $f_t(E_1 - W_t) = \{0\}$;
2) $g_t(d) = 1$, $g_t(E_2 - V_t) = \{0\}$.

The function g defined below is continuous (because the family $(W_t \times V_t)_{t \in T}$ is discrete in $E_1 \times E_2$):

$$g : E_1 \times E_2 \to [0, 1]$$
$$(x, y) \to 0 \quad \text{if } (x, y) \text{ does not belong to } \bigcup_{t \in T} W_t \times V_t$$
$$(x, y) \mapsto f_t(x)\, g_t(y) \quad \text{if} \quad (x, y) \in W_t \times V_t, \quad t \in T.$$

By the hypothesis, there are $U_1 \in \mathfrak{U}_1$ and $U_2 \in \mathfrak{U}_2$, such that $(x, y) \in U_1$ and $(u, v) \in U_2$ imply $|g(x, u) - g(y, v)| < \frac{1}{4}$. But this is not possible, because then $U_2[d]$ would be contained in $\bigcap_{t \in T} V_t$. ($(a_t, a_t) \in U_1$ and $(u, d) \in U_2$ imply $g(a_t, u) \geqq \frac{3}{4}$, so $u \in V_t$.) The proof is completed; notice that $m_1 = m_2$.

Remark 1. Suppose $m_1 = m_2 = p$. If E_1 and E_2 are paracompact and every locally finite open covering of E_i $(i = 1, 2)$ has cardinality less than p, then $E_1 \times E_2$ is paracompact and, further, every locally finite open covering of $E_1 \times E_2$ has cardinality less than p. So if $\mathfrak{U}_1$ and $\mathfrak{U}_2$ are the universal uniformities in E_1 and E_2, $\mathfrak{U}_1 \otimes \mathfrak{U}_2$ is the universal uniformity in $E_1 \times E_2$.

The next theorem follows easily.

Theorem 2. *Let E be a nondiscrete paracompact space and m the index of E. The following conditions are equivalent:*

1) *there is a uniformity $\mathfrak{U}$ in E such that every continuous function of $E \times E$ (with the product topology) into $[0, 1]$ is a uniformly continuous map of $(E \times E, \mathfrak{U} \otimes \mathfrak{U})$ into $([0, 1], \mathfrak{U}')$;*

2) *every locally finite open covering of E has cardinality less than m;*

3) *there is a uniformity $\mathfrak{U}$ in E such that $\mathfrak{U} \otimes \ldots \otimes \mathfrak{U}$ (n times) is the universal uniformity in the product topological space E^n, $n = 2, 3, \ldots$;*

4) *there is a uniformity $\mathfrak{U}$ in E such that $\mathfrak{U} \otimes \mathfrak{U}$ is the universal uniformity in the product topological space $E \times E$.*

Hint. If E satisfies the condition 2), then E^n is a paracompact space for each $n = 2, 3, \ldots$ On the other hand, it is well-known that if X is a paracompact space, the set $\{ \bigcup_{Y \in \alpha} Y \times Y \mid \alpha \text{ is a locally finite open covering of } X \}$ is a basis of the universal uniformity in X.

Remark. The implication 2) $\Rightarrow$ 4) is a particular case of Theorem 35 ([7], p. 137).

For topological groups we have the following theorem ([2]):

Theorem 3. *Suppose E is a paracompact topological group. E satisfies the condition* 2 *of Theorem* 2 *if and only if the right uniformity in E is the universal uniformity in E.*

For other results in the same area see, for instance, [3], [5], [7] and [8]. Professor L. Nachbin also investigated a similar question for metric spaces.

References

[1] *O. T. Alas:* Paracompact topological groups and uniform continuity. (To appear.)

[2] *O. T. Alas:* Normal topological groups and universal uniformities. An. Acad. Brasil. Ci. *42* (1970), 411–413.

[3] *M. Atsuji:* Uniform continuity of continuous functions on uniform spaces. Canad. J. Math. *13* (1961), 654–663.

[4] *N. Bourbaki:* Éléments de Mathématique. Topologie Générale. Livre 3, chapitres 1 et 2. Hermann, Paris, 1965.

[5] *W. W. Comfort and K. A. Ross:* Pseudocompactness and uniform continuity in topological groups. Pacific J. Math. *16* (1966), 483–496.

[6] *R. Engelking:* Outline of General Topology. North-Holland Publishing Co, Amsterdam, 1968.

[7] *J. R. Isbell:* Uniform spaces. Math. Surveys *12*. Amer. Math. Soc., 1964.

[8] *N. Onuchic:* On the Nachbin uniform structure. Proc. Amer. Math. Soc. *11* (1960), 177–179.

SOME TIETZE TYPE EXTENSION THEOREMS

R. A. ALÒ

Pittsburgh

In the historical development of the separation axioms in set-theoretical topology, the normal topological spaces received appropriate recognition for their capability to determine, set-theoretically, the existence of non-trivial continuous real valued functions. In fact they are precisely the class of topological spaces in which every closed subset is C-embedded (or C^*-embedded). This is recorded for us in the famous Tietze-Urysohn Extension Theorem.

Since the study of continuous real valued functions was at the core of the early developments of topology, this class of spaces satisfied, then, the desires of many. On the other hand, however, normal spaces do not yield to many of the topological operations we often like to perform on classes of spaces. For example, the category of normal topological spaces is neither closed under finite products nor hereditary for arbitrary subspaces. The standard examples of the Sorgenfrey plane and the Tychonoff plank are appropriate here.

Thus the question as to when the product set of two normal spaces is again normal in its Tychonoff product topology has been of some interest. In particular, in homotopy theory one often likes to know when the product space, $X \times I$, is normal where I is the closed unit interval of real numbers and where X is some normal topological space. In dimension theory the normality of the product $X \times Y$ often appears in the dimension product theorem for any of the various concepts of dimension that occur in spaces which are not separable metric spaces.

The normality of $X \times I$ was settled by C. H. Dowker in [6]. He showed that for any compact metric space A, $X \times A$ is a normal (respectively, collectionwise normal) Hausdorff space if and only if X is a countably paracompact, normal, (respectively, collectionwise normal) Hausdorff space. Also M. E. Rudin in [11] has shown the existence of a collectionwise normal (and therefore normal) Hausdorff space which is not countably paracompact. H. Tamano in [14] has shown that $X \times A$ is normal Hausdorff for any compact Hausdorff space A if and only if X is a paracompact Hausdorff space. K. Morita in [10] has demonstrated that $X \times A$ is normal for any metric space A if and only if X is a normal P-space. Not so easy to describe are the conditions on X, as recently discovered by Y. Katuta in [9], for the equivalent formulation of $X \times A$ being normal when A is a paracompact Hausdorff space.

In another direction the strengthening of normality to collectionwise normality in [5] by R. H. Bing assisted in resolving the metrization problem. It has also been shown to be useful in extension theory. In fact, a space X is collectionwise normal if and only if every closed subset is P-embedded (that is, every continuous pseudometric on the subset extends to a continuous pseudometric on the entire space) in the space (see [7] and [12]).

The concept of P-embedding is definitely stronger than that of C-embedding which in turn is used to characterize the class of normal spaces. Also every paracompact Hausdorff space is collectionwise normal. Consequently compact subsets of Tychonoff spaces are P-embedded.

In [4], R. Arens began the serious consideration of P-embedding. Some characterizations of it were given, reminiscent of the Tietze-Urysohn extension theorem for normal spaces. In [1], L. I. Sennott and the author showed the following.

Theorem 1. *Let S be a subset of a non-empty topological space X. Then S is P-embedded in X if and only if every continuous function from S into a bounded, closed, convex subset of a Banach space extends to a continuous function on X.*

Thus, as a corollary, collectionwise normal spaces can be characterized in the sense of Tietze-Urysohn as shown in (2) of the following corollary.

Corollary 1.1. *For a non-empty topological space X, the following statements are equivalent:*

(1) *The space X is collectionwise normal.*

(2) *For every closed subset F of X, every continuous function from F into a bounded closed convex subset B of a Banach space can be extended continuously to X (and the range of the extension is contained in B).*

(3) *For every closed subspace F of X, the product space $F \times X$ is C^*-embedded in $F \times \beta X$, where βX is the Čech-Stone extension of X.*

The third statement in this corollary is due to H. Tamano in [15].

In [3], Arens asked whether a continuous function from a closed subset of a normal space into a bounded closed convex subset C of a Banach space could be extended continuously to the whole space with values still in the subset C. But in [5], Bing gave an example of a normal space which is not collectionwise normal. Thus the corollary gives a negative reply to Arens' original query.

Looking at Corollary 1.1, one asks if a similar statement may be made regarding normal spaces. Such was already done in [3]. However in [1], the authors were able to give a stronger version of this by first proving the following formulation of C-embedding.

Theorem 2. *The following statements are equivalent for a non-empty subspace S of a topological space X.*

(1) *The subspace S is C-embedded (respectively C^*-embedded) in X.*

(2) *Given a complete convex metrizable subset M of a locally convex topological vector space L, every continuous function f on S with $f(S)$ contained in M and separable (respectively, totally bounded) has a continuous extension f^* to X with $f^*(X)$ contained in M.*

(3) *Every continuous function f from S to a Fréchet space (that is, a complete, metrizable, locally convex topological vector space) such that $f(S)$ is separable (respectively, totally bounded) extends to a continuous function on X.*

Thus the sharpened version of the result in [3], reads as

Corollary 2.1. *A non-empty topological space X is normal if and only if for every closed subset F of X, and for any continuous function f from F into a closed, convex metrizable subset M of a locally convex topological vector space L such that $f(F)$ is contained in M and is separable, there is a continuous extension f^* of f to X where $f^*(X)$ is contained in M.*

The third statement in Theorem 2 says that

Corollary 2.2. *The Hewitt realcompactification υX of a Tychonoff space X is that unique realcompactification of X for which every continuous function f from X into a Fréchet space, such that $f(X)$ is separable, can be extended to a continuous function on υX.*

In [12], it was shown that if the cardinality of a dense C-embedded subset S of a Tychonoff space X is non-measurable, then S is P-embedded in X. From this 2.2 can be restated for Tychonoff spaces of non-measurable cardinality by dropping the requirement that $f(X)$ be separable.

As Corollary 2.2 characterized the Hewitt realcompactification, the Čech-Stone compactification may be characterized by considering the equivalent formulations for C^*-embedding in Theorem 2.

Corollary 2.3. *The Čech-Stone compactification of a Tychonoff space X is that unique compactification βX of X for which every continuous function f from X to a Fréchet space, such that $f(X)$ is totally bounded, can be extended to a function on βX.*

Let us point out here that in the proof of statement (1) implies that of (2) in Theorem 2, one may adapt the proof to show the following (see [1] for details) which is an improvement of a result in [8].

Theorem 3. *If S is a non empty subspace of a uniform space X and if L is any Fréchet space, then every uniformly continuous function from S into L can be extended to a continuous function on X.*

Let us say that a subset S is *strongly P-embedded* (respectively, *strongly C-embedded*) in the space X if for every σ-locally finite (respectively, every countable) open cover $\mathscr{U}$ of S there is a locally finite cozero set cover $\mathscr{V}$ of X such that $\mathscr{V} \mid S = \{V \cap S : V \in \mathscr{V}\}$ refines $\mathscr{U}$.

In [13], it was shown that a subset S is P-embedded in X if and only if every σ-locally finite cozero set cover $\mathscr{U}$ of S has a locally finite cozero set cover $\mathscr{V}$ of X such that $\mathscr{V} \mid S$ refines $\mathscr{U}$. Consequently strongly P-embedding is definitely stronger than P-embedding.

In [2], we have shown that a strongly P-embedded subset is strongly C-embedded and that a strongly C-embedded subset is C-embedded. Also none of the implications are reversible.

Countably paracompact spaces have been mentioned above regarding the question of normality of the product. Also P-embedding, C-embedding and C^*-embedding have been considered. Let us now put these together with countable paracompactness.

Theorem 4. *For T_1 spaces X, the following statements are equivalent.*

(1) *The space X is collectionwise normal and countably paracompact.*
(2) *Every closed subset F of X is strongly P-embedded in X.*
(3) *The product set $X \times I$ is collectionwise normal.*

Corollary 4.1. *Compact subsets of Tychonoff spaces are strongly P-embedded.*

Statement (2) of Theorem 4 is shown in [2] as is statement (2) of the following.

Theorem 5. *For T_1 spaces X, the following statements are equivalent.*

(1) *The space X is normal and countably paracompact.*
(2) *Every closed subset F is strongly C-embedded in X.*
(3) *The product set $X \times I$ is normal.*

The author is grateful to Prof. Giovanni Boetti, Il Direttore, Instituto di Matematica, Università degli Studi di Siena, for his assistance.

References

[1] *R. A. Alò and L. I. Sennott:* Extending linear space-valued functions. Math. Ann. *191* (1971), 79—86.

[2] *R. A. Alò and H. L. Shapiro:* Countably paracompact, normal, and collectionwise normal spaces. Indag. Math. (to appear).

[3] *R. Arens:* Extensions of functions on fully normal spaces. Pacific J. Math. *2* (1952), 11—22.

[4] *R. Arens:* Extensions of coverings, of pseudometrics, and of linear-space-valued mappings. Canad. J. Math. *5* (1953), 211—215.

[5] *R. H. Bing:* Metrization of topological spaces. Canad. J. Math. *3* (1951), 175—186.
[6] *C. H. Dowker:* On countably paracompact spaces. Canad. J. Math. *3* (1951), 219—224.
[7] *C. H. Dowker:* On a theorem of Hanner. Ark. Mat. *2* (1952), 307—313.
[8] *T. E. Gantner:* Extensions of uniformly continuous pseudometrics. Trans. Amer. Math. Soc. *132* (1968), 147—157.
[9] *Y. Katuta:* On the normality of the product of a normal space with a paracompact space. General Topology and its Applications (to appear).
[10] *K. Morita:* Products of normal spaces with metric spaces. Math. Ann. *154* (1964), 365—382.
[11] *M. E. Rudin:* A normal space X for which $X \times I$ is not normal. Bull. Amer. Math. Soc. *77* (1971), 246.
[12] *H. L. Shapiro:* Extensions of pseudometrics. Canad. J. Math. *18* (1966), 981—998.
[13] *H. L. Shapiro:* More on extending continuous pseudometrics. Canad. J. Math. *27* (1970), 984—993.
[14] *H. Tamano:* On paracompactness. Pacific J. Math. *10* (1960), 1043—1047.
[15] *H. Tamano:* On compactifications. J. Math. Kyoto Univ. *1* (1961/62), 161—193.

UNIVERSITÀ DEGLI STUDI DI SIENA
CARNEGIE-MELLON UNIVERSITY, PITTSBURGH, PENNSYLVANIA

SOME OPEN QUESTIONS IN INFINITE-DIMENSIONAL TOPOLOGY

R. D. ANDERSON

Baton Rouge

Since the Second Prague Symposium five years ago, there have been many worthwhile results in infinite-dimensional point set topology. Open embedding theorems for manifolds modeled on Hilbert space, l_2, or on any of many other linear spaces have been established and such manifolds have been topologically classified by homotopy type. Various useful σ-compact structures for such manifolds as well as for the Hilbert cube, Q, and for Hilbert cube manifolds have been identified and studied. General homeomorphism extension theorems have been proved. And many product and factor theorems have been established and used. As reported in Mardešić's paper in these Proceedings, Chapman has recently established the equivalence of the concept of shape of compacta with the existence of homeomorphisms of certain subsets of the Hilbert cube, thus enabling the homeomorphism theory of Q to be used in settling questions of shape. Brief summaries of known results in various directions are included in [3], in [16] and in the introduction to [5].

Three working sessions identifying and listing open problems in infinite-dimensional topology have been held in Ithaca (January, 1969), in Baton Rouge (December, 1969) and in Oberwolfach (September, 1970). Based on problem lists prepared at these sessions, an extensive list of open problems [5] has been published by (and is available from) the Mathematisch Centrum, Amsterdam.

The purpose of the present paper is to identify and discuss one general and five special areas of open questions which appear both interesting and promising for future research. Of necessity, we omit many intriguing questions, e.g. questions about bundle maps over l_2-manifolds suggested by Wong's results about bundles over polyhedra cited in his paper in these Proceedings [18]. See [5] for a much more extensive list of open problems. First we give a few definitions and some notation.

In this paper, all spaces will be separable metric. Hilbert space, l_2, is the space of all square-summable sequences of reals with the norm topology, i.e., $l_2 = \{(x_i) \mid x_i$ is real and $\sum x_i^2 < \infty\}$ with $d((x_i), (y_i)) = \sqrt{\sum (x_i - y_i)^2}$. The Hilbert cube, Q, is defined as $\prod_{j>0} I_j$ where $I_j = [-1, 1]$ and $s \subset Q$ is defined as $s = \prod_{j>0} I_j^0$ where $I_j^0 = (-1, 1)$. An E-manifold, i.e., manifold modeled on a (homogeneous) space E, is a space admitting a cover by open sets homeomorphic to open subsets of E. For a compactum M, $H(M)$ is the space of homeomorphisms of M onto itself with the metric $d(h_1, h_2) = \sup_{x \in M} d(h_1(x), h_2(x))$. The symbol "$\cong$" denotes "is homeomorphic to". A set K in a space X is a *Z-set* in X if for each non-empty homotopically trivial open set $U \subset X$, $U \setminus K$ is non-empty and homotopically trivial. A σ-Z-set is a countable union of Z-sets.

For $Y_1 \subset X_1$ and $Y_2 \subset X_2$, the pair $(X_1, Y_1) \cong (X_2, Y_2)$ if there exists a homeomorphism h of X_1 onto X_2 such that $h(Y_1) = Y_2$. If X_1 and Y_1 are E-manifolds and Y_1 is a closed subset of X_1, we call (X_1, Y_1) a manifold pair.

A *Q-factor* is a space X such that $X \times Q \cong Q$.

The two dominant "building blocks" in infinite-dimensional topology are the complete but nowhere locally compact, l_2, and the compact Q. Since it is known, [1], that $l_2 \cong s$ we may regard Q as a compactification of l_2. An "almost true" meta-conjecture states that if a separable metric space is not obviously different from l_2 (or Q) then it must be homeomorphic to l_2 (or Q). The general question is to find broader and more useful characterizations of l_2 and Q. Indeed while many useful characterizations of l_2 and Q are known (we list several below), most of the specific questions which follow can be regarded as further attempts to identify (properties of) l_2 and Q or to recognize l_2 or Q as spaces arising in contexts other than the usual ones. The following theorems are among the basic characterization and representation theorems for l_2 and Q.

$Q - 1$. Every compact convex infinite-dimensional subset of l_2 (or of any linear metric space) is homeomorphic to Q, [12] and [13].

$Q - 2$. Any countable infinite product of non-degenerate Q-factors is homeomorphic to Q (and, for example, all compact contractible polyhedra or cell-complexes are Q-factors), [15] and [16].

$Q - 3$. Any contractible compactum which is locally Q is Q, [7].

$l_2 - 1$. Every separable infinite-dimensional Banach (or Fréchet) space is homeomorphic to l_2 (or to s), [1] and [11].

$l_2 - 2$. Any product of l_2 by a Q-factor or by a Q-factor slash any σ-Z-set of it is homeomorphic to l_2.

$l_2 - 3$. Any contractible separable metric space which is locally l_2 is l_2, [10].

$l_2 - 4$. Any complement of a σ-Z-set in l_2 is homeomorphic to l_2 (and, for example, every compactum in l_2 is a Z-set in l_2), [1] and [2].

$l_2 - 5$. For any two compact polyhedra K and L having no isolated points, the space of maps of K into L is homeomorphic to l_2 provided L is contractible, [9].

Problems on Spaces of Homeomorphisms

Perhaps the most far-reaching open question in the area of infinite-dimensional topology is the following:

Question. *For M any compact M-manifold, is $H(M)$ an l_2-manifold?*

An affirmative answer would permit the application or the powerful infinite-dimensional theory to questions about homeomorphisms of finite-dimensional manifolds. Many researchers in the area consider it very probable that the answer is affirmative. The affirmative answer is known (and is easy to prove) for M one-dimensional. It is also known by Geoghegan, [9], that $H(M) \times l_2 \cong H(M)$, a property that gives a partial coordinate structure to $H(M)$ and a property that is known to be shared by all l_2-manifolds. For $n \geqq 2$, one approach to the problem of whether $H(M)$ is an l_2-manifold is to attempt (1) to show that $H(M)$ is an ANR and (2) to show that if $H(M)$ is an ANR, then $H(M)$ is an l_2-manifold. For $n = 2$, Luke and Mason [14] have shown that $H(M)$ is an ANR but their argument is delicate and depends on special properties of the plane. It does not appear to generalize easily.

Other approaches to the problem have been tried but generally have led to formulations which are no more intuitively evident and perhaps no easier than the original problem. For example, one may try to identify a space of maps or embeddings which contains the space of homeomorphisms as a suitable dense subset or is contained in such space as a suitable dense subset.

Problems on Q-manifolds

A second major problem is the topological classification of Q-manifolds, i.e., manifolds modeled on the Hilbert cube.

Question. *Are all Q-manifolds characterized by proper homotopy type?* (Here "proper homotopy type" is defined in the same way as homotopy type but with all maps required to be proper.)

It is not even known whether all compact Q-manifolds are characterized by homotopy type (which for compacta is equivalent to proper homotopy type). However, by omitting compacta of various shapes from an endslice of Q, it is easy to exhibit continuumwise many Q-manifolds which are contractible and are thus of the same homotopy type but for which no two are of the same proper homotopy type.

It is interesting to note that the intuitively easier (and necessarily locally compact) Q-manifolds are not yet classified with respect to homeomorphism type whereas the obviously nowhere locally compact l_2-manifolds are homeomorphically classified by homotopy type. Perhaps the reason for this anomaly is that the local "holes" in l_2 permit many more homeomorphisms to exist and to be exhibited thus permitting many more spaces to be observed to be homeomorphic to each other (see l_2-4 above, for example).

In the proof by Henderson [10] that l_2-manifolds are characterized by homotopy type, a vital step was the open embedding theorem, namely that any l_2-manifold may be embedded as an open subset of l_2 thus yielding a single coordinate structure for the manifold. It is immediate that for a circle C, $Q \times C$ cannot be openly embedded in Q (since the boundary of the image of $Q \times C$ cannot be empty). Therefore, given a Q-manifold, M, we cannot guarantee an open embedding in Q. An affirmative answer to the following question would guarantee a (partial) coordinate structure for M and thus provide a possible substitute for the open embedding theorem. A subset of Q which contains an open set and is the product of closed non-degenerate subintervals of the coordinate intervals of Q is called a *basic closed set* in Q. A collection is star-finite if each element intersects at most finitely many elements of the collection.

Question. *For any Q-manifold M, does there exist an embedding f of M into Q such that $f(M)$ is the union of the elements of a star-finite collection of basic closed sets of Q?*

Perhaps the best two existing theorems about Q-manifolds are due to Chapman, [7], who proved (1) for any Q-manifold, M, $M \times [0, 1)$ can be openly embedded in Q and (2) for any two Q-manifolds, M_1 and M_2, $M_1 \times [0, 1) \cong M_2 \times [0, 1)$ iff $M_1 \times [0, 1)$ and $M_2 \times [0, 1)$ are of the same homotopy type. (Here $[0, 1)$ denotes the half-open interval.) The $[0, 1)$ factor kills the "proper" distinctions made by proper homotopy type for Q-manifolds.

Problems on Manifold Pairs

There are a number of interesting questions on manifold pairs which are open even for l_2-manifolds. If in the l_2-manifold pair (M, K), K is a Z-set in M then K plays the role of an abstract boundary of M since for any $p \in K$, there is an open embedding h of $l_2 \times [0, 1)$ into M such that $p \in h(l_2 \times \{0\})$ and $K \cap h(l_2 \times [0, 1)) = h(l_2 \times \{0\})$. And, conversely, a closed abstract boundary must be a Z-set.

Question. *Let (M, K) be an l_2-manifold pair with K a Z-set in M. Under what conditions can M be embedded in l_2 such that K is the topological boundary under the embedding?*

It should be remarked that for any two l_2-manifolds M_1 and M_2, there are open embeddings h_1 and h_2 of M_1 and M_2 in l_2 such that $\mathrm{Cl}\, h_1(M_1)$, $\mathrm{Cl}\, h_2(M_2)$, $\mathrm{Cl}\, h_1(M_1) \cap \mathrm{Cl}\, h_2(M_2)$ are l_2-manifolds, $l_2 = \mathrm{Cl}\, h_1(M_1) \cup \mathrm{Cl}\, h_2(M_2)$, and $h_1(M_1)$ and $h_2(M_2)$ are disjoint. Thus weird embeddings of manifold pairs in l_2 are possible.

In the l_2-manifold pair (M, K), K is said to be of finite local deficiency n at $p \in K$ if there is an open embedding h of $l_2 \times (-1, 1)^n$ into M such that $p \in l_2 \times \{0\}^n$ and $K \cap h(l_2 \times (-1, 1)^n) = h(l_2 \times \{0\}^n)$.

Question. *Given the l_2-manifold pair (M, K) with K of local deficiency n at every point of K except possibly at the points of a subset K' of K which is a Z-set both in M and in K. Is K of local deficiency n at every point?*

For $n \neq 2$ this question is open even for K' compact or consisting of a single point. For $n = 2$, Kuiper has given a counterexample using knots where K' can be a single point (or a finite cell). The answer is known to be affirmative for the analagous questions for $n = \infty$ or for $n = 0$ (using the abstract boundary $[0, 1)$ definition for $n = 0$) and indeed by the Z-set theory, the cases for $n = \infty$ and $n = 0$ are known to be different formulations of the same question.

Problems on Q-factors

In his paper in these Proceedings, West has reviewed the status of the impressive recent research on Q-factors, i.e., spaces X such that $X \times Q \cong Q$.

Question. *Characterize the Q-factors.*

Since it is clear that every Q-factor must be a compact absolute retract, we would have a characterization of Q-factors if we could answer the following question affirmatively.

Question. *Is every compact* AR *a Q-factor*?

The results cited in West's paper give many partial results and thus characterize Q-factors for certain classes of compacta, e.g., a polyhedron or a cell-complex is a Q-factor iff it is contractible.

For the general case of compact AR's, it is not intuitively clear what answer should be expected, e.g., there are compact AR's which are not local AR's, indeed, by Borsuk, [6], there is an AR, M, and a point $p \in M$ such that no open set containing p lies in an AR which is a proper subcompactum of M. If there does exist a compact AR which is not a Q-factor, then a solution of the Q-factor characterization problem might lead to interesting new classes of compacta.

Problems on Upper Semi-Continuous Decompositions of Q

A compactum K in Q is point-like if $Q \setminus K \cong Q \setminus \{pt\}$. As discussed in Mardešić's paper in these Proceedings, Chapman, [8], has proved that for any two Z-sets K_1 and K_2 in Q, $Q \setminus K_1 \cong Q \setminus K_2$ iff K_1 and K_2 have the same shape (as defined by Borsuk). Thus a Z-set in Q is point-like iff it has the shape of a point. However, compacta in Q may be point-like even though they are not Z-sets and there exist wild arcs in Q, i.e., arcs which are not point-like. In Euclidean spaces, point-like upper semi-continuous decompositions have been extensively studied and many interesting examples and theorems are known. Frequently a condition that the elements, e.g., arcs, are "tame" replaces the more general "point-like".

It seems likely that comparable studies for Q would include the hypothesis that the elements be Z-sets of trivial shape since homeomorphic Z-sets in Q are necessarily equivalently embedded and, for example, all compacta in s or in $Q \setminus s$ are Z-sets. Indeed any point-like decomposition of s is a point-like Z-set decomposition.

Let G be an upper semi-continuous decomposition of Q into Z-sets of trivial shape with hyperspace X; equivalently, let h be a map of Q onto X such that G is the collection of point-inverses under h and the elements of G are point-like Z-sets. Let H be the collection of non-degenerate elements of G. Let H^* denote the union of the elements of H.

Question. *Under what conditions can we conclude that* $X \cong Q$?

It seems likely that there exists a "dogbone" example in Q, i.e., an upper semi-continuous decomposition of Q into points and Z-set arcs such that $h(H^*)$ is a topological Cantor set and such that $X \not\cong Q$. A possible candidate for such an example is to consider two disjoint Wong-type [17] wild Cantor sets C_1 and C_2 in Q and a continuous collection of arcs joining the points of C_1 with the points of C_2 such that each arc lies except possibly for its endpoints in s (and thus is a Z-set). Let this collection of arcs be the collection H of all non-degenerate elements of G. Then, noting some special properties of Wong's construction, it seems unlikely that the elements of H can be shrunk to points in Q, i.e., that the hyperspace of G is homeomorphic to Q.

Three special cases of the general question appear interesting.

(1) If H^* lies in a Z-set (or equivalently in an endslice of Q), can we conclude that $X \cong Q$?

(2) If H^* lies in a countable union of Z-sets (or, equivalently, in $Q \setminus s$), can we conclude that $X \cong Q$?

(3) If H^* lies in s, can we conclude that $X \cong Q$?

A substantial partial result for (1) is known whereas no significant additional results are known for (2) and (3). It should be noted that for the candidate for the dogbone example above, H^* cannot lie topologically in s since then Wong's "wild" Cantor sets would be "tame".

The result known for (1) is the following:

Theorem. *If H^* is finite-dimensional, then $X \cong Q$ (and $h(H^*)$ lies in a Z-set in X)* [4].

The finite-dimensionality is probably an unnecessary hypothesis but the known proof uses the condition strongly. Notice that if for a dogbone construction with H^* zero-dimensional and the dogbone in a Z-set, then the theorem implies that $Q \cong X$. An immediate corollary of the theorem and of Chapman's characterization of shape is the following:

Corollary. *For any compactum A and map f of A onto a compactum B such that the image under f of the union of the non-degenerate point-inverses is finite-dimensional and for each $b \in B$, $f^{-1}(b)$ has trivial shape, then shape A = shape B.*

A result like this corollary has reportedly recently been obtained independently by Koslowski and by Sher. The argument for the corollary involves (1) embedding A in a Z-set (or endslice) in Q, (2) employing the theorem for the U.S.C. decomposition which consists of the given U.S.C. decomposition of (the image of) A and the set of individual points not in (the image of) A, and then (3) employing Chapman's characterization since the decomposition map carries Q onto Q and $Q \setminus A$ homeomorphically onto $Q \setminus B$. If the finite dimensionality condition can be omitted from the theorem, then it also can be omitted from the corollary.

References

[1] *R. D. Anderson:* Hilbert space is homeomorphic to the countable infinite product of lines. Bull. Amer. Math. Soc. *72* (1966), 515–519.
[2] *R. D. Anderson:* On topological infinite deficiency. Michigan Math. J. *14* (1967), 365–383.
[3] *R. D. Anderson:* Homeomorphisms on infinite-dimensional manifolds. Proc. Int. Cong. of Math., Nice, 1970 (to appear).
[4] *R. D. Anderson:* Point-like decompositions of the Hilbert cube. (In preparation.)
[5] *R. D. Anderson, T. A. Chapman, and R. M. Schori* (Editors): Problems in the topology of infinite-dimensional spaces and manifolds. Mathematisch Centrum Report ZW 1/71,2^e Boerhaavestraat 49, Amsterdam.
[6] *K. Borsuk:* Theory of Retracts. Polish Scientific Publishers, Warsaw, 1967.
[7] *T. A. Chapman:* On the structure of Hilbert cube manifolds. (Preprint.)
[8] *T. A. Chapman:* On some applications of infinite-dimensional manifolds to the theory of shape. To be published in Fund. Math.
[9] *R. Geoghegan:* On spaces of homeomorphisms, embeddings and functions. (Preprint.)
[10] *D. W. Henderson:* Infinite-dimensional manifolds are open subsets of Hilbert space. Bull. Amer. Math. Soc. *75* (1969), 759–762.
[11] *M. I. Kadec:* A proof of topological equivalence of all separable infinite-dimensional Banach spaces. Funkcional. Anal. i Priložen. *1* (1967), 53–62.
[12] *O. H. Keller:* Die Homoiomorphie der kompakten konvexen Mengen in Hilbertschen Raum. Math. Ann. *105* (1931), 748–758.
[13] *V. L. Klee:* Some topological properties of convex sets. Trans. Amer. Math. Soc. *78* (1955), 30–45.
[14] *R. Luke and W. K. Mason:* The space of homeomorphisms on a compact two-manifold is an absolute neighborhood retract. (Preprint.)
[15] *J. E. West:* Infinite products which are Hilbert cubes. Trans. Amer. Math. Soc. *150* (1970), 1–25.
[16] *J. E. West:* Identifying Hilbert cubes: General methods and their application to hyperspaces by Schori and West. These Proceedings.
[17] *R. Y. T. Wong:* A wild Cantor set in the Hilbert cube. Pacific J. Math. *24* (1968), 189–193.
[18] *R. Y. T. Wong:* On homeomorphisms of infinite-dimensional bundles. These Proceedings.

LOUISIANA STATE UNIVERSITY, BATON ROUGE, LOUISIANA

[illegible] Corollary [illegible] that in [illegible] of the [illegible] and show that [illegible]

A [illegible] that [illegible] if and only if [illegible] decomposition [illegible] the unique [illegible] in the [illegible]

Reference

[illegible] Math. Soc. [illegible]

[illegible] Math. [illegible] Proc. [illegible]

[illegible] Holland, Amsterdam, 1966, [illegible]

[illegible] (1966). [illegible]

[illegible] (1960), 359–366.

[illegible] Math. Soc. 150 (1970), [illegible]

[11] R. [illegible] Hilbert cube [illegible] of the Hilbert cube, Trans. Amer. Math. Soc. 150 (1970), 1–25.

[16] J. E. West, [illegible] Hilbert cube [illegible] general methods and their application to hyperspaces [illegible] Proceedings.

[17] R. Y. T. Wong, [illegible] J. Math. 21 (1968), 189–193.

[18] R. Y. T. Wong, On homeomorphisms of [illegible] bundles, These Proceedings.

LOUISIANA STATE UNIVERSITY, BATON ROUGE, LOUISIANA

ON CARDINAL INVARIANTS

A. V. ARCHANGELSKIJ

Moskva

We call cardinal invariants such topological invariants which assume values in the class of all cardinal numbers. These invariants play a very important role in all branches of general topology. The first of them appeared at the earliest stages of development of general topology and showed their importance at once. For example, the second axiom of countability is the basic condition in the classical metrization criteria of Urysohn. The weight of a space X (denoted $w(X)$), the character of a point x in the space X (denoted $\chi(x, X)$) are other examples of classical cardinal invariants. The cardinal-valued topological invariants are very essential in many theorems on classification of topological spaces and continuous mappings. Even the bicompactness property may be considered from this point of view – the definition of bicompactness in terms of complete accumulation points shows it clearly.

The constant and reasonable use of cardinal invariants in different areas of general topology stimulated the appearance of many new interesting invariants of this kind. So it became quite necessary to clarify the interrelations between them – in other words, to classify them, to develop their theory.

Many good results were obtained in this direction in the course of the last five years. I do not intend to give a full survey. This would need a book. In fact, such a book has been written recently by a well-known specialist I. Juhasz. I will present mainly some of my results.

The notation follows that in [3]. In particular, I write $c(X)$ for the Suslin number of the space X, i.e. for the least infinite cardinal τ such that the cardinal of every disjoint family γ of nonempty open subsets of the space X is $\leqq\tau$. I put $cc(X) = \sup\{c(Y) : Y \subset X\}$. All spaces considered are Hausdorff.

1. Here I describe some circumstances in which the condition $c(X) \leqq \aleph_0$ implies paracompactness or even metrizability of X.

1.1. Theorem. *If X is a Čech-complete σ-metacompact and $c(X) = \aleph_0$, then X is Lindelöf.*

(Following F. D. Tall I call a space X σ-metacompact if each open covering of this space has a σ-point finite open refinement. Hence all metacompact and all screenable spaces are σ-metacompact.)

The crucial role in the proof of this theorem is played by the following:

1.2. Lemma. *If X is Čech-complete, $c(X) = \aleph_0$ and γ is a point finite family of open sets in X, then $|\gamma| \leqq \aleph_0$.*

The following result constitutes an essential step in the proof of this lemma.

1.3. Lemma. *If $c(X) = \aleph_0$, k is a natural number and γ is a family of open sets in X such that no point of X belongs to more than k elements of γ, then $|\gamma| \leqq \aleph_0$.*

Notice that we do not suppose here the space X to be Čech-complete.

Besides Theorem 1.1, other interesting conclusions follow from Lemma 1.2. Let us show now some of them.

1.4. Theorem. *If X is metacompact, locally Čech-complete and has locally a countable Suslin number, then X is strongly paracompact.*

1.5. Theorem. *If X is σ-metacompact locally bicompact perfectly normal space, then X is strongly paracompact.*

1.6. Theorem. *If X is a Čech-complete space with a σ-point finite base $\mathscr{B}$ and $c(X) \leqq \aleph_0$, then the base $\mathscr{B}$ is countable (and X is metrizable).*

Pixley and Roy constructed (in an appropriate model of the set theory) an example of a non-metrizable completely regular space X with the uniform base $\mathscr{B}$ (in the sense of P. S. Alexandrov) such that $c(X) = \aleph_0$. As each uniform base $\mathscr{B}$ is σ-point finite it is not possible to extend Theorem 1.6 to all non-Čech complete spaces. (Maybe such an extension is possible in some models.)

Of course, each σ-point finite base is point countable. So in connection with Theorem 1.6 the following question seems to be very interesting.

1.7. Problem. Let X be a Čech-complete space with a point countable base $\mathscr{B}$ and $c(X) \leqq \aleph_0$. Is it true then that the base $\mathscr{B}$ is countable?

2. The remaining part of this article is devoted to an exposition of results on interrelations between cardinal-valued invariants. In the end the reader will find himself completely surrounded by a host of unsolved problems.

I would start with the following example of a result typical for this area.

2.1. Theorem. (A. Hajnal and I. Juhasz, [8]). *If X is first countable and $c(X) = \aleph_0$, then $|X| \leqq 2^{\aleph_0}$.*

An interesting and important generalization of the notion of first countable space is the notion of sequential space. A space X is called sequential if all sequentially closed subsets of X are closed (see [17]). A space X is called Fréchet-Urysohn space (FU-space) if for each $z \in X$ and each $A \subset X$ such that $x \in [A]$ there exists a sequence $\{a_n : n = 1, 2, \ldots\}$ of points belonging to A which converges to x. Evidently, each FU-space is sequential but not conversely. Even a bicompactum exists which is sequential, but not an FU-space [17].

Obviously, each subspace of an FU-space is again an FU-space. Hence each subspace of an FU-space is sequential. But a subspace of a sequential space need not be sequential. It is easy to see that a sequential spece X is an FU-space if and only if each subspace of the space X is sequential. In fact, a stronger result holds: X is an FU-space iff each subspace of X is a k-space [5]. It is worth noticing that the Σ-product of an arbitrary family ξ of real lines (or, more generally, of complete separable metric spaces) is an FU-space ([18]). As the product has a countable Suslin number and the Σ-product is dense in the product, we conclude that the Suslin number of the Σ-product is also countable. So we have exhibited an FU-space X such that $c(X) \leqq \aleph_0$ and $|X| \geqq |\xi|$. And the cardinality of ξ may be as high as we wish. It is easy to see that $\chi(x, X) \geqq |\xi|$ for each $x \in X$. Hence we have proved the following:

2.2. Assertion. *For each cardinal $\tau \geqq \aleph_0$ there exists an FU-space X_τ such that $c(X) \leqq \aleph_0$, $|X| \geqq \tau$ and $\chi(x, X) \geqq \tau$ for all $x \in X$.*

This means that Theorem 2.1 cannot be extended to the class of all Fréchet-Urysohn spaces. But the situation miraculously changes to the best if we concentrate our attention on the class of all bicompact Hausdorff spaces. Of course, for each cardinal τ there exists a bicompactum X such that $|X| \geqq \tau$ and $c(X) = \aleph_0$ – take for example $\mathscr{D}^\tau$ or I^τ (where $I = [0, 1]$ and $\mathscr{D} = \{0, 1\}$). On the other hand, the (Alexandroff's) one point bicompactification A_τ of a discrete space of cardinality $\tau \geqq \aleph_0$ enjoys the following properties: A_τ is first countable at all points but one; A_τ is sequential; A_τ is not homogeneous; $|A_\tau| = \tau$; $c(A_\tau) = \tau$; A_τ is a bicompact Hausdorff space. Results which follow show us that these properties bunched together not by chance.

2.3. Theorem. *If $2^{\aleph_0} = \aleph_1$ and X is a bicompact (Hausdorff) sequential space, then the set of points in which X is first countable is dense in X* [3].

The crucial point in the proof is the following:

2.4. Lemma. *If X is a bicompact sequential space and U is a nonempty open subset in X, then there exists a non-empty closed set $P \subset X$ such that:* (i_1) $P \subset U$; (i_2) $\chi(P, X) \leqq \aleph_0$; ($\mathrm{i}_3$) $|P| \leqq 2^{\aleph_0}$ [3].

Using this lemma and the well known Ramsey type theorem proved by Erdös and Radó, we arrive also to the following result:

2.5. Theorem. *If X is a bicompact sequential space and $c(X) \leqq \aleph_0$, then $|X| \leqq 2^{\aleph_0}$* [3].

It is useful to look now once more at the spaces A_τ.

2.6. Theorem. *If X is a sequential bicompact homogeneous space, then either $|X| = 2^{\aleph_0}$ or $|X| < \aleph_0$.*

Let us sketch the proof of this theorem to reveal the close interrelations between the results mentioned above.

From 2.4 it follows that $\chi(x, X) \leqq 2^{\aleph_0}$ for some $x \in X$. As X is homogeneous, $\chi(x, X) \leqq 2^{\aleph_0}$ for all $x \in X$. In [2] was proved that if X is a bicompact sequential Hausdorff space such that $\chi(x, X) \leqq 2^{\aleph_0}$ for all $x \in X$, then $|X| \leqq 2^{\aleph_0}$. If X has an isolated point, then X is finite. If there are no isolated points in X, then $|X| \geqq 2^{\aleph_0}$ [1]. So in this case $|X| = 2^{\aleph_0}$.

With the aid of GCH (which means: $2^\tau = \tau^+$ for each $\tau \geqq \aleph_0$) we can prove the following

2.7. Theorem. *If X is a homogeneous bicompactum, then $|X|$ is an isolated cardinal number* [3].

Now, we turn to some generalizations naturally connected with the notion of a sequential space.

3.3.1. Let X be a topological space and τ a cardinal number different from zero. For any $A \subset X$ put $[A]_\tau = \bigcup\{[B] : B \subset A \text{ and } |B| \leqq \tau\}$.

It is easy to see that always $[[A]_\tau]_\tau = [A]_\tau$; moreover, the operator $[\ ]_\tau$ is a closure operator on X for some topology on X which will be denoted by $\mathscr{T}_\tau$ provided the given topology on X is denoted by $\mathscr{T}$.

3.2. We say that $A \subset X$ is a G_τ-set (in X) if there is a family γ of open subsets of X such that $\bigcap\{U : U \in \gamma\} = A$ and $|\gamma| \leqq \tau$.

Put $[A]^\tau = \{x \in X$: if $x \in Q$ and Q is a G_τ-set, then $Q \cap A \neq \Lambda\}$. Obviously, $[[A]^\tau]^\tau = [A]^\tau$. The operator $[\ ]^\tau$ is a closure operator for some topology on X which will be denoted by $\mathscr{T}^\tau$. Evidently, $\mathscr{T}_\tau \supset \mathscr{T}$ and $\mathscr{T}^\tau \supset \mathscr{T}$. Usually, $\mathscr{T}^\tau$ and $\mathscr{T}_\tau$ are not comparable and in general, the formula $\mathscr{T}^\tau \cap \mathscr{T}_\tau = \mathscr{T}$ does not hold.

3.3. For example, let $H = (I, \mathscr{T})$ be the space which we obtain by declaring all countable subsets of the segment $I = [0, 1]$, as well as the subsets of I which are closed in the usual topology, to be closed. F. Hausdorff was the first who considered this space. I list some important properties of H: 1) $cc(H) \leqq \aleph_0$; 2) $S(H) > \aleph_0$; 3) $\psi(x, H) = \aleph_0$ for each $x \in H$; 4) $[A]_{\aleph_0} = A$ for each $A \subset H$; 5) $[A]^{\aleph_0} = A$ for each $A \subset H$; 6) $\mathscr{T}_{\aleph_0} = \mathscr{T}^{\aleph_0}$ – discrete topology, hence 7) $\mathscr{T}_{\aleph_0} \cap \mathscr{T}^{\aleph_0} \neq \mathscr{T}$ (because H is not discrete).

3.4. Theorem. [4] *If X is bicompact, then for each $A \subset X$ and each $\tau > 0$*

$$[[A]_\tau]^\tau = [A].$$

I think that this is really an important formula. The proof of this formula, if not long, is not trivial. I wish also to underline that it is extremely general. To derive some valuable consequences from the formula we need two definitions.

3.5. Let X be a space. The least cardinal number τ such that $[A]_\tau = [A]$ for each $A \subset X$ is called the tightness of X and is denoted by $t(X)$. Of course, $t(X) \leqq \chi(X) \leqq w(X)$ for each X. If X is sequential, $t(X) \leqq \aleph_0$. But the converse to the last assertion is not true: If X is countable, then clearly $t(X) \leqq \aleph_0$. However, it is easy to construct a countable space X which is not sequential: take for X the set $N \cup \xi \subset \beta N$, where $\xi \in \beta N \setminus N$ and $N \cup \xi$ is considered as a subspace of βN (βN is the Stone-Čech bicompactification of a countable infinite discrete space N).

3.6. Let X be a space and τ a cardinal number. Suppose that for each ordinal $\alpha < \tau$ a point $x_\alpha \in X$ is chosen. Then we say that $\xi = \{x_\alpha : \alpha < \tau\}$ is a free sequence of length τ if for each $\beta < \tau$ the following condition holds:

$$[\{x_\alpha : \alpha < \beta\}] \cap [\{x_\alpha : \beta \leqq \alpha\}] = \Lambda .$$

Obviously, points of a free sequence in X constitute a discrete subspace of X. Of course this subspace may not be closed. If a closed discrete subspace Y of X is given, any minimal well ordering on Y makes Y a free sequence. Obviously, not each discrete subspace of X can be represented as the set of all points of a free sequence. But each countable discrete subspace is the set of all points of some free sequence.

3.7. Theorem. *Let X be a bicompactum. Then $t(X) = \sup\{\tau :$ there is a free sequence of the length τ in $X\}$.*

The proof of Theorem 3.7 heavily depends on Theorem 3.4.

3.8. Corollary. *If X is a bicompactum, then $t(x) \leqq cc(X)$.*

B. Shapirovskij – a student of mine – was the first to formulate and prove this last assertion. A few days later, knowing the Shapirovskij's result but not its proof I gave an independent proof of 3.8. Working on this proof I found Theorems 3.4 and 3.7. The assertion 3.8 may be easily generalized in the following way:

3.9. Corollary. *If X is a k-space, then $t(x) \leqq cc(X)$.*

Besides the results mentioned above and the following theorem almost nothing is known about the tightness – even in the case of bicompact Hausdorff spaces. Hence more attention should be paid to the following positive fact.

3.10. Theorem. *If X is a dyadic bicompactum, then* $t(X) = w(X)$.

The proof of this assertion, given in [19], makes use of the notion of Dante's space (see [19]) and is based essentially on the Hewitt-Marczewski-Pondiszeri's theorem.

3.11. Corollary. *If X is a dyadic bicompactum, then* $cc(X) = w(X)$.

To prove it, we simply combine 3.8 and 3.10. The original Efimov's proof of 3.11 is rather complicated (see [7]).

It is remarkable that Theorem 3.10 includes also the following well known assertions:

3.12. Corollary (A. S. Esenin-Vol'pin). *If X is a dyadic bicompactum, then* $\chi(X) = w(X)$.

3.13. Corollary (A. V. Arhangel'skij). *If a dyadic bicompactum X is a factor-space of some metric space, then X is metrizable.*

Another series of corollaries of 3.8 holds in a special model of the set theory.

4. There exists (see [14], [10]) a model of set theory in which the following assertion is true:

4.1. (TAM) If X is a space such that $c(X) = \aleph_0$, then for each uncountable family ξ of non-empty open subsets of X there exists an uncountable subfamily $\eta \subset \xi$ which is centered.

4.2. Theorem. *If* (TAM) *holds and X is a bicompactum such that* $cc(X) \leqq \aleph_0$, *then each subspace of X is separable.*

In other words, if (TAM) holds then bicompact Hausdorff space is hereditary separable if and only if each discrete subspace of this space is countable. The proof of the fact is non-trivial. We essentially use the relation $t(X) \leqq cc(X) \leqq \aleph_0$, which holds by 3.8 (see [5]).

4.3. I. Juhasz proved that if (TAM) holds and X is a first countable bicompact Hausdorff space such that $c(X) = \aleph_0$, then X is separable [10].

There exist such models of the set theory in which not only (TAM) is true but the following relations also hold: $2^{\aleph_0} = 2^{\aleph_1} = \aleph_2$. In any of these models the following theorem may be proved:

4.4. Theorem. *If X is a sequential bicompactum such that* $c(X) = \aleph_0$, *then X is separable.*

In the proof, Lemma 2.4 plays a very important role. D. Kurepa [11] proved that the product X of an arbitrary family $\{X_\alpha : \alpha \in A\}$ of spaces such that $c(X) = \tau \geqq \aleph_0$ for each A has a Suslin number not greater than 2^τ (see also [9]). In his other very elegant work [20] D. Kurepa proved that if X is a topological space, its topology being induced by a linear order on X, and $c(X \times X) = \aleph_0$, then $s(X) \leqq \aleph_0$.

Now from the arguments in [15] (see also [12]), it is clear that the following very remarkable theorem holds:

4.5. Theorem. *If* (TAM) *is true,* $X = \prod\{X_\alpha : \alpha \in A\}$ *and* $c(X_\alpha) = \aleph_0$ *for each* $\alpha \in A$, *then* $c(X) = \aleph_0$.

The technique developed to prove the results of the last group yields some conclusions of absolute character – which are true in all reasonable models.

We shall use now the following notation: $ss(X) = \sup\{s(Y) : Y \subset X\}$. Here $s(Y)$ stands for the density of Y.

4.6. Theorem. *Let* X *be a bicompactum,* τ *a cardinal number and, for each closed subspace* Y *of the space* X, $s(Y) \leqq \tau$. *Then* $ss(X) \leqq \tau$.

We sketch here the proof of 4.6. From the assumptions about X it follows that $cc(X) \leqq \aleph_0$. By 3.8 we have then $t(X) \leqq \aleph_0$. It is sufficient now to use the following simple

4.7. Lemma. *If* X *is a space,* Y *a subspace of* X, τ *a cardinal number and* (a) $[Y] = X$; (b) $t(X) \leqq \tau$; (c) $s(X) \leqq \tau$; *then* $s(Y) \leqq \tau$.

Although the following part is devoted mainly to problems, it involves also a discussion of problems and in its course we also mention some new interesting results.

5.5.1. Problem. Is there a bicompactum X such that $|X| \leqq 2^{\aleph_0}$, $c(X) = \aleph_0$ and $s(X) > \aleph_0$? It is easy to prove that if (TAM) holds and $|X| = \aleph_1$, $c(X) = \aleph_0$, then $s(X) \leqq \aleph_0$. But as far as I know it is not known whether (TAM) is consistent with $2^{\aleph_0} = \aleph_1$. A. Hajnal and I. Juhasz have proved recently that if (TAM) holds and X is a bicompactum such that π-weight of X is less than or equal to $\aleph_1$ and $c(X) = \aleph_0$, then X is separable [13].

5.2. Problem. Let X be a Lindelöf space in which each point is a G_δ-set. Is it true then that $|X| \leqq 2^{\aleph_0}$?

In [2] I showed that the answer is yes if X is first countable Lindelöf space. Now I have the following result:

5.3. Theorem. *If X is a regular Lindelöf space such that $\{x\}$ is a G_δ-set for each $x \in X$ and $t(X) \leqq \aleph_0$, then $|X| \leqq 2^{\aleph_0}$.*

Is it possible to extend this result from regular to Hausdorff spaces?

Problem 5.2 would be settled in the positive way if we could prove one of the following two hypotheses:

5.4. Hypothesis. If X is a Hausdorff space and each point of X is a G_δ-set in X, then there exists a first countable Hausdorff space Y which is a one-to-one continuous image of the space X.

5.5. Hypothesis. If X is a regular Lindelöf space and each point of X is a G_δ-set in X, then there exists a first countable Hausdorff space Y which is a one-to-one continuous image of the space X.

5.7. Problem (A. Hajnal and I. Juhasz). Let X be a bicompactum such that $ss(X) \leqq \aleph_0$. Is it true then that $|X| \leqq 2^{\aleph_0}$?

5.8. Problem (A. Hajnal and I. Juhasz). The same question as in 5.7 for an arbitrary Hausdorff space X.

5.9. Problem (B.A. Efimov). Is it true that for each infinite bicompact Hausdorff space X one of the following two alternatives holds:

1. X contains a non-trivial convergent sequence of points;
2. X contains a topological copy of βN?

Problems 5.7 and 5.9 are closely connected with the following group of problems.

5.10. Problems. Let X be a bicompactum such that $t(X) \leqq \aleph_0$. Which of the following assertions 5.10.1 – 5.10.5 are then true?

5.10.1. X is sequential.

5.10.2. If $2^{\aleph_0} = \aleph_1$, then X is first countable at some point. (See 2.3.)

5.10.3. If $c(X) = \aleph_0$, then $|X| \leqq 2^{\aleph_0}$.

Notice that if $ss(X) \leqq \aleph_0$, then $t(X) \leqq \aleph_0$ and $c(X) = \aleph_0$. Look now at 5.7. See also 2.5.

5.10.4. If $|X| \geqq \aleph_0$, then X contains a non-trivial convergent sequence of points.

5.10.5. If (TAM) and $2^{\aleph_0} = 2^{\aleph_1} = \aleph_2$ holds, and $c(X) = \aleph_0$, then X is separable. (See 4.4.)

5.11. Problem. Let X be a bicompactum and $cc(X) \leqq \aleph_0$. Is it true then that $s(X) \leqq \aleph_0$? (See 4.2.)

5.12. Problem. Is it true that if for each closed subspace Y of a completely regular space X, $s(Y) \leqq \tau$, then $s(Z) \leqq \tau$ for each subspace Z of the space X? (See 4.6 and 3.3.)

5.13. Problem. Let X be a completely regular space such that $cc(X) \leqq \tau$. Is there a bicompact Hausdorff extension bX of X such that $cc(bX) \leqq \tau$?

5.14. Problem. The same question as in 5.13 for X such that $s(Y) \leqq \tau$ for each closed subspace Y of the space X. (We do not seek for a stronger conclusion.)

5.15. Problem. Which spaces of countable tightness can be realized as subspaces of a sequential spaces? (Clearly each subspace of a sequential space has countable tightness; more generally, $t(Y) \leqq t(X)$ as soon as Y is a subspace of X. Let us remark that not each countable space is a subspace of a sequential space.)

5.16. Problem. Which spaces are homeomorphic to subspaces of bicompact Hausdorff spaces of countable tightness, of sequential bicompact Hausdorff spaces, respectively?

It is true that if X is a bicompactum such that $c(X) \leqq \aleph_0$ and $t(X) \leqq \aleph_0$, then $|X| \leqq 2^{2^{\aleph_0}}$ (look now once more at 5.10.3). Hence not every FU-space has a bicompactification with countable tightness (for the proof take an appropriate Σ-product — see the arguments on page 39).

5.17. Problem (F. D. Tall). Suppose (TAM) holds and let X be a Čech-complete first countable space such that $c(X) = \aleph_0$. Is it true then that X is separable? (See 4.3 and 1.7.)

5.18. Problem. Is the product of countably many spaces of countable tightness again a space of countable tightness? V. Malychin proved that the answer is yes if all the factors are bicompact.

References

[1] *P. S. Alexandrov et P. S. Urysohn:* Mémoire sur les espaces topologiques compacts. Nederl. Akad. Wetensch. Proc. Ser. A *14* (1929), 1—96.

[2] *A. V. Arhangel'skij:* The power of bicompacta with first axiom of countability. Dokl. Akad. Nauk SSSR *187* (1969), 967—970 (Soviet Math. Dokl. *10* (1969), 951—955).

[3] *A. V. Arhangel'skij:* The Suslin number and cardinality. The characters of points in sequential bicompacta. Dokl. Akad. Nauk SSSR *192* (1970), 255—258.

[4] *A. V. Arhangel'skij:* On very k-spaces. Czechoslovak Math. J. *18(93)* (1968), 392—395.

[5] *A. V. Arhangel'skij:* On bicompacta which satisfy the Suslin condition hereditarily. Tightness and free sequences. Dokl. Akad. Nauk SSSR *199* (1971), 1227—1230.

[6] *E. Čech et B. Pospíšil:* Sur les espaces compacts. Publ. Fac. Sci. Univ. Masaryk *258* (1938), 1—14.

[7] *B. Efimov:* Dyadic bicompacta. Trudy Moskov. Mat. Obšč. *14* (1965), 211—247.

[8] *A. Hajnal and I. Juhasz:* Discrete subspaces of topological spaces I—II. Indag. Math. *29* (1967), 343—356; *31* (1969), 18—30.

[9] *Z. Hedrlín:* An application of Ramsey's theorem to the topological products. Bull. Acad. Polon. Sci. Sér. Sci. Math. Astronom. Phys. *14* (1966), 25—26.

[10] *I. Juhasz:* Cardinal functions in Topology. (In collaboration with A. Verheek and N. S. Kronenberg.) Mathematical Centre Tracts *34,* Amsterdam.

[11] *D. Kurepa:* The cartesian multiplication and the cellularity number. Publ. Inst. Math. (Beograd) *2* (1962), 121—139.

[12] *R. Engelking:* An Outline of General Topology. North-Holland Publ. Comp., Amsterdam, 1968.

[13] *A. Hajnal and I. Juhasz:* A consequence of Martin's Axiom. The University of Calgary, Department of Mathematics, Research Paper 110.

[14] *A. Martin and R. Solovay:* Internal Cohen Extensions. Ann. Math. Logic (to appear).

[15] *K. A. Ross and A. H. Stone:* Products of separable spaces. Amer. Math. Monthly *71* (1964), 398—403.

[16] *F. D. Tall:* Set-theoretic consistency results and topological problems concerning the normal Moore space conjecture and related problems. Thesis, University of Wisconsin, Madison, 1969.

[17] *S. P. Franklin:* Spaces in which sequences suffice II. Fund. Math. *61* (1967), 51—56.

[18] *N. Noble:* Ph. D. Thesis.

[19] *A. V. Arhangel'skij:* On approximation of the theory of dyadic bicompacta. Dokl. Akad. Nauk SSSR *184* (1969), 767—770.

[20] a) *G. Kurepa:* La condition de Souslin et une propriété caractéristique des nombres réels. C. R. Acad. Sci. Paris Sér. A—B *231* (1950), 1113—1114.
b) *D. Kurepa:* Sur une propriéte caractéristique du continu lineaire et le problème de Suslin. Publ. Inst. Math. (Beograd) *4* (1952), 97—108.

POINT-COUNTABLE BASES AND QUASI-DEVELOPMENTS

C. E. AULL

Blacksburg

1. Introduction

An important class of topological spaces is the one with point-countable bases. (We will refer to these as point-countable spaces).

In this paper we discuss some properties of these spaces and of some spaces with more restrictive base properties and relate these to spaces with corresponding type covering properties.

In a recent survey paper of the author [1], some of the important properties of point-countable spaces are listed. By using standard techniques for minimal properties of topological spaces as in [5], one can show that spaces minimal with respect to being Hausdorff and having a point-countable base are semiregular and feebly compact. It would be interesting to know if such spaces are necessarily H-closed. Corson and Michael [7] showed that countably compact point-countable spaces have a countable base. Modifying their techniques one can show that a topological space that is H-closed, Hausdorff, semiregular and has a base such that every point is in the closure of only a countable number of members of the base has a countable base. However there is an example of Miščenko [11] of a point countable Hausdorff, H-closed space that does not have a countable base. This space is not semiregular nor is each point in the closure of just a countable number of members of any base.

2. Quasi-developments

Recently E. E. Grace and H. R. Bennett [3] have introduced a generalization of the developable space.

Definition 1. A sequence $\mathscr{G}_1, \mathscr{G}_2, \ldots$ of collections of open sets of a topological space $(X, \mathscr{T})$ is called a *quasi-development* for $(X, \mathscr{T})$ provided that if $x \in T \in \mathscr{T}$, there exists n and G such that $x \in G \in \mathscr{G}_n$, and if $x \in G \in \mathscr{G}_n$ then $G \subset T$. We will refer to each $\mathscr{G}_k$ as a *collection of the quasi-development*.

We note that a quasi-developable space is a weak σ-space (follows from Lemma 3 below)[1] and a point-countable weak σ-space is quasi-developable. From the latter result follows the result of Okuyama [13] that a collectionwise normal T_1 σ-space is metrizable iff it has a point-countable base and the result of Heath [9] that a T_3 stratifiable space is metrizable iff it has a point-countable base.

Definition 2. A topological space $(X, \mathscr{T})$ is a *weak σ-space* if it has a σ-disjoint network where each disjoint family is discrete with respect to some open subspace containing all members of the family.

As the developable space has the property of converting certain covering properties to corresponding base properties, the quasi-developable space has the property of converting certain hereditary covering properties to corresponding base properties. This will be clarified in Theorem 4.

Bennett and Lutzer [4] have showed the equivalence of a space being quasi-developable with having a θ-base. The concept of θ-base was introduced by Worrell and Wicke [14].

The following definitions will also be useful.

Definition 3. A family of sets will be said to be[2])

1. σ-0 if it is σ-disjoint.
2. σ-1 if it is σ-relatively locally finite, i.e., if the union of a denumerable number of families each locally finite in its union.
3. σ-2 if it is σ-point finite.
4. σ-3 if it is point-countable.

A topological space will be said to be *σ-k refinable* if every open cover has a σ-k open refinement.

Definition 4. A topological space is said to be *CN-k* if for every discrete family $\{D_a\}$ there is a σ-k family $\{G_a\}$ such that $D_a \subset G_a$ and $D_b \cap G_a = \emptyset$ for $a \neq b$. *HCN-k* will be used to indicate that every subspace is *CN-k*.

We note that a metacompact space is *CN*-2 and a point-countable space is *CN*-3. The first result is in Krajewski and Smith [9].

Theorem 1. *A countably paracompact (countably metacompact) σ-k refinable space is paracompact (metacompact) for $k = 0, 1$ (for $k = 2$).*

We note the result for $k = 0$ is due to Nagami [12].

Theorem 2. *A perfectly normal T_1 space with a σ-k base is metrizable for $k = 0, 1$.*

Proof. For $k = 0$ it is proved in [1] and the other case is similar.

---

1) Result obtained independetly by E. S. Berney.

2) In subsequent statements $k = 0, 1, 2$ or 3 unless specifically indicated otherwise.

Theorem 3. *Let $(X, \mathcal{T})$ be quasi-developable and satisfy HCN-k, then $(X, \mathcal{T})$ has a σ-k base and is hereditary σ-k refinable.*

The proof is a modification of the proof of Theorem 2 in [2] in which the θ-base replaces the σ-point finite base. The following lemma is needed and its proof is similar to Lemma 1 in [2].

Lemma 3. *Let $\mathscr{V}$ be a family of open subsets of a topological space $(X, \mathcal{T})$. Then there exists a σ-disjoint family $\mathscr{M} = \bigcup_{1}^{\infty} \mathscr{M}_n$ such that*

(1) *Each $\mathscr{M}_n$ is discrete with respect to some open subspace G_n of X. (It is understood that if $M \in \mathscr{M}_n$ then $M \subset G_n$.)*

(2) *If x is contained in a finite number of members of $\mathscr{V}$ and if $x \in V \in \mathscr{V}$ then there exists $M \in \mathscr{M}$ such that $x \in M \subset V$.*

(3) *If x is not contained in at least one member of $\mathscr{V}$ or if x is contained in an infinite number of members of $\mathscr{V}$ then there does not exist $M \in \mathscr{M}$ such that $x \in M$.*

Corollary 3. *A CN-0 (Bing) or a CN-1 normal Moore space is metrizable.* See [6].

Theorem 4. *In a quasi-developable space hereditary σ-k refinability is equivalent to having a σ-k base.*

3. σ-locally countable bases[3])

To the author's knowledge not much is known about spaces with σ-locally countable basis. However Fedorčuk [8] proved that paracompact Hausdorff spaces with this base property are metrizable. Like developable spaces it seems to have the property of converting certain covering properties into corresponding base properties. For instance it can be proved that if such spaces are σ-*n* refinable (weakly θ-refinable [4]) they have a σ-*n* base (θ-base).

References

[1] *C. E. Aull:* Some base axioms for topology involving enumerability. General Topology and its Relations to Modern Analysis and Algebra (Proc. Kanpur Topological Conf., 1968). Academia, Prague, 1971, 54—61.

[2] *C. E. Aull:* Topological spaces with a σ-point finite base. Proc. Amer. Math. Soc. *29* (1971), 411—416.

[3]) The author is indebted to F. D. Tall for pointing out an error in this section.

[3] *H. R. Bennett:* Quasi-developable spaces. Topology Conference, Arizona State University, 1967. Tempe, 1968, 314—317.

[4] *H. R. Bennett and D. J. Lutzer:* A note on weak θ-refinability. General Topology and its Applications *2* (1972), 49—54.

[5] *M. P. Berri, J. R. Porter and R. M. Stephenson:* A survey of minimal topological spaces. General Topology and its Relations to Modern Analysis and Algebra (Proc. Kanpur Topological Conf., 1968). Academia, Prague, 1971, 93—114.

[6] *R. H. Bing:* Metrization of topological spaces. Canad. J. Math. *3* (1951), 175—186.

[7] *H. Corson and E. Michael:* Metrizability of certain countable unions. Illinois. J. Math. *8* (1964), 351—360.

[8] *V. V. Fedorčuk:* Ordered sets and the product of topological spaces. (Russian.) Vestnik Moskov. Univ. Ser. I Mat. Meh. *21* (4) (1966), 66—71.

[9] *R. W. Heath:* On spaces with point-countable bases. Bull. Acad. Polon. Sci. Sér. Sci. Math. Astronom. Phys. *13* (1965), 393—395.

[10] *L. Krajewski and J. C. Smith:* Expandability and collectionwise normality. Trans. Amer. Math. Soc. *160* (1971), 437—451.

[11] *A. Miščenko:* Spaces with point-countable base. Soviet Mathematics *3* (1962), 855—858.

[12] *K. Nagami:* Paracompactness and strong screenability. Nagoya Math. J. *8* (1955), 83—88.

[13] *A. Okuyama:* σ-spaces and closed mappings I. Proc. Jap. Acad. *44* (1968), 427—477.

[14] *J. M. Worrell, Jr. and H. H. Wicke:* Characterizations of developable spaces. Canad. J. Math. *17* (1965), 820—850.

VIRGINIA POLYTECHNIC INSTITUTE AND STATE UNIVERSITY, BLACKSBURG, VIRGINIA

INSTITUTE FOR ADVANCED STUDY, PRINCETON, NEW JERSEY

ON PROFINITE UNIVERSAL ALGEBRAS

B. BANASCHEWSKI

Hamilton

This paper deals with a certain part of the somewhat uncharted terrain of universal topological algebra. Singling out profiniteness for study was mainly motivated by the fact that it provides a strong tie to the properties of the underlying algebras so that universal algebraic conditions on the latter could readily be expected to have interesting consequences.

The terminology used here follows, in the main, Grätzer [5] for universal algebra, Mitchell [11] for category theory, and Bourbaki [2] for topology; in particular compactness includes Hausdorffness. The presentation does not include any proofs, beyond the occasional hint; the omitted details are expected to appear elsewhere.

Some of the general material in the first section is contained in, or related to, the doctoral dissertation of my student D. E. Eastman [3] whose interest in universal topological algebra did a great deal to stimulate my own.[1])

1. Compact and pro-C algebras. The topological algebras to be discussed here are particular compact algebras; we therefore begin with some general remarks about the latter.

A topological (universal) algebra is an object $A = (X, (f_\alpha)_{\alpha\in I}, \mathfrak{D})$ where X is a set, $(f_\alpha)_{\alpha\in I}$ a family of maps $f_\alpha\colon X^{n_\alpha} \to X$, n_α the *arity* of the *operation* f_α, and $\mathfrak{D}$ a topology on X such that each f_α maps the product space $(X, \mathfrak{D})^{n_\alpha}$ *continuously* to the space $(X, \mathfrak{D})$. For such A, $(X, (f_\alpha)_{\alpha\in I})$ is called the *underlying algebra,* and $(X, \mathfrak{D})$ the *underlying space,* of the topological algebra A. Between algebras of the same arity type $\tau = (n_\alpha)_{\alpha\in I}$ one has the familiar notion of homomorphism; when speaking of *topological* algebras, homomorphisms are, unless specified otherwise, understood to be homomorphisms between the underlying algebras which, in addition, map the underlying spaces *continuously*. Clearly, any given class of topological algebras is then the class of objects of a category, the maps (=morphisms) being the homomorphisms between the members of the class; we refer to such categories usually by merely naming their objects.

The most useful classes of topological algebras appear to be those defined by

---

[1]) Financial assistance from the National Reseach Council of Canada, making attendance at this conference possible, is gratefully acknowledged.

specifying a particular algebraic property for the underlying algebras and a topological one for the underlying spaces, e.g. metrizable topological groups, or locally compact topological rings. Of especial significance in the present context are the classes **KA** of topological algebras whose underlying spaces are *compact* and whose underlying algebras belong to a given *equational class* **A** (of algebras) such as, for instance the class **G** of all groups, **Ab** of all abelian groups, *R* **Mod** of all (left) modules over a given ring *R*, or **Ann** of all commutative rings with unit.

Also associated with an equational class **A** one has the class **FA** of all finite algebras belonging to **A**; since every finite algebra becomes a compact topological algebra when given the discrete topology of its underlying set, provided all its operations have finite arity[2]), we shall (par abus de langage) also regard **FA** as a subclass of **KA** in that case. Finally, there is the subclass $\mathbf{K_0A}$ of **KA** consisting of all those $A \in \mathbf{KA}$ whose underlying spaces are *zero-dimensional*, and one has $\mathbf{FA} \subseteq \mathbf{K_0A}$ for finite arities.

For any equational class **A**, **KA** as a category has *products*, namely the obvious cartesian ones, and any *closed topological subalgebra B* of an $A \in \mathbf{KA}$, i.e. a topological algebra whose underlying algebra is a subalgebra, and whose underlying space is a *closed* subspace, of that of *A*, again belongs to **KA**. The latter, together with the fact that the set of points on which two homomorphisms $f, g\colon A \to C$ in **KA** coincide determines a closed subalgebra of *A*, provides **KA** with equalizers. It follows that **KA** is a complete category.

In addition, any closed congruence Θ on an $A \in \mathbf{KA}$, i.e. a congruence on the underlying algebra of *A* which is a closed subset of $A \times A$, determines in the usual way operations and a topology derived from those of *A* on the associated quotient set which make up a topological quotient algebra A/Θ [3]), evidently belonging to **KA**. The existence of these quotients in **KA** readily provides coequalizers. Further, one shows **KA** also has coproducts, and hence it is a cocomplete category. Finally, the functors from **KA** to the category **Ens** of sets by passing to underlying sets, and to the category of, say, completely regular Hausdorff spaces by passing to underlying spaces, have left adjoints.

The latter assertions can be proved directly via a number of topological lemmas, where the construction of coproducts and free objects (over either sets or spaces) is carried out by the completion of certain algebras with respect to suitable totally bounded uniformities. On the other hand, they can also be established, albeit in a somewhat less explicit way, by categorical arguments involving the Adjoint Functor Theorem and the notion of tripleability (Manes [10]). For a further approach to coproducts, see also Golema [4].

Finally, we note about the category **KA** that its monomorphisms are exactly

[2]) In general, this no longer holds for infinitary algebras.

[3]) The argument here employs compactness; whether the same is true for arbitrary *A* we do not know.

its one-to-one maps and hence the same as its embeddings, i.e. the maps which provide isomorphisms with the image, and dually, its coequalizers are exactly its onto maps.

The subcategory $\mathbf{K_0A}$ of $\mathbf{KA}$ is evidently productive and hereditary, and hence epireflective, in $\mathbf{KA}$ (Herrlich [7]). Moreover, the reflection $\mathbf{KA} \rightarrow \mathbf{K_0A}$ is provided in a very natural topological manner: One proves that, on any topological algebra, the connected component relation is a congruence, and hence one can take quotients in the present setting; passage from $A \in \mathbf{KA}$ to the resulting algebra of its connected components then provides the reflection to $\mathbf{K_0A}$, the quotient maps giving the adjunction.

We now turn to the topological algebras we are specifically concerned with here. In the following, $\mathbf{A}$ will always be a fixed equational class, and $\mathbf{C}$ a subclass of $\mathbf{FA}$ which is hereditary and finitely productive, tacitly assumed to be non-trivial, i.e. to contain algebras with more than one element. With this, let $\mathfrak{R}A$, for any $A \in \mathbf{KA}$, be the set of those closed congruences Θ on A for which $A/\Theta \in \mathbf{C}$, and $\mathbf{ProC} \subseteq \mathbf{KA}$ the subclass of those A for which the intersection of all $\Theta \in \mathfrak{R}A$ is trivial, i.e. the diagonal Δ of $A \times A$. Note that, in a more categorical way, $A \in \mathbf{KA}$ belongs to $\mathbf{ProC}$ iff for any distinct maps $f, g: B \rightarrow A$ in $\mathbf{KA}$ there exists a map $h: A \rightarrow C$ for some $C \in \mathbf{C}$ such that $hf \neq hg$. The $A \in \mathbf{ProC}$ will be called the pro-$\mathbf{C}$ algebras.

Proposition 1. (1) $\mathbf{ProC}$ *is hereditary and productive in* $\mathbf{KA}$.

(2) $\mathbf{ProC} \subseteq \mathbf{K_0A}$.

(3) $A \in \mathbf{ProC}$ *iff* $\mathbf{A}$ *is a closed topological subalgebra of a product of algebras in* $\mathbf{C}$.

(4) $A \in \mathbf{ProC}$ *iff A is a projective limit of algebras in* $\mathbf{C}$.

This can be proved most directly from the definition of $\mathbf{ProC}$ by means of the $\mathfrak{R}A$, but it can also be obtained as a formal consequence of certain categorical properties of $\mathbf{KA}$.

It follows from this proposition that $\mathbf{ProC}$ is epireflective in $\mathbf{KA}$, the reflection being provided by passing from A to its quotient modulo the congruence $\bigcap\Theta(\Theta \in \mathfrak{R}A)$; moreover, $\mathbf{ProC}$ is actually the epireflective *hull* of $\mathbf{C}$ in $\mathbf{KA}$.

Concerning $\mathfrak{R}\mathbf{A}$ one has, for any $A \in \mathbf{ProC}$: (1) Each $\Theta \in \mathfrak{R}A$ is open, and thus open-closed, in $A \times A$. (2) $\mathfrak{R}A$ is closed under finite intersections and hence a basis for the uniformity of A. (3) If $\mathbf{C}$ is also closed under quotients then $\mathfrak{R}A$ consists of all closed congruences of finite index on A, and these are exactly the congruences on A open-closed in $A \times A$.

Possibly the most typical examples of pro-$\mathbf{C}$ algebras are those where $\mathbf{C} = \mathbf{FA}$, such as the class $\mathbf{ProFG}$ of profinite groups, $\mathbf{ProFAb}$ of profinite abelian groups, or $\mathbf{ProFAnn}$ of profinite commutative rings with unit. However, one also has naturally occuring cases where $\mathbf{C} \subset \mathbf{FA}$, such as the class $p\mathbf{G}$ of (finite) p-groups leading to the pro-p groups.

Regarding **ProFA**, there are instances where this is topologically characterized in **KA** as $\mathbf{K}_0\mathbf{A}$; some examples are groups, semigroups, Boolean lattices, and associative rings. We have no general result concerning the identity $\mathbf{ProFA} = \mathbf{K}_0\mathbf{A}$ apart from the obvious remark that any equational subclass of an equational class with this property inherits it; on the other hand, there are the following counterexamples: $\beta\mathbf{N}$ together with the continuous extension of the successor function $k \rightsquigarrow k + 1$ is a zero-dimensional compact algebra, of type (1), but not profinite since it has only countably many open-closed congruences and $\beta\mathbf{N}$ is not metrizable. Similarly, for any finitely generated infinite abelian group G, the Stone-Čech compactification of its underlying set, made into a topological G-set by extending the left translations of G continuously, is a counterexample for the same reason. Another such example, with countably many unary operations on the one-point compactification of **N**, is given in Eastman [3].

2. Properties transferred from C to ProC. Here we collect a number of properties, primarily of a categorical nature, which are invariant under passage from **C** to **ProC**.

Proposition 2. *If all epimorphisms are onto in* **C** *then this also holds in* **ProC**.

Examples of classes **C** to which this applies are the **FA** for the equational classes **A** consisting of the following kinds of algebras: Boolean algebras; commutative rings with unit satisfying the equations $x^p = x$, $px = 0$ for some prime p; commutative rings with unit satisfying the equation $x^n = x$ for some n; groups; abelian groups. Non-equationally defined such **C** are given by the p-groups, the nilpotent groups, and the finite semi-prime commutative rings with unit. On the other hand, finite distributive lattices and finite semigroups have epimorphisms which are not onto.

That epimorphisms in **ProC** are onto whenever they are so in **C** is not a purely categorical consequence of the fact that **ProC** is the epireflective hull of **C** in some category: the class of discrete spaces and its epireflective hull in all Hausdorff spaces provide a counterexample.

In the equational class **A**, the monomorphisms are exactly the one-to-one maps, as one readily sees from the existence of free algebras in **A**. Furthermore, this carries over to **C** as well as to **ProC**. Equalizers are special monomorphisms, and it is noteworthy when these two coincide in a category.

Proposition 3. *If all monomorphisms are equalizers in* **C** *then this also holds in* **ProC**.

Here, finite groups, abelian groups, and M-sets for any monoid M provide examples of classes **C** with the stated property, and finite semigroups and distributive lattices counterexamples. Also, this proposition rests on more than the fact that **ProC** is an epireflective hull of **C**, as can again be seen by looking at discrete spaces in all Hausdorff spaces.

The most significant of the additional hypothesis on **A** which will be used here is described as follows: Two congruences Θ and Λ, on any algebra, are called *permutable* iff $\Theta \circ \Lambda = \Lambda \circ \Theta$, and if this is so then $\Theta \vee \Lambda = \Theta \circ \Lambda$, the join $\vee$ referring to the lattice of all congruences on the algebra. The equational class **A** will be called *congruence permutable* iff any two congruences on any algebra in **A** are permutable. The importance of this condition for universal topological algebra was first noted in Malcev [9]. We recall that congruence permutability is equivalent to the existence of a ternary polynomial p, i.e. an element in the absolutely free algebra, of the same arity type as **A**, with a three element basis $\{x, y, z\}$, such that the equations $p(x, x, z) = z$ and $p(x, z, z) = x$ hold in the class **A**, and that it implies modularity for the congruence lattice of each algebra in **A** (Malcev [8]). Natural examples of congruence permutable equational classes are given by groups, rings, modules, and Boolean lattices.

Proposition 4. *For congruence permutable* **A**, *if* **C** *is closed under quotients then* **ProC** *is also closed under quotients (in* **KA**).

For any congruence permutable **A**, **FA** has the required property and thus all such **ProFA** are closed under quotients. Other examples for such **C** are given by the p-groups, nilpotent groups, abelian p-groups, and the p-rings, i.e. the finite rings whose additive group is a p-group.[4]) That this proposition does not hold in general is shown by the example **A** = **Ens** ("algebras" without operations): Here **ProFA** is the class of zero-dimensional compact Hausdorff spaces and clearly not closed under quotients.

Next, for two subclasses **D** and **E** $\supseteq$ **D** of **A** (and, analogously, of **KA**), call **D** extensive in **E** iff any $B \in$ **E** already belongs to **D** whenever there exists a non-void $A \subseteq B$ and a congruence Θ on B such that $\Theta(A) = A$, and A and B/Θ belong to **D**.

Proposition 5. *For congruence permutable* **A** *and* **C** *closed under quotients, if* **C** *is extensive in* **FA** *then* **ProC** *is extensive in* **ProFA**.

Examples for this situation are provided by the p-groups, abelian p-groups, and p-rings.

With additional hypotheses on **A** one can reach the stronger conclusion that **ProC** is extensive in **KA**: this is the case whenever **A** is congruence regular, i.e. the algebras in **A** have the property that the only congruence with a singleton class is the trivial one, and **ProFA** = $\mathbf{K}_0$**A**. The above examples, of course, all come under this.

Given a category **K**, we shall mean by a *(reflective) factorization* of **K** a family $(\mathbf{K}_i)_{i \in J}$ of reflective subcategories $\mathbf{K}_i$ of **K** such that the adjunctions ϱ_i: $A \to R_i A$, R_i the functor reflecting **K** into $\mathbf{K}_i$, provide an isomorphism $\varrho : A \to \prod R_i A$ for each

[4]) This term is sometimes used with a different meaning, and these rings are also called the finite p-torsion rings.

$A \in \mathbf{K}$. Further, we shall call such a factorization *disjoint* iff the only maps $A \to B$ where $A \in \mathbf{K}_i$ and $B \in \mathbf{K}_j$ for $i \neq j$ are those that factor through the terminal object of $\mathbf{K}$. It is clear that such factorizations reduce the structure of $\mathbf{K}$ to a good extent to that of the product category $\prod \mathbf{K}_i$.

In the following, let $(\mathbf{C}_i)$ be a family of subclasses of $\mathbf{C}$, each hereditary and finitely productive and hence reflective with onto maps as adjunctions; then the latter also holds for the subcategories $\mathbf{ProC}_i$ of $\mathbf{ProC}$, and one has:

Proposition 6. *If* $(\mathbf{C}_i)$ *is a disjoint factorization of* $\mathbf{C}$ *then* $(\mathbf{ProC}_i)$ *is a disjoint factorization of* $\mathbf{ProC}$.

Examples of classes $\mathbf{C}$ with disjoint factorizations are given by the finite abelian groups, the nilpotent groups, and the finite rings: for the first and second, the subclasses consist of the p-groups, in the original class, for each prime p, and for the third of the p-rings. A ready way of proving disjointness for a factorization $(\mathbf{C}_i)$ of $\mathbf{C}$ is to show that the $\mathbf{C}_i$ are closed under quotients and $\mathbf{C}_i \cap \mathbf{C}_j$, for $i \neq j$, contains only trivial (i.e. singleton) algebras.

We call an algebra $A \in \mathbf{C}$ *semi-simple* iff the intersection of its maximal (proper) congruences is trivial, and use the term analogously for $A \in \mathbf{ProC}$, the congruences to be admitted being the open ones. Of course, a maximal open congruence on an $A \in \mathbf{ProC}$ is, in fact, a maximal congruence.

Proposition 7. *If all* $C \in \mathbf{C}$ *are semi-simple then so are all* $A \in \mathbf{ProC}$.

Examples of classes $\mathbf{C}$ all whose members are semi-simple are evidently easy to obtain: In general, whenever $\mathbf{FA}$ contains any non-trivial algebras, it also contains non-trivial semi-simple ones, and they form a hereditary and finitely productive class. Particular instances are given by all finite abelian groups of squarefree index, all finite distributive lattices, and all finite commutative rings with unit satisfying the equation $x^n = x$ for some n.

3. Completions. This section deals with the construction of pro-$\mathbf{C}$ algebras from algebras in $\mathbf{A}$ by means of uniform space completion and a number of applications of this process.

We shall call an algebra $A \in \mathbf{A}$ $\mathbf{C}$-separated iff its congruences Θ for which $A/\Theta \in \mathbf{C}$ have trivial intersection. The class $\mathbf{TC}$ of all $\mathbf{C}$-separated $A \in \mathbf{A}$ is then hereditary and productive in $\mathbf{A}$ and consists of all algebras isomorphic to subalgebras of products of algebras in $\mathbf{C}$. Consequently, $\mathbf{TC}$ is reflective in $\mathbf{A}$, the reflection R: $\mathbf{A} \to \mathbf{TC}$ again given by passage to quotients, i.e. $RA = A/\Lambda$ where Λ is the intersection of all congruences Θ on A such that $A/\Theta \in \mathbf{C}$, and the adjunction by the corresponding quotient maps. Note, incidentally, that the underlying algebra of any $A \in \mathbf{ProC}$ belongs to $\mathbf{TC}$.

Now consider, on a given $A \in \mathbf{TC}$, any filter basis $\mathfrak{F}$ of congruences Θ such that $A/\Theta \in \mathbf{C}$ which has trivial intersection. Then $\mathfrak{F}$ is a basis of a separated uniformity on the underlying set of A, and all operations of A are uniformly continuous with respect to this since congruences are subalgebras of $A \times A$. This uniformity is, moreover, totally bounded, and hence the completion of the resulting uniform space is compact. The operations of A, by their uniform continuity, then have continuous extensions to the completed space, and whatever equations they satisfy in A the latter will also satisfy, by the familiar principle of extension of identities. It follows that one obtains a topological algebra $\tilde{A} \in \mathbf{KA}$ in this manner.

For more information about these $\tilde{A}$, note that the quotient map $\nu: A \to A/\Theta$, where $\Theta \in \mathfrak{F}$, has a continuous extension $\tilde{\nu}: \tilde{A} \to A/\Theta$ by uniform continuity (discrete uniformity on A/Θ), and therefore $\tilde{A}/\tilde{\Theta} \cong A/\Theta$ for the kernel congruence $\tilde{\Theta}$ of $\tilde{\nu}$. Since $\tilde{\Theta}$ is also the closure of Θ in $\tilde{A}$, $\bigcap \tilde{\Theta} (\Theta \in \mathfrak{F})$ is trivial. It follows that all $\tilde{\Theta}$ for $\Theta \in \mathfrak{F}$ belong to $\mathfrak{R}\tilde{A}$ and $\tilde{A} \in \mathbf{ProC}$. We call $\tilde{A}$ the *$\mathfrak{F}$-completion* of A. For the set of *all* Θ with $A/\Theta \in \mathbf{C}$ this completion will be called the **C**-*adic completion* of A, and the uniformity involved the **C**-*adic uniformity* of A.

Note that $\tilde{A} = \varprojlim_{\Theta \in \mathfrak{F}} \tilde{A}/\tilde{\Theta} = \varprojlim_{\Theta \in \mathfrak{F}} A/\Theta$, the inverse systems involved being the obvious ones, so that $\tilde{A}$, for given $\mathfrak{F}$, could also be described without reference to uniformities and completions; there does, however, seem to be some technical advantage in making use of the latter.

For any $A \in \mathbf{TC}$, let PA be its **C**-adic completion. Then, any $h: A \to B$ in **TC**, being evidently uniformly continuous with respect to the **C**-adic uniformities, extends uniquely to a homomorphism $Ph: PA \to PB$ of topological algebras, and the correspondence $A \rightsquigarrow PA$ and $h \rightsquigarrow Ph$ clearly provides a functor $P: \mathbf{TC} \to \mathbf{ProC}$.

Proposition 8. *The functor P of* **C**-*adic completion is left adjoint to the underlying algebra functor* **ProC** → **TC**.

Recall that the inclusion functor **TC** → **A** has the reflection R: **A** → **TC** as left adjoint. If one defines the generalized **C**-adic completion by the functor PR: **A** → → **ProC** one has, for purely formal reasons:

Corollary 1. *The underlying algebra functor* **ProC** → **A** *has the functor PR of generalized* **C**-*adic completion as left adjoint.*

In similar manner it follows that the underlying set functor **ProC** → **Ens** has PRF as left adjoint, F: **Ens** → **A** being the free algebra functor. Note that there are pro-**C** algebras of arbitrarily large cardinality, so that the front adjunction is one-to-one.

Corollary 2. *For any set X there exists a free pro-***C** *algebra with basis X, provided by the generalized* **C**-*adic completion of the free algebra in* **A** *with basis X.*

In some cases, the free algebras in **A** are known to belong to **TC**, which somewhat simplifies the description of the free pro-**C** algebras. For instance, this holds for both, **FG** and *p***G** in **G**, for **FAb** and *p***Ab** in **Ab**, and for **FAnn** and *p***Ann** in **Ann**.

Let $(\mathbf{C}_i)$ now be a factorization of $\mathbf{C}$, R_i: $\mathbf{A} \to \mathbf{TC}_i$ the reflections, P_i: $\mathbf{TC}_i \to \mathbf{ProC}_i$ the $\mathbf{C}_i$-adic completion functor. Then one has:

Corollary 3. *For any $A \in \mathbf{A}$, PRA is isomorphic to $\prod P_i R_i A$, natural in A.*

Similarly, if E and E_i are the free pro-$\mathbf{C}$, and free pro-$\mathbf{C}_i$ algebra functors on **Ens**:

Corollary 4. *For any set X, EX is isomorphic to $\prod E_i X$, natural in X.*

Thus, one has that the free profinite abelián group on X is the product of the free pro-p abelian groups on X, and the free pro-nilpotent group on X is the product of the free pro-p groups on X, and similar consequences.

Consider now any zero-dimensional Hausdorff space X. If F_X is the **TC**-reflection of the algebra in **A** with the underlying set of X as basis (which may be assumed to contain that set), let $\mathfrak{F}_X$ be the filter basis of those congruences Θ on F_X for which $F_X/\Theta \in \mathbf{C}$, and which induce on X an open-closed decomposition. One proves that $\bigcap \Theta(\Theta \in \mathfrak{F}_X)$ is trivial and hence has the $\mathfrak{F}_X$-completion GX of F_X. Futhermore, any continuous map u: $X \to Y$ between zero-dimensional Hausdorff spaces induces a homomorphism $\bar{u}$: $F_X \to F_Y$ which is uniformly continuous for the uniformities given by $\mathfrak{F}_X$ and $\mathfrak{F}_Y$, respectively, and hence extends to a homomorphism Gu: $GX \to GY$ in **ProC**. The result is a functor G into **ProC**.

Proposition 9. *The functor G is left adjoint to the underlying space functor from* **ProC**, *and the front adjunction provides embeddings.*

Of course, for a factorization $(\mathbf{C}_i)$ of $\mathbf{C}$ one again has $GX \cong \prod G_i X$, analogous to the case of free pro-**C** algebras over sets.

Since any zero-dimensional Hausdorff space X can be regarded as a subspace of the pro-**C** algebra GX one may ask what its closure is in GX, this providing a particular compactification procedure for all zero-dimensional Hausdorff spaces. The answer is simple: It is just the maximal zero-dimensional compactification of X.

Corollary. *If A is a free pro-*__C__ *algebra with basis space X and B the closed subalgebra of A generated by a subspace Y of X then B is a free pro-*__C__ *algebra on Y iff every finite open-closed partition of Y is the restriction of such a partition of X.*

Completions can also be used to obtain coproducts: Given any family (A_λ) of pro-**C** algebras, let F be the coproduct in **A** of their underlying algebras, with algebra homomorphism u_λ: $A_\lambda \to F$ as the canonical maps. Then consider the filter basis $\mathfrak{F}$ of congruences Θ on F for which $F/\Theta \in \mathbf{C}$ and $(u_\lambda \times u_\lambda)^{-1}(\Theta) \in \mathfrak{R}A_\lambda$ for all λ. For $\Lambda = \bigcap \Theta(\Theta \in \mathfrak{F})$, let $F_0 = F/\Lambda$, $\mathfrak{F}_0$ the filter basis of congruences Θ_0 induced modulo Λ on F_0 by the $\Theta \in \mathfrak{F}$, A the $\mathfrak{F}_0$-completion of F_0, and j_λ: $A_\lambda \to A$ the map resulting from u_λ.

Proposition 10. *A is the coproduct of* (A_λ) *in* **ProC** *with* (j_λ) *as its family of canonical maps.*

Note that the existence of coproducts in **ProC** and of the adjoint functors considered above could also be obtained from general principles (reflection from **KA**, Adjoint Functor Theorem), but this would yield rather less information than the constructions by means of completion provide.

We conclude this section with a result concerning the structure of certain individual completions. Let Σ be a set of filter bases $\mathfrak{F}$ of congruences on some algebra $A \in$ **TC** such that $A/\Theta \in$ **C** for each $\Theta \in \mathfrak{F}$ and $\bigcap\Theta(\Theta \in \mathfrak{F}) = \Delta$. Σ will be called independent iff $\Theta_0 \circ (\Theta_1 \cap \ldots \cap \Theta_n) = \nabla$, the congruence with a single class, for any $\Theta_i \in \mathfrak{F}_i \in \Sigma$ where all $\mathfrak{F}_i$ are distinct.

Proposition 11. *For independent* Σ, *the product of the* $\mathfrak{F}$-*completions of* A, *for* $\mathfrak{F} \in \Sigma$, *is isomorphic to the* $\mathfrak{G}$-*completion of* A, $\mathfrak{G}$ *consisting of all* $\Theta_1 \cap \ldots \cap \Theta_n$ *where* $\Theta_i \in \mathfrak{F}_i \in \Sigma$, *the isomorphism being the extension, to the* $\mathfrak{G}$-*completion, of the diagonal map of* A *into the product.*

4. Projectives. In the present context, projectivity is always understood with respect to onto maps. Since all topological algebras under consideration are compact, this makes the general theory given in Banaschewski [1] applicable, and one has the following:

(1) $A \in$ **ProC** *is projective iff every onto map* $B \to A$ *in* **ProC** *has a right inverse iff every coessential onto map* $B \to A$ *in* **ProC** *is an isomorphism, a map being coessential onto iff it is onto and maps proper closed subalgebras to proper subalgebras.*[5])

(2) *For any* $A \in$ **ProC** *there exists an essentially unique coessential onto map* h: $B \to A$ *with projective* B. B, *or sometimes* h, *is called a projective cover of* A.

(3) *For any onto map* h: $B \to A$ *in* **ProC**: B *is a projective cover of* A *iff* h *is coessential onto and any map* g *in* **ProC** *for which* hg *is coessential onto is an isomorphism iff any onto map* g *in* **ProC** *with projective codomain and a factorization* $h = fg$ *is an isomorphism.*

Our aim here is to obtain statements about the projectives in **ProC** on the basis of algebraic conditions on **A** and **C**. We make one blanket hypothesis: **A** is taken to be congruence permutable, **C** closed under quotients, and for the congruences in **A** it is further assumed that for any algebras A, B and C in **A** such that $B \subseteq C \subseteq A$, and any congruence Θ on A, $\Theta(B) = B$ implies $\Theta(C) = C$. Note that these conditions, as far as they concern **A**, are satisfied by any equational class each of whose algebras

[5]) In Banaschewski [1], "essential" was used in place of "coessential"; the latter seems preferable in the context of projectivity.

has a group as a reduct, typical instances being groups themselves, modules over a ring, rings, and algebras over a ring (associative or not).

A preliminary result, which only depends on the first two parts of the present assumption, is:

Proposition 12. *$A \in$ **ProC** is semi-simple iff it is a product of simple $C \in$ **C**.*

The proof uses the fact that congruence permutability implies the modularity of the congruence lattices for all algebras in **A**.

In dealing with projectivity it is desirable to have a characterization of the coessential maps. To this end, one first proves that (i) any closed proper subalgebra of an $A \in$ **ProC** is contained in a maximal closed subalgebra, and (ii) for any maximal closed subalgebra M of A there exist congruences $\Theta \in \mathfrak{R}A$ such that $\Theta(M) = M$ and among these a largest one, say Θ_M. The intersection of all Θ_M will then be called the *Frattini congruence* ΦA of A. Its significance is as follows:

Lemma. *In* **ProC**, *an onto map h: $A \to B$ is coessential iff* $\operatorname{Ker}(h) \subseteq \Phi A$.

Corollary 1. *For projective A, $B \in$ **ProC**, $A \cong B$ iff $A/\Phi A \cong B/\Phi B$.*

Corollary 2. *A, $B \in$ **ProC** have isomorphic projective covers iff $A/\Phi A \cong B/\Phi B$.*

Note also: $\Phi(A/\Phi A)$ is trivial for any $A \in$ **ProC**.

Applications of the Frattini congruence, in the case of profinite groups, occur in Banaschewski [1] and in Gruenberg [6].

Proposition 13. *If all $C \in$ **C** are semi-simple, and each simple $S \in$ **C** has a proper subalgebra then all $A \in$ **ProC** are projective.*

The proof of this and the following proposition proceeds by showing that the hypotheses imply $\Phi A = \Delta$ for all A, which eliminates all non-trivial coessential maps.

Instances of classes **C** to which this is applicable are given by the finite abelian groups of square free index and analogous classes of finite modules over a given ring.

Any algebra $C = A/\Theta_M$ is a finite algebra containing a maximal subalgebra D such that the only congruence Θ on C for which $\Theta(D) = D$ is the trivial one. We call such algebras *compressed*. Note that any simple algebra which has proper subalgebras is compressed, but one can easily give example of compressed algebras which are not simple. In the other direction, let an algebra S be called *strongly simple* iff the diagonal $\Delta \subseteq S \times S$ is a maximal subalgebra (rather than merely a maximal congruence, which means simplicity). One then has, for finite S, $S \times S$ is compressed whenever $S \in$ **A** is strongly simple and **A** congruence regular.

Proposition 14. *If **A** is congruence regular and all $C \in$ **C** are powers of a single strongly simple $S \in$ **C** then all non-trivial $A \in$ **ProC** are projective.*

A particular instance of this, which does not fall under the previous proposition, is the fact that all non-trivial profinite commutative rings satisfying the equations $x^p = x$, $px = 0$ for some prime p are projective (in their category) which was proved by an entirely different method in Banaschewski [1].[6])

We now turn to the relationship between projective and free pro-**C** algebras. Clearly, any pro-**C** algebra free on a set is projective (the maps admitted are onto), whereas, conversely, the last two propositions readily lead to cases in which not every projective is free on a set, or, for that matter, on a space. Rather more surprising is:

Proposition 15. *Any pro-***C** *algebra free on some zero-dimensional Hausdorff space is projective.*

The proof of this depends strongly on the assumption of congruence permutability and is related to the corresponding proof for profinite groups in Serre [12].

Since a projective pro-**C** algebra A is determined by $A/\Phi A$ one expects information on projectives if one has some on the pro-**C** algebras of the form $A/\Phi A$. Take, for instance, **C** to have, up to isomorphism, only one compressed algrebra S, and assume further that S is simple. This makes each $A/\Phi A$ isomorphic to some power $S^{\mathfrak{m}}$, $\mathfrak{m}$ uniquely determined by A. We call $\mathfrak{m}$ the *colength* of A and note that a projective pro-**C** algebra is then simply determined by its colength. Considering the colengths of free pro-**C** algebras one obtains:

Proposition 16. *If all compressed algebras in* **C** *are simple and isomorphic to each other then the projective pro-***C** *algebras with infinite colength* $\mathfrak{m}$ *are exactly the pro-***C** *algebras free on the one-point compactification of a discrete space of cardinality* $\mathfrak{m}$.

It should be added that not all projectives of infinite colength are free on some set; the latter are exactly those of colength $2^{\mathfrak{m}}$.

What the situation is regarding the projectives of finite colength we do not know. A priori it would seem possible that only certain natural numbers occur as the colengths of free pro-**C** algebras, but this is not the case for pro-p groups, abelian pro-p groups, and the categories of profinite modules analogous to the latter, to which the above applies: there, the colength of the free pro-**C** algebra on a finite set of n elements is n, and the equation free = projective holds completely (Banaschewski [1]).

We conclude with a result concerning factorizations and projectives.

Proposition 17. *If* $(\mathbf{ProC}_i)$ *is a factorization of* **ProC**, *resulting from a factorization* $(\mathbf{C}_i)$ *of* **C**, *with reflections* R_i: $\mathbf{ProC} \to \mathbf{ProC}_i$, *then* $A \in \mathbf{ProC}$ *is projective iff all* R_iA *are projective in* $\mathbf{ProC}_i$.

---

[6]) Erratum: The qualification "non-trivial" was omitted there, but the trivial ring is evidently not projective here. In general, the status of the trivial $A \in \mathbf{ProC}$ depends on **C**: for any $\mathbf{C} \subseteq \mathbf{G}$, for instance, the trivial groups are projective in **ProC** since they are initial besides being terminal objects.

As an immediate consequence of this and Proposition 14 one has, for instance, that all non-trivial profinite rings, for any equational class of rings generated by a set of finite prime fields, are projective. Similarly, but with some additional arguments, one obtains that projective pronilpotent groups are exactly the products of projective pro-p groups.

References

[1] *B. Banaschewski:* Projective covers in categories of topological spaces and topological algebras. General Topology and its Relations to Modern Analysis and Algebra (Proc. Kanpur Topological Conf., 1968). Academia, Prague, 1971, 63—91.
[2] *N. Bourbaki:* General Topology I. Herrman, Paris, and Addison-Wesley, Reading, Mass., 1966.
[3] *D. E. Eastman:* Universal topological and uniform algebra. Doctoral dissertation, McMaster University, 1970.
[4] *K. Golema:* Free products of compact general algebras. Colloq. Math. *13* (1965), 165—166.
[5] *G. Grätzer:* Universal Algebra. Van Nostrand, Princeton, N. J., 1968.
[6] *K. W. Gruenberg:* Projective profinite groups. J. London Math. Soc. *42* (1967), 155—165.
[7] *H. Herrlich:* Topologische Reflexionen und Coreflexionen. Springer Lecture Notes *78*, 1968.
[8] *A. I. Malcev:* On the general theory of algebraic systems. (Russian.) Mat. Sb. *35* (*77*) (1954), 3—20.
[9] *A. I. Malcev:* Free topological algebras. A.M.S. Translations (2) *17* (1961), 173—200.
[10] *E. Manes:* A triple theoretic construction of compact algebras. Springer Lecture Notes *80*, 1969, 91—118.
[11] *B. Mitchell:* Theory of Categories. Academic Press, New York and London, 1965.
[12] *J. P. Serre:* Cohomologie Galoisienne. Springer Lecture Notes *5*, 1965.

CONTRACTION OF SOME SPACES OF HOMEOMORPHISMS

W. BARIT

Amsterdam

Let $Q = \prod_{i>0} I_i$ and $s = \prod_{i>0} I^o$ where $I_i = [-1, 1]$ and $I_i^o = (-1, 1)$. Let l_2 denote separable Hilbert space. Following Anderson [1], we say a set K in X is a Z-set if K is closed and for each non-empty homotopically trivial open set U in X, $U \setminus K$ is non-empty and homotopically trivial. Some examples of Z-sets in l_2 are closed σ-compact subsets and closed sets whose projection in infinitely many directions is a point. Let $H(X)$ be the space of homeomorphisms of X onto X with the compact-open topology. Let $H_K(X) = \{h \in H(X) \mid h/K = \text{id}\}$. The main result is the following:

Theorem 1. *Let $X = Q$, s, or l_2, and let K be a compact Z-set in X. Then $H_K(X)$ is contractible.*

As background to this theorem, Wong [4] showed that any homeomorphism of X is isotopic to the identity. Renz [3] observed that this process is continuous and in fact contracts $H(X)$. In a later paper [5] Wong showed that any homeomorphism of X which is the identity on a compact Z-set K, is isotopic to the identity with each level of the isotopy being the identity on K. The proof of Theorem 1 requires a non-trivial modification of Wong's technique and the use of a canonical homeomorphism extension theorem due to Chapman [2]. We also obtain the following theorem.

Theorem 2. *Let $X = s$ or l_2, and let K be a Z-set in X. If h is a homeomorphism of X such that $h/K =$ id, then h is isotopic to the identity via $\{H_t\}_{t\in[0,1]}$ where for each t, $H_{t/K} =$ id.*

Theorem 2 shows that the compactness condition is not required for K, and thus answers a question posed in Wong's paper [5]. The methods used here do not show, however, that $H_K(s)$ is contractible for K a non-compact Z-set, and this question is still open. The requirement that K be a Z-set is necessary, and some examples are given.

References

[1] *R. D. Anderson:* On topological infinite deficiency. Michigan Math. J. *14* (1967), 365—383.

[2] *T. A. Chapman:* Canonical Extensions of Homeomorphisms. (Preprint.)

[3] *P. Renz:* The contractibility of the homeomorphism group of some product spaces by Wong's method. Math. Scand. (to appear).

[4] *R. Y. T. Wong:* On homeomorphism of certain infinite-dimensional spaces. Trans. Amer. Math. Soc. *28* (1967), 140—153.

[5] *R. Y. T. Wong:* Stationary isotopes of infinite-dimensional spaces. (Preprint.)

NON-CYCLIC TRANSFORMATIONS AND UNIFORM CONVERGENCE OF PICARD SEQUENCES

A. BELLEN and A. VOLČIČ

Trieste

Definition. A continuous transformation of a topological space S into itself is said to be *non-cyclic* iff $f(x) \neq x$ implies $f^2(x) \neq x$.

In a recent paper, extending a theorem due to S. C. Chu and R. D. Moyer, the second author proved that, if f is a continuous transformation of a compact and connected space S into itself, whose topology is deduced from a total ordering, then all the Picard sequences converge iff f is non-cyclic (and this last property is proved to be equivalent to five other properties).

In this paper the authors characterize (under the same hypotheses on S) the non-cyclic transformations for which the Picard sequences converge uniformly with respect to $x \in S$. Besides, some partial answers to the same problem in a more general setting are given.

Notations: $F(f)$ indicates the set of all fixed points of f. $F^*(f)$ indicates the set $\bigcap_N f^n(S)$.

Obviously we have $F^*(f) \supset F(f)$.

Theorem. *If S is totally ordered, connected and compact in the order topology, and f a non-cyclic transformation of S into itself, then the following properties are equivalent:*

(a) *$F(f)$ is connected;*
(b) $F^*(f) = F(f)$;
(c) *the convergence of the Picard sequences is uniform.*

Published in the Rend. Ist. di Matem. Univ. Trieste, vol. IV, fasc. I (1972), 1—7.

UNIVERSITY OF TRIESTE

RECENT RESULTS IN THE FUNCTIONAL ANALYTIC INVESTIGATIONS OF CONVERGENCE SPACES

E. BINZ

Mannheim

In the past fifteen years various generalizations of the notion of a topology have appeared in connection with several branches of mathematics, mainly with functional analysis, e.g. [16]. The generalization I have in mind is that of the so-called convergence structure introduced in [13] and [15].

During the Symposium Dr. Simon brought my attention to the following two papers of Katětov: On continuity structures and spaces of mappings, Comment. Math. Univ. Carolinae 6 (1965), 257–278; Convergence structures, Proceedings of the Second Prague Topological Symposium, 1966, 207–216. Those papers are indeed closely related to several papers listed in the References. I therefore would like to thank Dr. Simon very much.

A *convergence structure* Λ is a map from a non-empty set X into the power set $P(F(X))$ of the set of all filters (in the sense of Bourbaki [9]) $F(X)$ of X assigning to each point $p \in X$ a collection of filters $\Lambda(p)$ satisfying:

(i) The filter $\dot{p}$ generated by $\{p\}$ belongs to $\Lambda(p)$,
(ii) every filter Ψ finer than a member Φ of $\Lambda(p)$ belongs to $\Lambda(p)$ and finally
(iii) the infimum $\Phi \wedge \Psi$ of any two filters Φ and Ψ of $\Lambda(p)$ belongs to $\Lambda(p)$.

The filters in $\Lambda(p)$ are called the filters converging to p with respect to Λ or simply the filters converging to p. The set X together with Λ is called a convergence space or simply a space.

Every *topology* T on the set X will be interpreted as a convergence structure in the following way: For any point $p \in X$ let $T(p)$ be the set of all filters converging to p. Hence we say that the topology T is a convergence structure. We call a convergence structure *topological* if it is a topology.

Let me now construct an example of a space which is not a topological space. A map from a space X into a space Y is said to be continuous if for each point $p \in X$ the filter $f(\Phi)$ converges to $f(p)$ for any filter Φ converging to p. The collection of all continuous real-valued functions of the space X is denoted by $C(X)$. Now we endow $C(X)$ with the continuous convergence structure Λ_c defined as follows: For any function $f \in C(X)$ the set $\Lambda_c(f)$ consists of all filters Θ for which the filter $\Theta(\Phi)$ generated by

$$\{T(F) \mid T \in \Theta \text{ and } F \in \Phi\}$$

converges to $f(p)$ for any point $p \in X$ and any filter Φ converging to p. The set $C(X)$ together with Λ_c is denoted by $C_c(X)$. The latter space is, under the pointwise defined operations, an **R**-*convergence algebra*, meaning that the operations are continuous. For some basic facts on $C_c(X)$ one may consult [5]. For a completely regular topological space X, the continuous convergence structure Λ_c is a topology iff X is locally compact, in which case Λ_c coincides with the topology of compact convergence. Hence to give an example of a space which is not topological we choose in $C_c(X)$ the space X to be the rationals with the usual topology.

In this lecture I would like to present some of the recent results concerning the investigations of the relationship between a certain type of space X and $C_c(X)$. The class of spaces we restrict ourselves to is that of all c-embedded spaces. Let me briefly explain the notion of a *c-embedded space*. For any space X let $\mathrm{Hom}_c\, C_c(X)$ denote the collection of all real-valued continuous **R**-algebra-homomorphisms from $C(X)$ onto **R**, endowed with the continuous convergence structure. The map

$$i_X : X \to \mathrm{Hom}_c\, C_c(X)$$

defined by $i_X(p) = f(p)$ for all $p \in X$ and all $f \in C_c(X)$ is a continuous surjection [6]. A space X now is said to be c-embedded if i_X is a homeomorphism, i.e. a bicontinuous bijection. As examples of c-embedded spaces let me give $C_c(Y)$ for any space Y, any completely regular topological space (satisfying T_1) and any subspace of a c-embedded space.

The reason we restrict ourselves to the class of c-embedded spaces is the fact that any two c-embedded spaces X and Y are homeomorphic iff $C_c(X)$ and $C_c(Y)$ are bicontinuously isomorphic.

Now let me present the first result [17]:

Theorem 1. *For any c-embedded space X the following conditions are equivalent*:

(i) *X is locally compact*;
(ii) *$C_c(X)$ is topological.*

If $C_c(X)$ is topological (X being c-embedded), then Λ_c is the topology of compact convergence.

What means locally compact? A space is said to be *compact* if every ultrafilter converges to a unique point. We call a space X locally compact, if every convergent filter contains a compact subset of X.

For compact c-embedded spaces we have [8]:

Theorem 2. *For any c-embedded space X the following conditions are equivalent*:

(i) *X is compact;*
(ii) *X is compact and topological;*
(iii) $C_c(X)$ *carries a norm topology.*

In a c-embedded convergence space any compact subspace is topological. Hence any c-embedded locally compact space is the inductive limit (in the category of convergence spaces) of compact topological spaces. We shall soon meet an example of a c-embedded locally compact space which is not topological.

The next two results concern completely regular topological spaces. We would like to convert the topological term "normal" of a completely regular topological space X into a functional analytic term of $C_c(X)$. The space X is normal iff the restriction map

$$r : C(X) \to C(A)$$

is surjective for any (non-empty) closed subset A of X. Let $I(A)$ denote the ideal in $C(X)$ of all functions vanishing on A. Hence we have the following commutative diagram:

$$\begin{array}{ccc} C_c(X) & \xrightarrow{\ r\ } & C(A) \\ {\scriptstyle\pi}\downarrow & \nearrow{\scriptstyle\bar{r}} & \\ C(X)/I(A) & & \end{array}$$

where π denotes the canonical projection and $\bar{r}$ the map induced by r. Hence r is surjective iff $\bar{r}$ is surjective. Now let us endow $C(X)/I(A)$ with the finest [3] of all convergence structures for which π is continuous. This space will be denoted by $C_c(X)/I(A)$. Then $\bar{r}$ is a homeomorphism onto a subspace of $C_c(A)$.

For any space Y the convergence algebra $C_c(Y)$ is complete [7], i.e. every Cauchy-filter (in the obvious sense) converges. Now there is a Stone-Weierstrass theorem [7] saying that for any completely regular topological space Y any complete subalgebra of $C_c(Y)$ inducing the topology and containing the constants is all of $C(Y)$.

Since $\bar{r}(C_c(X)/I(A))$ induces the topology of A and contains the constants, $\bar{r}$ is surjective iff $C_c(X)/I(A)$ is complete. Since any closed proper ideal in $C_c(X)$ is of the form $I(A)$ for some closed (non-empty) subset $A \subset X$, we have:

Theorem 3. *Let X be a completely regular topological space. Then X is normal iff $C_c(X)/I$ is complete for any closed (proper) ideal in $C_c(X)$.*

The next two theorems are due to W. A. Feldman [12].

Theorem 4. *Let X be a completely regular topological space. The following conditions are equivalent:*

(i) *X is metrizable and separable;*
(ii) $C_c(X)$ *is second countable.*

Second countable means the following:

There is a system S of at most countably many subsets of $C_c(X)$ such that to each filter in $C_c(X)$ there exists a coarser one (still convergent) with a basis of members of $\mathbf{S}$.

To find a functional analytic equivalent of the term *metric* one may consult [2].

Theorem 5. *Let X be a c-embedded space. The following conditions are equivalent*:

(i) *X is Lindelöf;*
(ii) *$C_c(X)$ is first countable.*

The notion of Lindelöf is based on the notion of a *covering system.* A system S of subsets of X is a covering system if in every convergent filter in X there is a member of S. The space X is Lindelöf if to every covering system S there is a countable covering system S' refining (defined in the obvious way) S.

First countable means simply that to any convergent filter there exists a coarser one (still converging) with a countable basis.

We now turn our attention to some functional analytic properties of $C_c(X)$, namely to those of *duality*. Let $L_cC_c(X)$ be the c-dual space, i.e. the space of all continuous linear real-valued functions carrying the continuous convergence structure. The next three theorems are due to H. P. Butzmann [10], [11].

Theorem 6. *For any convergence space X the canonical map*

$$j : C_c(X) \to L_cL_cC_c(X)$$

(defined by $j(f) = l(f)$ for all $l \in L_cC_c(X)$ and all $f \in C(X)$) is a bicontinuous isomorphism, i.e. $C_c(X)$ is c-reflexive.

This theorem is based on the following two theorems:

Theorem 7. *Let X be a c-embedded space. The locally convex topology on $C(X)$ generated by all continuous seminorms of $C_c(X)$ is the topology of compact convergence.*

Theorem 8. *For any space X the* **R**-*vector space generated by $i_X(X)$ in $L_cC_c(X)$ is dense in $L_cC_c(X)$.*

The theory of c-duality for general convergence spaces is, except for certain special classes of such spaces [14], not developed at all. With the intention to develop such a theory my assistents Dr. H. P. Butzmann, Dr. K. Kutzler and myself began to study the c-dual spaces of topological **R**-vector spaces. Here are some of the results the proof of which can be found in [4] and [11].

Theorem 9. *A convergence* **R**-*vector space F is a c-dual space of some topological vector space iff the following conditions hold*:

(i) *F is locally compact;*
(ii) *all compact subsets in F are topological and*
(iii) *F has point-separating continuous linear functionals.*

Now we can easily give an example of a locally compact c-embedded convergence space which is not topological. Theorem 9 applied to any infinite dimensional locally convex separated vector space E asserts that the c-dual L_cE is locally compact. Clearly L_cE is c-embedded and not topological.

Theorem 10. *For any topological* **R**-*vector space E the canonical map*

$$j : E \to L_c L_c E$$

maps E homeomorphically onto a dense subspace of the complete locally convex **R**-*vector space* L_cL_cE *iff E is locally convex.*

Another branch of our studies is devoted to an extension of *Pontryagin's duality theory* for locally compact Abelian groups. The main result, due to H. P. Butzmann, links the c-dual space of an **R**-convergence vector space E of a certain type with the group $\Gamma_c E$ (carrying the continuous convergence structure) of all continuous group homomorphisms of E into the unit circle T. It allows us to describe an extension of Pontryagin's duality theory for certain groups:

Theorem 11. *Let E be an* **R**-*convergence vector space in which for any filter* Φ *in E converging to zero the filter* $[-1,1]\cdot\Phi$ *generated by* $\{[-1,1]\cdot F \mid F \in \Phi\}$ *converges to zero, too. Then the canonical projection* $\pi : \mathbf{R} \to T$, *sending each* $\lambda \in \mathbf{R}$ *into* $e^{2\pi i\lambda}$ *induces a bicontinuous group isomorphism*

$$\pi^* : L_cE \to \Gamma_cE\,,$$

defined by $\pi^*(l) = \pi \circ l$ *for all* $l \in L_cE$. *Moreover, the canonical group homomorphism*

$$j'_E : E \to \Gamma_c\Gamma_cE$$

(defined by $j'_E(p)(g) = g(p)$ *for all* $p \in E$ *and all* $g \in \Gamma_cE$*) is a bicontinuous group isomorphism iff E is c-reflexive. Hence* $j'_{C_c(X)}$ *is a bicontinuous group isomorphism for any space X.*

For a given topological **R**-*vector space E the group homomorphism* j'_E *is a bicontinuous bijection iff E is locally convex and complete.*

References

[1] *E. Binz and H. H. Keller:* Funktionenräume in der Kategorie der Limesräume. Ann. Acad. Sci. Fenn. Ser. A I. *383* (1966), 1—21.

[2] *E. Binz and K. Kutzler:* Über metrische Räume und $C_c(X)$. Ann. Scuola Norm. Sup. Pisa *26* (1) (1972), 197—223.

[3] *E. Binz and W. A. Feldman:* A functional analytic description of normal spaces. Canad. J. Math. *24* (1) (1972), 45—49.

[4] *E. Binz, H. P. Butzmann and K. Kutzler:* Über den *c*-Dual eines topologischen Vektorraumes. Math. Z. *127* (1972), 70—74.

[5] *E. Binz:* Convergence spaces and convergence function algebras. Proc. Internat. Sympos. on Topology and its Applications (Herceg-Novi, 1968). Savez Društava Mat. Fiz. i Astronom., Belgrade, 1969, 87—92.

[6] *E. Binz:* Zu den Beziehungen zwischen *c*-einbettbaren Limesräumen und ihren limitierten Funktionenalgebren. Math. Ann. *181* (1969), 45—52.

[7] *E. Binz:* Notes on a characterization of function algebras. Math. Ann. *186* (1970), 314—326.

[8] *E. Binz:* Kompakte Limesräume und limitierte Funktionenalgebren. Comment. Math. Helv. *43* (1968), 195—203.

[9] *N. Bourbaki:* Topologie générale. Chapitre I, 3ème ed. Act. Sci. Ind. 1142, Paris, 1961.

[10] *H. P. Butzmann:* Dualitäten in $C_c(X)$. Ph. D. Thesis, University of Mannheim, W. Germany.

[11] *H. P. Butzmann:* Über die *c*-Reflexivität von $C_c(X)$. Comment. Math. Helv. *47* (1972), 92—101.

[12] *W. A. Feldman:* Topological spaces and their associated convergence function algebras. Ph. D. Thesis, Queen's Univ., Kingston, Canada.

[13] *H. R. Fischer:* Limesräume. Math. Ann. *137* (1959), 269—303.

[14] *H. Jarchow:* Duale Charakterisierung der Schwartz-Räume. Math. Ann. *196* (1972), 85—90.

[15] *H. J. Kowalsky:* Limesräume und Komplettierung. Math. Nachr *12* (1954), 301—340.

[16] *G. Marinescu:* Espaces vectoriels pseudotopologiques et théorie des distributions. Deutsch. Verlag Wissensch., Berlin, 1963.

[17] *M. Schroder:* Continuous convergence in a Gelfand theory for topological algebras. Ph. D. Thesis, Queen's Univ., Kingston, Canada.

FOUR GENERALIZATIONS OF STRATIFIABLE SPACES

C. R. BORGES

Davis

1. Introduction

Recently various authors have introduced and studied various generalizations of stratifiable spaces, whose definitions appear in the next section. In particular, we have the (*completely*) *monotonically normal* spaces of Zenor, the *elastic* spaces of Tamano and Vaughan, and the *linearly stratifiable* spaces of Vaughan. In attempting to decipher the common denominator of these spaces we will introduce one more class of spaces, which we shall call *stratonormal* spaces. We will also state and prove a characterization of monotonically normal spaces which will play a critical role in the following results, the proofs of which will appear elsewhere:

(a) The class of monotonically normal spaces equals the class of completely monotonically normal spaces (this answers questions of Heath, Lutzer and Zenor).

(b) "Linearly stratifiable" implies "elastic" implies "stratonormal" implies "monotonically normal".

2. Preliminary results

For the sake of completeness, let us begin with the definitions of elastic and stratonormal spaces, linearly stratifiable spaces and monotonically normal spaces.

Definition 2.1 (see [3]). Let X be a space. Then

(a) Let $\mathscr{U}$ be any collection of subsets of X and let $\mathscr{R}$ be a relation on $\mathscr{U}$ (i.e., $\mathscr{R} \subset \mathscr{U} \times \mathscr{U}$). (We shall often write $U \leqq V$ instead of $(U, V) \in \mathscr{R}$. The relation $\mathscr{R}$ is said to be a *framing of* $\mathscr{U}$ provided that, for every $U, V \in \mathscr{U}$ with $U \cap V \neq \emptyset$, either $U \leqq V$ or $V \leqq U$.

(b) A collection $\mathscr{U}$ is said to be *framed in a collection* $\mathscr{V}$ *with frame map* f: $\mathscr{U} \to \mathscr{V}$ provided that there exists a framing $\mathscr{R}$ of $\mathscr{U}$ such that for every $\mathscr{U}' \subset \mathscr{U}$ which has an $\mathscr{R}$-upper bound we get that $\overline{\bigcup \mathscr{U}'} \subset \bigcup f(\mathscr{U}')$.

(c) If $\mathscr{U}$ is framed in $\mathscr{V}$ and $\mathscr{R}$ is a transitive relation then $\mathscr{U}$ is said to be *elastic* in $\mathscr{V}$.

(d) A pair-base $\mathscr{P}$ for X (i.e., $\mathscr{P}$ is a collection of pairs $P = (P_1, P_2)$ of subsets of X such that P_1 is open, $P_1 \subset P_2$ and, for each $x \in X$ and neighborhood U of x, there exists $(P_1, P_2) \in \mathscr{P}$ with $x \in P_1 \subset P_2 \subset U$) is said to be an *elastic* (*framed*) *base* if there is a framing of $\mathscr{P}_1 = \{P_1 \mid (P_1, P_2) \in \mathscr{P}\}$ such that $\mathscr{P}_1$ is elastic (framed) in $\mathscr{P}_2 = \{P_2 \mid (P_1, P_2) \in \mathscr{P}\}$ with respect to the map $f : \mathscr{P}_1 \to \mathscr{P}_2$ defined by $f(P_1) = P_2$.

(e) A T_1-space with an elastic (framed) base is called an *elastic* (*stratonormal*) *space.*

Definition 2.2 (see [4]). A T_1-space X is said to be *linearly stratifiable* provided that there exists some infinite cardinal number α (we assume that cardinal numbers are ordinal numbers) such that to each open $U \subset X$ one can assign a family $\{U_\beta\}_{\beta<\alpha}$ of open subsets of X such that

(a) $U_\beta^- \subset U$ for all $\beta < \alpha$,
(b) $\bigcup\{U_\beta \mid \beta < \alpha\} = U$,
(c) $U_\beta \subset V_\beta$ whenever $U < V$,
(d) $U_\gamma \subset U_\beta$ whenever $\gamma < \beta < \alpha$.

Definition 2.3 (see [2]). For any space X, let $\mathscr{D}_X = \{(A, B) \mid A$ and B are disjoint closed subsets of $X\}$, $\mathscr{S}_X = \{(A, B) \mid A, B \subset X$ and $A \cap \bar{B} = \emptyset = B \cap A^-\}$. The T_1-space X is said to be *monotonically normal* (respectively *completely monotonically normal*) provided that to each $(A, B) \in \mathscr{D}_X$ (respectively $(A, B) \in \mathscr{S}_X$) one can assign an open subset $G(A, B)$ of X such that

(a) $A \subset G(A, B) \subset G(A, B)^- \subset X - B$,
(b) $G(A, B) \subset G(A', B')$ whenever $A \subset A'$ and $B' \subset B$.

The function G is called a *monotone normality operator.*

We will now state and prove various characterizations of monotonically normal spaces. The techniques developed in the following result permit rather elementary proofs of most results announced in the introduction.

Theorem 2.4. *The following are equivalent*[1]):

(a) *X is completely monotonically normal.*
(b) *X is monotonically normal.*
(c) *To each pair (A, U) of subsets of X, with A closed, U open and $A \subset U$, we can assign an open set U_A such that*

(i) *$U_A \subset V_B$ whenever $A \subset B$ and $U \subset V$,*
(ii) *$U_A \cap (X - A)_{X-U} = \emptyset$.*

[1]) P. Zenor informed us at the Third Prague Topological Symposium that R. Heath and D. Lutzer have also obtained the equivalence of (a) and (b). This has been confirmed by a letter from D. Lutzer, without proofs.

(d) *For each open* $U \subset X$ *and* $x \in U$ *there exists an open neighborhood* U_x *of* x *such that* $U_x \cap V_y \neq \emptyset$ *implies* $x \in V$ *or* $y \in U$.

(e) *There exists a base* $\mathscr{B}$ *for* X *such that, for each* $B \in \mathscr{B}$ *and* $x \in B$ *there exists an open neighborhood* B_x *of* x *such that*

$$B_x \cap C_y \neq \emptyset \quad \textit{implies} \quad x \in C \textit{ or } y \in B.$$

Proof. Clearly (a) implies (b). Therefore we first prove that (b) implies (c): By Lemma 5.1 of [2], X has a monotone normality operator G such that $G(A, B) \cap G(B, A) = \emptyset$ for each $(A, B) \in \mathscr{D}_X$ (indeed, the proof consists of observing that, for any monotone normality operator H on X, letting $G(A, B) = H(A, B) - H(B, A)^-$ will do the trick). Now, for each pair (A, U) with A closed, U open and $A \subset U$, let

$$U_A = G(A, X - U).$$

It is quite easy to see that

(i) $U_A \subset V_B$ whenever $A \subset B$ and $U \subset V$,
(ii) $U_A \cap (X - A)_{X-U} = \emptyset$.

Next we prove (c) implies (d): For each open $U \subset X$ and $x \in U$, let

$$U_x = U_{\{x\}}.$$

Suppose $U_x \cap V_y \neq \emptyset$, $x \notin V$ and $y \notin U$. Then $U_x \cap (X - \{x\})_{X-U} = \emptyset$ and $V_y \subset (X - \{x\})_{X-U}$. Consequently $U_x \cap V_y = \emptyset$, a contradiction. This proves that $U_x \cap V_y \neq \emptyset$ implies $y \in U$ or $x \in V$.

Clearly (d) implies (e). So we complete the proof by showing that (e) implies (a): For each $(A, B) \in \mathscr{S}_X$, let

$$G(A, B) = \bigcup\{U_x \mid x \in A,\ U \subset X - B,\ U \in \mathscr{B}\}.$$

Clearly $G(A, B) \subset G(A', B')$ whenever $A \subset A'$, $B' \subset B$ and (A, B), $(A', B') \in \mathscr{S}_X$; each $G(A, B)$ is open and $A \subset G(A, B)$. Therefore we only need to prove that $G(A, B)^- \subset X - B$. Actually we prove the stronger result that $G(A, B) \cap G(B, A) = \emptyset$: Assume there exists $w \in G(A, B) \cap G(B, A)$. Then $w \in U_x$ for some $x \in A$ and some $U \subset X - B$ $(U \in \mathscr{B})$ and $w \in V_y$ for some $y \in B$ and some $V \subset X - A$ $(V \in \mathscr{B})$. Then $U_x \cap V_y \neq \emptyset$ which implies that $y \in U$ or $x \in V$, a contradiction (for example, $y \in U \subset X - B$ contradicts the fact that $y \in B$).

3. Results and questions

Among the various results we have so far obtained we mention the following:

1. *Every elastic space is monotonically normal* (indeed we have constructed a monotone normality operator for elastic spaces which satisfies much stronger properties than those in Theorem 2.4 (c)).

2. *The closed continuous image of a monotonically normal space is monotonically normal.* (The proof depends on Theorem 2.4 (d).)

3. *A monotonically normal space X is paracompact if and only if each open cover of X has a* (not necessarily open) *σ-cushioned refinement.*

4. Let X be monotonically normal. *To each pair (A, U) of subsets of X, with A closed, U open and $A \subset U$, one can assign a continuous function $f_{U,A}: X \to I$ such that $f_{U,A}(A) = 0$, $f_{U,A}(X - U) = 1$ and $f_{U,A} \geqq f_{V,B}$ whenever $A \subset B$ and $U \subset V$.* (The proof is essentially the same as the proof of Theorem 5.1 of [1], because of Theorem 2.4 (c).)

5. *Every monotonically normal space is hereditarily monotonically normal.* (This is immediate from Theorem 2.4 (d) and answers a question of Heath and Lutzer.)

6. (Theorem 3.3 of [2].) *Any linearly ordered topological space $(X, \leqq, \tau)$ is completely monotonically normal.* (While the original proof of this result by Heath and Lutzer is quite long, our Theorem 2.4 (e) allows the following very elementary proof: Without loss of generality, we assume that X has no $\leqq$-first and no $\leqq$-last point (if necessary, add rays and apply 5.). Let $\mathscr{B}$ be the base for X which consists of all open intervals and well-order X by the order W. For each $B \in \mathscr{B}$ and $x \in B$, let

$$B_x =]l(x, B), r(x, B)[$$

where $l(x, B)$ $(r(x, B))$ is the W-first element of B such that $l(x, B) < x$ $(x < r(x, B))$. It is quite easy to see that the B_x just defined satisfy Theorem 2.4 (e).)

The following two questions appear intriguing: Is the closed continuous image of an elastic (stratonormal) space an elastic (stratonormal) space? (Tamano and Vaughan [3] conjecture a positive answer, for the first case.)

References

[1] *C. R. Borges:* On stratifiable spaces. Pacific J. Math. *17* (1966), 1—16.

[2] *R. Heath, D. Lutzer and P. Zenor:* A note on monotone normality. Trans. Amer. Math. Soc. (to appear).

[3] *H. Tamano and J. E. Vaughan:* Paracompactness and elastic spaces. Proc. Amer. Math. Soc. *28* (1971), 299—303.

[4] *J. E. Vaughan:* Linearly stratifiable spaces. Pacific J. Math. (to appear).

UNIVERSITY OF CALIFORNIA, DAVIS, CALIFORNIA

SOME REMARKS CONCERNING THE THEORY OF SHAPE IN ARBITRARY METRIZABLE SPACES

K. BORSUK

Warszawa

The classical notion of the homotopy type (introduced by W. Hurewicz, [4], p. 125) allows to classify the spaces from the point of view of their most important global topological properties, called homotopy properties. Let us recall the definition of this notion:

Two spaces X, Y have the same *homotopy type* (notation: $X \underset{h}{\simeq} Y$) if there exist two maps (i.e. continuous functions) $f: X \to Y$ and $g: Y \to X$ such that $gf: X \to X$ and $fg: Y \to Y$ are homotopic to the identities $i_X: X \to X$ and $i_Y: Y \to Y$ respectively.

If only the relation $fg \simeq i_Y$ holds, then one says (following J. H. C. Whitehead, [6], p. 1133) that X *homotopically dominates* Y and we write $X \underset{i}{\geq} Y$.

Both notions of the homotopy type and of the homotopy domination are important tools in the study of global properties of spaces with a rather regular local topological structure as polyhedra or — more generally — as absolute neighborhood retracts for metric spaces, i.e. ANR$(\mathfrak{M})$-spaces. However, for spaces with a more complicated local structure, the family of maps of one space into another can be too limited to give a basis for a reasonable classification of spaces from the point of view of their global properties.

The endeavor to avoid this difficulty is the origin of the theory of shape. Instead of the notion of maps, one uses in it a more elastic notion of the fundamental sequences, which in the most important case of compacta (that is of metric compact spaces) may be defined as follows:

Let X be a compactum lying in a space $M \in$ AR$(\mathfrak{M})$ and Y be a compactum lying in another space $N \in$ AR$(\mathfrak{M})$. By a fundamental sequence $\boldsymbol{f}: X \to Y$ one understands a triple $\{f_k, X, Y\}_{M,N}$ consisting of a sequence of maps $f_k: M \to N$ and of the compacta X, Y satisfying the following condition:

(I) *For every neighborhood V of Y (in N) there is a neighborhood U of X (in M) such that $f_k/U \simeq f_{k+1}/U$ in V for almost all k.*

In particular, if $X = Y$, $M = N$ and if f_k denotes the identity map $i: M \to M$ for every $k = 1, 2, \ldots$, then $\{f_k, X, X\}_{M,M}$ is a fundamental sequence, which we denote by $\boldsymbol{i}_{X,M}$.

If $\boldsymbol{f} = \{f_k, X, Y\}_{M,N}$ and $\boldsymbol{g} = \{g_k, Y, Z\}_{N,P}$ are two fundamental sequences, then $\boldsymbol{gf} = \{g_k f_k, X, Z\}_{M,P}$ is also a fundamental sequence. Two fundamental sequences

$f = \{f_k, X, Y\}_{M,N}$ and $f' = \{f'_k, X, Y\}_{M,N}$ are said to be *homotopic* (notation: $f \simeq f'$) if

(II) *For every neighborhood V of Y (in N) there is a neighborhood U of X (in M) such that* $f_k/U \simeq f'_k/U$ *in V for almost all k.*

By the *shape* Sh (X) of a compactum X one understands the collection of all compacta Y such that there exist two fundamental sequences $f: X \to Y$, $g: Y \to X$ such that $gf \simeq i_{X,M}$ and $fg \simeq i_{Y,N}$. We write Sh $(X) \geqq$ Sh (Y) if there exist two fundamental sequences $f: X \to Y$ and $g: Y \to X$ such that $fg \simeq i_{Y,N}$. One sees easily that the choice of spaces $M, N \in \mathrm{AR}(\mathfrak{M})$, and also the choice of the embeddings of X and of Y into M and N respectively is immaterial. The properties of a compactum X depending only on the shape Sh (X) are called *shape invariants.* One shows that the most important global topological properties of compacta, as the homology groups (in the sense of Vietoris or of Čech, but not the singular homology groups) are shape invariants. The homotopy groups (in the classical sense) do not belong to shape invariants, but by a slight modification of their definitions, one obtains the so called *fundamental groups* (see [1], p. 251) which are shape invariants. Also cohomology groups ([5], p. 54) and cohomotopy groups ([3], p. 81) belong to shape invariants.

Let us add that for compact ANR-spaces, the notion of the shape is the same as the notion of the homotopy type, i.e. if $X, Y \in$ ANR then Sh $(X) =$ Sh (Y) if and only if $X \underset{h}{\simeq} Y$, and Sh $(X) \geqq$ Sh (Y) if and only if X homotopically dominates Y.

The theory of shape of compacta has been already developed, due to work of T. A. Chapman, S. Godlewski, D. Henderson, W. Holsztyński, S. Mardešić, M. Moszyńska, H. Patkowska, J. Segal and others. However, the attempts to extend it onto arbitrary metrizable spaces are only at the beginning. The purpose of this note is to point out some difficulties which appear in this way.

By a theorem of K. Kuratowski and M. Wojdysławski ([7], p. 186), every metric space X can be embedded as a closed subset of an AR $(\mathfrak{M})$-space M. This fact allows to replace in the definition of the fundamental sequences and in the definition of their homotopy given by conditions (I) and (II) the hypothesis that the considered spaces X and Y are compacta by a weaker one, viz. that they are arbitrary metric spaces. However, it seems, that the definitions obtained in this way are not satisfactory, because they do not constitute a sufficient base for the proof that the homology groups (in the sense of Vietoris, with compact carriers) belong to shape invariants. But this difficulty is not serious, because for compacta, conditions (I) and (II) are equivalent to the following conditions:

(I′) *For every compactum* $A \subset X$ *there is a compactum* $B \subset Y$ *such that for every neighborhood V of B in N there exists a neighborhood U of A in M such that* $f_k/U \simeq f_{k+1}/U$ *in V for almost all k.*

(II′) *For every compactum* $A \subset X$ *there is a compactum* $B \subset Y$ *such that for every neighborhood V of B in N there exists a neighborhood U of A in M such that* $f_k/U \simeq f'_k/U$ *in V for almost all k.*

One can show by examples that conditions (I) and (I′), though equivalent for compacta, are independent from each other in the case of arbitrary metric spaces. In this general case, the condition (I′) seems to be more important than the condition (I) because, by the definition of the shape based on it, the homology groups, and also the fundamental groups are shape invariants.

In one of my notes [2] I develop the theory of shape for arbitrary metrizable spaces, starting from the definition of the fundamental sequences as triples $\{f_k, X, Y\}_{M,N}$ satisfying both conditions (I), (I′) and using the definition of the homotopy based on both conditions (II) and (II′). Let us call the theory of shape obtained in this way the *strong shape theory*, or shortly, *S-shape theory*, and let us call the fundamental sequences in this sense – the *S-sequences*, their homotopy – the *S-homotopy* and the corresponding notion of shape of a space X, the *S-shape* $\mathrm{Sh}_S(X)$. One shows that in this S-shape theory (which for compacta is the same as the usual shape theory) many results extend from compacta onto arbitrary metrizable spaces. In particular, one shows that if $X, Y \in \mathrm{ANR}(\mathfrak{M})$ then the relation $\mathrm{Sh}_S(X) = \mathrm{Sh}_S(Y)$ is equivalent to the homotopy equivalence of X and Y, and the relation $\mathrm{Sh}_S(X) \geqq \mathrm{Sh}_S(Y)$ is equivalent to the homotopy domination of Y by X.

Unfortunately the S-shape theory is not always adequate to our intuition. The following simple example, due to S. Nowak, shows that the S-shape of the Cartesian product $X \times Y$ of two spaces X and Y is not determined by the S-shapes of X and of Y. In fact, let X denote the space of all natural numbers (with the usual, discrete topology) and let Y_1 be a circle and Y_2 be the so called "Polish circle" (that is the union of the closure C of the diagram of the function $y = \sin \pi/x$ for $0 < x < 1$ and of an arc with endpoints (0, 0) and (1, 0) and the interior disjoint with C). It is clear that $\mathrm{Sh}_S(Y_1) = \mathrm{Sh}_S(Y_2)$, but it is easy to show that $\mathrm{Sh}_S(X \times Y_1) \neq \mathrm{Sh}_S(X \times Y_2)$.

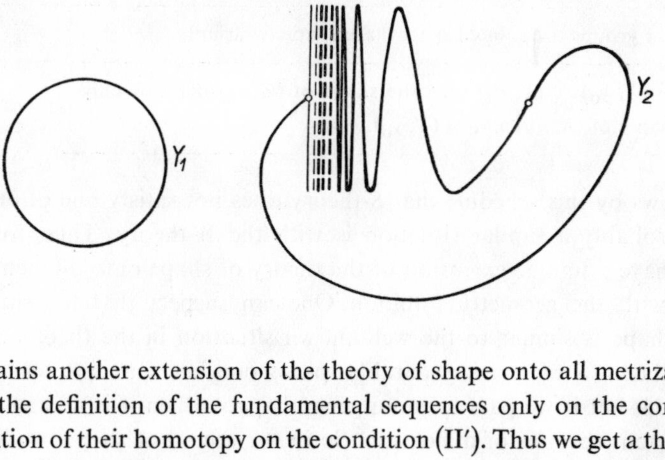

One obtains another extension of the theory of shape onto all metrizable spaces if one bases the definition of the fundamental sequences only on the condition (I′) and the definition of their homotopy on the condition (II′). Thus we get a theory which we call the *weak shape theory*, or shortly the *W-shape theory*, dealing with the corresponding notion of the *W-shape*. For compacta the W-theory is the same as the usual shape theory and one obtains in it generalizations of many theorems concerning the shapes, from compacta onto arbitrary metrizable spaces. In particular, one proves that the W-shape of the Cartesian product $X \times Y$ depends only on the W-shapes of X and of Y. As I have already mentioned, an analogous statement is not true in the S-theory.

Thus the W-theory has some advantages in comparison with the S-theory. However, it is rather doubtful if the W-theory satisfies all our expectations. In particular, we do not know if it is a direct extension of the classical homotopy theory (limited to ANR $(\mathfrak{M})$-spaces), because the problem whether two ANR $(\mathfrak{M})$-spaces with the same W-shape are necessarily homotopically equivalent, remains still open. As I have already mentioned, an analogous proposition holds in the S-theory.

Let us confront some features of both theories: the S-theory and the W-theory, with some postulates which one may expect from a reasonable extension of the theory of shape onto all metrizable spaces:

Some postulates for an extension of the theory of shape onto all metrizable spaces	S-theory	W-theory
Coincidence with the usual shape-theory for compacta.	+	+
If $X, Y \in$ ANR $(\mathfrak{M})$ then X and Y have the same shape iff $X \underset{h}{\simeq} Y$.	+	?
Homology groups (of Vietoris with compact carriers) are shape-invariants.	+	+
Shape of $X \times Y$ depends only on the shape of X and of Y.	—	+
Shape of the suspension of X depends only on the shape of X.	+	+
Fundamental groups (of pointed spaces) are shape invariants.	+	+
If $(Z, z_0) = (X, x_0) \underset{\text{top}}{+} (Y, y_0)$ then the shape of (Z, z_0) depends only on the shapes of (X, x_0) and of (Y, y_0).	+	+

It follows by this schedule that S-theory does not satisfy one of our postulates, and that probably a similar situation is with the W-theory. Thus, for the present we do not have a unique extension of the theory of shape onto all metrizable spaces consistent with the geometric intuition. One can suspect that the situation in the theory of shape is similar to the well-known situation in the theory of dimension, where for the class of metric separable spaces one has a unique reasonable theory of dimension, but if we pass to more general spaces then it is necessary to develop several different theories of dimension. So in the theory of shape, we have a unique theory for compacta, but for arbitrary metric spaces it seems to be necessary to develop several different theories, in particular the weak and the strong theory of shape.

References

[1] *K. Borsuk:* Concerning homotopy properties of compacta. Fund. Math. *62* (1968), 223—254.

[2] *K. Borsuk:* On the concept of shape for metrizable spaces. Bull. Acad. Polon. Sci. Sér. Sci. Math. Astronom. Phys. *18* (1970), 127—132.

[3] *S. Godlewski:* Homotopy dependence of fundamental sequences, relative fundamental equivalence of sets and a generalization of cohomotopy groups. Fund. Math. *69* (1970), 63—91.

[4] *W. Hurewicz:* Beiträge zur Topologie der Deformationen III. Klassen und Homologietypen von Abbildungen. Proc. Ak. Amsterdam *39* (1936), 117—125.

[5] *S. Mardešić and J. Segal:* Shapes of compacta and ANR-systems. Fund. Math. *72* (1971), 41—59.

[6] *J. H. C. Whitehead:* On the homotopy type of ANR's. Bull. Amer. Math. Soc. *54* (1948), 1133—1145.

[7] *M. Wojdysławski:* Rétractes absolus et hyperespaces des continus. Fund. Math. *32* (1939), 184—192.

ON THE FIXED POINT PROPERTY FOR SET-VALUED MAPPINGS OF HEREDITARILY DECOMPOSABLE CONTINUA

J. J. CHARATONIK

Wrocław

Let X and Y be two topological spaces. We say that $F: X \to Y$ is a closed set-valued mapping from X into Y if $F(x)$ is a non-empty closed subset of Y. A closed set-valued mapping $F: X \to Y$ is said to be upper (lower) semi-continuous if $\{x \in X : F(x) \cap A \neq \emptyset\}$ is closed (open) in X whenever A is closed (open) in Y. F is said to be continuous if it is both upper and lower semi-continuous. If $F(x)$ is connected for each $x \in X$, then F is called continuum-valued.

Let $\mathfrak{C}$ be a class of closed set-valued mappings of a topological space X into itself. We say that X has the fixed point property for $\mathfrak{C}$ (the F.p.p. for $\mathfrak{C}$) if, for each $F \in \mathfrak{C}$, there exists $x \in X$ such that $x \in F(x)$.

Three conditions for metric continua X are considered in the paper:

(I) X has the F.p.p. for upper semi-continuous, continuum-valued mappings;

(II) X is hereditarily unicoherent;

(III) X has the F.p.p. for continuous, closed set-valued mappings.

Main problems:

Problem 1. Characterize all continua X with property (I);

Problem 2. Characterize all continua X with property (III);

have only some partial solutions. It follows from results of A. D. Wallace ([5], Theorem A, p. 757), R. L. Plunkett ([4], Theorems 1 and 2, p. 161 and 162) and L. E. Ward, Jr. ([7], Lemma 4, p. 162 and Theorem 3, p. 164) that

Theorem 1. *If a continuum X is locally connected, then*

$$(\mathrm{I}) \Leftrightarrow (\mathrm{II}) \Leftrightarrow (\mathrm{III}).$$

L. E. Ward, Jr. proved ([7], Corollary, p. 163, and [6], Theorem 2, p. 926) the following

Theorem 2. *If a continuum X is arcwise connected, then*

$$(\mathrm{I}) \Leftrightarrow (\mathrm{II}) \Rightarrow (\mathrm{III}).$$

The problem if (III) implies (II) for arcwise connected continua X was posed for the first time in [7], p. 160. There is a conjecture suggesting that the answer is affirmative ([8], p. 92).

The aim of the paper is to prove

Theorem 3. *If a continuum X is hereditarily decomposable, then*

$$(I) \Rightarrow (II) \Rightarrow (III).$$

The proof of the first implication is patterned after Ward's proof of the same implication in Theorem 2, using a result of H. C. Miller ([3], Theorem 2.6, p. 187). The second implication follows from results of H. Cook ([1], Theorem 1, p. 20), S. Mardešić and J. Segal ([2], Theorem 1*, p. 148) and P. O. Wheatley ([9], p. 546).

It is natural to ask if both the inverse implications to those in Theorem 3 hold for hereditarily decomposable continua X. I conjecture that the answer is yes.

References

[1] *H. Cook:* Tree-likeness of dendroids and λ-dendroids. Fund. Math. *68* (1970), 19—22.

[2] *S. Mardešić and J. Segal:* ε-mappings onto polyhedra. Trans. Amer. Math. Soc. *71* (1951), 151—182.

[3] *H. C. Miller:* On unicoherent continua. Trans. Amer. Math. Soc. *69* (1950), 179—194.

[4] *R. L. Plunkett:* A fixed point theorem for continuous multi-valued transformations. Proc. Amer. Math. Soc. *7* (1956), 160—163.

[5] *A. D. Wallace:* A fixed point theorem for trees. Bull. Amer. Math. Soc. *47* (1941), 757—760.

[6] *L. E. Ward, Jr.:* A fixed point theorem for multi-valued functions. Pacific J. Math. *8* (1958), 921—927.

[7] *L. E. Ward, Jr.:* Characterization of the fixed point property for a class of set-valued mappings. Fund. Math. *50* (1961), 159—164.

[8] *L. E. Ward, Jr.:* Set-valued mappings on partially ordered spaces. Lecture Notes in Mathematics *171*, Springer Verlag, 1970, 88—99.

[9] *P. O. Wheatley:* Inverse limits and the fixed point property for set-valued maps. Notices Amer. Math. Soc. *18* (1971), Abstract 685-G3, 546.

INSTITUTE OF MATHEMATICS OF THE WROCŁAW UNIVERSITY

REALIZATIONS OF CLOSURE SPACES BY SET SYSTEMS

J. CHVALINA and M. SEKANINA

Brno

Let $\mathscr{C}$ be the category of all topological spaces in the sense of [1] with continuous mappings as morphisms, i.e., an object of $\mathscr{C}$ is a pair $\langle P, u\rangle$ where $u : \exp P \to \exp P$ fulfils the following axioms:

$$u\emptyset = \emptyset\,, \quad X \subset P \Rightarrow X \subset uX\,, \quad X \subset Y \subset P \Rightarrow uX \subset uY\,.$$

Let us remind that $f: P \to Q$ is a continuous mapping from $\langle P, u\rangle$ into $\langle Q, v\rangle$ iff $f(uX) \subset vf(X)$ for all $X \subset P$. Similarly, a map $f : P \to Q$ is called inversely continuous if $f(uX) \supset vf(X)$. If $f(uX) = vf(X)$, the map f is called closed. $\mathscr{C}'$ or $\mathscr{C}''$ will be the categories with the same objects as $\mathscr{C}$ has, but with inversely continuous mappings or closed mappings respectively as morphisms.

$\mathscr{A}$ will be the full subcategory of $\mathscr{C}$ formed by all topological spaces $\langle P, u\rangle$ from $\mathscr{C}$ for which $u(X \cup Y) = uX \cup uY$ for all X, $Y \subset P$, (the theory of such spaces is developed in [2]), and with continuous mappings as morphisms. $\mathscr{B}$ means the full subcategory of $\mathscr{A}$ formed by all topological spaces $\langle P, u\rangle$ from $\mathscr{A}$ with $u(uX) = uX$ for all $X \subset P$ (morphisms – continuous mappings). Let $\mathscr{A}'$, $\mathscr{A}''$, $\mathscr{B}'$, $\mathscr{B}''$ be defined in similar way as $\mathscr{C}'$, $\mathscr{C}''$ are for $\mathscr{C}$.

Now, let $\mathscr{S}^-$ be the category defined in the following way. Objects of $\mathscr{S}^-$ are pairs $\langle P, S\rangle$ where P is a set and $S \subset \exp P$. Morphisms from $\langle P, S\rangle$ into $\langle Q, T\rangle$ are mappings $f : P \to Q$ for which $X \in T \Rightarrow f^{-1}(X) \in S$. If instead of this condition $X \in S \Rightarrow f(X) \in T$ holds we get the category $\mathscr{S}^+$. $\mathscr{S}$ will mean the intersection of $\mathscr{S}^-$ and $\mathscr{S}^+$.

A full embedding of one category into another is defined as in [3], i.e., it is a full functor true for morphisms and objects. If $\mathscr{K}$ means some of categories $\mathscr{C}$, $\mathscr{A}$, $\mathscr{B}$ or their subcategories then by realization of $\mathscr{K}$ in $\mathscr{S}^-$ such a full embedding Φ is meant for which

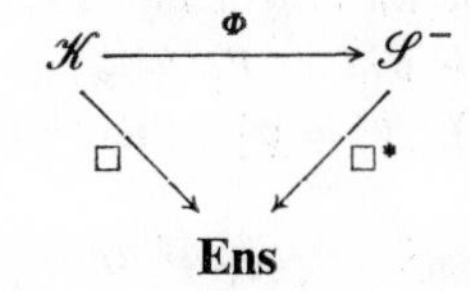

where **Ens** is the category of all sets together with all mappings as morphisms and $\Box$, $\Box^*$ are forgetful functors $[\Box\langle P, u\rangle = P = \Box^*\langle P, S\rangle]$ (see e.g. [4]).

It is clear that the systems of open (closed) sets for an object from $\mathscr{B}$ induce two realizations Φ_1, Φ_2 of $\mathscr{B}$ into $\mathscr{S}^-$, similarly systems of closed sets induce the realizations Φ_3, Φ_4 of $\mathscr{B}'$ into $\mathscr{S}^+$, or $\mathscr{B}''$ into $\mathscr{S}$ respectively. It is not difficult to prove (probably this is a known result) the following

Proposition 1. *Let Φ be a full embedding of $\mathscr{B}$ into $\mathscr{S}^-$ such that if $\Phi\langle X, u\rangle = \langle Y, S\rangle$ then $\emptyset, Y \in S$. Then Φ is equivalent either to Φ_1 or to Φ_2.*

An immediate consequence is

Proposition 1'. *Φ_1, Φ_2 are the only realizations of $\mathscr{B}$ into $\mathscr{S}^-$.*

Similarly one can prove the following

Proposition 2. *Φ_3 is the only realization of $\mathscr{B}'$ into $\mathscr{S}^+$*

Up to now, for $\mathscr{B}''$, the authors can only deduce that for every realization $\Phi : \mathscr{B}'' \to \mathscr{S}$ for all $\langle P, u\rangle$ the following assertion is valid: If $\Phi\langle P, u\rangle = \langle P, S\rangle$ and X is a closed set in $\langle P, u\rangle$, then $X \in S$.

It is a natural question to ask what can be said about realizations of $\mathscr{C}$ or $\mathscr{A}$ in $\mathscr{S}^+$. A detailed investigation of the system of three-point spaces of $\mathscr{A}$, and not quite trivial transfer of some results to infinite case, give the following negative answers.

Proposition 3. *Let $\mathscr{K}$ be a full subcategory of $\mathscr{A}$ ($\mathscr{A}'$) for which there exists a set X, $\operatorname{card} X \geq 3$, such that all topological spaces from $\mathscr{A}$ of the form $\langle X, u\rangle$ are objects of $\mathscr{K}$. Then there is no realization of $\mathscr{K}$ into $\mathscr{S}^-$ ($\mathscr{S}^+$).*

Proposition 4. *The analogous assertion to that in Proposition 3 is valid for $\mathscr{A}''$ and $\mathscr{S}$ under the condition that $\operatorname{card} X = 3$ or X is infinite.*

It remains an open question, whether finite X with $\operatorname{card} X \geq 4$ can be allowed, too[1]).

There exist, of course, embeddings of $\mathscr{C}$ in $\mathscr{S}^-$ defined by various set theoretical functors. E.g., one can prove

Proposition 5. *Let $\langle P, u\rangle$ be an object in $\mathscr{C}$. Put $\mathscr{S}_{\langle P, u\rangle}(\emptyset) = \{\emptyset\}$, $\mathscr{S}_{\langle P,u\rangle}(M) = \{\exp P\}$ for all $M \subset P$ with $uM = P$, $\mathscr{S}_{\langle P,u\rangle}(M) = \{\{X \cup Y \mid X \in \mathscr{N}, Y \subset M, Y \neq \emptyset\} \mid \mathscr{N} \in \exp\exp(P - uM), \mathscr{N} \neq \emptyset\}$ otherwise. Let $\mathscr{S}\langle P,u\rangle = \bigcup_{M \subset P} \mathscr{S}_{\langle P,u\rangle}(M)$. Put*

$$\Phi\langle P, u\rangle = \langle \exp P, \mathscr{S}\langle P, u\rangle\rangle .$$

---

[1]) Added in proofs. The answer is positive.

For a continuous map $f : \langle P, u \rangle \to \langle Q, v \rangle$ *put* $\Phi(f) : \exp P \to \exp Q$ *with* $\Phi(f)(X) = f(X)$ *for all* $X \subset P$. *Then* Φ *is an embedding of* $\mathscr{C}$ *in* $\mathscr{S}^-$.

Or

Proposition 6. *Let* $\langle P, u \rangle$ *be an object in* $\mathscr{C}$. *For every* $x \in P$ *and every neighborhood* $\mathscr{U}$ *of* x *in* $\langle P, u \rangle$, *put* $\langle x, \mathscr{U} \rangle = \{\langle x, y \rangle \mid y \in \mathscr{U}\}$. *Let* $\mathscr{S}\langle P, u \rangle$ *be an additive hull of the system of all* $\langle x, \mathscr{U} \rangle$. *For a continuous map* $f : \langle P, u \rangle \to \langle Q, v \rangle$ *define* $f \times f : P \times P \to Q \times Q$ *as usual by* $f \times f\langle x_1, x_2 \rangle = \langle f(x_1), f(x_2) \rangle$. *Put*

$$\Phi\langle P, u \rangle = \langle P \times P, \mathscr{S}\langle P, u \rangle \rangle, \quad \Phi(f) = f \times f .$$

Then Φ *is an embedding of* $\mathscr{C}$ *in* $\mathscr{S}^-$.

(Notice that, if $\langle P, u \rangle$ is in $\mathscr{A}$, then $\Phi\langle P, u \rangle$ is, in fact, in $\mathscr{B}$.)

Proofs and other details of the above propositions will be published partly in a common paper, partly in the first author's Thesis.

References

[1] *E. Čech:* Topologické prostory. Časopis Pěst. Mat. a Fys. *66* (1937), D 225—D 264.

[2] *E. Čech:* Topological spaces. Academia, Prague, 1966.

[3] *B. Mitchell:* Theory of categories. Academic Press, New York and London, 1965.

[4] *A. Pultr:* On selecting of morphisms among all mappings between underlying sets of objects in concrete categories and realisations of these. Comment. Math. Univ. Carolinae *8* (1), (1967), 53—83.

J. E. PURKYNĚ UNIVERSITY, BRNO

CONTINUOUS FUNCTIONS ON PRODUCTS WITH STRONG TOPOLOGIES

W. W. COMFORT (1) and S. NEGREPONTIS (2)

Middletown and Montréal

This paper is organized as follows. First, the necessary definitions, notation and terminology are introduced; next, our result is stated and proved; and last, various special cases of the theorem are mentioned together with references to the literature where these appear. This indicates the extent to which we are indebted to other authors both for ideas and motivation.

Throughout, we assume that we are given a nonvoid family $\{X_i\}_{i\in I}$ of nonvoid (completely regular, Hausdorff) spaces. We write $X_J = \prod_{i\in J} X_i$ for each nonvoid subset J of I, so that in particular $X_I = \prod_{i\in I} X_i$. We set

$$\mathscr{P}^*(I) = \{J \subset I : J \neq \emptyset\},$$

and

$$\mathscr{P}_\varkappa(I) = \{J \in \mathscr{P}^*(I) : |J| < \varkappa\}$$

for each cardinal number $\varkappa$.

When $\varkappa$ is infinite and $J \in \mathscr{P}^*(I)$, the $\varkappa$-box topology on X_J is that topology which has as base all sets of the form $U = \prod_{i\in J} U_i$ with U_i open in X_i for each i in J and with $|R(U)| < \varkappa$ (where we have set $R(U) = \{i \in I : U_i \subsetneqq X_i\}$). Thus the ω-box topology on X_J is the usual product topology on X_J. The set X_J with the $\varkappa$-box topology is denoted $(X_J)_\varkappa$. When $\varkappa$ is regular the space $(X_J)_\varkappa$ has the property that each intersection of fewer than $\varkappa$ of its open sets is open if and only if each of the spaces X_i $(i \in J)$ has this property.

When $J \in \mathscr{P}^*(I)$ the map π_J is the projection from X_I onto X_J. It is easily verified that for each $\varkappa$ with $\omega \leqq \varkappa$, the map π_J when considered as a map from $(X_I)_\varkappa$ to $(X_J)_\varkappa$ is continuous. For x in X_I, we write x_J for $\pi_J(x)$.

For $\omega \leqq \varkappa \leqq \alpha$, a space is said to be pseudo-$(\alpha, \varkappa)$-compact if, given any α of its open sets, each neighborhood of one of its points meets at least $\varkappa$. A pseudo-(α, ω)-compact is said, simply, to be pseudo-α-compact; and the pseudo-ω-compact spaces – i.e., those spaces with no infinite, locally finite family of open sets – are the familiar pseudocompact spaces.

Finally, if $Y \subset X_I$ and f is a function with domain Y, then f is said to depend on fewer than α coordinates if there exists $J \in \mathscr{P}^*_\alpha(I)$ for which $f(x) = f(y)$ whenever $x \in Y$ and $y \in Y$ and $x_J = y_J$.

Theorem. *Assume the following*: (1) $\omega \leqq \varkappa < \alpha$, α *is regular, and* $\beta^\lambda < \alpha$ *whenever* $\beta < \alpha$ *and* $\lambda < \varkappa$; (2) $Y \subset (X_I)_\varkappa$ *and* $\pi_J(Y) = X_J$ *for each* J *in* $\mathscr{P}^*_\alpha(I)$; *and* (3) $(X_J)_\varkappa$ *is pseudo-*$(\alpha, \varkappa)$*-compact for each* J *in* $\mathscr{P}^*_\varkappa(I)$. *Then* (a) Y *is pseudo-*$(\alpha, \varkappa)$*-compact*; (b) *each continuous function from* Y *to a metric space depends on fewer than* α *coordinates*; *and* (c) *each such function extends continuously over* $(X_I)_\varkappa$.

Proof. To prove (a) we will show that if $\mathscr{U}$ is a family of open subsets of $(X_I)_\varkappa$ with $|\mathscr{U}| = \alpha$, then there is a point p of Y each of whose neighborhoods meets at least $\varkappa$ elements of $\mathscr{U}$. (This will suffice. For, Y being dense in $(X_I)_\varkappa$ by (2), any pair of intersecting open subsets of $(X_I)_\varkappa$ have a point in common in Y.) We may suppose that each element U of $\mathscr{U}$ has the form $U = \prod_{i \in I} U_i$ with each U_i open in X_i and with $|R(U)| < \varkappa$. According to a combinatorial theorem of Erdös and Rado [9], there exist a subset $\mathscr{V}$ of $\mathscr{U}$ with $|\mathscr{V}| = \alpha$ and a (possibly empty) subset J of I for which $R(U) \cap R(V) = J$ whenever U and V are distinct elements of $\mathscr{V}$. If $J = \emptyset$ then for each point p of Y it is true that each neighborhood of p meets each member of $\mathscr{V}$, with fewer than $\varkappa$ exceptions. (Indeed, given a $\varkappa$-box basic neighborhood $W = \prod_{i \in I} W_i$ of p, each i in $R(W)$ belongs to $R(U)$ for at most one U in $\mathscr{V}$. Thus $R(W) \cap R(U) = \emptyset$ for all U in $\mathscr{V}$ with fewer than $\varkappa$ exceptions, and $W \cap U \neq \emptyset$ for each such U.) If $J \neq \emptyset$, i.e. if $J \in \mathscr{P}^*_\varkappa(I)$, then from hypothesis (3) and the fact that α is regular it follows that there is a point x of $(X_J)_\varkappa$ with the property that if W_J is a neighborhood in $(X_J)_\varkappa$ of x then there exists $\mathscr{V} \subset \mathscr{U}$ with $|\mathscr{V}| = \varkappa$ for which $W_J \cap \pi_J(U) \neq \emptyset$ whenever $U \in \mathscr{V}$. Then for p we choose any point of Y for which $\pi_J(p) = x$. (The existence of such a point p requires not all of hypothesis (2) but the weaker condition that $\pi_J(Y) = X_J$ whenever $J \in \mathscr{P}^*_\varkappa(I)$; (2) can be weakened to this, if only conclusion (a) is wanted.) Given a $\varkappa$-box neighborhood $W = W_J \times W_{I \setminus J}$ of p, where W_J and $W_{I \setminus J}$ are $\varkappa$-box open in $(X_J)_\varkappa$ and $(X_{I \setminus J})_\varkappa$ respectively, let $\mathscr{V} \subset \mathscr{U}$ with $|\mathscr{V}| = \varkappa$ and with $W_J \cap \pi_J(U) \neq \emptyset$ for each U in $\mathscr{V}$. Since $|R(W_{I \setminus J})| < \varkappa$ and each element of $R(W_{I \setminus J})$ lies in $R(U)$ for at most one element U of $\mathscr{V}$, there are $\varkappa$ elements U of $\mathscr{V}$ for which $R(W_{I \setminus J}) \cap R(U) = \emptyset$. Given such U, let $q_i \in W_J \cap \pi_J(U)$ if $i \in J$; let $q_i \in W_i$ if $i \in R(W_{I \setminus J})$; let $q_i \in U_i$ if $i \in R(U) \setminus J$; and let q_i be any point of X_i if $i \in I \setminus (J \cup R(W) \cup R(U))$. Thus $q \in W \cap U$, and it follows that W meets (at least) $\varkappa$ elements of $\mathscr{V}$. The proof of (a) is complete.

When (1) is given, statement (b) holds for any dense, pseudo-$(\alpha, \varkappa)$-compact subspace Y of $(X_I)_\varkappa$. It suffices to show that if f is continuous on Y to a metric space (M, ϱ), then for each $\varepsilon > 0$ there exists J in $\mathscr{P}^*_\alpha(I)$ for which $\varrho(f(x), f(y)) \leqq \varepsilon$ whenever x and y are in Y and $\pi_J(x) = \pi_J(y)$. Assuming the contrary, one argues by recursion to produce, for $0 \leqq \xi < \alpha$, points x^ξ and y^ξ of Y and neighborhoods U^ξ and V^ξ of x^ξ and y^ξ respectively, basic in $(X_I)_\varkappa$, for which:

(i) $R(U^\xi) = R(V^\xi)$;

(ii) $x^\xi_i = y^\xi_i$ whenever $i \in \bigcup_{\zeta < \xi} R(U^\zeta)$;

(iii) $\varrho(f(x), f(y)) > \varepsilon$ whenever $x \in U^\xi \cap Y$ and $y \in V^\xi \cap Y$;
(iv) $U_i^\xi \cap V_i^\xi = \emptyset$ whenever $U_i^\xi \neq V_i^\xi$.

This having been done, one uses the fact that Y is pseudo-$(\alpha, \varkappa)$-compact to find a point p in Y each of whose neighborhoods meets $\varkappa$ of the sets U^ξ. It then follows from the fact that for each i in I the relation $U_i^\xi = V_i^\xi$ is valid for all but at most one ξ that each neighborhood in $(X_I)_\varkappa$ of p meets (for some ξ) both $U^\xi \cap Y$ and $V^\xi \cap Y$. This contradicts the continuity of f at p and concludes the proof of (b).

To prove (c) we show that if (2) holds and $\omega \leqq \varkappa \leqq \alpha$, then one can extend continuously over $(X_I)_\varkappa$ any continuous function f from Y (to any space whatever) for which there exists J in $\mathscr{P}_\alpha^*(I)$ with $f(x) = f(y)$ whenever $x \in Y$, $y \in Y$, and $x_J = y_J$. Given such f and J, choose for each x in X a point $\bar{x}$ in Y for which $x_J = \bar{x}_J$ and define $g(x) = f(\bar{x})$. After checking that g is well-defined and extends f, one verifies the continuity of g at each point x of X as follows. Given a neighborhood W of $g(x)$, find neighborhoods U and V of $\bar{x}_J$ and $\bar{x}_{I\setminus J}$ in $(X_J)_\varkappa$ and $(X_{I\setminus J})_\varkappa$ respectively for which $f[(U \times V) \cap Y] \subset W$ and let z be any point in the neighborhood $\pi_J^{-1}(U)$ of x. Because $J \cup R(V) \in \mathscr{P}_\alpha^*(I)$ there exists z' in Y for which $z'_{J\cup R(V)} = z_{J\cup R(V)}$, and then $z' \in U \times V$ and $g(z) = f(z') \in W$.

We remark that in the presence of (1), hypothesis (3) can be replaced by the hypothesis that each of the spaces X_i has a dense subspace with fewer than α elements. For in this case, from (1), each $(X_J)_\varkappa$ with $J \in \mathscr{P}_\varkappa^*(I)$ also has such a subspace, hence is even pseudo-(α, α)-compact.

When $\varkappa = \omega$ and $Y = X$ hypotheses (1) and (2) are automatically fulfilled and (3) is the statement that each space X_F, with F a finite, nonvoid subset of I, is pseudo-α-compact. The argument we used for (b), a straightforward generalization of one presented by Glicksberg [10], shows that in this case there is for each $\varepsilon > 0$ and each continuous f from X_I to (M, ϱ) a finite subset F of I for which $\varrho(f(x), f(y)) \leqq \varepsilon$ whenever $x \in X_I$ and $y \in X_I$ and $x_F = y_F$; thus each such f depends on countably many coordinates. When each X_i is compact, this last assertion is due to Mibu [13] and Bishop [1]; when each is separable and metrizable, to Mazur [12] (with an additional cardinality hypothesis, but for sequentially continuous functions) and Corson and Isbell [5]; when each has Knaster's property (K), to Ross and Stone [16]; when each is separable, to Gleason (see Ross and Stone [16] or Isbell [11], pages 130–132); when each finite product of the X_i is a Lindelöf space, to Engelking [7]; and finally, when X (or each X_F with $F \in \mathscr{P}_\omega^*(I)$) is pseudo-$\omega^+$-compact, to Noble and Ulmer [15]. Some of these authors consider functions defined on subspaces of $\prod_{i\in I} X_i$, and some impose hypotheses weaker than metrizability on the range space.

The argument used to prove (a) was developed by Ulmer [17] to treat (in the case $\varkappa = \omega$) the case where Y is a Σ-space in the sense of Corson [4] or a Σ_α-space (the natural generalization); other applications are given by Noble and Ulmer (loc. cit.). More consequences and equivalents to the Erdös-Rado theorem appear in our work [2]. Results related to those of the present paper and [2], dealing with families of disjoint, open sets in various $\varkappa$-box topologies on a product, appear in Engelking-Karłowicz [8], Engelking [6], and Mostowski [14], Theorem 13.3.1.

The case $\varkappa = \omega$, $\alpha = \omega^+$, of the present theorem was announced by the authors in an abstract which appeared in the Notices of the American Mathematical Society *18* (1971), page 669. Detailed proofs and additional references will appear in [3].

References

[1] *E. Bishop:* A minimal boundary for function algebras. Pacific J. Math. *9* (1959), 629—642.
[2] *W. W. Comfort and S. Negrepontis:* On families of large oscillation. Fund. Math. *75* (1972), 275—290.
[3] *W. W. Comfort and S. Negrepontis:* Manuscript in preparation.
[4] *H. H. Corson:* Normality in subsets of product spaces. Amer. J. Math. *81* (1959), 785—796.
[5] *H. H. Corson and J. R. Isbell:* Some properties of strong uniformities. Quart. J. Math. Oxford Ser. *11* (2) (1960), 17—33.
[6] *R. Engelking:* Cartesian products and dyadic spaces. Fund. Math. *57* (1965), 287—304.
[7] *R. Engelking:* On functions defined on Cartesian products. Fund. Math. *59* (1966), 221—231.
[8] *R. Engelking and M. Karłowicz:* Some theorems of set theory and their topological consequences. Fund. Math. *57* (1965), 275—285.
[9] *P. Erdős and R. Rado:* Intersection theorems for systems of sets (II). J. London Math. Soc. *44* (1969), 467—479.
[10] *I. Glicksberg:* Stone-Čech compactifications of products. Trans. Amer. Math. Soc. *90* (1959), 369—382.
[11] *J. R. Isbell:* Uniform Spaces. Math. Surveys *12*, Amer. Math. Soc., Providence, Rhode Island, 1964.
[12] *S. Mazur:* On continuous mappings in product spaces. Fund. Math. *39* (1952), 229—238.
[13] *Y. Mibu:* On Baire functions on infinite product spaces. Proc. Japan Acad. *20* (1944), 661—663.
[14] *A. Mostowski:* Constructible sets with applications. Studies in Logic and the Foundations of Mathematics, North-Holland Publishing Co., Amsterdam, 1969.
[15] *N. Noble and M. Ulmer:* Factoring functions on cartesian products. Trans. Amer. Math. Soc. *163* (1972), 329—339.
[16] *K. A. Ross and A. H. Stone:* Products of separable spaces. Amer. Math. Monthly *71* (1964), 398—403.
[17] *M. Ulmer:* Continuous functions on product spaces. Doctoral Dissertation, Wesleyan University, 1970.

(1) WESLEYAN UNIVERSITY, MIDDLETOWN, CONNECTICUT
(2) MCGILL UNIVERSITY, MONTRÉAL, QUÉBEC

H-CLOSED EXTENSIONS OF TOPOLOGICAL SPACES

K. CSÁSZÁR

Budapest

P. S. Alexandroff and P. S. Urysohn [1] introduced the following

Definition 1. A Hausdorff (or T_2) space E is *H-closed* if it is closed in every T_2-space E' in which it is contained.

The same authors gave the following characterization of H-closed spaces:

Theorem 1. (Alexandroff-Urysohn) *A T_2-space E is H-closed if and only if from every open covering $E = \bigcup_{i \in I} G_i$ a finite system can be selected such that $E = \bigcup_{j=1}^{n} \bar{G}_{i_j}$.*

The latter condition may be formulated for any topological space, whether T_2 or not, and is fulfilled in particular for every compact space. Therefore, let us introduce

Definition 2. A topological space E is *almost compact* if in each open covering $E = \bigcup_{i \in I} G_i$ there is a finite subsystem $G_{i_1}, \ldots, G_{i_n}$ such that $E = \bigcup_{j=1}^{n} \bar{G}_{i_j}$.

The almost compact spaces have many properties analogous to those of compact spaces. We mention only the following ones:

A topological space E is almost compact if and only if

1) every open filter has a cluster point, or

2) every maximal open filter is convergent (a filter is called open if it has an open base).

Every almost compact regular space is compact.

We see from Theorem 1 that almost compact spaces are generalizations of H-closed T_2-spaces.

However, it is interesting to formulate a direct generalization of the definition of H-closed T_2-spaces, equivalent to the condition of almost compactness. This may be done by means of

Definition 3. ([2]) Let E' be a topological space and $E \subset E'$ a subspace of E'. The space E' is said to be T_2 *with respect to* E if arbitrary two points $x \in E' - E$, $x \neq y \in E'$ have disjoint neighbourhoods.

We can now formulate the definition of H-closedness for a topological space, whether T_2 or not:

Definition 4. A topological space E is *H-closed* if it is closed in every space $E' \supset E$, E' being T_2 with respect to E.

It is easy to see that, if E is T_2, Definition 4 and Definition 1 are equivalent. Moreover, Theorem 1 can be generalized as follows:

Theorem 2. *A topological space E is almost compact if and only if it is H-closed.*

Concerning the extensions of a topological space E, Alexandroff and Urysohn asked whether every T_2-space E has an extension E' which is T_2 and H-closed. M. H. Stone [3] gave a positive answer to this question. Since then, a number of authors: A. D. Alexandroff [4], S. Fomin [5], N. Shanin [6], M. Katětov [7], [8], J. Flachsmeyer [9] etc. have investigated H-closed T_2-extensions of T_2-spaces.

A direct generalization of the problem of H-closed T_2-extensions of T_2-spaces would be the question whether an arbitrary topological space has H-closed extensions. However, the question is obvious in this form, because each topological space possesses e.g. compact extensions. In order to formulate an adequate generalization of the problem of H-closed T_2-extensions of T_2-spaces, we need the following

Definition 5. E' is an *ordinary extension* of the topological space E if it is T_2 with respect to E.

Now we look for *ordinary H-closed extensions* of a topological space E.

If E itself is T_2, an ordinary H-closed extension is the same as an H-closed T_2-extension. It turns out that the theory of ordinary H-closed extensions is very similar to that of H-closed T_2-extensions. E.g., the construction of Flachsmeyer [9] may be transferred with slight modifications to general topological spaces and permits to construct a number of ordinary extensions. For this purpose, let $\mathfrak{P}$ be a base in E such that $\mathfrak{P}$ is a lattice and $P \in \mathfrak{P}$ implies $E - \bar{P} \in \mathfrak{P}$. A filter in E is said to be a $\mathfrak{P}$-filter if it has a base composed of sets belonging to $\mathfrak{P}$. Let us take a set $E' \supset E$ such that there exists a one-to-one map $\mathfrak{S}$ from $E' - E$ onto the set of all non-convergent maximal $\mathfrak{P}$-filters. Further, for $x \in E$ let us denote by $\mathfrak{S}(x)$ the neighbourhood filter of x in E. Now, there exist topologies on E' such that the trace in E of the neighbourhood filter of $x \in E'$ coincides with $\mathfrak{S}(x)$. Among these topologies, there is the coarsest one denoted by $\sigma(\mathfrak{P})$ and the finest one denoted by $\tau(\mathfrak{P})$. The set E', equipped with either $\sigma(\mathfrak{P})$ or $\tau(\mathfrak{P})$ or an arbitrary topology between $\sigma(\mathfrak{P})$ and $\tau(\mathfrak{P})$ is an ordinary H-closed extension of E.

The above construction is far from yielding all possible ordinary H-closed extensions. However, it furnishes a lot of important ordinary H-closed extensions. E.g. E', equipped with $\sigma(\mathfrak{P})$, is an ordinary H-closed extension having a base such that the boundary of its elements is contained in E, and conversely, each extension of this kind is obtained by this construction. In particular, if $\mathfrak{P} = \mathfrak{G}$ (the system of all open sets of E), then E' equipped with $\sigma(\mathfrak{G})$ is the *Fomin extension* of E. It is characterized by the properties of being an ordinary H-closed strict and hypercombinatorial extension; by a strict extension of E, we understand an extension E' such that the closures of subsets of E constitute a base for the closed sets in E', and E' is a hypercombinatorial extension if $\bar{A} \cap \bar{B} = A \cap B$ whenever A and B are closed and $A \cap B$ is nowhere dense in E.

Another important particular case is E' equipped with $\tau(\mathfrak{G})$, called the *Katětov extension* of E. It is characterized by being an ordinary H-closed, hypercombinatorial extension such that $E' - E$ is a discrete closed subset of E'.

It can be shown that the Katětov extension E' is the *finest* ordinary H-closed extension of the given space E in the sense that an arbitrary ordinary H-closed extension of E is a continuous image of E' under a map coinciding on E with the identity.

With the help of Flachsmeyer's method we can examine other types of H-closed extensions too. E.g., if E is *semi-regular* (it possesses a base, composed of interiors of closed sets), then it has a semi-regular ordinary H-closed extension, namely E' equipped with $\sigma(\mathfrak{P})$ where $\mathfrak{P}$ is the system of the interiors of all closed subsets of E.

Finally, let us mention an open question:

Which spaces E have an ordinary compactification E'?

If E is a T_2-space, then it has to be completely regular (a Tychonoff space).

If E is not T_2, a necessary condition is that in E the closure of a compact set has to be compact. A sufficient condition is that E is compact, or that each point of E has a compact closed neighbourhood. However, I do not know a necessary and sufficient condition.

References

[1] *P. S. Alexandroff et P. S. Urysohn:* Mémoire sur les espaces topologiques compacts. Verh. Akad. Wetensch. Amsterdam *14* (1929), 1—96.

[2] *K. Császár:* Untersuchungen über Trennungsaxiome. Publ. Math. Debrecen *14* (1967), 353—364.

[3] *M. H. Stone:* Application of the theory of Boolean rings to general topology. Trans. Amer. Math. Soc. *41* (1937), 375—481.

[4] *A. D. Alexandroff:* Über die Erweiterung eines Hausdorffschen Raumes zu einem *H*-abgeschlossenen. Dokl. Akad. Nauk SSSR *37* (1942), 138—141.

[5] *S. Fomin:* Extensions of topological spaces. Ann. of Math. *44* (1943), 471—480.

[6] *N. Shanin:* On special extensions of topological spaces. Dokl. Akad. Nauk SSSR *38* (1943), 6—9.

[7] *M. Katětov:* Über *H*-abgeschlossene und bikompakte Räume. Časopis Pěst. Mat. Fys. *69* (1940), 36—49.

[8] *M. Katětov:* On *H*-closed extensions of topological spaces. Časopis Pěst. Mat. Fys. *72* (1947), 17—31.

[9] *J. Flachsmeyer:* Zur Theorie der *H*-abgeschlossenen Erweiterungen. Math. Z. *94* (1966), 349—381.

ON NORMS AND SUBSETS OF LINEAR SPACES

J. DANEŠ

Praha

The subject of the communication are certain results concerning the existence, in normed linear spaces, of nonequivalent norms satisfying certain properties as well as the existence of absolutely convex finitely open bounded sets, which are nowhere dense. The results have been published under the same title in Comment. Math. Univ. Carolinae 12 (1971), 835–844.

CHARLES UNIVERSITY, PRAHA

ONE RESULT ON INVERSE LIMITS AND HYPERSPACES

R. DUDA

Wrocław

Hyperspaces, i.e. families consisting of non-empty subsets of a given space X and provided with a topology connected with that of X, are becoming nowadays a subject of intense research. In the present note we restrict our attention to compact Hausdorff spaces only and by a hyperspace we shall mean any family consisting of compact Hausdorff subspaces of a given compact Hausdorff space X. Our aim is to show that some "hyperfunctors" transforming compact Hausdorff spaces into some of its hyperspaces commute with the inverse limits, i.e.,

$$\mathfrak{H}(\lim \boldsymbol{X}) = \lim \mathfrak{H}(\boldsymbol{X}),$$

where $\mathfrak{H}$ is a hyperfunctor and $\boldsymbol{X}$ is an inverse system.

Let $\mathfrak{A}$ be the category consisting of all compact Hausdorff spaces and all continuous mappings, and let $\mathfrak{B}$ be a subcategory of $\mathfrak{A}$. A covariant functor $\mathfrak{H} : \mathfrak{B} \to \mathfrak{A}$ will be called a *hyperfunctor* of $\mathfrak{B}$ if, for each $(f : X \to Y) \in \mathfrak{B}$, $\mathfrak{H}(X)$ is a hyperspace of X, $\mathfrak{H}(Y)$ is a hyperspace of Y, and the induced mapping $\mathfrak{H}(f) : \mathfrak{H}(X) \to \mathfrak{H}(Y)$ is defined by $(\mathfrak{H}(f))(A) = f[A]$ for each $A \in \mathfrak{H}(X)$.

A subcategory $\mathfrak{B}$ will be called *closed with respect to hyperfunctor* $\mathfrak{H}$ if, for each inverse system $\boldsymbol{X} = \{X_\alpha, \pi^\alpha_\beta, \Sigma\}$ in $\mathfrak{B}$, it satisfies the following two conditions:

(i) the limit of $\boldsymbol{X}$ does exist and belongs to $\mathfrak{B}$ together with all projections $\pi_\alpha : \lim \boldsymbol{X} \to X_\alpha$,

(ii) for each "partial" inverse system $\boldsymbol{A} = \{A_\alpha, \pi^\alpha_\beta \mid A_\alpha, \Sigma\}$ such that $A_\alpha \in \mathfrak{H}(X_\alpha)$ and $\pi^\alpha_\beta \mid A_\alpha : A_\alpha \to A_\beta$ are onto, the limit of $\boldsymbol{A}$ does exist and belongs to $\mathfrak{H}(\lim \boldsymbol{X})$.

Some of the well known examples of hyperfunctors are Comp, C and Conv. If X is a compact Hausdorff space, then Comp (X) consists of all non-empty compact subsets of X, $C(X)$ consists of all non-empty compact and connected subsets (i.e., of all subcontinua) of X, and in the case of X being metric, Conv (X) consists of all non-empty compact convex subsets of X. If the topology in hyperspaces is that of Vietoris[1]), it is also known that the category $\mathfrak{B}_1 = \mathfrak{A}$ is closed with respect to Comp

[1]) A base of the Vietoris topology in a hyperspace $\mathfrak{H}(X)$ consists of all sets of the form

$$\langle U_1, \ldots, U_n; \mathfrak{H}(X)\rangle = \{A \in \mathfrak{H}(X) : A \subset \bigcup_{i=1}^{n} U_i,\ A \cap U_i \neq \emptyset \text{ for } i = 1, 2, \ldots, n\},$$

where $U_1, \ldots, U_n$ are open in X and n is any integer.

(cf. [4], Propositions 4.9.2 and 5.10.1, and [3], Theorem 3.2.10), the full subcategory $\mathfrak{B}_2$ of all continua is closed with respect to C (cf. [3], Theorem 3.1.5), and the subcategory $\mathfrak{B}_3$ of all convex subcontinua of the Hilbert cube and mappings preserving convexity[2]) is closed with respect to Conv (cf. [1], Theorem 5.1 (d), and [2], Corollary 1).

Theorem. *Let $\mathfrak{B}$ be a subcategory of $\mathfrak{A}$, closed with the respect to a hyperfunctor $\mathfrak{H}$. If*

$$X = \{X_\alpha, \pi^\alpha_\beta, \Sigma\} \tag{1}$$

is an inverse system in $\mathfrak{B}$, then

$$\mathfrak{H}(X) = \{\mathfrak{H}(X_\alpha), \mathfrak{H}(\pi^\alpha_\beta), \Sigma\} \tag{2}$$

is an inverse system in $\mathfrak{A}$ and

$$\mathfrak{H}(\lim X) = \lim \mathfrak{H}(X). \tag{3}$$

Proof. Consider the diagram

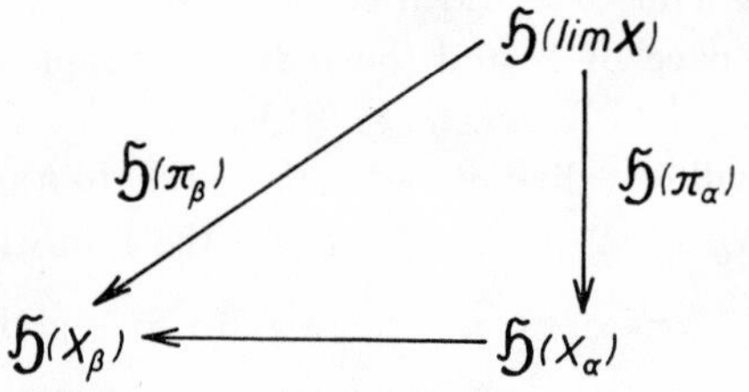

obtained by imposing the functor $\mathfrak{H}$ upon the system (1) augmented with its limit. Since $\mathfrak{H}$ is a functor, the diagram commutes and so (2) is an inverse system.

Since $\mathfrak{H}$ is a hyperfunctor, (2) is an inverse system in the category $\mathfrak{A}$. Hence its limit $\lim \mathfrak{H}(X)$ exists (cf. [3], Theorem 3.2.10) and so there exists a unique continuous mapping

$$h : \mathfrak{H}(\lim X) \to \lim \mathfrak{H}(X)$$

---

[2]) A continuous mapping $f: X \to Y$ from a metric space X onto a metric space Y *preserves convexity* if it satisfies the following two conditions:

(i) if $K \subset X$ is a segment, then $f[K]$ is a segment or a point,
(ii) if $L \subset Y$ is a segment or a point, then $f^{-1}[L]$ is convex.

such that the diagram

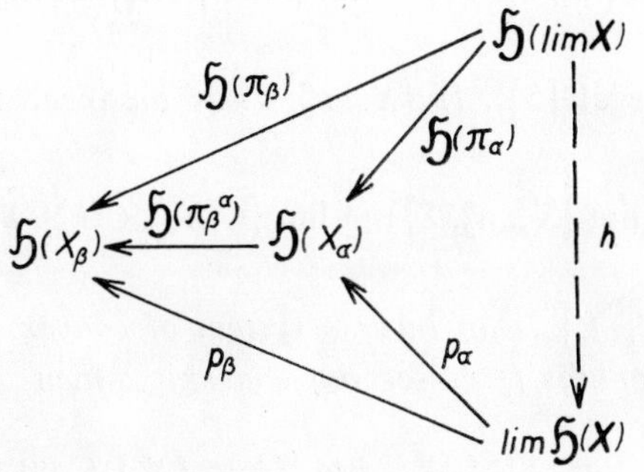

commutes, p_α and p_β being projections.

To see what h is like, take $A \in \mathfrak{H}(\lim X)$ and put $A_\alpha = [\mathfrak{H}(\pi_\alpha)](A)$ for each $\alpha \in \Sigma$. Since $[\mathfrak{H}(\pi_\beta^\alpha)](A_\alpha) = [\mathfrak{H}(\pi_\beta)](A) = A_\beta$, the set $\{A_\alpha\}_{\alpha\in\Sigma}$ is a thread of the system (2), and since $p_\alpha h(A) = [\mathfrak{H}(\pi_\alpha)](A) = A_\alpha$, we have $h(A) = \{A_\alpha\}_{\alpha\in\Sigma}$.

Now we show that h is one-to-one. Assume, on the contrary, that there are two sets $B, C \in \mathfrak{H}(\lim X)$ such that $B \setminus C \neq \emptyset$ and

$$B_\alpha = C_\alpha \quad \text{for each} \quad \alpha \in \Sigma. \tag{4}$$

Since $B \setminus C \neq \emptyset$, there is a thread $\{b_\alpha\}_{\alpha\in\Sigma}$ which is in B and not in C. In view of the equality $[\mathfrak{H}(\pi_\alpha)](B) = \pi_\alpha[B]$ we have $b_\alpha \in B_\alpha$ for each $\alpha \in \Sigma$, which together with the assumption (4) implies that $b_\alpha \in C_\alpha$ for each $\alpha \in \Sigma$. And by virtue of the last relation there exists, for each $\gamma \in \Sigma$, a thread $t_\gamma = \{c_\alpha^\gamma\}_{\alpha\in\Sigma} \in C$ such that $c_\alpha^\gamma = b_\alpha$ for all $\alpha \leqq \gamma$. The set of threads t_γ is a Moore-Smith sequence convergent to $\{b_\alpha\}_{\alpha\in\Sigma}$. In view of the compactness of C, there must be $\{b_\alpha\}_{\alpha\in\Sigma} \in C$ – a contradiction.

Hence h, being defined on a compact space $\mathfrak{H}(\lim X)$ and being one-to-one and continuous, must be a homeomorphism.

It remains to show that h is onto. For that purpose take $A \in \lim \mathfrak{H}(X)$. Then $\{p_\alpha(A)\}_{\alpha\in\Sigma}$ is a thread of the inverse system (2) and, consequently, $\mathbf{A} = \{p_\alpha(A), \pi_\beta^\alpha \mid p_\alpha(A), \Sigma\}$ is an inverse system consisting of spaces $p_\alpha(A) \in \mathfrak{H}(X_\alpha)$ and mappings $\pi_\beta^\alpha \mid p_\alpha(A)$ being onto, and so, since $\mathfrak{B}$ is closed with respect to $\mathfrak{H}$, its limit $\lim \mathbf{A}$ belongs to $\mathfrak{H}(\lim X)$. Therefore $p_\alpha(A) = [\mathfrak{H}(\pi_\alpha)](\lim \mathbf{A}) = p_\alpha h(\lim \mathbf{A})$ for each $\alpha \in \Sigma$ and, consequently, $h(\lim \mathbf{A}) = A$.

Thus the proof of Theorem is completed.

Three examples of hyperfunctors Comp, C and Conv and of subcategories $\mathfrak{B}_1$, $\mathfrak{B}_2$ and $\mathfrak{B}_3$ yield three corollaries. The first two are known, since they have been proved directly by S. Sirota and J. Segal (however, the second for metric continua only). Recall that the topology in hyperspaces is here that of Vietoris.

Corollary 1 (cf. Sirota [6]). *If $\{X_\alpha, \pi^\alpha_\beta, \Sigma\}$ is an inverse system of compact Hausdorff spaces, then*

$$\mathrm{Comp}\,(\lim\,\{X_\alpha, \pi^\alpha_\beta, \Sigma\}) = \lim\,\{\mathrm{Comp}\,(X_\alpha), \mathrm{Comp}\,(\pi^\alpha_\beta), \Sigma\}\,.$$

Corollary 2 (cf. Segal [5]). *If $\{X_\alpha, \pi^\alpha_\beta, \Sigma\}$ is an inverse system of Hausdorff continua, then*

$$C(\lim\,\{X_\alpha, \pi^\alpha_\beta, \Sigma\}) = \lim\,\{C(X_\alpha), C(\pi^\alpha_\beta), \Sigma\}\,.$$

Corollary 3. *If $\{X_n, f_n\}$ is an inverse system of convex subcontinua X_n of the Hilbert cube and mappings f_n preserving convexity, then*

$$\mathrm{Conv}\,(\lim\,\{X_n, f_n\}) = \lim\,\{\mathrm{Conv}\,(X_n), \mathrm{Conv}\,(f_n)\}\,.$$

References

[1] *R. Duda:* On convex metric spaces V. Fund. Math. *68* (1970), 87—106.
[2] *R. Duda:* Inverse limits and hyperspaces. Colloq. Math. *23* (1971), 231—238.
[3] *R. Engelking:* An Outline of General Topology. Warsaw, 1968.
[4] *E. Michael:* Topologies on spaces of subsets. Trans. Amer. Math. Soc. *71* (1951), 152—182.
[5] *J. Segal:* Hyperspaces of the inverse limit space. Proc. Amer. Math. Soc. *10* (1959), 706—709.
[6] *С. Сирота:* О спектральном представлении пространств замкнутых подмножеств бикомпактов. Докл. Акад. Наук СССР *181* (1968), 1069—1072. English translation: Soviet Math. Dokl. *9* (1968), 997—1000.

ON THE IMBEDDING OF EXTREMALLY DISCONNECTED SPACES INTO BICOMPACTA

B. A. EFIMOV

Moskva

1. The formulation of the problem

The class of bicompacta decomposes into two classes $\mathcal{R}_1$ and $\mathcal{R}_2$. The first class includes the bicompacta which contain βN – the Stone-Čech compactification of a countable discrete space – and the second includes the bicompacta which do not contain βN. We note that the class $\mathcal{R}_1$ contains all the infinite quasi-extremal bicompacta (the closure of any open set of the F_σ type is open) and all the dyadic bicompacta of weight $\geqq \mathfrak{c} = \exp \aleph_0$. (A bicompactum X is called dyadic if X is a continuous image of the Cantor space D^τ for some $\tau \geqq \aleph_0$.) The class $\mathcal{R}_2$ contains all the hereditarily normal spaces (and consequently all the linearly ordered bicompacta) and all the sequential bicompacta.

1.1. A necessary and sufficient condition for a bicompactum X to contain βN is that X can be mapped onto a Tyhonov cube $I^{\mathfrak{c}}$ with weight $\mathfrak{c}$.

Note that βN is an extremally disconnected space (the closure of any open set is open). Moreover, βN contains all the extremally disconnected spaces with weight $\leqq \mathfrak{c}$. Naturally there arises a question as to which bicompacta contain extremally disconnected bicompacta having sufficiently large weights?

Let $\tau \geqq \mathfrak{t} \geqq \aleph_0$ and denote $\tau^{1/\mathfrak{t}} = \min\{\mathfrak{m}, \mathfrak{m}^{\mathfrak{t}} \geqq \tau\}$. The cardinal number $\mathfrak{m}$ will be called admissible if $\mathfrak{m}^{\aleph_0} = \mathfrak{m}$.

1.2. Theorem. *If a bicompactum X can be mapped onto a Tyhonov cube I^τ, then X contains all the extremally disconnected spaces with weight $\leqq \tau$. Conversely, if a bicompactum X contains an infinite extremally disconnected bicompactum Y, with weight τ, and Souslin number $\mathfrak{t}$, then X can be mapped onto $I^{\mathfrak{m}}$, $\mathfrak{m} = \tau^{1/\mathfrak{t}} + \mathfrak{k}$ where $\mathfrak{k} = \exp \mathfrak{t}$ if $\mathfrak{t}$ is accessible or else $\mathfrak{k} = \exp \sigma$ for any $\sigma < \mathfrak{t}$, if $\mathfrak{t}$ is a weakly inaccessible cardinal number.*

The last estimate follows from [1].

1.3. Theorem. *Every infinite extremally disconnected bicompactum X with weight τ which satisfies the Souslin condition and for which the cardinal number $\mathfrak{c} + \tau^{1/\aleph_0}$ is $\aleph_0$-admissible, can be mapped onto I^τ.*

1.4. Theorem ([2]). *Any infinite bicompactum X with weight* τ *and Souslin number* $\mathfrak{t}$, *where* $\tau \geqq \exp\exp\exp \mathfrak{t}$, *contains all extremally disconnected spaces with weight* $\leqq (\exp \tau)^+$.

In [2] a special cardinal number invariant called the strength eX of a topological space X is defined.

1.5. Theorem ([2]). *If a bicompactum X can be mapped onto* I^τ, *then* $eX \geqq \tau$. *Conversely, if* $eX > \tau$, *then X can be mapped onto* $I^{\log(\tau^+)}$.

2. The imbedding of extremally disconnected spaces as nowhere dense subsets into dyadic bicompacta and their absolutes

An absolute $\mathrm{p}X$ of a topological space X in the Gleason-Ponomarev sense is an irreducible perfect extremally disconnected preimage of X. According to V. I. Ponomarev [3] the class of completely regular spaces decomposes into classes of spaces which are co-absolute to one another. X and Y are called co-absolute if $\mathrm{p}X \underset{\text{top}}{=} \mathrm{p}Y$.

2.1. Theorem. *If a bicompactum X with weight* τ *and satisfying Souslin's condition can be mapped onto* I^τ, *then there exists a nowhere dense closed* $F \subset \mathrm{p}X$ *homeomorphic with* $\mathrm{p}X$.

A topological space X is called τ-dispersed if for every closed $F \subset X$ there exists a point $x \in F$ such that $\chi(x, F) < \tau$, where $\chi(x, F)$ denotes the character of the point x in F. If, for example, $\tau = \aleph_0$, then the usual definition of a dispersed space is obtained, i.e. every subset $F \subset X$ is not dense in itself.

2.2. Theorem. *In order that it may be possible to map a dyadic bicompactum X with weight* τ *onto* I^τ *it is necessary and sufficient that X be not* τ*-dispersed.*

2.3. Theorem [4]. *The Continuum Hypothesis is equivalent to the following statement: Every non-metrizable dyadic bicompactum contains* βN.

Denote by (γ) the following hypothesis:

$$(\gamma) \quad (\forall\tau)\,(cf(\tau) = \aleph_0)\,\&\,(\forall\mathfrak{n})\,(\mathfrak{n} < \tau) \Rightarrow (\exp \mathfrak{n} \leqq \tau)$$

2.4. Theorem. *Let* (γ) *hold. The absolute of any* τ*-dispersed dyadic bicompactum X with weight* τ *can be mapped onto* $I^{\exp\tau}$. *Moreover, in order that a mapping of* $\mathrm{p}I^\tau$ *onto* $I^{\exp\tau}$ *may exist it is necessary that* $\tau^{\aleph_0} = \exp \tau$ *and it is sufficient that* $cf(\tau) = \aleph_0$.

Note that no τ-dispersed dyadic bicompactum with weight τ can be mapped even onto I^τ.

2.5. Theorem. *The absolute* pX *of every non-*τ*-dispersed dyadic bicompactum with weight* τ *contains a closed nowhere dense* $F \subset \mathrm{p}X$ *homeomorphic with* pX. *An analogous statement is true for the absolutes of* τ*-dispersed dyadic bicompacta provided* (γ) *holds.*

3. Classes of non-homogeneous extremally disconnected bicompacta

A topological space X is called homogeneous if for any two points $x, y \in X$ there exists a homeomorphism $\varphi : X \xrightarrow{\text{onto}} X$ such that $\varphi(x) = y$.

As A. V. Arhangelskii has shown [5] every extremally disconnected bicompactum with weight $\mathfrak{c}$ is non-homogeneous. Z. Frolík [6] proved that if the Continuum Hypothesis is true or if there are cardinal numbers in between $\mathfrak{c}$ and $\exp \mathfrak{c}$ then every infinite extremally disconnected bicompactum is non-homogeneous. The author proved [2] that if cX is weakly accessible and $cX \geqq \log(\pi w X)$, then the extremally disconnected bicompactum X is non-homogeneous. Here we show the non-homogeneity of some new classes of extremally disconnected bicompacta in a number of cases without using any special set theory hypotheses. These results are a consequence of the above theorems and the following Frolík's result [7]: If E is a closed nowhere dense subspace of an extremally disconnected bicompactum X and if E contains X (in particular, if E is homeomorphic to X) then E is non-homogeneous.

3.1. Theorem.[1]) *Any extremally disconnected bicompactum* X *satisfying one of the below conditions is non-homogeneous*:

1) *X satisfies Souslin's condition and the cardinal* $\mathfrak{c} + (wX)^{1/\aleph_0}$ *is* $\aleph_0$*-admissible.*

2) *X is the absolute of the dyadic bicompactum Y with weight τ where Y is not* τ*-dispersed, in particular, if* $cf(wY) \geqq \aleph_1$.

3) *X is the absolute of a* τ*-dispersed bicompactum with weight* τ *provided* (γ) *holds.*

4) *X is the absolute of an ordered bicompactum Y, wY being weakly accessible.*

4. The dependence of the power of a bicompactum on its weight

Consider $|X|$ as a function which assigns to each space X of weight $\leqq \tau$ the cardinality of X. If $|X|$ is defined on the set of all infinite metric compacta then it can assume at most two values, viz. $\aleph_0$ or $\mathfrak{c}$. This fact is proved independently of the

---

[1]) Editor's Note. K. Kunen announced in his preliminary report On the compactification of the integers, Notices Amer. Math. Soc. *17* (1970), p. 299, that the usual orderings on the types of ultrafilters are not linear. Now by another theorem in [6] it follows that no infinite extremally disconnected compact space is homogeneous.

Continuum Hypothesis. If $|X|$ is defined on the set of all bicompacta with weight τ then $\tau \leqq |X| \leqq \exp \tau$. A question naturally arises: Is it again possible to show independently of the Generalized Continuum Hypothesis that the function $|X|$ assumes at most two values? In general, the answer is negative. Namely, in the model M of the set theory ZF in which 1) $\mathfrak{c} = \aleph_{\omega_1}$, 2) $\exp \mathfrak{c} = \aleph_{\omega_2}$, 3) $(\forall \mathfrak{n})(\mathfrak{n} < \mathfrak{c}) \Rightarrow \Rightarrow (\exp \mathfrak{n} < \exp \mathfrak{c})$ there exists a bicompactum with weight $\mathfrak{c}$ for which $\mathfrak{c} < |X| < < \exp \mathfrak{c}$. The existence of such models M is proved by Cohen's method. However, in the case of dyadic bicompacta a positive answer can be given without the GCH or any of its analogues.

4.1. Theorem. *The cardinality of any dyadic bicompactum with weight τ which is not τ-dispersed equals* $\exp \tau$. *The cardinality of any τ-dispersed dyadic bicompactum equals* $\mathfrak{e} = \sum_{k<\omega_0} \exp \mathfrak{n}_k$ *for some countable sequence* $\mathfrak{n}_1 < \mathfrak{n}_2 < \ldots < < \mathfrak{n}_k < \ldots$ *of cardinal numbers, where* $\sum_{k<\omega_0} \mathfrak{n}_k = \tau$.

Note that the cardinal number $\mathfrak{e}$ is independent of the choice of the sequence $\{\mathfrak{n}_k\}$.

4.2. By πwX we shall denote the π-weight [3] of a topological space X, i.e. the least cardinal number of a system of open subsets of X cofinal with the system of all open subsets of X ordered according to inclusion.

4.3. Theorem. *The cardinality of every bicompactum X co-absolute with a dyadic bicompactum which is not (πwX)-dispersed satisfies the inequalities*

$$\exp(\pi wX) \leqq |X| \leqq \exp(wX).$$

and

$$\pi wX \leqq wX \leqq (\pi wX)^{\aleph_0}.$$

If a bicompactum X co-absolute with a dyadic bicompactum is (πwX)-dispersed then the cardinality of X satisfies the inequality

$$\sum_{k<\omega_0} \exp \mathfrak{n}_k \leqq |X| \leqq \sum_{k<\omega_0} \exp \exp \mathfrak{n}_k$$

for any sequence of cardinal numbers $\mathfrak{n}_1 < \mathfrak{n}_2 < \ldots < \mathfrak{n}_k < \ldots$ such that $\sum_{k<\omega_0} \mathfrak{n}_k = = \pi wX$.

Since any two compactifications of a space X are co-absolute the following Corollary gives a negative answer to A. V. Arhangelskii's question as to whether any countable completely regular space S has a compactification of power $\mathfrak{c}$.

4.4. Corollary. *There exists a countable completely regular space S (for instance a countable dense subset of $I^{\mathfrak{c}}$) such that its every compactification has the power* $\exp \mathfrak{c}$.

References

[1] *D. A. Vladimirov and B. A. Efimov:* On the power of extremally disconnected spaces and complete Boolean algebras. Dokl. Akad. Nauk SSSR *194* (6) (1970), 1247—1250.

[2] *B. A. Efimov:* Extremally disconnected bicompacta and absolutes. Trudy Moskov. Mat. Obšč. *23* (1970), 235—276.

[3] *V. I. Pomomarev:* On space co-absolute with metric spaces. Uspehi Mat. Nauk *21* (4) (1966), 101—132.

[4] *B. A. Efimov:* Subspaces of dyadic bicompacta. Dokl. Akad. Nauk SSSR *187* (1) (1969), 21—24.

[5] *A. V. Arhangelskii:* An extremally disconnected bicompactum having weight $\mathfrak{c}$ is non-homogeneous. Dokl. Akad. Nauk SSSR *175* (4) (1967), 751—754.

[6] *Z. Frolík:* Homogeneity problems for extremally disconnected spaces. Coment. Math. Univ. Carolinae *8* (4) (1967), 757—763.

[7] *Z. Frolík:* Fixed points of maps of extremally disconnected spaces. Proc. Internat. Sympos. on Topology and its Applications (Herceg-Novi, 1968). Savez Društava Mat. Fiz. i Astronom., Belgrade, 1969, 164—167.

CEMI AS USSR, MOSCOW

NORMAL AND CATEGORY MEASURES ON TOPOLOGICAL SPACES

J. FLACHSMEYER

Greifswald

The general theory of topological spaces can be built without using real numbers. But many parts of the advanced topology have a substantial relation to real numbers. One of these relations turns up in the investigation of measures on topological spaces. Here we are interested in the case when topologically thin sets are thin in a measure theoretic sense too. Several authors have considered such questions with more or less restrictive assumptions on the underlying space. It is not immediately evident how significant these assumptions are. The purpose of this paper is to make some steps in an interesting program to clarify the relevance of a special topological structure for topological measures.

1. The space of regular Borel measures

Let X be an arbitrary topological space. A bounded Borel measure μ on X is a positive countably additive function on the Borel field of X (the least σ-field containing the open sets) with $\mu(X) < +\infty$. By regularity of a Borel measure μ we mean that for every Borel set B in X:

$$\mu(B) = \inf \{\mu(U) \mid U \text{ open in } X \text{ and } U \supset B\}.$$

A bounded signed Borel measure μ is called regular if its positive and negative parts μ^+ and μ^- are regular Borel measures.

The system $\mathcal{M}(X)$ of all bounded regular signed Borel measures with pointwise addition and scalar multiplication and pointwise ordering is a vector lattice. By definition

$$\|\mu\| := |\mu|(X) = \mu^+(X) + \mu^-(X).$$

$\mu \rightsquigarrow \|\mu\|$ is a norm for $\mathcal{M}(X)$ and makes it a Banach lattice.

Now a Banach lattice L is called a Kakutani-L-space if the norm has the additivity property for positive elements:

$$x \geqq 0, \quad y \geqq 0, \quad \|x + y\| = \|x\| + \|y\|.$$

Every Kakutani-L-space is an order complete vector lattice (cf. Day [2]). From the results mentioned in Dunford-Schwartz [4], Chap. III § 7, it can be seen that:

The system $\mathcal{M}(X)$ of all bounded signed regular Borel measures on an arbitrary topological space forms in a natural way a Kakutani-L-space.

2. The space of normal measures

The notion of a normal Borel measures is due to J. Dixmier [3] for the special case of extremally disconnected compact Hausdorff spaces. Some writers gave an extension of this notion to more general spaces (cf. Heider [8] for Boolean spaces, Knowles [11] for completely regular spaces, Hebert and Lacey [7] for compact Hausdorff spaces).

Definition. Let X be an arbitrary topological space. A bounded regular Borel measure $\mu \in \mathcal{M}^+(X)$ (the set of positive measures) is called a *normal Borel measure* if every nowhere dense Borel set has measure zero. Roughly speaking, a normal positive Borel measure is a measure for which topological null-sets are measure null-sets.

A *signed normal Borel measure* is a measure $\mu \in \mathcal{M}(X)$ such that μ^+, μ^- are normal Borel measures.

Theorem 1. (*The structure of the system of all normal Borel measures.*)

Suppose X is an arbitrary topological space. Then the system $\mathcal{N}(X)$ of all bounded signed normal Borel measures on X with respect to pointwise addition and scalar multiplication, and pointwise ordering is a vector lattice.

For the positive part μ^+ and the negative part μ^- of $\mu \in \mathcal{N}(X)$ it holds

$$\mu^+(B) = \sup\{\mu(A) \mid A \text{ Borel set with } A \subset B\}$$
$$\mu^-(B) = -\inf\{\mu(A) \mid A \text{ Borel set with } A \subset B\} .$$

By the total variation $\|\mu\| := |\mu|(X)\ (|\mu| = \mu^+ + \mu^-)$ of the measure $\mu \in \mathcal{N}(X)$ the vector lattice $\mathcal{N}(X)$ becomes a Banach lattice, this means complete with respect to the norm $\|.\|$ and the order and the norm structures are related through

$$|\mu| \leqq |\nu| \Rightarrow \|\mu\| \leqq \|\nu\| .$$

Moreover, the norm has the additivity property on $\mathcal{N}(X)$:

$$\mu \geqq 0, \quad \nu \geqq 0, \quad \|\mu + \nu\| = \|\mu\| + \|\nu\| .$$

Summarizing all these properties:

The space $\mathcal{N}(X)$ of all signed normal Borel measures on the topological space X is a Kakutani-L-space. $\mathcal{N}(X)$ is a convex lattice-subspace of $\mathcal{M}(X)$.

Remark. For the case of Boolean spaces X and normal Baire measures this theorem is stated in Heider [8], for the case of compact Hausdorff spaces and normal Borel measures it is remarked in Hebert-Lacey [7], p. 116.

Proof. The difference of two positive normal Borel measures μ_1, μ_2 must be normal, since $\mu = \mu_1 - \mu_2 = \mu^+ - \mu^-$ implies $\mu_1 \geqq \mu^+ \geqq 0$, $\mu_2 \geqq \mu^- \geqq 0$ and therefore μ^+ and μ^- are normal. Thus we have $\mathcal{N}(X)$ as a subspace of $\mathcal{M}(X)$. This subspace is a l-convex (a solid) subspace, namely for $\mu \in \mathcal{N}(X)$ and $\nu \in \mathcal{M}(X)$ $|\nu| \leqq |\mu|$ implies $\nu \in \mathcal{N}(X)$. To prove the norm completeness of the normal vector lattice $\mathcal{N}(X)$ we have only to follow the reasoning by which the norm completeness of $\mathcal{M}(X)$ is proved in Dunford-Schwartz [4], Chap. III § 7. There the equivalence of the total variation norm $\|.\|$ in $\mathcal{M}(X)$ and the supremum norm $\|\mu\|^{\bullet} := \sup \{|\mu(B)| \mid B \text{ Borel set in } X\}$ is shown. But for the norm $\| \ \|^{\bullet}$, $\mathcal{N}(X)$ is a closed subspace of $\mathcal{M}(X)$.

Examples. 1. *For every separable metrizable space X without isolated points there exists only the trivial normal Borel measure $\mu \equiv 0$. Thus $\mathcal{N}(X)$ is here the null space.*

Particularly, this can be established for a regular Borel measure $\mu(X) > 0$ with $\mu(\{x\}) = 0$ for each $x \in X$ by finding a dense open subset of arbitrary small measure using a cover of a countable dense subset (see for this question also Marczewski and Sikorski [12]).

2. *Let X be a T_1-space. Every bounded signed regular Borel measure on X is a normal Borel measure iff X is discrete.* ($\mathcal{M}(X) = \mathcal{N}(X)$.)

For this it is only needed to consider the point-measures (Dirac measures) δ_x, $x \in X$.

Recall that a *Freudenthal-unit* in a vector lattice is a positive element $\nu > 0$ such that $\inf(|x|, \nu) = 0$ implies $x = 0$.

The L-space $\mathcal{N}(X)$ for a discrete space X has a Freudenthal-unit iff card $X = \aleph_0$. *A positive measure on X is a Freudenthal-unit iff each point has strict positive measure. Of course $\mathcal{N}(X)$ $(=\mathcal{M}(X))$ is order isomorphic and isometric to the space $l_1(X)$ of absolute summable real "sequences" on X.*

3. Apart from its intrinsic interest, the relevance of normal Borel measures arises from the well known situation of representation of every Kakutani-L-space by a space of all normal measures on a suitable compact extremally disconnected Hausdorff space X (cf. Kelley and Namioka [15]). There the space X can be constructed as the structure space of the dual space L^* which is isometrically isomorphic to the space of all continuous real valued functions $\mathscr{C}(X)$ on a topologically uniquely determined compact Hausdorff space X. The canonical embedding of L in its second

dual L^{**} gives L, after identifying L^{**} with the space $\mathcal{M}(X)$ of all regular Borel measures on X, as the closed linear subspace $\mathcal{N}(X)$ of $\mathcal{M}(X)$.

According to this representation theorem every L-space $\mathcal{M}(X)$ of all bounded signed regular Borel measures on a topological space X is isometric isomorphic to the space $\mathcal{N}(Y)$ of all normal Borel measures on a space Y.

The space $\mathcal{M}(\alpha\mathbf{N})$ of all bounded signed regular Borel measures on the one-point compactification of the space $\mathbf{N}$ of natural numbers is isometric isomorphic to the space $\mathcal{N}(\mathbf{N})$ of all normal measures on $\mathbf{N}$ resp. isometric isomorphic to the space $\mathcal{N}(\beta\mathbf{N})$ of all normal Borel measures on the Stone-Čech compactification of the natural numbers.

By the way we can ask for which spaces X there is a "nice" correspondence to spaces Y such that $\mathcal{M}(X)$ is isometric isomorphic to $\mathcal{N}(Y)$.

3. Supports of normal measures

The support for $\mu \in \mathcal{M}^+(X)$ is defined as the following

$$\operatorname{supp} \mu := X \setminus \bigcup U \mid (U \text{ open in } X \text{ with } \mu(U) = 0).$$

For arbitrary elements $\mu \in \mathcal{M}(X)$, the support is understood to be the union of the supports of the positive part μ^+ and the negative part μ^-.

Definition. An arbitrary topological space X is said to have a *rich system of normal Borel measures* iff every nonvoid open subset contains the support of a non trivial normal Borel measure on X.

Remark. This concept applied to extremally disconnected Hausdorff spaces (sometimes called Stonian spaces) is equivalent to Dixmier's notion of hyperstonian spaces. According to Dixmier, in a hyperstonian space the union of supports of normal measures is dense in the space.

Theorem 2. *Suppose X is an arbitrary topological space. The support of a normal Borel measure on X is a regular-closed subset of X, which is a Baire space: If $\mu \in \mathcal{N}(X)$ then for $F := \operatorname{supp} \mu$, $F = \overline{\operatorname{Int} F}$, and every relatively open subset of F is not meager. Let X be a space with a rich system of normal Borel measures. Then every meager set is nowhere dense and X must be a Baire space in which every non empty open set contains a non empty regular closed set.*

Remark. For compact Hausdorff spaces X which are extremally disconnected (sometimes called Stonian spaces) this is the contents of two propositions by Dixmier [3]. The last part is a generalization of Theorem 3 in Oxtoby's paper [14]. Measures with the whole space as the support are considered there. Spaces that contain in every non empty open set a regular closed set are called quasi-regular by Oxtoby.

Proof. We can consider positive measures because the union of two regular closed sets is regular closed again. Let $\mu \in \mathcal{N}^+(X)$, $F := \operatorname{supp} \mu$. For the regular-closed kernel $\hat{F} := \overline{\operatorname{Int} F}$ of F, $F \setminus \hat{F}$ is a nowhere dense Borel set. Therefore $\mu(F \setminus \hat{F}) = 0$. Then $\hat{F}$ must be the support of μ, i.e. $F = \hat{F}$. μ restricted to its support is a normal Borel measure on its support. Then every nonvoid open subset of supp μ has strictly positive measure but meager open sets are null-sets with respect to normal measures. To prove the second part it is sufficient to consider Borel sets and show for meager Borel sets B, $\operatorname{Int} \bar{B} = \emptyset$. Each meager Borel set is for every normal measure a null set. Then $\operatorname{Int} B = \emptyset$, since for $\operatorname{Int} B \neq \emptyset$ there exists a positive Borel measure with $\mu(B) \geqq \mu(\operatorname{Int} B) > 0$. From $\bar{B} = (\bar{B} \setminus \operatorname{Int} B) \cup \operatorname{Int} B$ it follows $\mu(\bar{B}) = 0$ for each normal Borel measure. Thus we have

$$\operatorname{Int} \bar{B} = \emptyset .$$

Corollary. *Let X be a T_1-space separable in the sense of Fréchet (that means X has a countable dense subset) without isolated points. Then there is no nontrivial normal Borel measure on X.*

Proof. Let μ be a nontrivial normal Borel measure on X. Supp μ is regular closed in X. Then supp μ is a T_1-space separable in the sense of Fréchet without isolated points. Thus we may assume supp $\mu = X$. Then X is a space with a rich system of normal Borel measures. The countable dense set is meager but not nowhere dense. It follows $\mu \equiv 0$.

Remark. Dixmier [3] constructs an extremally disconnected compact Hausdorff space which is not hyperstonian in the sense of Dixmier but for which every meager set is nowhere dense.

Definition. Let X be an arbitrary topological space. A positive normal measure $\mu \in \mathcal{N}^+(X)$ is said to be a *category Borel measure* iff its support is the whole space X.

Remark. Oxtoby [14] called a positive bounded measure μ on the class of sets having the property of Baire (the σ-field generated by the open sets and the meager sets) a category measure for the topological space X if $\mu(A) = 0$ means A is meager. Every category measure is the completion of a Borel measure. It is easy to see that a category Borel measure in our sense gives a regular category measure of Oxtoby by completion and conversely a category measure of Oxtoby restricted to the class of Borel sets is a Borel category measure. In other words, a category Borel measure means a regular Borel measure such that the Borel null-sets are identical with the Borel meager sets (Borel sets of the first category).

Theorem 3. *Let X be a topological space with a rich system of normal Borel measures. Then the Kakutani-L-space $\mathcal{N}(X)$ of all normal Borel measures has the following properties:*

1. *The Freudenthal-units coincide with the Borel category measures on X.*

2. *There exist Freudenthal-units in $\mathcal{N}(X)$ iff the space X has the Souslin-property that every open disjoint family is countable.*

Remark. Without any assumption on the space X each Borel category measure in $\mathcal{N}(X)$ is a Freudenthal-unit in $\mathcal{N}(X)$, and the converse does not hold.

Proof. Let $\nu \in \mathcal{N}^+(X)$. If $\operatorname{supp} \nu \neq X$ then $X \setminus \operatorname{supp} \nu$ contains a support of a nontrivial normal Borel measure μ. Then $\inf(\nu, |\mu|) = 0$. Therefore Freudenthal-units must have X for its support. Conversely, for a category measure ν let $\inf(\nu, |\mu|) = 0$ with $\mu \in \mathcal{N}(X)$. If $U \neq \emptyset$ is open, then $\nu(U) > 0$. Therefore $|\mu|(U) = 0$ and $\operatorname{supp} \mu = \emptyset$. For a positive normal Borel measure μ with support X every disjoint open family must be countable.

Now let X be a space with a rich system of normal Borel measures and such that it has the Souslin-property. We consider a disjoint family in the vector lattice $\mathcal{N}(X)$. This means $(\mu_i)_{i \in I}$, $0 \neq \mu_i \in \mathcal{N}(X)$, with $\inf(|\mu_i|, |\mu_j|) = 0$ for $i \neq j$. Then since $\operatorname{Int}(\operatorname{supp} \mu_i)$ is a disjoint family of open sets, it must be a countable family, i.e. $\operatorname{card} I \leqq \aleph_0$. With the help of Zorn's Lemma we can find a maximal disjoint family of positive normal measures. This family is countable: $\mu_1, \mu_2, \ldots, \mu_n, \ldots$. Hence

$$\mu := \sum_{n=1}^{\infty} \frac{1}{2^n} \frac{\mu_n}{\|\mu_n\|}$$

is a normal measure. We have $\operatorname{supp} \mu = X$, since in the other case we find a positive normal measure ν with $\operatorname{supp} \nu \subset X \setminus \operatorname{supp} \mu$, which contradicts the maximality of the family $\mu_1, \mu_2, \ldots$.

4. Normal measures and their induced measures on the Boolean algebra of regular open sets

The regular open sets of a topological space X form (with respect to the inclusion as an ordering) a complete Boolean algebra. This Boolean algebra is isomorphic to the Boolean algebra of regular closed sets.

Definition. Let $\mathbf{B}$ be a complete Boolean algebra. A positive additive function $\mu : \mathbf{B} \to \mathbf{R}$ is said to be *σ-additive* if for every countable disjoint family $(a_n)_{n \in \mathbf{N}}$ of $\mathbf{B}$

$$\mu(\sup a_n) = \sum_{n=1}^{\infty} \mu(a_n),$$

completely additive if for every arbitrary disjoint family $(a_i)_{i \in I}$

$$\mu(\sup a_i) = \sum \mu(a_i) = \sup \{ \sum_{i \in E} \mu(a_i), E \text{ finite subset of } I \}.$$

Of course the first and the second condition are equivalent respectively to the conditions

$$\mu(\sup a_n) = \sup \mu(a_n) \text{ by } a_n \nearrow a$$

and

$$\mu(\sup a_i) = \sup \mu(a_i) \text{ by } a_i \nearrow a\,.$$

These things are well known.

Example 4. Let $X = [\![0, \Omega]\!]$ be the space of ordinals less than or equal to the first uncountable ordinal Ω. The point measure δ_Ω is a regular Borel measure on X. We see that this measure induces on the Boolean algebra $\mathfrak{R}_0(X)$ of all regular open sets a σ-additive measure, because every countable family of regular open neighborhoods of the limit point Ω has a non empty regular open infimum. But this measure on $\mathfrak{R}_0(X)$ is not completely additive, since the infimum of all regular open neighborhoods of Ω is zero in $\mathfrak{R}_0(X)$ and the measure of all these neighborhoods is identically 1.

Theorem 4. *Let μ be a positive normal Borel measure on an arbitrary topological space X. Then the function induced by μ on the Boolean algebra of regular open sets is a completely additive measure. Two distinct normal Borel measures have distinct completely additive traces on the Boolean algebra of regular open sets. Every complete additive measure on the Boolean algebra of regular open sets is the trace of a normal Borel measure on X if the space X is a quasi-regular Baire space.*

Remark. This statement extends a theorem of Oxtoby [14] (cf. also Mibu [13]) to the case of normal measures while Oxtoby considers category measures. That every normal measure induces a σ-additive measure on $\mathfrak{R}_0(X)$ is also remarked in Hebert-Lacey [7].

Proof. Let $\mu \in \mathcal{N}^+(X)$; we restrict μ to its regular closed support. Then the trace of μ on $\mathfrak{R}_0(\operatorname{supp} \mu)$ is a σ-additive strictly positive measure. Therefore it is completely additive on $\mathfrak{R}_0(\operatorname{supp} \mu)$. But then the extension to $\mathfrak{R}_0(X)$ remains completely additive. For normal measures $\mu \in \mathcal{N}^+(X)$ we have $\mu(U) = \mu(\operatorname{Int} \overline{U})$, U open in X. By regularity of μ for a Borel set B:

$$\mu(B) = \inf \{\mu(H) \mid H \supset B,\ H \text{ regular open}\}\,.$$

Thus a normal measure is uniquely determined by its trace on $\mathfrak{R}_0(X)$.

The last part is a consequence of Theorem 2 in Oxtoby [14].

Theorem 5. *Let X, Y be regular Baire spaces and $f : X \to Y$ a proper irreducible map from X onto Y (this means that f is continuous, closed with compact fibres, and no closed subset $F \neq X$ is mapped onto Y). Then f gives in a natural way an isometric isomorphism from the space of normal measures on X onto the space of normal measures on Y.*

Remark. This statement generalizes the investigation of Hebert-Lacey on normal measures and the projective resolution.

Proof. It follows from Theorem 4 and our investigation about the behavior of regular open (closed) sets under proper irreducible maps [5], [6].

References

[1] *G. Birkhoff:* Lattice theory. 2nd ed., New York, 1948.

[2] *M. M. Day:* Normed linear spaces. Springer Verlag, 1958.

[3] *J. Dixmier:* Sur certains espaces considérés par M. H. Stone. Summa Brasil. Math. *2* (1951), 151—181.

[4] *N. Dunford and J. T. Schwartz:* Linear operators. Part I, General theory. Interscience Publishers, 1958.

[5] *J. Flachsmeyer:* Topologische Projektivräume. Math. Nachr. *26* (1963), 57—66.

[6] *J. Flachsmeyer:* On the system of regular open (closed) sets. (Russian.) Dokl. Akad. Nauk SSSR *156* (1964), 32—34.

[7] *D. J. Hebert and H. E. Lacey:* On supports of regular Borel measures. Pacific J. Math. *27* (1968), 101—118.

[8] *L. J. Heider:* A representation theory for measures on Boolean algebras. Michigan Math. J. *5* (1958), 213—221.

[9] *S. Kakutani:* Concrete representation of abstract (*L*)-spaces and the mean ergodic theorem. Ann. of Math. *42* (1941), 523—537.

[10] *S. Kakutani:* Concrete representation of abstract (*M*)-spaces. Ann. of Math. *42* (1941), 994—1024.

[11] *J. D. Knowles:* Measures on topological spaces. Proc. London Math. Soc. *17* (1967), 139—156.

[12] *E. Marczewski and R. Sikorski:* Remarks on measure and category. Colloq. Math. *2* (1951), 13—19.

[13] *Y. Mibu:* Relations between measure and topology in some Boolean space. Proc. Japan Acad. *20* (1944), 454—458.

[14] *J. C. Oxtoby:* Spaces that admit a category measure. J. Reine Angew. Math. *205* (1960/61), 156—170.

[15] *J. L. Kelley and I. Namioka:* Linear topological spaces. Van Nostrand, 1963.

ERNST — MORITZ — ARNDT — UNIVERSITÄT, GREIFSWALD

TOPOLOGICAL SPACES WHICH ADMIT A COMPATIBLE COMPLETE QUASI-UNIFORMITY

P. FLETCHER (1) and W. F. LINDGREN (2)

Blacksburg and Slippery Rock

1. Introduction

The concept of a quasi-uniformity on a set X was introduced by L. Nachbin [13]. A *quasi-uniformity on a set* X is a filter $\mathscr{U}$ on $X \times X$ with the following two properties

i) if $U \in \mathscr{U}$, then $\Delta = \{(x, x) : x \in X\} \subset U$
ii) if $U \in \mathscr{U}$, there exists $V \in \mathscr{U}$ such that $V \circ V \subset U$.

If $\mathscr{U}$ is a quasi-uniformity on a set X, then $\mathscr{T}_{\mathscr{U}} = \{A \subset X : \text{if } x \in A, \text{ there is } V \in \mathscr{U}$ with $V(x) \subset A\}$ is a topology for X. If X is a set, $\mathscr{U}$ is a quasi-uniformity on X and $\mathscr{T}$ is a topology on X, then $\mathscr{U}$ is *compatible* with $(X, \mathscr{T})$ provided that $\mathscr{T} = \mathscr{T}_{\mathscr{U}}$. It is known that every topological space admits a compatible quasi-uniformity. In this paper we consider the question, does every topological space admit a compatible complete quasi-uniformity.

Throughout this paper, if $\mathscr{C}$ is an open cover of a topological space $(X, \mathscr{T})$ and $x \in X$ then $A_x^{\mathscr{C}}$ denotes $\bigcap\{C \in \mathscr{C} : x \in C\}$.

2. Compatible quasi-uniformities

In 1955, V. S. Krishnan proved essentially the following result.

Theorem 2.1 [12]. *Let $(X, \mathscr{T})$ be a topological space and let U be the collection of all upper semi-continuous functions on $(X, \mathscr{T})$. For each $\varepsilon > 0$ and each $f \in U$, let $U_{(f,\varepsilon)} = \{(x, y) \in X \times X : f(y) - f(x) < \varepsilon\}$. The quasi-uniformity $\mathscr{U}$ generated by $\{U_{(f,\varepsilon)} : f \in U, \varepsilon > 0\}$ is compatible with $(X, \mathscr{T})$.*

The quasi-uniformity of Theorem 2.1 is called the upper semi-continuous quasi-uniformity and is denoted by $\mathscr{USC}$. A *Q-cover* of a topological space $(X, \mathscr{T})$ is an open cover $\mathscr{C}$ of X such that if $x \in X$, then $A_x^{\mathscr{C}} = \bigcap\{C \in \mathscr{C} : x \in C\} \in \mathscr{T}$ [15].

Theorem 2.2. [6]. *Let $\mathscr{A}$ be the collection of all Q-covers of a topological space $(X, \mathscr{T})$. For each $\mathscr{C} \in \mathscr{A}$,* let *$U_{\mathscr{C}} = \bigcup\{\{x\} \times A_x^{\mathscr{C}} : x \in X\}$ and let $\mathscr{B} = \{U_{\mathscr{C}} : \mathscr{C} \in \mathscr{A}\}$. Then $\mathscr{B}$ is a base for a compatible quasi-uniformity $\mathscr{FT}$ for $(X, \mathscr{T})$.*

A quasi-uniformity $\mathscr{U}$ is *transitive* provided that there is a base $\mathscr{B}$ for $\mathscr{U}$ with the property that if $B \in \mathscr{B}$, then $B \circ B = B$. The quasi-uniformity of Theorem 2.2 is denoted by $\mathscr{F}\mathscr{T}$ because it is always the finest compatible transitive quasi-uniformity for a given space $(X, \mathscr{T})$ [6, Corollary to Theorem 2]. The authors have been unable to find an example of a space for which $\mathscr{F}\mathscr{T}$ is not in fact the fine quasi-uniformity. Consequently, it may well be that the problem discussed herein may reduce to the problem of deciding for which spaces $\mathscr{F}\mathscr{T}$ is complete.

3. Completeness

Definition [14]. Let $(X, \mathscr{U})$ be a quasi-uniform space. A filter $\mathscr{F}$ on X is *$\mathscr{U}$-Cauchy* provided that if $U \in \mathscr{U}$, there is $p \in X$ such that $U(p) \in \mathscr{F}$. The quasi-uniform space $(X, \mathscr{U})$ is *complete* provided that every $\mathscr{U}$-Cauchy filter converges.

It is natural to ask if there exists a non-realcompact Tychonoff space which admits a compatible complete quasi-uniformity. This question was answered by J. Carlson and T. Hicks who exhibited a compatible complete quasi-uniformity for $(\mathcal{O}, \Omega)$, the space of ordinals less than the first uncountable ordinal [3]. Actually, as we now show, it follows from known results that any linearly ordered space admits a compatible complete quasi-uniformity.

Definition [15]. A topological space $(X, \mathscr{T})$ is *orthocompact* (has property Q in terms of [15]) provided that every open cover has an open Q-refinement.

Definition [5]. A cover $\mathscr{R}$ of a space $(X, \mathscr{T})$ is a *fundamental* cover provided that if $x \in X$, then there is $U \in \mathscr{T}$ such that $x \in U \subset A_x^{\mathscr{R}}$. A topological space *has property F* provided that every open cover has a fundamental open refinement.

We note that in the literature the concept of orthocompactness [15] predates the concept of having property F [5].

Lemma. *A topological space is orthocompact if and only if it possesses property F.*

Proof. Let $\mathscr{C}$ be an open cover of X and let $\mathscr{R}$ be a fundamental open refinement. For each $x \in X$, let $U_x = \operatorname{int}(A_x^{\mathscr{R}})$ and let $\mathscr{S} = \{U_x : x \in X\}$. It is clear that $\mathscr{S}$ is an open refinement of $\mathscr{C}$. We now show that $\mathscr{S}$ is a Q-cover. Let $y \in X$ and let $p \in A_y^{\mathscr{S}}$. Let $q \in U_p$ and let $x \in X$ such that $y \in U_x$. Then $p \in U_x \subset A_x^{\mathscr{R}}$ and $q \in U_p \subset A_p^{\mathscr{R}}$ so that $q \in A_x^{\mathscr{R}}$. Thus $U_p \subset A_x^{\mathscr{R}}$. Since U_p is open, $U_p \subset \operatorname{int}(A_x^{\mathscr{R}}) = U_x$. Thus $U_p \subset U_x$ for each x such that $y \in U_x$. That is $U_p \subset A_y^{\mathscr{S}}$.

Lemma [6, Theorem 3]. *Every orthocompact topological space has a compatible complete transitive quasi-uniformity.*

Lemma [5, Theorem 3]. *Every linearly ordered topological space has property F.*

Theorem 3.1. *Every linearly ordered topological space has a compatible complete transitive quasi-uniformity.*

Definition [14]. A quasi-uniform space $(X, \mathscr{U})$ is *precompact* provided that if $U \in \mathscr{U}$, there is a finite subset F of X such that $U(F) = X$.

The classic proof of J. Dieudonné that $(\mathcal{O}, \Omega)$ does not possess a compatible complete uniformity does not depend upon Shirota's theorem; rather it makes use of a generalization of the Niemytski-Tychonoff theorem. A principal justification for defining completeness with respect to quasi-uniform spaces is that this same generalization still obtains for quasi-uniform spaces.

Theorem 3.2 [14, Theorem 2.2]. *Let $(X, \mathscr{U})$ be a quasi-uniform space. Then $(X, \mathscr{T}_{\mathscr{U}})$ is compact if and only if $\mathscr{U}$ is complete and precompact.*

We recall the argument of J. Dieudonné: One first shows that $(\mathcal{O}, \Omega)$ has only one compatible uniformity which must be precompact. By the above theorem, this uniformity cannot be complete since $(\mathcal{O}, \Omega)$ is not compact. This argument suggests the searching out of a non-compact topological space for which every compatible quasi-uniformity is precompact.

Theorem 3.3. *A topological space $(X, \mathscr{T})$ is compact if and only if $\mathscr{F}\mathscr{T}$ is precompact.*

Proof. By [7, Proposition 4.2], a necessary and sufficient condition that $\mathscr{F}\mathscr{T}$ be precompact is that every Q-cover have a finite subcover. Evidently every well-ordered (by set inclusion) open cover is a Q-cover, and it is known that a space is compact if and only if every well-ordered open cover has a finite subcover [1, Theorem 7.1].

In light of the above theorem we now seek a topological space whose fine quasi-uniformity is neither complete nor precompact.

Definition [9]. A quasi-uniform space $(X, \mathscr{U})$ is *almost precompact* provided that if $U \in \mathscr{U}$, there is a finite subset F of X such that $\overline{U(F)} = X$; and $(X, \mathscr{U})$ is *almost complete* provided that every open Cauchy filter has a cluster point.

Theorem 3.4 [2, Theorem 3.3]. *The upper semi-continuous quasi-uniformity $\mathscr{U}\mathscr{S}\mathscr{C}$ for a topological space $(X, \mathscr{T})$ is precompact if and only if $(X, \mathscr{T})$ is countably compact.*

Definition [10]. A filter $\mathscr{F}$ has the countable *subcollection intersection property* provided that if $\mathscr{C}$ is a countable subcollection of $\mathscr{F}$, then $\bigcap\{\overline{F} : F \in \mathscr{C}\} \neq \emptyset$. A space

$(X, \mathcal{T})$ is *almost realcompact* provided that every open ultrafilter with the countable subcollection intersection property converges.

The concept of almost realcompactness was defined in 1961 by Z. Frolík. Only recently has the following internal characterization of almost realcompactness been obtained.

Theorem 3.5 [7, Corollary to Theorem 5.1]. *A topological space is almost realcompact if and only if $\mathcal{USC}$ is almost complete.*

Corollary. *Let $(X, \mathcal{T})$ be a topological space which is neither almost realcompact nor countably compact. Then $\mathcal{USC}$ is neither complete nor precompact.*

Recall that a Tychonoff space is pseudocompact if and only if it is countably almost compact (=countably H-closed), and that very countably almost compact almost realcompact space is almost compact [10, Theorem 2].

Example. The space Ψ [11, page 79] is pseudocompact but not countably compact. Consequently $\mathcal{USC}$ is neither complete nor precompact. It is easily seen that Ψ is orthocompact so that $\mathcal{FT}$ is complete.

We do not know of an example of a normal almost precompact quasi-uniform space which is not precompact; however if $(X, \mathcal{T})$ is any pseudocompact Tychonoff space which is not countably compact, then $\mathcal{USC}$ is an almost precompact quasi-uniformity, which is not precompact.

4. Questions

The authors believe that the following questions are related to the problem of deciding which topological spaces admit a compatible complete quasi-uniformity.

For which topological spaces is $\mathcal{FT}$ the fine quasi-uniformity?

For which topological spaces is $\mathcal{USC}$ the fine quasi-uniformity?

References

[1] *P. Alexandroff and P. Urysohn:* Mémoire sur les espaces topologiques compacts. Verh. Akad. Wetensch. *14* (1929), 1—96.

[2] *C. Barnhill and P. Fletcher:* Topological spaces with a unique compatible quasi-uniform structure. Arch. Math. *21* (1970), 206—209.

[3] *J. Carlson and T. Hicks:* Some quasi-uniform space examples. J. Math. Anal. Appl. (to appear).

[4] *J. Dieudonné:* Un exemple d'espace normal non susceptible d'une structure d'espace complet. C. R. Acad. Sci. Paris Sér. A—B, *209* (1939), 145—147.

[5] *W. M. Fleischman:* On fundamental open coverings. Proc. Internat. Sympos. on Topology and its Applications (Herceg-Novi, 1968). Savez Društava Mat. Fiz. i Astronom., Belgrade, 1969, 154—155.

[6] *P. Fletcher:* On completeness of quasi-uniform spaces. Arch. Math. *22* (1971), 200—204.

[7] *P. Fletcher and W. F. Lindgren:* Quasi-uniformities with a transitive base. Pacific J. Math. (to appear).

[8] *P. Fletcher and W. F. Lindgren:* Transitive quasi-uniformities. J. Math. Anal. Appl. (to appear).

[9] *P. Fletcher and S. A. Naimpally:* Almost complete and almost precompact quasi-uniform spaces. Czechoslovak Math. J. *21* (1971), 383—390.

[10] *Z. Frolík:* A generalization of realcompact spaces. Czechoslovak Math. J. *13* (1963), 127—137.

[11] *L. Gillman and M. Jerison:* Rings of continuous functions. D. Van Nostrand Company, Princeton, N. J., 1960.

[12] *V. S. Krishnan:* A note on semi-uniform spaces. J. Madras Univ. B. *25* (1969), 273—274.

[13] *L. Nachbin:* Topology and Order. D. Van Nostrand Company, Princeton, N. J., 1965.

[14] *J. L. Sieber and W. J. Pervin:* Completeness in quasi-uniform spaces. Math. Ann. *158* (1965), 79—81.

[15] *M. Sion and R. C. Willmott:* Hausdorff measures on abstract spaces. Trans. Amer. Math. Soc. *123* (1966), 275—309.

(1) VIRGINIA POLYTECHNIC INSTITUTE AND STATE UNIVERSITY, BLACKSBURG, VIRGINIA

(2) SLIPPERY ROCK STATE COLLEGE, SLIPPERY ROCK, PENNSYLVANIA

SEQUENTIAL ENVELOPE AND SUBSPACES OF THE ČECH-STONE COMPACTIFICATION

R. FRIČ

Praha

A convergence $\mathfrak{L}$ for a non-empty set L is a set of pairs $(\{x_n\}, x)$ where $\{x_n\}$ is a sequence of points of L and $x \in L$, satisfying the axioms of Fréchet:

($\mathscr{L}_0$) If $(\{x_n\}, x) \in \mathfrak{L}$ and $(\{x_n\}, y) \in \mathfrak{L}$, then $x = y$.

($\mathscr{L}_1$) If $x_n = x$ for each positive integer n, then $(\{x_n\}, x) \in \mathfrak{L}$.

($\mathscr{L}_2$) If $(\{x_n\}, x) \in \mathfrak{L}$ and $\{x_{n_i}\}$ is a subsequence of $\{x_n\}$, then $(\{x_{n_i}\}, x) \in \mathfrak{L}$.

A convergence $\mathfrak{L}$ can be enlarged to a convergence $\mathfrak{L}^*$ for L satisfying the axiom of Urysohn:

($\mathscr{L}_3$) If each subsequence $\{x_{n_i}\}$ of a sequence $\{x_n\}$ contains a subsequence $\{x_{n_{i_j}}\}$ such that $(\{x_{n_{i_j}}\}, x) \in \mathfrak{L}$, then $(\{x_n\}, x) \in \mathfrak{L}$.

For each $A \subset L$ denote

$$\lambda A = \{x : (\{x_n\}, x) \in \mathfrak{L},\ \bigcup(x_n) \subset A\}.$$

Then $(L, \mathfrak{L}, \lambda)$ or briefly (L, λ) is a closure space and it is called a convergence space. Notice that $\mathfrak{L}$ and $\mathfrak{L}^*$ induce the same convergence closure λ; operator λ^{ω_1} defined inductively for $\xi \leqq \omega_1$ by $\lambda^\xi A = \bigcup_{\eta < \xi} \lambda(\lambda^\eta A)$, $A \subset L$, is a topology for L and $\lambda^{\omega_1} A$ is the smallest sequentially closed set in L containing A. If $\lambda^{\omega_1} A = L$, then we say that A is sequentially dense in L.

If $\langle (L_\alpha, \mathfrak{L}_\alpha, \lambda_\alpha), \alpha \in I \rangle$ is a non-empty family of convergence spaces, then $(L, \mathfrak{L}, \lambda)$, where $L = \prod_{\alpha \in I} L_\alpha$, $(\{\langle x(\alpha), \alpha \in I \rangle_n\}, \langle x(\alpha), \alpha \in I \rangle) \in \mathfrak{L}$ whenever $(\{x_n(\alpha)\}, x(\alpha)) \in \mathfrak{L}^*_\alpha$, $\alpha \in I$ (coordinatewise star convergence), and λ is the induced convergence closure for L, is a convergence space (cf. [5]). We shall call it a *star convergence Cartesian product space.*

In what follows the set of all continuous functions on (L, λ) is denoted by $\mathscr{C} = \mathscr{C}(L)$, the subset of all bounded continuous functions by $\mathscr{C}^*$, and a subset of $\mathscr{C}$ by $\mathscr{C}_0$.

Definition. We say that a convergence space (L, λ) has the *property p* with respect to $\mathscr{C}_0$ if

(p) For each two sequences $\{x_n\}$, $\{y_n\}$ of points of L such that

$$(\lambda \bigcup(x_n)) \cap (\lambda \bigcup(y_n)) = \emptyset$$

there is a function $f \in \mathscr{C}_0$ such that

$$\lim f(x_n) = \lim f(y_n)$$

does not hold.

Note 1. If in (p) we have $y_n = y$, $n = 1, 2, \ldots$, then we speak about the $\mathscr{C}_0$ *sequential regularity* (Cf. [6]).

Note 2. The property p is topological.

Theorem 1. *The properties p with respect to* $\mathscr{C}$ *and* $\mathscr{C}^*$ *are equivalent.*[1])

Theorem 2. *The property p is star convergence Cartesian productive and hereditary with respect to sequentially closed subspaces.*

Theorem 3. *A convergence space associated*[2]) *with a normal or a realcompact topological space has the property p.*

Theorem 4. *A convergence space has the property p if and only if it is homeomorphic to a sequentially closed subspace of a star convergence Cartesian product of the real lines.*

Theorem 5. *A* $\mathscr{C}_0$ *sequentially regular convergence space* (L, λ) *has the property p with respect to* $\mathscr{C}_0$ *if and only if it cannot be* $\mathscr{C}_0$ *embedded*[3]) *as a sequentially dense proper subspace in any sequentially regular convergence space.*

It is well known that the Čech-Stone compactification (Hewitt realcompactification) of a completely regular topological space is the maximal completely regular space in which the original one is $\mathscr{C}^*$ embedded ($\mathscr{C}$ embedded) dense subspace. The sequential envelope of a sequentially regular convergence space introduced by J. Novák ([4]) is the maximal sequentially regular convergence space in which the original space is $\mathscr{C}^*$ embedded sequentially dense subspace. More generally, following [6]:

---

[1]) We speak simply of the property p in this case.

[2]) A sequence converges to a point whenever it is eventually in each topological neighborhood of the point.

[3]) A subspace (L, λ) of a convergence space (L', λ') is $\mathscr{C}_0(L)$ embedded in (L', λ') if each $f \in \mathscr{C}_0$ has a continuous extension onto the whole space.

A convergence space (S, σ) is called a $\mathscr{C}_0$ sequential envelope of a $\mathscr{C}_0$ sequentially regular convergence space (L, λ) if

1° (L, λ) is a sequentially dense subspace of (S, σ).

2° (L, λ) is $\mathscr{C}_0$ embedded in (S, σ) and (S, σ) is $\overline{\mathscr{C}}_0(S)$ sequentially regular, where

$$\overline{\mathscr{C}}_0(S) = \{f : f \in \mathscr{C}(S),\ f/L \in \mathscr{C}_0\}.$$

3° There is no convergence space (S', σ') containing (S, σ) as a proper subspace and fulfilling 1° and 2° with regard to (L, λ) and (S', σ').

Theorem 6. *A convergence space (S, σ) is a $\mathscr{C}_0$ sequential envelope of a $\mathscr{C}_0$ sequentially regular convergence space (L, λ) if and only if*

(i) *(L, λ) is a sequentially dense $\mathscr{C}_0$ embedded subspace of (S, σ).*

(ii) *(S, σ) has the property p with respect to $\overline{\mathscr{C}}_0(S)$.*

The Čech-Stone compactification and the Hewitt realcompactification can substantially differ, e.g. $\beta N \neq N = \upsilon N$. In comparison with this fact Theorem 1 yields the following

Theorem 7. *$\mathscr{C}^*$ sequential and $\mathscr{C}$ sequential envelopes of a sequentially regular convergence space are homeomorphic and the homeomorphism leaves the original space pointwise fixed.*

In the sequel let (L, λ) be a sequentially regular convergence space, $\tilde{\lambda}$ the completely regular modification[4] of λ, and (P, u) the Čech-Stone compactification of $(L, \tilde{\lambda})$, where P is regarded as the set of all z-ultrafilters on $(L, \tilde{\lambda})$. For every infinite cardinal $\aleph_\alpha$ let $P_{\aleph_\alpha} \subset P = P_{\aleph_0}$ be the set of all z-ultrafilters having the $\aleph_\alpha$-intersection property (i.e. the intersections of less than $\aleph_\alpha$ members of the z-ultrafilter are non-empty). For example $(P_{\aleph_1}, u/P_{\aleph_1})$ is the Hewitt realcompactification of $(L, \tilde{\lambda})$ (cf. [2]). Observe that if (P, π) is the convergence space associated with (P, u), then $(Q, \pi/Q)$ is associated with $(Q, u/Q)$ for each non-empty subset Q of P.

V. Koutník pointed out ([3]) that the $\mathscr{C}^*$ sequential and hence, by Theorem 5, the $\mathscr{C}$ sequential envelope of (L, λ) is the smallest sequentially closed subset in the Čech-Stone compactification of $(L, \tilde{\lambda})$ which contains L endowed with the associated convergence and convergence closure.

Theorem 8. *Denote $S = \pi^{\omega_1}L$. Then $P_{\aleph_0} \supset P_{\aleph_1} \supset S$ and $(S, \pi/S)$ is a sequential envelope of (L, λ).*

There are examples such that $P_{\aleph_1} \neq S$ and $S - P_{\aleph_2} \neq \emptyset$. Consequently, Theorem 8 is in this direction the best possible result. On the other hand, it is easy

[4]) $\tilde{\lambda}$ is the finest of all completely regular topologies for L coarser than λ (cf. [3]).

to verify that if $2^{\text{card}L} = \aleph_\alpha$ then $P_{\aleph_{\alpha+1}} = L$. Hence the following problems might be of interest.

Problem 1. Find the least cardinal $\aleph_\alpha$ depending on card L such that $S \supset P_{\aleph_\alpha}$.

Problem 2. Is there the least cardinal $\aleph_\alpha$ such that $S \supset P_{\aleph_\alpha}$ whenever (L, λ) is a sequentially regular convergence space?

References

[1] *E. Čech:* Topological spaces. Prague, 1966.
[2] *L. Gillman and M. Jerison:* Rings of continuous functions. Princeton, 1960.
[3] *V. Koutnik:* On convergence topologies. General Topology and its Relations to Modern Analysis and Algebra, II (Proc. Second Prague Topological Sympos., 1966). Academia, Prague, 1967, 226—228.
[4] *J. Novák:* On the sequential envelope. General Topology and its Relations to Modern Analysis and Algebra (I) (Proc. (First) Prague Topological Sympos., 1961). Publishing House of the Czechoslovak Academy of Sciences, Prague, 1962, 292—294.
[5] *J. Novák:* On convergence spaces and their sequential envelopes. Czechoslovak Math. J. *15* (1965), 74—100.
[6] *J. Novák:* On some problems concerning the convergence spaces and groups. General Topology and its Relations to Modern Analysis and Algebra (Proc. Kanpur Topological Conf., 1968). Academia, Prague, 1971, 219—229.

INSTITUTE OF MATHEMATICS OF THE CZECHOSLOVAK ACADEMY OF SCIENCES, PRAHA

TOPOLOGICAL METHODS IN MEASURE THEORY AND THE THEORY OF MEASURABLE SPACES

Z. FROLÍK

Praha

There has been a growing interest in topological problems which have applications in measure theory and the theory of measurable spaces. The most important material seems to be related to the existence of projective limits in various categories of measure spaces; for a short expository account we refer to [F3] (the basic technique goes back to Kolmogorov, Bochner and Marczewski, generalized by Choksi and Metivier).

I do not intend to give a systematic survey, but merely a short selection of results and trends which interest me personally. In other words I will talk about the material discussed in my seminars (leaving out projective limits).

In the first two paragraphs some background material is presented. Paragraphs 3 and 4 concern the spaces of measures. The paragraphs 5 and 6 serve as an introduction to § 8 on uniform methods in the theory of measurable spaces. § 7 gives an important example on the construction introduced in § 6, and at the same time provides an exposition of some surprising recent results in non-separable descriptive theory. In conclusion we discuss quite mysterious Borel complete and Baire-Borel complete spaces.

1. Baire and Borel sets

The category of measurable spaces has measurable spaces (pairs $\mathscr{X} = \langle X, \mathscr{B} \rangle$ where $\mathscr{B}$ is a σ-algebra of subsets of X) for the objects, and measurable mappings for the morphisms. Denote by **Bo** the functor which assigns to each topological space X the Borel space of X, i.e. **Bo** X is the set X endowed with the smallest σ-algebra containing the open sets in X. The symbol **Bo** X will be used for the Borel space of X as well as for the structure of the Borel space, i.e. for the σ-algebra of Borel sets. In measure theory the functor **Bo** is commonly used; in many questions another functor is needed, namely the functor **Ba** which assigns to each X the Baire space of X which is defined to be the set X endowed with the smallest σ-algebra such that each continuous (real-valued) function on X is measurable. Again **Ba** X stands for both the Baire space of X and the structure of the Baire space, namely the σ-algebra of all Baire sets in X. Recall that **Ba** X is the smallest σ-algebra which contains all exact open (=co-zero) sets in X. If X is perfectly normal (in particular, if X is metrizable) then $\mathbf{Ba}\,X = \mathbf{Bo}\,X$; it is not known if every normal X such that $\mathbf{Ba}\,X = \mathbf{Bo}\,X$ is necessarily perfectly normal (this is a problem of M. Katětov).

It seems that $\mathbf{Ba}\,X$ more closely reflects the properties of X than $\mathbf{Bo}\,X$ does. For example, notice that

if K is compact Hausdorff, and if $\mathbf{Ba}\,K$ is countably generated then K is metrizable.

It is not known whether **Ba** may be replaced by **Bo** in this statement; the answer in affirmative would be an important theorem. I guess that the problem will be answered in negative.

The second example on close attachment of $\mathbf{Ba}\,X$ to X is the following non-trivial theorem [F 5] which says, roughly speaking, that a Baire measurable mapping of an analytic space into a metrizable space can be regarded continuous, and the image is analytic.

Theorem. *If A is analytic, M is metrizable and if $f : \mathbf{Ba}\,A \to \mathbf{Ba}\,M$ is measurable then there exists an analytic space A' such that $\mathbf{Ba}\,A' = \mathbf{Ba}\,A$ and $f : A' \to M$ is continuous. Hence $f[A]$ is analytic.*

It should be remarked that one can take the graph of f in $A \times M$ for A'.

I don't know of any survey of **Bo** and **Ba**. On the other hand an exposition of many classical results on abstract measurable spaces (mostly countably generated) is provided by B. V. Rao's thesis.

2. Baire and Borel measures

All topological spaces we consider are assumed to be completely regular. Let $\mathbf{C}_b(X)$ denote the Banach space of all bounded continuous functions on X. Everybody knows that the dual of $\mathbf{C}_b(X)$ can be represented as the space of all finitely additive regular signed measures on $\mathbf{Ba}\,X$; regular means that for each B in $\mathbf{Ba}\,X$ and each $r > 0$ there is a zero set $Z \subset B$ such that

$$B' \subset B - Z\,, \quad B' \in \mathbf{Ba}\,X \quad \text{implies} \quad |\mu B'| < r\,.$$

This representation of the dual of $\mathbf{C}_b(X)$ is usually denoted by $\mathbf{M}(X)$; here we shall use this symbol just for the positive cone, i.e. for the set of all non-negative measures (this will be essential in § 3 and § 4!). There are two distinguished topologies on $\mathbf{M}(X)$, the norm topology (the norm of μ is μX), and the weak topology ($\{\mu_a\}$ converges to μ if and only if $\mu_a(f)$ converges to $\mu(f)$ for each f in $\mathbf{C}_b(X)$).

Consider the following subspaces of $\mathbf{M}(X)$:

$\mathbf{P}(X)$: the probabilities (i.e. $\mu X = 1$).

$\mathbf{P}^2(X)$: the probabilities which assume two values only.

$\mathbf{M}_\sigma(X)$: σ-additive measures.

$\mathbf{M}_\downarrow(X)$: strongly continuous measures (i.e., if $\mu(f_a)$ converges to 0 whenever $\{f_a\}$ is a net in $\mathbf{C}_b(X)$, which pointwise decreases to zero).

$\mathbf{M}_t(X)$: tight or Radon measures (the functionals continuous w.r.t. the compact-open topology on $\mathbf{C}_b(X)$; another description: for each B in $\mathbf{Ba}\, X$ and each $r > 0$ there exists a compact set $K \subset B$ such that $B' \subset B - K$ implies $\mu B' < r$).

The meaning of $\mathbf{P}_\sigma(X)$ and $\mathbf{P}_\downarrow(X)$ is obvious. It is easy to see that

$$\mathbf{M} \supset \mathbf{M}_\sigma \supset \mathbf{M}_\downarrow \supset \mathbf{M}_t\,,$$
$$\mathbf{P}^2 \supset \mathbf{P}^2_\sigma \supset \mathbf{P}^2_\downarrow = \mathbf{P}^2_t\,.$$

The following two results are due to E. Hewitt [1]:

$\mathbf{P}(X) = \mathbf{P}_\sigma(X) \Leftrightarrow \mathbf{M}(X) = \mathbf{M}_\sigma(X) \Leftrightarrow X$ is pseudocompact.

$\mathbf{P}^2_\sigma(X) = \mathbf{P}^2_t(X) \Leftrightarrow X$ is realcompact.

Definition. A space X is called *measure-compact* if $\mathbf{M}_\sigma(X) = \mathbf{M}_\downarrow(X)$, a *Radon space* if $\mathbf{M}_\sigma(X) = \mathbf{M}_t(X)$.

It has been proved by W. Moran that the product of two measure-compact spaces need not be measure-compact ($S \times S$ where S is the Sorgenfrey line), and that the class of Radon spaces is finitely productive. There is no known topological characterization of measure-compact nor Radon topological spaces.

Since $\mathbf{C}_b(X)$ and $\mathbf{C}_b(\beta X)$ are isomorphic, $\mathbf{M}(X)$ and $\mathbf{M}(\beta X)$ are also isomorphic, and if $\tilde{\mu} \in \mathbf{M}(\beta X)$ corresponds to $\mu \in \mathbf{M}(X)$, then

$$\tilde{\mu} B = \mu(B \cap X)$$

for each $B \in \mathbf{Ba}(\beta X)$. Since βX is compact,

$$\mathbf{M}(\beta X) = \mathbf{M}_t(\beta X)\,.$$

Now clearly $\mu \in \mathbf{M}_t(X)$ if and only if X is Caratheodory measurable in $\langle \beta X, \mathbf{Ba}\,(\beta X), \tilde{\mu}\rangle$. Thus we can say that X is a Radon space if and only if X is Caratheodory measurable for every $\mu \in \mathbf{M}_\sigma(X)$. For example, every analytic space is a Radon space [F 3], [S 1], a locally compact space need not be a Radon space (it is if and only if it is measure-compact).

3. Prochorov spaces

In this section we assume that $\mathbf{M}(X)$ is endowed with the weak topology. By a Prochorov space we mean a space X with the following property:

If C is compact subset of $\mathbf{P}_t(X)$, and if $r > 0$ then there exists a compact set $K \subset X$ such that $\mu(X - K) < r$ for each μ in C.

Certainly every compact space is a Prochorov space. About ten years ago Varadarajan asserted that every Radon space is a Prochorov space (a Prochorov space need not be Radon), but his proof gives just that every locally compact space is a Prochorov space (the mapping $\mathbf{C}_b(X) \times \mathbf{M}_t(X) \to \mathrm{R}$ is continuous where $\mathbf{C}_b(X)$ is given the compact-open topology). Varadarajan noticed that each G_δ subspace of a Prochorov space is a Prochorov space.

This year D. Preiss proved that the space Q of rational numbers is not a Prochorov space, and that an absolute Borel separable metrizable space is completely metrizable if and only if it does not contain a copy of Q as a G_δ subspace. Hence

Theorem (Preiss [1]). *The following three properties on a separable metrizable absolute Borel space X (more generally, coanalytic space) are equivalent:*

1. *X is completely metrizable.*
2. *X contains no G_δ copy of* Q.
3. *X is a Prochorov space.*

It should be remarked that V. Prochorov proved in 1956 that every separable completely metrizable space is a Prochorov space. For a connection of Prochorov spaces with null sets of vector measures we refer to G. Choquet.

In conclusion we should note that a separable metrizable Prochorov space need not be completely metrizable, and that it would be useful to complete the study of Prochorov spaces.

4. BC-spaces

Call a space X a BC-space if the following condition is fulfilled:

If T is a 0-dimensional compact space then for every continuous mapping $f: T \to \mathbf{P}_t(X)$ there exists a Radon probability μ on $\mathbf{C}(T, X)$ endowed with the compact-open topology such that

$$f = f_\mu$$

where $f_\mu t$ is the image of μ under the evaluation mapping e_t at t; by definition

$$f_\mu\, t(B) = \mu(e_t^{-1}[B]) .$$

It is easy to see that if T is compact and X is any space, then each $f_\mu : T \to \mathbf{P}_t(X)$ is continuous. R. M. Blumental and H. H. Corson [1] proved:

Theorem. *Every completely metrizable space is a* BC-*space.*

Every BC-space is a Prochorov space. Indeed if C is a compact subset of $\mathbf{P}_t(X)$ then we can take a continuous mapping f of a 0-dimensional compact space T onto C.

Let μ be a Radon probability on $\mathbf{C}(T, X)$ such that $f = f_\mu$. Given $r > 0$ take a compact subset K' in $\mathbf{C}(T, X)$ such that $\mu K' > 1 - r$. Consider the image K of $T \times K'$ in X under the mapping $T \times \mathbf{C}(T, X) \to X$. Clearly $ftK > 1 - r$ for each $t \in T$.

It follows that in the theorem of § 3 we can add one more equivalent condition: X is a BC space.

The statement "X is a BC-space" is equivalent to: if f is a continuous mapping of a 0-dimensional compact space into $\mathbf{P}(X)$ then there exists a Radon probability μ on $\mathbf{C}(T, X)$ such that the random function

$$\{e_t : \langle \mathbf{C}(T, X), \mu \rangle \to X\}$$

has the 1-dimensional distributions prescribed by f. With this motivation in mind, one is interested in the properties of the set of all probabilites representing a given mapping, and also in the existence theorems for T as general as possible. We mention another theorem by Blumenthal and Corson [2]:

Theorem. *Let $\mathbf{P}_\alpha(X)$ be the space of all $\mu \in \mathbf{P}(X)$ such that X is the support of μ. If X is a compact metrizable space then X is connected and locally connected if and only if each continuous mapping f of a compact metrizable space T into $\mathbf{P}_\alpha(X)$ can be represented by a Radon measure on $\mathbf{C}(T, X)$ (i.e. $f = f_\mu$ for some μ).*

5. Separable measurable spaces

In this paragraph let $\mathscr{X} = \langle X, \mathscr{B} \rangle$ be a measurable space.

A family $\{B_a \mid a \in A\}$ is called $\mathscr{B}$-preserving if the union of each subfamily $\{B_a \mid a \in A'\}$, $A' \subset A$, belongs to $\mathscr{B}$. A family $\{X_a \mid a \in A\}$ is called $\mathscr{B}$-discrete if there exists a $\mathscr{B}$-preserving disjoint family $\{B_a\}$ such that $B_a \supset X_a$ for each a. Hence, a disjoint $\mathscr{B}$-preserving family is $\mathscr{B}$-discrete, and a disjoint cover of X is $\mathscr{B}$-discrete if and only if it is $\mathscr{B}$-preserving. Observe that a disjoint cover $\{X_a \mid a \in A\}$ of X is $\mathscr{B}$-discrete if and only if the associated quotient mapping $f : \langle X, \mathscr{B} \rangle \to \langle A, \exp A \rangle$ is measurable ($f[X_a] = (a)$). A $\mathscr{B}$-discrete family $\{X_a \mid a \in A\}$ need not be $\mathscr{B}$-preserving even if $X_a \in \mathscr{B}$ for each a. For example, take a family $\{X_a \mid a < \omega_1\}$ such that X_a is a Baire set of class a in the unit interval I, and consider the subset $B' = \sum\{X_a\}$ in the sum space $X = \sum\{I \mid a < \omega_1\}$. Observe that X is metrizable, B' is no Baire set in X, but $\{(a) \times X_a \mid a \in A\}$ is $\mathbf{Ba}(X)$-discrete.

Definition. Call a measurable space $\langle X, \mathscr{B} \rangle$ *separable* if every $\mathscr{B}$-discrete cover of X is countable.

If $2^{\aleph_1} > 2^{\aleph_0}$ then every countably generated measurable space is separable. If there exists an uncountable separable metrizable space X such that $\exp X = \mathbf{Ba}\, X$, then $\langle X, \mathbf{Ba}\, X \rangle$ is countably generated but not separable.

On the other hand, a separable measurable space need not be countably generated. This follows from the following consequence of the theorem in § 1.

Theorem. *If A is analytic then* **Ba** *A is separable. If $\langle X, \mathscr{B}\rangle$ is a pseudoanalytic space then $\mathscr{B}$ is separable.*

In some spaces X there is a close connection between **Ba**-discreteness and topological discreteness. This will be shown in § 7.

6. Hyper-rocks

We have noticed that the union of a $\mathscr{B}$-discrete family in $\mathscr{B}$ need not belong to $\mathscr{B}$. We shall see later on that it is a nice property of $\mathscr{B}$ if the union of every $\mathscr{B}$-discrete family in $\mathscr{B}$ belongs to $\mathscr{B}$. We say that a σ-algebra with this nice property is **h**-closed.

Definition. If $\mathscr{B}$ is a σ-algebra on X, we denote by $\mathbf{h}\mathscr{B}$ the smallest σ-algebra $\mathscr{C} \supset \mathscr{B}$ with the property: the union of each $\mathscr{B}$-discrete family in $\mathscr{C}$ belongs to $\mathscr{C}$. We denote by $\mathbf{H}\mathscr{B}$ the smallest σ-algebra $\mathscr{C} \supset \mathscr{B}$ with $\mathbf{h}\mathscr{C} = \mathscr{C}$.

Clearly $\mathbf{h}\mathscr{B} = \mathscr{B}$ if and only if $\mathscr{B}$ is **h**-closed, and $\mathbf{h}\mathscr{B} = \mathscr{B}$ is equivalent to $\mathbf{H}\mathscr{B} = \mathscr{B}$.

It is obvious that $\mathbf{H}\mathscr{B}$ is obtained by transfinite iteration of **h**, i.e.

$$\mathbf{H}\mathscr{B} = \bigcup\{h_\alpha\mathscr{B}\}$$

where $h_\alpha\mathscr{B} = \mathbf{h}\mathscr{C}$ with $\mathscr{C}$ being the smallest σ-algebra which contains $\bigcup\{h_\beta\mathscr{B} \mid \beta < \alpha\}$.

Evidently $\mathbf{h}\mathscr{B}$ can be obtained by iterating the operations involved. The operation of taking σ-discrete unions is very important, and therefore some details may be in place.

Definition. Let $\mathscr{B}$ be a σ-algebra on X. Denote by $\mathbf{d}_{\mathscr{B}}$ the map from exp (exp X) into itself which assigns to each $\mathscr{C} \subset \exp X$ the set of all unions of $\mathscr{B}$-discrete families in $\mathscr{C}$.

The operation $\mathbf{d}_{\mathscr{B}}$ need not be idempotent; however, obviously, it is if $\mathscr{B}$ is **h**-closed.

The other operation involved is ϱ, which assigns to each $\mathscr{C} \subset \exp X$ the smallest $\mathscr{D} \supset \mathscr{C}$ with $\mathscr{D}_\sigma = \mathscr{D}_\delta = \mathscr{D}$. The elements of $\varrho\mathscr{D}$ are called the *rocks* over $\mathscr{D}$. The operation ϱ is idempotent, and $\varrho\mathscr{D}$ is a σ-algebra if $\mathscr{D}$ is.

Now we may write

$$\mathbf{h}\mathscr{B} = \bigcup\{\mathscr{B}_\alpha\},$$

where $\mathscr{B}_0 = \mathscr{B}$, and $\mathscr{B}_\alpha = \mathbf{d}_{\mathscr{B}}\mathscr{B}'_\alpha$ or $\mathscr{B}_\alpha = \varrho\mathscr{B}'_\alpha$ according to as α is odd or even, with $\mathscr{B}'_\alpha = \bigcup\{\mathscr{B}_\beta \mid \beta < \alpha\}$.

Remark. Sometimes the following situation occurs: We are given a σ-algebra $\mathscr{C} \subset \mathscr{B}$, and we consider the smallest σ-algebra $\mathscr{D} \supset \mathscr{C}$ with $\mathbf{d}_{\mathscr{B}}\mathscr{D} = \mathscr{D}$. This σ-algebra is denoted by $\mathbf{h}_{\mathscr{B}}\mathscr{C}$. This is the situation when one investigates completeness of bi-measurable spaces, see § 10.

There are two more functors of measurable spaces into itself. The first assigns to each σ-algebra the smallest σ-algebra with the property that the meet of two discrete covers is discrete, and the other one assigns to each $\mathscr{B}$ the σ-algebra of all bi-Souslin sets over $\mathscr{B}$; for the definition see § 7.

7. Non-separable Baire sets

The main result says:

Theorem. *If $\mathscr{B}$ is the σ-algebra of all Baire sets in a completely metrizable space X then*

$$\mathscr{C} = \mathbf{H}\mathscr{B} = \mathbf{h}\mathscr{B} = \textbf{bi-Souslin}\ \mathscr{B} = \textbf{bi-Souslin}\ X,$$

and $\mathscr{C}$ is locally determined in X.

Recall that if $\mathscr{M}$ is a cover of a set Y, then **Souslin** $\mathscr{M}$ stands for the collection of all Souslin sets over $\mathscr{M}$, and **bi-Souslin** $\mathscr{M}$ stands for the collection of all $M \subset Y$ such that the two sets M and $Y - M$ belong to **Souslin** $\mathscr{M}$. If X is a space, then

$$\textbf{Souslin}\ X = \textbf{Souslin}\ \mathscr{M}, \quad \textbf{bi-Souslin}\ X = \textbf{bi-Souslin}\ \mathscr{M}$$

where $\mathscr{M}$ is the collection of all closed sets in X. The Souslin sets over $\mathscr{M}$ are the sets of the form

$\mathbf{S}M = \bigcup\{\bigcap\{Ms \mid s \in S, s < \sigma\} \mid \sigma \in \Sigma\}$ where S is the set of the finite sequences in N, Σ is the set of all sequences in N, $<$ indicates restriction, and $Ms \in \mathscr{M}$ for each s.

The crucial step in the proof of Theorem is the following important and very new result by R. W. Hansell [1]:

Lemma. *Assume that $\{X_a\}$ is a disjoint family in a completely metrizable space X such that the union of each subfamily is a Souslin set in X. Then we can write $X_a = \bigcup\{X_{an} \mid n \in \mathsf{N}\}$, $a \in A$, where $\{X_{an} \mid a \in A\}$ is topologically discrete in X for each n.*

We may assume that each X_{an} is closed in X_a, and then $\{X_{an}\}$ is a σ-discrete refinement of $\{X_a\}$ by Souslin sets. It is easy to show that each σ-discrete family of Souslin sets has the property that the union of each subfamily is Souslin. Since

$$\mathscr{B} \subset \textbf{Souslin}\ \mathscr{B} \subset \textbf{Souslin}\ X$$

in any space X (with $\mathscr{B} = \mathbf{Ba}\,X$), in our setting we get

$$\mathbf{H}\mathscr{B} = \mathbf{h}\mathscr{B} \subset \mathbf{Souslin}\,X\,.$$

By another theorem of Hansell [3] (First principle type theorem)

$$\mathbf{bi\text{-}Souslin}\,X \subset \mathbf{h}\mathscr{B}\,.$$

In the next section we shall need another deep property of Baire-discrete covers in completely metrizable spaces.

Proposition. *Every Baire-discrete cover of a completely metrizable space is bounded, i.e. the supremum of the Baire classes of the members is countable.*

The proof depends on Hansell's Lemma.

Corollary. *Let X be a completely metrizable space, and let $\{X_a\}$, $\{Y_b\}$ be two Baire-discrete covers of X. Then $\{X_a \cap Y_b\}$ is Baire-discrete.*

Proof. Every σ-discrete (topologically) family of Baire sets of a bounded class is Baire-preserving.

Remark. The meet of two $\mathscr{B}$-discrete covers need not be $\mathscr{B}$-discrete. E.g. consider the measurable product $\langle Y, \mathscr{B}\rangle$ of $\langle X, \exp X\rangle$ by itself, and assume that $\mathscr{B} \neq \exp Y$. Clearly $\{(x) \times X \mid x \in X\}$ and $\{X \times (x) \mid x \in X\}$ are $\mathscr{B}$-discrete, but the meet $\{(\langle x, y\rangle) \mid x \in X,\, y \in X\}$ is not $\mathscr{B}$-discrete.

8. Basic functors into uniform spaces

It seems to me that an appropriate application of uniform methods to measurable spaces may be useful for the theory of measurable spaces as well as for the theory of uniform spaces. At least there are some highly non-trivial results the discovery of which has been stimulated by the viewpoint of uniform spaces. Note that it has been known for a long time that all non-trivial examples in the general theory of uniform spaces were "measurable".

Throughout this paragraph let $\mathscr{X} = \langle X, \mathscr{B}\rangle$ be a measurable space; it is convenient to assume that $\mathscr{B}$ is just an algebra, not necessarily a σ-algebra.

The proximity induced by $\mathscr{B}$ is designated by $\mathbf{p}_{\mathscr{B}}$, or simply $\mathbf{p}$, and is defined by setting

$$M_1 \mathbf{p} M_2 \Leftrightarrow M_1 \subset B\,, \quad M_2 \subset X - B \quad \text{for no} \quad B \in \mathscr{B}\,.$$

Evidently $\mathbf{p}$ has the following property: if M_1 and M_2 are distant then there exists a set B such that $M_1 \subset B$, $M_2 \subset X - B$, and B is distant to $X - B$. Such a proximity is called *measurable.* It is easy to see that $\{\mathscr{B} \to \mathbf{p}_{\mathscr{B}}\}$ is a bijection of algebras onto measurable proximities; if p is measurable then $p = \mathbf{p}_{\mathscr{B}}$ where $\mathscr{B} = \{Y \mid Y \text{ is distant to } X - Y\}$.

The proximities induced by σ-algebras are described as follows:

Theorem 1 [F 9]. *A proximity p on a set X is induced by a σ-algebra if and only if the following condition is fulfilled:*

If the union of a sequence $\{A_n\}$ is proximal to B then some A_n is proximal to B.

Thus one might say that p is induced by a σ-algebra if and only if p is σ-additive. Observe that σ-additivity implies measurability. The next result describes the compactifications associated with σ-additive proximities.

Theorem 2 [F 9], [Hager 1]. *Let $\langle X, p\rangle$ be a dense proximal subspace of a compact space K. Then p is σ-additive if and only if K is basically disconnected, X is G_δ-dense in K, and X is a P-space.*

It would be of some interest to find how to study measurable spaces by means of the corresponding compact spaces.

The precompact uniformity associated with $\mathbf{p}_{\mathscr{B}}$ is denoted by $\mathbf{u}_{\aleph_0}\mathscr{B}$, and the corresponding uniform space is denoted by $\mathbf{u}_{\aleph_0}\mathscr{X}$. Clearly $\{\mathscr{B} \to \mathbf{p}_{\mathscr{B}}\}$ defines a functor of measurable spaces into proximity spaces, and $\{\mathscr{B} \to \mathbf{u}_{\aleph_0}\mathscr{B}\}$ defines a functor into uniform spaces. It is easy to see that $\mathbf{u}_{\aleph_0}\mathscr{B}$ is projectively generated by all bounded $\mathscr{B}$-measurable functions. Let $\mathbf{u}_{\aleph_1}\mathscr{B}$ be the uniformity projectively generated by all measurable functions (not necessarily bounded).

Proposition. *The uniform partitions in $\mathbf{u}_{\aleph_1}\mathscr{X}$ form a basis for the uniform covers. If $\mathscr{U}$ is a separable uniformity with the precompact part $\mathbf{u}_{\aleph_0}\mathscr{X}$, then $\mathscr{U}$ is contained in $\mathbf{u}_{\aleph_1}\mathscr{B}$.*

Thus $\mathbf{u}_{\aleph_1}\mathscr{B}$ is proximally fine among separable uniformities. The property that the uniform partitions form a basis will be needed throughout, and we must agree on a short name for the uniformities with this property.▌

Definition. An *ultra-uniformity* is a uniformity with the property that the uniform partitions form a basis for the uniform covers.

An expository account of results and methods in separable uniform spaces with particular attention to measurable spaces is provided by A. Hager's papers listed in References, which are primarily concerned with the rings of functions.

Here we want to discuss the non-separable uniformities. There are some old problems connected with general uniformities, and therefore we restrict our attention to ultra-uniformities.

Definition. Let m be an infinite cardinal. Denote by $\mathbf{u}_m\mathscr{B}$ the uniformity on X which has all $\mathscr{B}$-discrete covers of cardinal less than m for a sub-basis of uniform covers. Put $\mathbf{u}\mathscr{B} = \bigcup\{\mathbf{u}_m\mathscr{B}\}$.

It is easy to see that this definition for $m = \aleph_0$, and $\aleph_1$ gives the previously defined uniformities.

One can say that $\mathbf{u}_m\mathscr{B}$ is projectively generated by all mappings

$$f : X \to M ,$$

where M is a uniformly discrete space of cardinal less than m, and $f : \langle X, \mathscr{B}\rangle \to \langle M, \exp M\rangle$ is measurable.

We noticed in § 7 that the meet of two $\mathscr{B}$-discrete covers need not be $\mathscr{B}$-discrete, and hence the uniformity $\mathbf{u}_m\mathscr{B}$ need not be proximally equivalent to $\mathbf{u}_{\aleph_0}\mathscr{B}$.

Definition. An algebra $\mathscr{B}$ is called *proximally fine* if $\mathbf{u}\mathscr{B}$ is proximally equivalent to $\mathbf{u}_{\aleph_0}\mathscr{B}$.

In other words, $\mathscr{B}$ is proximally fine if and only if the meet of any two $\mathscr{B}$-discrete covers is a $\mathscr{B}$-discrete cover.

Example. The Baire σ-algebra of a completely metrizable space is proximally fine. This is proved in the conclusion of § 7.

Theorem 3. *A σ-algebra $\mathscr{B}$ is* **h**-*closed if and only if it is proximally fine and* $\mathbf{u}\mathscr{B}$ *is locally fine.*

Recall that a uniform space is called locally fine if every uniformly locally uniform cover is uniform, see Isbell [1]. The proof of Theorem 3 is routine. Observe that in Theorem 3 one may replace locally fine by: every uniformly locally "finite uniform" cover is uniform.

It is shown in § 7 that the σ-algebra of Baire sets in a completely metrizable space X need not be **h**-closed, and it is a result of Hansell [3] that it is not if X is a Baire space, i.e. if $X = M^{\aleph_0}$ for some uncountable discrete space M.

It is easy to see that $\mathbf{uH}\mathscr{B}$ is the locally fine coreflection of $\mathbf{u}\mathscr{B}$. Thus the following diagram is commutative:

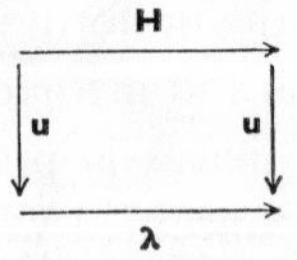

where **H** is a coreflection in measurable spaces defined in § 6, and λ is the coreflection of uniform spaces onto locally fine uniform spaces.

For the details and for the consequences in general theory of uniform spaces we refer to [F 8].

9. Another functors into uniform spaces

We get a scale of functors depending on an infinite cardinal m if we consider all partitions of cardinal less than m by measurable sets. This uniformity seems to be too fine, and I don't see anything interesting to be said about these functors.

There is another possibility, something between $\mathbf{u}_m$ and the latter one.

Definition. Let $\mathscr{X} = \langle X, \mathscr{B} \rangle$ be a measurable space. A *free* family in $\mathscr{B}$ is a family $\{B_a \mid a \in A\}$ such that there is a well ordering $\leqq$ on A such that

$$\bigcup \{B_a \mid a < b\} \in \mathscr{B}$$

for each b in A.

It is easy to check that the meet of two free partitions is a free partition. Thus we can define the functors $\mathbf{v}_m$ into ultra-uniform spaces, where the free partitions of cardinal less than m form a basis for uniform covers in $\langle X, \mathbf{v}_m \mathscr{B} \rangle$.

These functors may be quite interesting. I don't know of any deep theorem, however some examples I know are of certain interest.

10. Complete measurable spaces

A measurable space $\mathscr{X} = \langle X, \mathscr{B} \rangle$ is called complete if the uniform space $\mathbf{u}_{\aleph_1} \mathscr{X}$ is complete. It is immediate that $\mathscr{B}$ is complete if and only if every $\mathscr{B}$-ultrafilter with the countable intersection property (abb. CIP) is fixed.

It is proved in [F 7] that $\mathscr{B}$ is complete if and only if $\mathbf{h}\mathscr{B}$ is complete, and the cardinal of every $\mathscr{B}$-discrete partition is non-measurable. One may replace $\mathbf{h}$ by $\mathbf{H}$.

Examples. A topological space X is said to be *Baire complete* or *Borel complete* if, respectively, $\mathbf{Ba}\,X$ or $\mathbf{Bo}\,X$ is complete. By the Hewitt's theorem quoted in § 2 a space is Baire complete if and only if it is realcompact. Almost nothing is known about Borel complete spaces, see Hager [4] and [F 8].

Definition. Let $\mathscr{C}$ and $\mathscr{B}$ be σ-algebras on X, and let $\mathscr{C} \supset \mathscr{B}$. We say that the bi-measurable space $\langle X, \mathscr{B}, \mathscr{C} \rangle$ is *complete* if every $\mathscr{B}$-ultrafilter which extends to a $\mathscr{C}$-ultrafilter with CIP is fixed.

Clearly if $\mathscr{B}$ or $\mathscr{C}$ is complete then $\langle \mathscr{B}, \mathscr{C} \rangle$ is complete. A topological space X is called *Baire-Borel complete* if $\langle \mathbf{Ba}\,X, \mathbf{Bo}\,X \rangle$ is complete. The category of all Baire complete spaces is a simple epireflective category, the product of an uncountable number of at least two point spaces is not Borel complete. The category of Baire-Borel complete spaces is epireflective; is it the epireflective hull of the Borel complete spaces? For the details and other results we refer to [F 7, 8].

References

R. M. Blumenthal and H. H. Corson:

[1] On continuous collections of measures. Ann. Inst. Fourier (Grenoble).

[2] On continuous collections of measures. Proc. Sixth Berkeley Sympos. Math. Statist. and Probability.

J. R. Choksi: Inverse limits of measure spaces. Proc. London Math. Soc. *8* (3) (1958), 321–342.

G. Choquet: Sur les ensembles uniformément négligeables. Initiation à l'analyse, 9e année, 1969/70, n° 6.

H. H. Corson: see R. M. Blumenthal [1], [2].

Z. Frolík:

[1] A note on $C(P)$ and Baire sets in compact and metrizable spaces. Bull. Acad. Polon. Sci. Sér. Sci. Math. Astronom. Phys. *15* (1967), 779–784.

[2] Abstract analytic and Borelian sets. Bull. Soc. Math. France *97* (1969), 357–368.

[3] Projective limits of measures. Proc. Sixth Berkeley Sympos. Math. Statist. and Probability.

[4] Absolute Borel and Souslin sets. Pacific J. Math. *32* (1970), 663–683.

[5] A measurable map with analytic domain and metrizable range is quotient. Bull. Amer. Math. Soc. *76* (1970), 1112–1117.

[6] On capacity-compact measures. (To appear.)

[7] On prime filters with CIP. (To appear.)

[8] Baire-Borel complete spaces. (To appear.)

[9] Uniformities associated with measurable spaces. (To appear.)

P. Gänssler: Compactness and sequential compactness in spaces of measures. Z. Wahrscheinlichkeitstheorie und Verw. Gebiete *17* (1971), 124–146.

A. W. Hager:

[1] Measurable uniform spaces. (To appear.)

[2] A category of uniform spaces. (To appear.)

[3] Three classes of uniform spaces. These Proceedings.

A. W. Hager, G. D. Reynolds and M. D. Rice:

[4] Borel-complete topological spaces. Fund. Math. *75* (1972), 135–143.

R. W. Hansell:

[1] PhD dissertation. Rochester, 1970.

[2] On the nonseparable theory of Borel and Souslin sets. Bull. Amer. Math. Soc. *78* (1972), 236–241.

[3] On the non-separable theory of k-Borel and k-Souslin sets, I. General Topology and its Applications (1972).

A. Hayes: Alexander's theorem for realcompactness. Proc. Cambridge Philos. Soc. *64* (1968), 41–43.

E. Hewitt: Linear functionals on spaces of continuous functions. Fund. Math. *37* (1950), 161–189.

E. R. Lorch: Compactification, Baire functions, and Daniell integration. Acta Sci. Math. (Szeged) *24* (1963), 204–218.

E. Marczewski: On compact measures. Fund. Math. *40* (1953), 113–121.

M. Metivier: Limites projectives de mesures. Martingales. Applications. Ann. Mat. Pura Appl. *63* (4) (1963), 225–352.

W. Moran:

[1] The additivity of measures on completely regular spaces. J. London Math. Soc. *43* (1968), 633–639.
[2] Measures and mappings on topological spaces. Proc. London Math. Soc. (3) *19* (1969), 493–508.
[3] Measures on metacompact spaces. Proc. London Math. Soc. (3) *20* (1970), 507–524.

D. Preiss: Metric spaces in which Prohorov's theorem is not valid. Ann. Inst. Fourier (Grenoble) (1973) (see also his abstract in these Proceedings).
Yu. V. Prohorov: Convergence of random processes and limit theorems in probability theory. Theor. Probability Appl. *1* (1956), 157–214.
B. V. Rao: PhD dissertation. Calcutta, 1969.
G. D. Reynolds: see A. W. Hager [4].
M. D. Rice: see A. W. Hager [4].
M. Sion: Topological and measure theoretic properties of analytic sets. Proc. Amer. Math. Soc. *11* (1960), 769–776.

A. H. Stone:

[1] Non-separable Borel sets. Rozprawy Mat. *28*, Warszawa, 1962.
[2] Non-separable Borel sets II. General Topology and its Applications (1972).

F. Topsoe:

[1] Preservation of weak convergence under mappings. Ann. Math. Statist. *38* (6) (1967), 1661–1665.
[2] A criterion for weak convergence of measures with an application to convergence of measures on *D*[0, 1]. Math. Scand. *25* (1969), 97–104.
[3] Compactness in spaces of measures. Studia Math. *36* (1970), 195–212.

V. S. Varadarajan: Measures on topological spaces. Amer. Math. Soc. Transl., ser. II, *48* (1965), 161–228.

INSTITUTE OF MATHEMATICS OF THE CZECHOSLOVAK ACADEMY OF SCIENCES, PRAHA

PROXIMITY APPROACH TO TOPOLOGICAL PROBLEMS

M. GAGRAT (1) and S. A. NAIMPALLY (2)

Kanpur and Thunder Bay

In this paper we use the techniques of proximity spaces to investigate the properties of semi-metric and developable spaces. By a proximity δ on a nonempty set X, we mean any one of the following: (i) S-proximity, (ii) LO-proximity, (iii) R-proximity, or (iv) EF-proximity (see [6], [7]). Whenever X is a topological space, δ is assumed to be compatible with the topology of X. For a semi-metric space (X, d), we define $\delta(d)$ by: $A\,\delta(d)\,B$ iff $d(A, B) = 0$, where $d(A, B) = \text{Inf}\{d(a, b) : a \in A, b \in B\}$. Set $\mathcal{U}_d = \{U = U^{-1} \subset X \times X : V_{1/n} \subset U$ for some $n \in \mathbf{N}\}$, where $V_\varepsilon = \{(x, y) : d(x, y) < \varepsilon\}$. For a developable space (X, Σ), $\Sigma = \{G_n : n \in \mathbf{N}\}$ where each G_n is an open cover of X and $G_n \supset G_{n+1}$ we set $d(x, y) = \text{Inf}\{1/(n+1) : y \in \text{St}(x, G_n)\}$. The concept of an M-uniformity is defined in [7].

We now present our main results. The following theorem is analogous to the well-known result: A T_1-space has a compatible uniformity with a countable base iff it is metrizable (and consequently has a compatible metric d such that $\delta(d)$ is an EF-proximity).

Theorem 1. *A T_1-space has a compatible* M-*uniformity with a countable base if and only if it has a compatible semi-metric d such that $\delta(d)$ is a* LO-*proximity.*

We give below two new characterizations of developable spaces which can be used to give transparent proofs of some of the results already known in the literature.

Theorem 2. *A T_1-space X is developable if and only if it has a compatible semi-metric d such that $\mathcal{U}_d$ is an* M-*uniformity with a countable open base.*

Theorem 3. (Cf. Cook [5].) *A T_1-space X is developable if and only if it has a compatible upper-semi-continuous semi-metric.* (Cook proved that if the semi-metric is continuous then the space is developable.)

We now present two metrization theorems which are improvements of those of Arkhangel'skii [4] and Nedev [8] respectively.

Theorem 4. *A T_1-space X is metrizable if and only if it has a compatible semi-metric d such that $\delta(d)$ is an* R-*proximity.* (Arkhangel'skii required $\delta(d)$ to be an EF-proximity.)

Theorem 5. *A T_1-space X is metrizable if and only if it has a compatible semi-metric d such that for each closed set $A \subset X$, $d(A, x)$ is a lower-semi-continuous function of x.* (Nedev required $d(A, x)$ to be continuous.)

In the sequel we suppose that δ is a compatible proximity on a T_1-space X and that $f : X \to Y$ is a continuous function onto a T_1-space Y. We generalize the concepts: T_1-map, uniform map, completely uniform map, pseudo-open map etc. so as to include proximity spaces (see [1], [2], [3], [4]). For example f will be called *uniform* iff for each y in Y and each neighbourhood N_y of y, $f^{-1}(y)\, \bar{\delta}\, (X - f^{-1}(N_y))$. When δ is induced by a metric on X, the above definition coincides with the usual one. We also define a nearness relation δ' on Y as follows:

$$E\, \bar{\delta}'\, F \text{ iff } f^{-1}(E)\, \bar{\delta}\, f^{-1}(F).$$

Theorem 6. *δ' is a compatible* S*-proximity on Y in the following two cases:*

(i) *δ is an* S*-proximity, f is pseudo-open and uniform;*

(ii) *δ is an* R*-proximity, f is pseudo-open and compact.*

Theorem 7. *If δ is a* LO*-proximity on X and f is open uniform, then δ' is a compatible* LO*-proximity on Y.*

Theorem 8. *Suppose δ is an* S*-proximity and f is open uniform. Then f is completely uniform if and only if δ' is a compatible* R*-proximity on Y.*

Theorem 9. *If (X, d) is a semi-metric space and f is open and completely uniform, then Y is metrizable.* (The known result requires d to be a metric [1].)

Theorem 10. *If X is developable and f is open uniform, then Y is developable.*

The detailed paper with proofs will appear in the Pacific Journal of Mathematics.

References

[1] *P. S. Alexandroff:* On some results concerning topological spaces and their continuous mappings. General Topology and its Relations to Modern Analysis and Algebra (I) (Proc. (First) Prague Topological Sympos., 1961). Academia, Prague, 1962, 41—54.

[2] *P. S. Alexandroff:* Some results in the theory of topological spaces, obtained within the last twenty five years. Russian Math. Surveys *15* (1960), 23—83.

[3] *P. S. Alexandroff:* Some basic directions in general topology. Russian Math. Surveys *19* (1964), 1—19.

[4] *A. V. Arkhangel'skii:* Mappings and spaces. Russian Math. Surveys *21* (1966), 87—114.

[5] *H. Cook:* Cartesian products and continuous semi-metrics. Topology Conference, Arizona State University, 1967, 58—63.

[6] *D. Harris:* Regular-closed spaces and proximities. Pacific J. Math. *34* (1970), 675—685.

[7] *S. A. Naimpally and B. D. Warrack:* Proximity spaces. Cambridge Tracts in Mathematics No. *59*, Cambridge University Press, 1970.

[8] *S. I. Nedev:* Continuous and semi-continuous 0-metrics. Soviet Math. Dokl. *11* (1970), 975—978.

(1) INDIAN INSTITUTE OF TECHNOLOGY, KANPUR, U.P.

(2) INDIAN INSTITUTE OF TECHNOLOGY, KANPUR, U.P.
LAKEHEAD UNIVERSITY, THUNDER BAY, ONTARIO

ÜBER 2-BANACH-RÄUME

S. GÄHLER

Berlin

Ein 2-*normierter Raum* L ist ein linearer Raum der Dimension >1, über dessen Punktepaaren a, b eine reelle Funktion v, die 2-*Norm* von L, erklärt ist, die folgende Eigenschaften besitzt:

1. Es gilt $v(a, b) = 0$ genau dann, wenn a und b linear abhängig sind,
2. $v(a, b) = v(b, a)$,
3. $v(a, \beta b) = |\beta| \, v(a, b)$ für jede reelle Zahl β sowie
4. $v(a, b + c) \leqq v(a, b) + v(a, c)$.

Auf jedem 2-normierten Raum L erzeugt das System $\{\mu_b\}_{b \in L}$ aller durch $\mu_b(a) = v(a, b)$ definierter Halbnormen μ_b, $b \in L$, eine uniforme Struktur und damit eine Topologie. Ein 2-normierter Raum, der eine beschränkte Nullumgebung besitzt, heißt *lokal beschränkt*. Ein vollständiger 2-normierter Raum wird 2-*Banach-Raum* genannt. Der Vortrag befaßt sich mit den 2-Banach-Räumen näher. Es wird unter anderem gezeigt, daß vom Standpunkt der uniformen Struktur aus die Klasse der Banach-Räume der Dimension >1 mit der der lokal beschränkten 2-Banach-Räume übereinstimmt.

Ein 2-normierter Raum L^* heißt *Vervollständigung* eines 2-normierten Raumes L, wenn L ein linearer Teilraum von L^* ist, $v = v^*|L \times L$ gilt, die Topologie von L mit der durch die Topologie von L^* in L induzierten Topologie übereinstimmt, L in L^* dicht sowie L^* vollständig ist. Durch Beispiele läßt sich zeigen, daß ein 2-normierter Raum nicht immer eine Vervollständigung besitzt. Es existiert genau dann eine Vervollständigung L^* von L, wenn die folgenden drei Bedingungen erfüllt sind:

1. Für zwei Cauchysche Moore-Smith-Folgen $(a_i)_{i \in I}$, $(b_j)_{j \in J}$ von Punkten $a_i, b_j \in L$ ist $\lim_{i \in I} \lim_{j \in J} v(a_i, b_j) = 0$ äquivalent der Existenz zweier reeller Zahlen α und β mit $\alpha^2 + \beta^2 \neq 0$ sowie $(\alpha a_i)_{i \in I} \sim (\beta b_j)_{j \in J}$, wobei $\sim$ die bei der Konstruktion einer Vervollständigung eines uniformen Raumes üblicherweise benutzte Äquivalenzrelation bezeichnet.

2. Für beliebige Cauchysche Moore-Smith-Folgen $(a_i)_{i \in I}$, $(b_j)_{j \in J}$ von Punkten $a_i, b_j \in L$ gilt $\lim_{i \in I} \lim_{j \in J} \lim_{i^* \in I} v(a_i - a_{i^*}, b_j) = 0$.

3. Zu jeder Cauchyschen Moore-Smith-Folge $(a_i)_{i \in I}$ von Punkten $a_i \in L$ und jedem $\varepsilon^* > 0$ existieren Punkte $b_1, \ldots, b_m \in L$ sowie ein $\varepsilon > 0$, so daß $v(a, b_j) < \varepsilon$, $j = 1, \ldots, m$, stets $\lim_{i \in I} v(a, a_i) < \varepsilon^*$ nach sich zieht.

Literatur

[1] *S. Gähler:* Über 2-Banach-Räume. Math. Nachr. *42* (1969), 335–347.
[2] *S. Gähler:* Zum Problem der Vervollständigung 2-normierter Räume. Hausdorff-Gedenkband.
[3] *A. White:* 2-Banach spaces. Math. Nachr. *42* (1969), 43–60.

ON SOME PRETOPOLOGIES ASSOCIATED WITH A TOPOLOGY

K. M. GARG and S. A. NAIMPALLY

Edmonton and Thunder Bay

Given a set X and a family $\mathscr{U}$ of subsets of X, let $(X, \mathscr{U})$ be a *pretopological space* if $\mathscr{U}$ contains X and the empty set and is closed under arbitrary union. Various notions like the interior, closure, frontier, and the derived set operators as well as set properties like open, closed, dense, nowhere dense, dense-in-itself, connected, compact etc. can be defined in a pretopological space in analogy with a topological space. Many results of topological spaces remain valid in pretopological spaces, whereas some become false.

Let, for every set $A \subset X$, $A^\alpha = A^{-0}$, $A^\beta = A^{0-}$, $A^\gamma = A^{0-0}$, $A^\delta = A^{-0-}$, $A^\xi = A^\alpha \cap A^\beta$ and $A^\eta = A^\alpha \cup A^\beta$. ($A^-$ means the closure of A, A^0 the interior of A.) Let $\mathscr{U}_\lambda = \{A \subset X : A \subset A^\lambda\}$ for $\lambda = \alpha, \beta, \gamma, \delta, \xi$ and η. Each of these families of sets is again a pretopology on X and we have, in general,

$$\mathscr{U} \subset \mathscr{U}_\gamma = \mathscr{U}_\xi \subset \mathscr{U}_\alpha[\mathscr{U}_\beta] \subset \mathscr{U}_\eta \subset \mathscr{U}_\delta .$$

We call these pretopologies the *associated pretopologies* of $\mathscr{U}$. If X has no isolated points then $\mathscr{U}_d = \{A \subset X : A \subset A^d\}$, where A^d denotes the derived set of A, is also a pretopology on X. In case $\mathscr{U}$ is a topology, it is only $\mathscr{U}_\gamma$ which is always a topology and the rest are in general only pretopologies. This is why we investigate pretopological spaces.

Let a pretopological space $(X, \mathscr{U})$ be a *pre*topological* space if $(A \cap A^\alpha)^\alpha = A^\alpha$ for every set $A \subset X$. A topological space is pre*topological and all the pretopologies associated with a pre*topology are pre*topologies. The topological structure of the pretopologies associated with a pre*topology are investigated and some of the properties of the original space are determined in terms of these asociated pretopologies. The pretopologies associated with the associated pretopologies of a pre*-topology are obtained as one of the original associated pretopologies.

The frontier of a set can be decomposed into six parts in such a way that various properties of sets and spaces are characterized in terms of these parts. The decomposed frontiers relative to the associated pretopologies are obtained in terms of the original pretopology.

The notion of continuity has a natural extension to functions in pretopological spaces. Many results on continuity in topological spaces remain valid in pretopological spaces.

Given $f: X \to Y$, where $(X, \mathscr{U})$, $(Y, \mathscr{V})$ are pretopological spaces, let, for each of $\lambda = \alpha, \beta, \gamma, \delta, \xi, \eta$, f be *λ-weakly continuous* if $f^{-1}(V) \in \mathscr{U}_\lambda$ for every $V \in \mathscr{V}$. In topological spaces the quasi-continuity and near continuity are characterized by β-weak continuity and α-weak continuity, respectively. It follows from the established results that these two notions of continuity yield together θ-continuity, which in turn implies continuity in case Y is regular. Whereas it is known that dense sets are the stationary sets for quasi-continuity, it turns out that the stationary sets for near continuity are the sets with nowhere dense complements.

We further investigate continuity and openness of the homomorphisms between topological groups in terms of much weaker properties.

ON $\mathfrak{m}$-ADIC SPACES

J. GERLITS

Budapest

S. Mrówka [4] generalizing the notion of dyadicity, introduced the class of $\mathfrak{m}$-adic spaces. Denoting by $A_{\mathfrak{m}}$ the one-point compactification of a discrete space of cardinality $\mathfrak{m}$, a T_2-space X is said to be $\mathfrak{m}$-adic if it is a continuous image of a suitable topological power of $A_{\mathfrak{m}}$. It is not difficult to prove that a space is dyadic iff it is $\aleph_0$-adic.

S. Mrówka also proposed the following generalization of $\mathfrak{m}$-adicity: let us denote by $W(\xi + 1)$ the order-topological space of the ordinal numbers $<\xi + 1$ for an ordinal ξ. A Hausdorff space which is a continuous image of some topological power of $W(\xi + 1)$ will be called a ξ-adic space. This class of spaces is wider than that of $\mathfrak{m}$-adic spaces; indeed $A_{\mathfrak{m}}$ is a continuous image of $W(\xi + 1)$, where ξ is any ordinal of cardinality $\mathfrak{m}$.

S. Mrówka raised the following question as an open problem: is it true that an $\mathfrak{m}$-adic space with character $\leq \mathfrak{n}$ ($\mathfrak{n} \leq \mathfrak{m}$) is necessarily $\mathfrak{n}$-adic? Our aim is to give an affirmative answer to this question; indeed, the following more general theorem holds:

Theorem 1. *The weight and the character of a ξ-adic space are equal.*

The method of the proof is very similar to a method of N. A. Shanin (the "calibers" [5]).

Definition. Let $\mathfrak{n}$ denote an infinite cardinality. A topological space X is said to have the *property* $\boldsymbol{B}(\mathfrak{n})$ if for any family $\{G_\alpha; \alpha \in A\}$, $|A| = \mathfrak{n}$, of non-empty open subsets of X a set $B \subset A$, $|B| = \mathfrak{n}$, and a point $p \in X$ can be selected such that each neighbourhood of p meets almost all sets G_β in the sense that

$$|\{\beta \in B; V \cap G_\beta = \emptyset\}| < \mathfrak{n}$$

for each neighbourhood V of p.

Our main tool for the investigation of ξ-adic spaces is the following theorem:

Theorem 2. *An arbitrary product of spaces with property $\boldsymbol{B}(\mathfrak{n})$ has this property as well.*

The continuous image of a space with property $B(\mathfrak{n})$ also has this property and the spaces $W(\xi+1)$ obviously have the property $B(\mathfrak{n})$, hence we have

Corollary. *If the space X is ξ-adic then X has the property $B(\mathfrak{n})$ for each infinite cardinality $\mathfrak{n}$.*

Using this Corollary and some other theorems of R. Engelking [4] and R. Marty [3] Theorem 1 can be proved.

Our Corollary implies also some related theorems for ξ-adic spaces. The following results are direct generalizations of two theorems of R. Engelking and A. Pelczynski [2].

Theorem 3. *If the Stone-Čech compactification of a Tychonoff space T is ξ-adic for an ordinal ξ, then T is pseudocompact.*

Theorem 4. *There is no infinite extremally disconnected ξ-adic Hausdorff space.*

To prove these two theorems it is enough to apply our Corollary to the case $\mathfrak{n}=\aleph_0$.

Using a different method, applying an argument due to Efimov [1] for a more general situation, we obtain

Theorem 5. *Let X be a ξ-adic space. If X has a dense set each point of which has a character $\leqq \mathfrak{n}$ and $|\xi| \leqq \mathfrak{n}$, then the weight of $X \leqq \mathfrak{n}$.*

Corollary. *If the Tychonoff space X has a ξ-adic compactification αX for some ordinal ξ, then the weight of αX does not exceed the weight of X.*

Problem. Has each metrizable space M an $\mathfrak{m}$-adic compactification? (By Theorem 5, if such an $\mathfrak{m}$ exists, then it can be chosen as the weight of the space M.)

A detailed paper with proofs will appear in Periodica Math. Hungarica.

References

[1] *B. Efimov:* On dyadic compacta. (Russian.) Dokl. Akad. Nauk SSSR *149* (1963), 1011—1014.
[2] *R. Engelking and A. Pelczynski:* Remarks on dyadic spaces. Colloq. Math. *11* (1963), 55—63.
[3] *R. Marty:* On $\mathfrak{m}$-adic spaces. Acta Math. Acad. Sci. Hungar. (to appear).
[4] *S. Mrówka:* Mazur Theorem and $\mathfrak{m}$-adic spaces. Bull. Acad. Polon. Sci. Sér. Sci. Math. Astronom. Phys. *18* (1970), 299—305.
[5] *N. A. Shanin:* On the product of topological spaces. (Russian.) Trudy Mat. Inst. Steklov. *24* (1948).

MATHEMATICAL INSTITUTE OF THE HUNGARIAN ACADEMY OF SCIENCES

ON MONOTONE DECOMPOSITIONS OF SMOOTH CONTINUA

G. R. GORDH, JR.

Riverside – Lexington

The notion of smoothness of fans, dendroids, and hereditarily unicoherent continua has been discussed in [1], [4], and [6], respectively. We shall define a class of continua, called smooth, which contains the class of smooth hereditarily unicoherent continua, and we shall discuss some of the basic properties of such continua.

A *continuum* is a compact, connected, metric space. A continuum is said to be *hereditarily unicoherent at the point p* provided that the intersection of any two subcontinua, each of which contains p, is connected. Clearly a continuum M is hereditarily unicoherent at p if and only if given any point x in M there exists a unique subcontinuum which is irreducible between p and x. If the continuum M is hereditarily unicoherent at p, and q is a point of M, then pq will denote the unique subcontinuum which is irreducible between p and q.

A continuum M is said to be *smooth at the point p* if M is hereditarily unicoherent at p, and for each convergent sequence of points a_n of M the condition $\lim a_n = a$ implies that the sequence of continua pa_n is convergent and $\operatorname{Lim} pa_n = pa$. The set of points at which a continuum M is smooth is called the *initial set* of M and is denoted by $I(M)$. If $I(M) \neq \emptyset$, then M is said to be *smooth*.

Theorem 1. *If M is a smooth continuum then* (i) *M is locally connected at each point of $I(M)$,* (ii) *M is a dendrite if and only if $I(M) = M$,* (iii) *M is unicoherent, and* (iv) *every indecomposable subcontinuum of M has void interior.*

Theorem 2. *If M is a smooth continuum then there exists a decomposition D of M* (*called the canonical decomposition*) *such that* (i) *D is upper semicontinuous,* (ii) *the elements of D are continua,* (iii) *the decomposition space of D is arcwise connected, and* (iv) *if E is a decomposition satisfying* (i), (ii), *and* (iii) *then D refines E. Moreover, the decomposition space of D is a smooth dendroid and each element of D has void interior.*

The decomposition of Theorem 2 is similar to the decomposition obtained for λ-dendroids in [2]; however, the canonical decomposition of a λ-dendroid may consist of a single element [3] while the canonical decomposition of a smooth continuum is never degenerate.

For a detailed discussion of these results including generalizations to compact Hausdorff continua, see [5].

References

[1] *J. J. Charatonik:* On fans. Rozprawy Matematyczne *54*, Warszawa, 1967.
[2] *J. J. Charatonik:* On decompositions of λ-dendroids. Fund. Math. *67* (1970), 15—30.
[3] *J. J. Charatonik:* An example of a monostratiform λ-dendroid. Fund. Math. *67* (1970), 75—87.
[4] *J. J. Charatonik and C. Eberhart:* On smooth dendroids. Fund. Math. *67* (1970), 297—322.
[5] *G. R. Gordh, Jr.:* On decompositions of smooth continua. Fund. Math. *75* (1972), 51—60.
[6] *L. Mohler:* A characterization of smoothness in dendroids. Fund. Math. *67* (1970), 369—376.

UNIVERSITY OF CALIFORNIA, RIVERSIDE, CALIFORNIA
UNIVERSITY OF KENTUCKY, LEXINGTON, KENTUCKY

ON THE SATURATION OF A TOPOLOGICAL PARTIAL ALGEBRA WITH RESPECT TO A CONGRUENCE RELATION

G. GRIMEISEN

Stuttgart

This is a report on a paper [4] being in preparation, in which extensions of a given topological partial algebra to topological (full) algebras are discussed. The notion of the saturation of a topological partial algebra with respect to a congruence relation serves as a mean to consider topological quotient partial algebras and such extensions within a common framework.

1. Let $\mathfrak{A} = (A, (f_\gamma)_{\gamma<\beta})$ be a partial algebra (for the algebraic terminology, see Grätzer [1], especially p. 79) with the base set A and the n_γ-ary partial operations f_γ on A ($0 < n_\gamma < \omega$, β an ordinal >0), furthermore, let Θ be a congruence relation of $\mathfrak{A}$ (see [1], p. 82). For each set $M \subseteq A$, the set $\{x \mid x \in A$ and, for some $y \in M$, $x\Theta y\}$ is said to be the *Θ-saturation of M*. Define, for each $\gamma < \beta$ and each $(X_0, \ldots, X_{n_\gamma-1}) \in (\mathfrak{P}A)^{n_\gamma}$ ($=n_\gamma$-th Cartesian power of the power set $\mathfrak{P}A$ of A), the set $({}^\Theta f_\gamma)(X_0, \ldots, X_{n_\gamma-1})$ to be the Θ-saturation of the set $f_\gamma(X_0 \times \ldots \times X_{n_\gamma-1})$. Then, the algebra $(\mathfrak{P}A, ({}^\Theta f_\gamma)_{\gamma<\beta})$ with the n_γ-ary operations ${}^\Theta f_\gamma$ is called the *Θ-saturation ${}^\Theta\mathfrak{A}$ of $\mathfrak{A}$*.

Examples.

1. Let $\Theta = A \times A$. Then, for each $\gamma < \beta$, $({}^\Theta f_\gamma)(X_0, \ldots, X_{n_\gamma-1})$ is equal to A if $(\mathscr{D}f_\gamma) \cap (X_0 \times \ldots \times X_{n_\gamma-1}) \neq \emptyset$, and equal to $\emptyset$ otherwise. ($\mathscr{D}f_\gamma$ denotes the domain of f_γ.)

2. Let $\Theta = id_A$ (= identical mapping on A). Then, for each $\gamma < \beta$, $({}^\Theta f_\gamma)(X_0, \ldots, X_{n_\gamma-1}) = f_\gamma(X_0 \times \ldots \times X_{n_\gamma-1})$, and, in this case, we will call ${}^\Theta\mathfrak{A}$ the *power $\mathfrak{P}\mathfrak{A}$ of $\mathfrak{A}$* (not to be mixed up with the power set of the set $\mathfrak{A}$).

As an immediate consequence of the definitions, we have

Proposition 1. *The quotient partial algebra $\mathfrak{A}/\Theta$* (see [1], p. 82) *is the relative (partial) subalgebra* (see [1], p. 80) *of $\mathfrak{A}$ with A/Θ as the base set.*

2. Let (A, τ) be a topological space. The *(first) power $\mathfrak{P}\tau$ of the topology τ* is defined to be that topology on the power set $\mathfrak{P}A$ of A which is induced by the limit operator ("Limesoperator")

$$\mathfrak{P} \circ (\lim \inf_{\tau})_{\Phi_0(\mathfrak{P}A)}$$

(= composition of the mapping "limit inferior" $\lim \inf_{\tau}$, restricted to the set $\Phi_0(\mathfrak{P}A)$ of all filters on subsets of $\mathfrak{P}A$, with the mapping $\mathfrak{P}$ assigning to each set its power set (for the terminology, see [2], p. 244 and 245)). (One has to be careful not to mix up $\mathfrak{P}\tau$ with the power set of the set τ.)

A relation R between topological spaces (E, σ), (F, ϱ) is said to be (σ, ϱ)-*continuous* (see [5]) if and only if

for each $(x, y) \in R$, *there is a* $(\sigma_{\mathscr{D}R}, \varrho)$-*continuous mapping* φ *on* $\mathscr{D}R$ *into* F *such that* $(x, y) \in \varphi \subseteq R$.

Equivalently, R *is* (σ, ϱ)-*continuous if and only if* R *is the union of some set consisting of* $(\sigma_{\mathscr{D}R}, \varrho)$-*continuous mappings on* $\mathscr{D}R$ *into* F. ($\mathscr{D}R$ denotes the domain of R.)

Examples.

3. A mapping is continuous as a relation if and only if it is continuous in the usual sense. Especially, the identical mapping on A is (τ, τ)-continuous. Also, $A \times A$ is (τ, τ)-continuous.

4. Each congruence relation of each topological group is continuous. The same holds for topological rings.

5. The natural order relation on the real line (endowed with its usual topology) is continuous.

6. If (A, τ) is the Cartesian plane, then the relation R defined by "xRg if and only if $x \in g$ and g is a straight line" for all $(x, g) \in A \times \mathfrak{P}A$ is $(\tau, \mathfrak{P}\tau)$-continuous.

Proposition 2 ([3], "Satz 2"). *If* Θ *is a* (τ, τ)-*continuous equivalence relation on* A, *then the quotient topology* τ/Θ *and the trace* $(\mathfrak{P}\tau)_{A/\Theta}$ *of the topology* $\mathfrak{P}\tau$ *in the quotient set* A/Θ *coincide.*

3. In this section, we are going to combine Section 1 with Section 2, and we make free use of the agreements made there. Let $(\mathfrak{A}, \tau)$ be a topological partial algebra, i.e., let all partial operations f_γ of the given partial algebra $\mathfrak{A}$ be $((\tau^{n_\gamma})_{\mathscr{D}f_\gamma}, \tau)$-continuous mappings. ($\mathscr{D}f_\gamma$ denotes the domain of f_γ.)

Theorem. *Assume that* Θ *is a* (τ, τ)-*continuous congruence relation of* $\mathfrak{A}$ *and all domains* $\mathscr{D}f_\gamma$ *of the* f_γ *with* $\gamma < \beta$ *are* τ^{n_γ}-*open sets. Then,* $({}^{\Theta}\mathfrak{A}, \mathfrak{P}\tau)$ *is a super-compact* (see [2], p. 243) *topological algebra and* $(\mathfrak{A}/\Theta, \tau/\Theta)$ *is (a topological partial algebra, namely) the relative topological (partial) subalgebra of* $({}^{\Theta}\mathfrak{A}, \mathfrak{P}\tau)$ *with* A/Θ *as the base set,* i.e. $\mathfrak{A}/\Theta$ is the relative (partial) subalgebra of ${}^{\Theta}\mathfrak{A}$ with A/Θ as the base set, and $\tau/\Theta = (\mathfrak{P}\tau)_{A/\Theta}$.

Proof. Theorem 1 in [4], "Satz 3" in [2], Propositions 1 and 2.

Corollary 1. *If all domains* $\mathscr{D}f_\gamma$ *of the* f_γ *are open, then* $(\mathfrak{P}\mathfrak{A}, \mathfrak{P}\tau)$ *is a supercompact topological algebra and, up to an algebraic and topological isomorphism,* $(\mathfrak{A}, \tau)$ *is a relative topological (partial) subalgebra of* $(\mathfrak{P}\mathfrak{A}, \mathfrak{P}\tau)$.

Proof. Apply Theorem for $\Theta = id_A$ (see Examples 2 and 3) and use that $(\mathfrak{A}, \tau)$ is isomorphic to $(\mathfrak{A}/id_A, \tau/id_A)$.

Let E be the set $\{X \mid X \subseteqq A$ and card $X \leqq 1\}$ and define, for each $\gamma < \beta$, h_γ to be the restriction of the mapping ${}^{(id_A)}f_\gamma$ to the set E^{n_γ} ($= n_\gamma$-th Cartesian power of the set E). Then, $\mathfrak{E} = (E, (h_\gamma)_{\gamma<\beta})$ is an algebra; if $\mathfrak{A}$ is not an algebra itself, $\mathfrak{E}$ is the smallest subalgebra of $\mathfrak{P}\mathfrak{A}$ into which $\mathfrak{A}$ can be embedded isomorphically. (E has just one element more than A; cf. Grätzer [1], p. 80, proof of Theorem 1.) If we include in this statement the topological situation, we obtain

Corollary 2. *Assume that all domains* $\mathscr{D}f_\gamma$ *of the* f_γ *are open. Then:* $(\mathfrak{E}, (\mathfrak{P}_\tau)_E)$ *is a supercompact topological algebra, namely a topological subalgebra of* $(\mathfrak{P}\mathfrak{A}, \mathfrak{P}\tau)$*, and, up to an algebraic and topological isomorphism,* $(\mathfrak{A}, \tau)$ *is a relative topological (partial) subalgebra of* $(\mathfrak{E}, (\mathfrak{P}\tau)_E)$.

Of course, in the case that $\mathfrak{A}$ is not an algebra, $(\mathfrak{E}, (\mathfrak{P}\tau)_E)$ is the smallest topological subalgebra of $(\mathfrak{P}\mathfrak{A}, \mathfrak{P}\tau)$ which extends $(\mathfrak{A}, \tau)$ to a topological algebra (provided that all $\mathscr{D}f_\gamma$ are open).

References

[1] *G. Grätzer:* Universal Algebra. Van Nostrand, Princeton, N. J., 1968.
[2] *G. Grimeisen:* Topologische Räume, in denen alle Filter konvergieren. Math. Ann. *173* (1967), 241–252.
[3] *G. Grimeisen:* Die Quotiententopologie als Spur der Potenz einer Topologie. Math. Ann. *173* (1967), 260–262.
[4] *G. Grimeisen:* Extensions of topological partial algebras to topological algebras. (To appear.)
[5] *G. Grimeisen:* Continuous relations. Math. Z. *127* (1972), 35–44.

ON THE TOPOLOGICAL CHARACTERIZATION OF MANIFOLDS

J. DE GROOT

Amsterdam

1. Introduction

A manifold (without boundary) is defined as a (connected) locally Euclidean space. But when is a space locally Euclidean? Which axioms or which simple geometric properties are necessary and sufficient? Substituting the axiomatic characterization for the cube I^n — cf. [4] — we obtain such an axiomatic characterization. It depends on two axiomatic properties, 2-compactness and comparability, which are claimed for some suitable subbase. We indicate in § 2, 3 how one might axiomatically characterize manifolds by already one of these, namely comparability. A comparable subbase leads naturally to the notion of the incomparability number. Iff — locally — this topological invariant equals the dimension of the space we readily obtain a manifold.

In § 4 we discuss geometric properties which characterize S^n, I^n and $\mathbf{R}^n$. The results are partially known and still far from complete as is shown by the *conjectures* mentioned at the end.

2. For an n-dimensional compact metrizable space M there exists a finite number of continuous real-valued functions which separate points, (i.e. for distinct points in M the function values are different for at least one of the functions), because M can be embedded in E^{2n+1}.

The minimal number k of such a set of separating functions equals the dimension of the Euclidean space of lowest dimension in which M can be embedded – the *m.e.d.*, minimal embedding dimension. This is also clear, because these functions induce a one-to-one continuous, hence topological, mapping of M into the product E^k. Here k = m.e.d.

Now consider a locally compact metrizable space X in which every two points have homeomorphic neighborhoods (e.g. a manifold). Moreover, assume that each point has a neighborhood for which the minimal number k of separating functions equals the dimension n of X. Hence X can be locally embedded in $E^n = E^k$. But because X is n-dimensional in every point, X contains an open n-dimensional subset of E^n – according to Brouwer's theorem on the invariance of domain –, so X contains an n-dimensional cube as neighborhood of each point. Hence X is a manifold. So it is easy to prove the following

Manifold — lemma. *A finite dimensional X is a manifold (with or without boundary), if and only if the following conditions hold*

1. *X is locally compact metrizable,*
2. *every two points have homeomorphic neighborhoods,*
3. *"the local minimal separating number" equals the dimension of X (i.e. there exists a compact neighborhood of every point for which this minimal separating number equals the dimension).*

In the next paragraph we replace the minimal separating number by an interior invariant of the space.

3. Let X be a set and $\mathfrak{U} = \{U\}$ a family of subsets U of X. $U^{(c)}$ denotes either U or U^c, the complement $X \setminus U$ of U in X. Two elements U_1 and U_2 of $\mathfrak{U}$ are called *comparable*, if

$$U_1^{(c)} \subset U_2^{(c)}$$

is true for a suitable interpretation of $U_1^{(c)}$ and $U_2^{(c)}$. Otherwise U_1 and U_2 are called incomparable. If there exists in $\mathfrak{U}$ an incomparable set (i.e. no two are comparable) of k elements (k finite), but no incomparable set of more than k elements, we call k the incomparability of $\mathfrak{U}$

$$\text{inc}\, \mathfrak{U} = k\,.$$

We define inc $\mathfrak{U} = \infty$, if no such finite k ($k = 0, 1, 2, \ldots$) exists.

We apply this notion for a topological T_1-space X and $\mathfrak{U}$ a set of generators of X (i.e. a subbase of X). We shall always assume $\emptyset, X \notin \mathfrak{U}$. Actually, we subject $\mathfrak{U}$ to two conditions. If $\mathfrak{U}$ is an *open* subbase, we require

(i) $\mathfrak{U}$ *is a T_1-subbase*, i.e. for every $U \in \mathfrak{U}$ and every point $p \in U$, there exists an $U' \in \mathfrak{U}$ such that

$$U \cup U' = X\,, \quad p \notin U'\,.$$

(ii) *Comparability condition.* $U \cup U' = X$ and $U \cup U'' = X$ in $\mathfrak{U}$, imply $U' \subset U''$ (or conversely).

The second condition can be proved to imply the transitivity of the notion of comparability in $\mathfrak{U}$. Hence inc $\mathfrak{U}$ denotes the number of incomparability classes of $\mathfrak{U}$.

An open subbase $\mathfrak{U}$ satisfying (i) and (ii) is called a *comparable T_1-subbase*, if, moreover, X is T_1.

Now we define a topological invariant, the *incomparability of X* by

$$\text{inc}\, X = \min_{\mathfrak{U} \in \Gamma} \text{inc}\, \mathfrak{U}$$

(hence inc X is finite or infinite), where Γ denotes the class of all comparable T_1-subbases $\mathfrak{U}$.

Lemma. *For a (locally) compact metrizable space X we have*

$$\text{inc}\, X = \text{m.e.d.}\, X\,.$$

The proof if this lemma runs along the same lines as a major part of the proof in [4] or [5].

Combining the preceding two lemmas we obtain

Theorem 1. *A finite-dimensional topological space X is a manifold, iff*

1. *X is locally compact metrizable,*
2. *every two points have homeomorphic neighborhoods,*
3. $$\text{inc}\, X = \dim X\,,$$

i.e. there exists a neighborhood of a point for which inc = dim.

4. Geometric characterizations of the n-sphere S^n, the n-cell I^n and Euclidean n-space $\mathbf{R}^n$

Definitions. A *suspension* is defined as a double cone over a common base-space.

X is an *infinite* cone if X is homeomorphic to a space which is obtained from $Z \times [0, 1)$ – where Z is *compact* –, by identifying $Z \times \{0\}$ to one point. Observe that the one-point-compactification of X is a suspension over a base-space Z' with Z' homeomorphic to Z.

Theorem 2. *A compact connected manifold (with boundary) which is a suspension, is homeomorphic to a sphere (to a cell).*

A non-compact manifold which is an infinite cone, is homeomorphic to $\mathbf{R}^n$.

The last part of this theorem is known (Rosen), and follows e.g. from a stronger theorem of Brown [cf. 2].

The sphere-case in *this* form is not known to me, but a simple proof can be based on the generalized Schoenflies theorem [1], while the theorem immediately follows from the Doyle-Hocking characterization [3].

By the one-point compactification the $\mathbf{R}^n$-case is reduced to the sphere-case. So the only – to me at least – unknown case is the cell-case, while the unifying formulation of the several characterizations should also be observed.

The proof of the cell-case depends on a variation of the generalized Schoenflies theorem (spheres replaced by cells in a proper fashion). Because such a "Schoenflies" theorem is already in existence for the infinite-dimensional case (Wong [6]), the cell-case extends to: a compact Q-manifold (i.e. every point has a neighborhood homeomorphic to the Hilbert cube Q) which is a suspension, is homeomorphic to Q

(also: a Q-manifold which is a cone, is homeomorphic to Q, because even every compact contractible Q-manifold is homeomorphic to Q according to T. A. Chapman).

Finally, it should be remarked that Theorem 2 can be strengthened in several ways.

Open, however, seems

Question. *If a compact n-manifold with boundary is a cone, is it necessarily an n-cell?*

To obtain completely satisfactory geometric characterizations the *manifold condition should be dropped completely.*

Conjectures. If a compact metrizable space is a suspension in *every* pair of points (i.e. each pair is a pair of vertices of some suspension representation) or a cone in *every* point (i.e. each point can be considered as a vertex), it is homeomorphic to a sphere or a cell.

If a compact metrizable space is both a suspension and a cone (i.e. in every pair of points, in every point respectively) it is homeomorphic to the Hilbert cube Q.

Remark. In view of Theorem 1 one has *only* to prove the manifold condition to solve the case of the sphere and of Q. However, there still arise difficulties for S^k $(k > 2)$ and I^n $(n > 3)$. These difficulties are not trivial. Indeed, Bing [0] constructs a "bad space", the suspension of which becomes S^4.

References

[0] *R. H. Bing:* The Cartesian product of a certain non-manifold and a line is E^4. Bull. Amer. Math. Soc. *64* (1958), 82–84.

[1] *M. Brown:* A proof of the generalized Schoenflies theorem. Bull. Amer. Math. Soc. *66* (1960), 74–76.

[2] *M. Brown:* The monotone union of open *n*-cells is an open *n*-cell. Proc. Amer. Math. Soc. *12* (1961), 812–814.

[3] *P. H. Doyle and J. G. Hocking:* A characterization of Euclidean *n*-space. Michigan Math. J. *7* (1960), 199–200.

[4] *J. de Groot:* Topological characterization of metrizable cubes. To be published in "Hausdorff Gedenkschrift", 1972.

[5] *J. de Groot and P. S. Schnare:* A characterization of products of compact totally ordered spaces. General Topology and its Applications (to appear).

[6] *R. Y. T. Wong:* Extending homeomorphisms by means of collarings. Proc. Amer. Math. Soc. *19* (1968), 1443–1447.

UNIVERSITY OF AMSTERDAM

UNIVERSITY OF FLORIDA

THREE CLASSES OF UNIFORM SPACES

A. W. HAGER

Middletown

1. Introduction

The classes consist of separable uniform spaces which are, respectively, subfine, $\mathcal{M}$-fine, and measurable (defined below). They have enough in common, and there are enough interesting relationships between them, to warrant discussing them together.

The results on subfine spaces and duality are due largely to Isbell and collaborators, with embellishments in [4d]. $\mathcal{M}$-fine and measurable spaces are treated in [4b, c]. (For other references, see these three papers.) There is some overlap with recent results of H. Gordon. We shall mention several unpublished results of M. D. Rice [7]. I cannot over-emphasize the dependence of this paper on Isbell's work, particularly [5a] and the paper [3] with Ginsberg.

Definitions. A fine uniform space αX is a uniformizable topological space X with its finest compatible uniformity α. If μX is a uniform space, $T\mu X$ is the associated topological space. μX is subfine if it is a subspace of a fine space: $\mu X = \alpha Y/X$. Let $\mathcal{M}$ be the class of metric spaces. μX is $\mathcal{M}$-fine if uniform continuity of $\mu X \xrightarrow{f} \varrho M \in \mathcal{M}$ implies that of $\mu X \xrightarrow{f} \alpha T \varrho M$. μX is measurable if there is a σ-field $\mathcal{F} \subset 2^X$ such that the countable $\mathcal{F}$-covers (covers of X with members in $\mathcal{F}$) are a basis for μ; these are so-named because, if μX and νY are two, associated with σ-fields $\mathcal{F}$ and $\mathcal{G}$ then $\mu X \xrightarrow{f} \nu Y$ is uniformly continuous iff $f^{-}(G) \in \mathcal{F}$ for each $G \in \mathcal{G}$, i.e., f is "measurable". μX is called separable if μ has a basis of countable covers; equivalently, μ is weak generated by uniformly continuous maps to separable metric spaces.

2. Coreflective subcategories

Subcategory $\mathcal{D}$ of $\mathcal{A}$ is coreflective in $\mathcal{A}$ if to each object $A \in \mathcal{A}$ is associated $dA \in \mathcal{D}$ and map $dA \xrightarrow{i} A$ with all maps $\mathcal{D} \ni D \xrightarrow{f} A$ factoring uniquely, as $f = i \circ g$. Then d is functorial, called the coreflector. Examples: fine spaces in uniform spaces, with coreflector αT; "Shirota spaces" (i.e., separable reflections $e\alpha X$ of fine spaces) in separable spaces, with coreflector $e\alpha T$. Propositions (Kennison [6], Frolík [2]): if $\mathcal{A}$ is the category of (separable) uniform spaces, $\mathcal{D}$ is coreflective iff $\mathcal{D}$ is closed under quotients and sums $\sum$ ($e\sum$); and, coreflection maps i are one-to-one, and $d\mu X$ can be interpreted as the coarsest uniformity on X which is in $\mathcal{D}$ and finer than μ.

Let $\mathcal{U}$ be a class of uniform spaces, and o a "pre-coreflector", of $\mathcal{U}$ into uniform spaces, with map $oU \xrightarrow{1} U$. μX is called a $\mathcal{U}$-o space if maps $\mu X \xrightarrow{f} U \in \mathcal{U}$ factor uniquely, as $f = 1 \circ g$; that is, if uniform continuity of $\mu X \xrightarrow{f} U$ implies that of $\mu X \xrightarrow{f} oU$. If $o = \alpha T$, we say "$\mathcal{U}$-fine".

We shall consider $\mathcal{U}$-fine spaces, where $\mathcal{U}$ is $\mathcal{M}$, $\gamma\mathcal{M}$ = complete metric spaces, $\mathcal{I}$ = injective uniform spaces, $\{\varrho\mathcal{R}^{\aleph_0}\}$ where $\varrho\mathcal{R}^{\aleph_0}$ is the product of $\aleph_0$ real lines $\varrho\mathcal{R}$; and $\mathcal{U}$-b spaces, where $\mathcal{U}$ is either sep $\mathcal{M}$ or $\{\varrho\mathcal{R}\}$, and for $\varrho M \in$ sep $\mathcal{M}$, $b\varrho M$ is the measurable space associated with the Baire sets of M – note that $b\varrho \supset \alpha T\varrho$.

2.1. Theorem. *In general, $\mathcal{U}$-o spaces form a coreflective subcategory of uniform spaces, with coreflector extending o. The separable $\mathcal{U}$-o spaces form a coreflective subcategory of separable spaces (with coreflector extending o/sep $\mathcal{U}$) iff the coreflector preserves separability (equivalently, e preserves the $\mathcal{U}$-o property); this applies to the examples above.*

2.2. Theorem. *Subfine* = $\mathcal{I}$*-fine* (Isbell, Rice); *for separable spaces, subfine* = $\gamma\mathcal{M}$*-fine* (Isbell, Ginsburg, Corson); *separable subfine* = $\{\varrho\mathcal{R}^{\aleph_0}\}$*-fine and* $\varrho\mathcal{R}^{\aleph_0}$*-weak. Hence separable* $\mathcal{M}$*-fine* $\Rightarrow$ *subfine.*

For separable spaces, measurable = sep $\mathcal{M}$-b = $\{\varrho\mathcal{R}\}$-b. *Hence, measurable* $\Rightarrow$ $\Rightarrow$ $\mathcal{M}$*-fine.*

A non-separable $\mathcal{M}$-fine, or $\gamma\mathcal{M}$-fine, space need not be subfine. Concerning measurability in the non-separable case, the proper *definition* has not been found; this was noted also by Frolík in his lecture.

2.3. Theorem. *If all spaces in $o\mathcal{U}$ are complete, then the completion of a $\mathcal{U}$-o space is $\mathcal{U}$-o.*

We use 2.2 to apply this to: subfine, because injective spaces are complete; $\mathcal{M}$-fine because metric spaces are topologically complete; measurable because $b\varrho\mathcal{R}$ is complete (Marczewski-Sikorski). Rice has shown that if coreflective $\mathcal{D}$ has topology-preserving coreflector, then $D \in \mathcal{D}$ implies the completion $\gamma D \in \mathcal{D}$; an example shows the restriction is needed. This applies to subfine, $\mathcal{M}$-fine, and fine (which 2.3 does not), but not to measurable.

3. The functors

By § 2, subfine and $\mathcal{M}$-fine are coreflective in all uniform spaces, and measurable spaces in separable spaces; there are coreflectors l, m, b. Isbell describes l by embedding $\mu X \to I \in \mathcal{I}$ (injective), and taking the subspace $\alpha TI/X$. This section concerns concrete descriptions of b, and m on separable spaces. (More generally, Rice has described a $\mathcal{U}$-o coreflector by a transfinite process similar to that used in [3] for the "locally-fine" coreflector – which agrees with l in separable spaces.) In what follows, we shall occasionally refer to a general coreflective subcategory $\mathcal{D}$, with coreflector d.

Each space is supposed separable. coz $C(\mu X)$ is the set of all cozero sets of real-valued uniformly continuous functions; $\mathcal{Z}(C(\mu X))$ consists of the complements. If $\mathcal{A} \subset 2^X$, $\sigma(\mathcal{A})$ is the generated σ-field.

3.1. Proposition. *μX is ($\mathcal{M}$-fine; measurable) iff each countable (coz $C(\mu X)$; σ(coz $C(\mu X)$))-cover is in μ. Hence, for general μX: ($m\mu X$; $b\mu X$) has basis of all countable (coz $C(\mu X)$; σ(coz $C(\mu X)$))-covers.*

Real-valued functions. How are $C(\mu X)$ and $C(d\mu X)$ related? We describe this for m and b, using 3.1; with no analogue for l, we have no description of $C(l\mu X)$.

3.2. Theorem. (a) *$f \in C(m\mu X)$ iff $f^{-1}(G) \in$ coz $C(\mu X)$, for open $G \subset \mathcal{R}$.*

(b) *$C(m\mu X)$ is the uniform closure of $\{f/g : f, g \in C^*(\mu X),\ g$ never $0\}$.*

(c) *If $f \in C(m\mu X)$ and $f \geqq 0$, then there is a sequence (f_n) from $C(\mu X)$ with $f_n \uparrow f$ pointwise.*

3.3. Theorem. (a) *$f \in C(b\mu X)$ iff $f^{-1}(G) \in \sigma$(coz $C(\mu X)$), for open $G \subset \mathcal{R}$.*

(b) *$C^*(b\mu X)$ is the uniform closure of the σ(coz $C(\mu X)$) – simple functions.*

(c) *$C(b\mu X)$ is the least class containing $C(\mu X)$ and closed under pointwise convergence of sequences.*

The approximation theorems 3.2 (b) and (c) apply to all continuous functions when $m\mu X = \alpha T\mu X$. This occurs if $T\mu X$ is Lindelöf (below).

3.3 can be proved from results of Mauldin. 3.2 (c) shows that $C(m\mu X)$ is a subset of $\mathcal{B}_1$, the first Baire class of $C(\mu X)$. Maulding shows that $f \in \mathcal{B}_1$, iff $f^{-1}(G) \in \mathcal{Z}(C(\mu X))_\sigma$, for open $G \subset \mathcal{R}$; compare 3.2 (a).

Topology. It is easily seen that l, m, preserve topology. b does not: $Tb\mu X$ carries the coarsest P-space topology (G_δ's are open) finer than $T\mu$, because σ(coz $C(\mu X)$) is an open basis. $\mathcal{Z}(C(\mu X))$ is another, and $\mathcal{B}_1$ generates $Tb\mu$.

When does uniformizable X have unique uniformity in $\mathcal{D}$? Equivalently if fine spaces $\in \mathcal{D}$, when is $d\mu X = \alpha T\mu X$? Since precompact spaces are subfine, X has unique subfine uniformity iff X has unique compactification (or uniformity) (Doss-Hewitt): X is "almost compact" = of each pair of disjoint zero sets, one is compact. For m: X has unique $\mathcal{M}$-fine uniformity iff X is Lindelöf or almost compact (essentially, Henrikson-Johnson and Hager-Johnson). For b: X has unique measurable uniformity iff X is "almost Lindelöf" and P (essentially Frolík).

Subspaces. When is $d(\mu Y/X) = d\mu Y/X$? Rice has some results for $\mathcal{D} = \mathcal{U}$-$o$, but any general answer is far from clear. For l, the equation holds, as Isbell shows from his contruction of l [5b]. Likewise for b; this is the equality σ(coz $C(\mu Y/X)$) = σ(coz $C(\mu Y) \cap X$), and uses the Katětov extension theorem. For m, we have the following rather complicated result, with interesting corollaries.

3.4. Theorem. *$m(\mu Y/X) = m\mu Y/X$ iff $Z \in \mathcal{Z}(C(\mu Y))$ and $Z \cap X = \emptyset$ imply a $Z_1 \in \mathcal{Z}(C(\mu Y))$ with $Z_1 \supset X$ and $Z_1 \cap Z = \emptyset$.*

Taking μY a Shirota space $e\alpha Y$, 3.4 becomes: $e\alpha Y/X$ is $\mathcal{M}$-fine iff X is completely separated from every disjoint zero set. Now, it is easy to see that $m(e\alpha Y/X) = e\alpha X$ iff $\mathscr{Z}(C(X)) = \mathscr{Z}(C(Y)) \cap X$, i.e., X is "Z-embedded". Then, we get the equivalence of (a) $e\alpha Y/X = e\alpha X$; (b) X is Z-embedded and completely separated from every disjoint zero set; (c) X is C-embedded, i.e., $C(X) = C(Y)/X$. (a) $\Leftrightarrow$ (c) is due to Shapiro-Gantner; (b) $\Leftrightarrow$ (c) improves and clarifies a result of Gillman-Jerison.

In case X is dense in Y, the condition in 3.4 becomes: each G_δ in Y meets X, or, X is "G_δ-dense".

3.5. Corollary. *Let μX be separable subfine, $= e\alpha Y/X$. Then, μX is $\mathcal{M}$-fine iff X is G_δ-dense in $\overline{X}^Y$. Hence, if μX is precompact, μX is $\mathcal{M}$-fine iff X is G_δ-dense in the compactification (completion).*

3.6. Corollary (of 3.4). *μX is measurable iff μX is separable and hereditarily $\mathcal{M}$-fine.*

Completion. When is $d\gamma = \gamma d$? (γ is the completion functor.) Isbell observes that this holds if d preserves topology and subspaces, and that this applies to l [5b]. Neither m nor b preserves both.

3.7. Lemma. *The points of $\gamma m\mu X$ "are" the $\mathscr{Z}(C(\mu X))$-ultrafilters with cip; these are in one-to-one correspondence with the $\sigma(\text{coz } C(\mu X))$-ultrafilters with cip. Hence, $\gamma m\mu X$ and $\gamma b\mu X$ live on the same set; and $b\mu X$ is complete iff $m\mu X$ is (which occurs if μX is).*

3.8. Lemma. *$\gamma b\mu X$ is a subspace of $b\gamma\mu X$, living on the closure of X in $b\gamma\mu X$ (in the $b\gamma\mu$-topology); this is the G_δ-closure of X in $\gamma\mu X$.*

3.9. Theorem. *$m\gamma\mu X = \gamma m\mu X$, and $b\gamma\mu X = \gamma b\mu X$, iff X is G_δ-dense in $\gamma\mu X$.*

This also uses 3.4 and 2.3. 3.8 uses the results for b on topology and subspaces. The first part of 3.7 is rather standard, generalizing Hewitt and Shirota on υX vs. $e\alpha X$. The second part of 3.7 follows from results of Hays and Frolík generalizing Hewitt and Marczewski-Sikorski on realcompactness vs. completeness of $be\alpha X$.

4. Functions algebras and duality

Consider a family $A\colon X \to R$ of real-valued functions on X which separates the points of X, is an algebra and lattice in the pointwise operations, is closed under uniform convergence, and contains the constant 1. A is: closed under countable composition (cc) if $f_1, f_2, \ldots \in A$ and $f \in C(R^{\aleph_0})$ imply that $f \circ (f_i) \in A$; closed under inversion (ci) if $f \in A$ and f never 0 imply that $1/f \in A$; regular if $f \in A$ implies $g \in A$ with $f^2 g = f$. Then, regular $\Rightarrow$ ci $\Rightarrow$ cc; the latter is not obvious. Let $\mathscr{A}_i(X)$, $i = 1, 2, 3$, be the class of $A\colon X \to R$ which are, respectively, cc, ci, and regular. Let $\mathscr{S}_i$, $i = 1, 2, 3$, be the category of separable spaces which are, respectively, subfine, $\mathcal{M}$-fine, and measurable.

4.1. Theorem. *If* $\mu X \in \mathscr{S}_i$, *then* $C(\mu X) \in \mathscr{A}_i(X)$. *If* $A \in \mathscr{A}_i(X)$, *then there is unique* $\mu X \in \mathscr{S}_i$ *with* $C(\mu X) = A$.

For $i = 1$, this is, apart from the uniqueness, due to Isbell, Ginsberg, and Corson. The proofs of the first statements are fairly direct. For the existence in the second, we require three constructions: (cc) Embed $X \to P \equiv \prod\{R_f : f \in A\}$ in the usual way. Then A becomes $C(P)/X$, and $\mu X = \alpha P/X$. (ci) Essentially use 3.2 (b), in the form $C(\mu X) =$ uniform closure of $\{f/g : f,\ g \in A^*\}$. (regular) Prove that coz A is a σ-field, and A is the measurable functions. (In each case, μ turns out to be the coreflection of the A-weak uniformity.)

It suffices to have uniqueness for $i = 1$, and this is the theorem following. c is the functor (reflector) which assigns to μX the space $c\mu X$ generated by $C(\mu X)$.

4.2. Theorem. *If* $\mu X \in \mathscr{S}_1$, *then* $lc\mu X = \mu X$. *Hence,* $(c/\mathscr{S}_1)^{-1} = l/c(\mathscr{S}_1)$, *and* $c/\mathscr{S}_1$ *is an isomorphism of categories.*

The proof [4d] is quite difficult. (It is essentially the proof that $\mu X \in \mathscr{S}$ iff μX is $\{\varrho\mathscr{R}^{\aleph_0}\}$-fine and $\alpha\mathscr{R}^{\aleph_0}$-weak, of 2.2.) Not surprisingly, there are easy proofs, based on 3.1, of the weaker theorems that if $\mu X \in \mathscr{S}_2$ or $\mathscr{S}_3$ then $mc\mu X = \mu X$, or $bc\mu X = \mu X$. For $i = 2$, even the following is easy. p denotes the reflector into precompact spaces: $p\mu X$ has basis finite μ-covers (and is generated by $C^*(\mu X)$).

4.3. Theorem. *If* $\mu X \in \mathscr{S}_2$, *then* $mp\mu X = \mu X$. *Hence,* $(p/\mathscr{S}_2)^{-1} = m/p(\mathscr{S}_2)$, *and* $p/\mathscr{S}_2$ *is an isomorphism of categories.*

Both 4.1 and 4.2 are sharp: $p/\mathscr{S}_1$ is not one-to-one; c is not one-to-one on the category of separable $\{\varrho\mathscr{R}^{\aleph_0}\}$-fine spaces, which is only "slightly larger" than $\mathscr{S}_1$. But 4.1 and 4.2 hardly exhaust the subject: Smirnov has shown that p is one-to-one on metric spaces (hence so is c).

Consider Archimedean lattice-ordered algebras over R, with identity a weak order unit, the Φ-algebras of Henrikson-Johnson. If A is one, let $\mathscr{R}(A)$ be the set of real ideals of A, those M with $A/M = R$. If $\bigcap\mathscr{R}(A) = \{0\}$, then $A \ni a \to \bar{a}: \mathscr{R}(A) \to$ $\to R$ defined by $\bar{a}(M) = a + M$, defines an isomorphism of A onto the function algebra $\bar{A}$: $\mathscr{R}(A) \to R$. Henrikson, Johnson, and Isbell give algebraic definitions of cc, ci, and regular such that when $\bigcap\mathscr{R}(A) = \{0\}$, A has the property iff $\bar{A}$ does (in the sense already used). Let $\mathscr{A}_i$, $i = 1, 2, 3$, be the category of Φ-algebras A with $\bigcap\mathscr{R}(A) = \{0\}$, which are, respectively, cc, ci, and regular (with Φ-algebra homomorphisms; equivalently, ring homomorphisms preserving 1). Let $\gamma\mathscr{S}_i$ be the category of complete $\mathscr{S}_i$-spaces.

4.4. Theorem. $\gamma\mathscr{S}_i$ *and* $\mathscr{A}_i$ *are dual.*

This follows from 4.2 and 4.1, upon identifying X, when $\mu X \in \gamma\mathscr{S}_i$, with the "real ideal space" of $C(\mu X)$. This identification results from the following. Let A: $X \to R$ be a point-separating ring and lattice, with $1 \in A$. Let μ_A be the weak uniformity on X generated by A.

4.5. Proposition. *$\mu_A X$ is complete iff each real ideal in A is of the form $\{f \in A : f(x) = 0\}$ for some $x \in X$.*

4.6. Remarks. Let $\mathscr{S}$ be the category of separable spaces. The coreflectors l, m, b restricted to $\gamma c(\mathscr{S})$ induce by duality reflectors of some category of vector lattices, say $\mathscr{L}$, onto $\mathscr{A}_1$, $\mathscr{A}_2$, $\mathscr{A}_3$, respectively. $\mathscr{L}$ has not been indentified algebraically, though work of Fenstad and Császár is relevant; and the reflectors have not been studied carefully. But the coreflectors l, m, b, restricted to those objects of $\gamma c(\mathscr{S})$ associated with rings $C(\mu X)$ induce by duality reflectors of a category of Φ-algebras (which is not identified), and the reflectors have extensions to reflectors of all Φ-algebras (onto super-categories of the $\mathscr{A}_i$); these have been studied by Eleanor Aron. The category of uniform spaces for which $C(\mu X)$ is a ring is coreflective (Rice), and a construction of the coreflection exists (G. D. Reynolds); the category remains poorly understood.

See [5a] for a careful discussion of duality.

Added July 5, 1972. Isbell points out that 4.2. above follows from some results in [3], and that my work on $\mathscr{M}$-fine spaces overlaps with work of A. D. Alexandroff, Additive set functions in abstract spaces, Mat. Sb. *50* (1940), 307–348, *51* (1941), 563–628, *55* (1943), 169–238. The latter point is discussed in (revised) [4b].

There is a theorem similar to 4.2 and 4.3 implicit in [4c], concerning the category $\mathscr{S}_3$ and the Samuel compactification $s : s/\gamma\mathscr{S}_3$ *is an isomorphism of categories* (though $bs\mu X = \mu X$ rarely holds).

References

[1] *H. H. Corson and J. R. Isbell:* Some properties of strong uniformities. Quart. J. Math. Oxford Ser. *11* (2) (1960), 17–33.

[2] *E. Čech:* Topological Spaces. (Revised by *Z. Frolík* and *M. Katětov.*) Prague and New York, 1966.

[3] *S. Ginsberg and J. R. Isbell:* Some operators on uniform spaces. Trans. Amer. Math. Soc. *93* (1959), 145–168.

[4] *A. W. Hager:* (a) An approximation technique for real-valued functions. General Topology and its Applications *1* (1971), 127–133.
(b) Some nearly fine uniform spaces. (To appear.)
(c) Measurable uniform spaces. (To appear.)
(d) Subfine uniform spaces and the functor c. (To appear.)

[5] *J. R. Isbell:* (a) Algebras of uniformly continuous functions. Ann. Math. *68* (1958), 96–125.
(b) Uniform Spaces. Providence, 1964.

[6] *J. Kennison:* Reflective functors in general topology and elsewhere. Trans. Amer. Math. Soc. *188* (1965), 303–315.

[7] *M. D. Rice:* Thesis, Wesleyan University. (To appear.)

WESLEYAN UNIVERSITY, MIDDLETOWN, CONNECTICUT

DIRECT LIMITS OF HAUSDORFF SPACES

D. W. HAJEK (1) and G. E. STRECKER (2)

Gainesville and Pittsburgh

It is well-known that for topological spaces direct limits have generally very poor preservation properties. Indeed, the direct limit of compact separable metrizable spaces can be an infinite indiscrete space [2, p. 422]. However if all of the bonding maps are injective, the direct limit of T_1 spaces will be T_1. Likewise if all of the bonding maps are closed embeddings, the direct limit of a countable well-ordered spectrum of T_4 spaces will be T_4. (This preservation fails, however, if the countability condition is dropped.) Recently Herrlich [3] has shown that the direct limit of a countable well-ordered spectrum of completely regular Hausdorff spaces with closed embedding bonding maps may fail to be Hausdorff. In this paper we will exhibit sufficient conditions for the Hausdorff property to be preserved under direct limits.

In [1] Banaschewski considered, for any T_0 space, X, a space consisting of all open filters on X. To achieve our direct limit results, we use essentially the same type of construction.

Definition 1. If X is a space, then an *open filter* on X is a nonempty collection F of open sets of X such that:

(i) $\emptyset \notin F$,

(ii) $U, V \in F \Rightarrow U \cap V \in F$, and

(iii) $W \supseteq U \in F \Rightarrow W \in F$.

For any space X, let γX denote the set of all open filters on X. If U is an open set in X, let $U^{\#} = \{F \in \gamma X \mid U \in F\}$. Then $\{U^{\#} \mid U \text{ is open in } X\}$ is a base for a topology on γX. Henceforth γX will denote this space.

Definition 2. If X and Y are spaces and $f\colon X \to Y$ is a continuous function, then

(1) f is said to be *dense* provided that $f[X]$ is dense in Y; and in this case (f, Y) is called a *range* of X.

(2) f is said to be *relatively open* provided that for each open set U of X, $f[U] = W \cap f[X]$ for some open set W of Y.

(3) f is said to be an *embedding* provided that it is injective and relatively open.

(4) (f, Y) is called an *extension* of X provided that f is a dense embedding.

(5) (f, Y) is called a C-*distinguishable range* of X provided that it is a range of X and for each $y \in Y$ and each closed subset A of Y not containing y, there is an open neighborhood U of y such that $f^{-1}[U] \nsupseteq f^{-1}[V]$ for any open set V of Y that meets A.

The following proposition shows that if X is a T_0-space, then γX is a T_0-compactification of X which can be thought of as a "universal" C-distinguishable extension of X.

Proposition 1. *For every space X, γX is a compact T_0-space with the following properties:*

(1) *If (f, Y) is a range of X, then there exists a continuous function $\hat{f}: Y \to \gamma X$ defined by: $\hat{f}(y) =$ the open filter generated by $\{f^{-1}[U] \mid U$ is an open neighborhood of $y\}$.*

(2) *If (f, Y) is a T_0 C-distinguishable range of X, then $\hat{f}$ is an embedding.*

(3) *$\hat{1}_X: X \to \gamma X$ is an embedding if and only if X is a T_0-space.*

(4) *If X is a T_0-space, then the set of all subspaces of γX that contain $\hat{1}_X[X]$ is (up to homeomorphisms) precisely the class of all T_0 C-distinguishable extensions of X.*

(5) *If (f, Y) and (g, Z) are ranges of X and $h: Y \to Z$ is an embedding for which $g = h \circ f$, then $\hat{f} = \hat{g} \circ h$; i.e., if the left-hand triangle of the diagram*

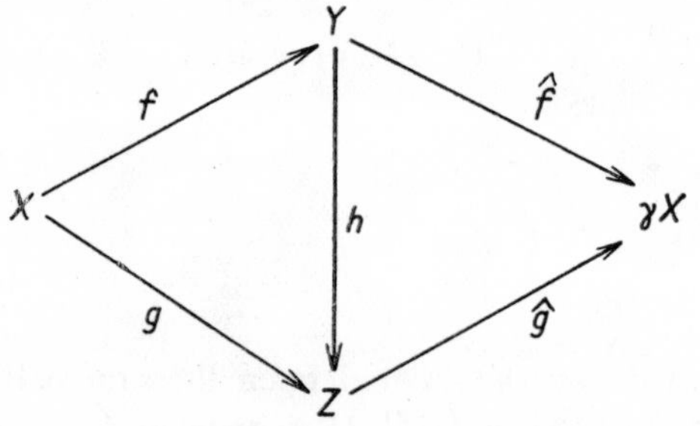

commutes, then so does the right-hand triangle.

Definition 3. Let I be a directed set and let $(X_i, g_{ij})_I$ be a direct spectrum of spaces and bonding maps over I.

(1) $(Y, f_i)_I$ is called a *natural source* for the spectrum provided that:

(i) Y is a space,

(ii) for each $i \in I$, $f_i: Y \to X_i$ is a continuous function, and

(iii) if $i, j \in I$ and $i \leq j$, then $g_{ij} \circ f_i = f_j$.

Dually,

(2) $(f_i, Y)_I$ is a *natural sink* for the spectrum provided that $f_i : X_i \to Y$ and $f_i = f_j \circ g_{ij}$ whenever $i \leqq j$.

(3) $(f_i, Y)_I$ is a *direct limit* of the spectrum provided that it is a maximal natural sink for it.

We will denote the direct limit of the spectrum by $(\mu_i, \underrightarrow{\text{Lim}}\, X_i)_I$.

Theorem 1. *Let $(X_i, g_{ij})_I$ be a direct spectrum of Hausdorff spaces and bonding maps over a directed set I. If each g_{ij} is an embedding and if there exists a natural source $(Y, f_i)_I$ for the spectrum, where Y is any space and each f_i is dense, then $\underrightarrow{\text{Lim}}\, X_i$ is a Hausdorff space.*

Proof. By Proposition 1, for all $i, j \in I$ there exist continuous functions $\hat{f}_i : X_i \to \gamma Y$ and $\hat{f}_j : X_j \to \gamma Y$ such that $\hat{f}_i = \hat{f}_j \circ g_{ij}$. Hence $(\hat{f}_i, \gamma Y)_I$ is a natural sink for the spectrum, so that by the definition of direct limit there is a (unique) continuous function g: $\underrightarrow{\text{Lim}}\, X_i \to \gamma Y$ such that for all $i, j \in I$ the diagram

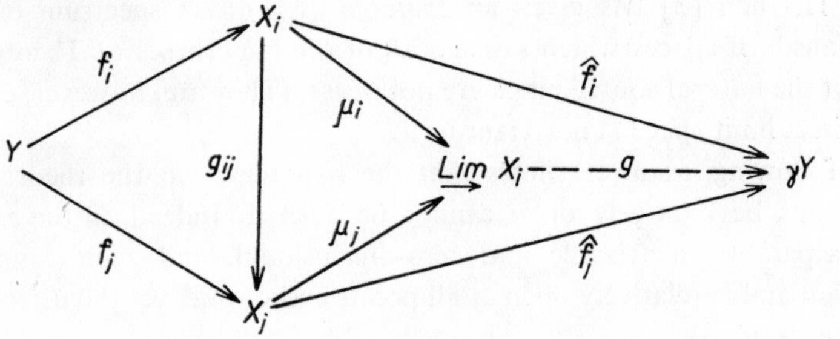

commutes.

If x and y are distinct points in $\underrightarrow{\text{Lim}}\, X_i$, then, since I is directed, there is some $k \in I$ and points $a, b \in X_k$ such that $\mu_k(a) = x$ and $\mu_k(b) = y$. Since X_k is Hausdorff, there are disjoint open sets U and V such that $a \in U$ and $b \in V$. Then since $\emptyset$ cannot belong to any filter, $(f_k^{-1}[U])^{\#}$ and $(f_k^{-1}[V])^{\#}$ are disjoint open sets in γY; so that $g^{-1}[(f_k^{-1}[U])^{\#}]$ and $g^{-1}[(f_k^{-1}[V])^{\#}]$ are disjoint open neighborhoods of x and y, respectively.

Lemma 1. *Let Y be a Hausdorff space and let $f : X \to Y$ and $h : Y \to Z$ be continuous functions for which f is dense, h is relatively open, and $h \circ f$ is a dense embedding. Then h is an embedding.*

Theorem 2. *Let $(X_i, g_{ij})_I$ be a direct spectrum of Hausdorff spaces and bonding maps over a directed set I. If each g_{ij} is relatively open and if there exists* *a natural source $(Y, f_i)_I$ for the spectrum, where Y is any space and each f_i is a dense embedding, then $\varinjlim X_i$ is a Hausdorff space.*

Proof. By the lemma, each g_{ij} must be an embedding. Apply Theorem 1.

Corollary 1. *If $(X_i, g_{ij})_I$ is a direct spectrum of Hausdorff spaces and dense embedding bonding maps over a directed set I, then $\varinjlim X_i$ is a Hausdorff space.*

Proof. If $I = \emptyset$, then $\varinjlim X_i$ is the empty space which is Hausdorff. If $I \neq \emptyset$, pick $\hat{\imath} \in I$ and let $J = \{i \in I \mid i \geqq \hat{\imath}\}$. Then J is cofinal in I; so $\varinjlim_J X_j \cong \varinjlim_I X_i$. But $(X_{\hat{\imath}}, g_{\hat{\imath}j})_J$ is a natural source for $(X_i, g_{ij})_J$; so that, by Theorem 2, $\varinjlim_J X_j$ is Hausdorff.

It seems difficult to weaken the hypotheses of the above theorems. Indeed, Dugundji [2, p. 422] has given an example of a direct spectrum of spaces (each homeomorphic to the unit circle) which satisfies all the hypotheses of Theorem 1, except that the bonding maps are not injective. They are, however, relatively open, so that the same example satisfies all of the hypotheses of Theorem 2, except that the natural source maps are not injective. Yet the direct limit space is infinite and indiscrete.

Also Herrlich [3] has given an example of a direct spectrum of completely regular Hausdorff spaces which satisfies all of the hypotheses of Theorems 1 and 2 except that the natural source maps are not dense. (They are, however, embeddings.) But the direct limit space is not Hausdorff.

The following example shows that the hypotheses in the theorems that the bonding maps be relatively open cannot be deleted. Indeed, in the example each space is separable, metrizable and zero-dimensional, and each connecting map is a bijection and is relatively open at all points except one, yet the direct limit space is not Hausdorff.

Example 1. For each positive integer k, let X_k be the rational numbers in the closed interval $[0, 1]$ with the points $\{1/2^n \mid n > k\}$ removed and the points $\{1/2^i \mid 1 \leqq \leqq i \leqq k\}$ identified. For each $n < m$, let g_{nm} be the obvious continuous bijection from X_n to X_m and let $(\mu_n, \varinjlim X_n)$ be the direct limit of the spectrum $(X_n, g_{nm})_N$. Then the points $\mu_1(0)$ and $\mu_1(\frac{1}{2})$ do not have disjoint neighborhoods in $\varinjlim X_n$.

It is also natural to ask whether or not the conclusions of the above theorems can be strengthened; i.e., whether or not stronger separation properties will be preserved by direct limits when the bonding maps are all dense embeddings. The following example shows that stronger separation properties are not preserved since the direct limit of a direct spectrum of separable, metrizable, zero-dimensional spaces with dense embedding bonding maps may fail to be Urysohn; i.e., there may exist distinct points which do not have disjoint closed neighborhoods.

Example 2. For each positive integer k, let X_k be the rational numbers in the closed interval $[-1, 1]$ with the points $\{1 - 1/2^n, -1 + 1/2^n \mid n \geqq k\}$ removed, and for $1 < i < k$ each pair of points $(1 - 1/2^i, -1 + 1/2^i)$ identified. For each $n < m$, let g_{nm} be the obvious inclusion map from X_n to X_m. Then $(X_n, g_{nm})_N$ is a direct spectrum with dense embedding bonding maps. However if $(\mu_n, \underrightarrow{\mathrm{Lim}}\, X_n)$ is the direct limit of the spectrum, then $\mu_1(1)$ and $\mu_1(-1)$ do not have disjoint closed neighborhoods in $\underrightarrow{\mathrm{Lim}}\, X_n$.

References

[1] *B. Banaschewski:* Extensions of topological spaces. Canad. Math. Bull. *7* (1964), 1—22.
[2] *J. Dugundji:* Topology. Allyn and Bacon, Inc., Boston, 1966.
[3] *H. Herrlich:* Separation axioms and direct limits. Canad. Math. Bull. *12* (1969), 337—338.

(1) UNIVERSITY OF FLORIDA, GAINESVILLE, FLORIDA

(2) UNIVERSITY OF PITTSBURGH, PITTSBURGH, PENNSYLVANIA

ON INTERNAL CHARACTERIZATIONS OF COMPLETE REGULARITY AND WALLMAN-TYPE COMPACTIFICATIONS

P. HAMBURGER

Budapest

To give an internal characterization of Tychonoff spaces, O. Frink [2] generalized the method introduced by Wallman [7] to provide Hausdorff compactifications for Tychonoff spaces. His procedure uses a *normal base* of closed sets instead of the family of all closed sets employed by Wallman. A base for the closed sets is a *normal base* if it is closed under the operations of finite unions and finite intersections, and satisfies the following conditions:

(i) for any element H of the base and any $x \in X \setminus H$ there are two elements H_1, H_2 of the base such that

$$H_1 \cup H_2 = X\,, \quad x \notin H_2\,, \quad H_1 \cap H = \emptyset\,,$$

(ii) for any two disjoint elements H_1, H_2 of the base there are two elements H', H'' of the base such that

$$H' \cup H'' = X\,, \quad H_1 \subset X \setminus H'\,, \quad H_2 \subset X \setminus H''\,.$$

Frink raised the following questions:

Let X be a compact Hausdorff space and Y a dense subset of X; is there any normal base $\mathfrak{B}$ of the closed sets of Y such that the Wallman-type compactification of Y, $\omega(Y, \mathfrak{B})$ is homeomorphic to X?

He also asked whether $\mathfrak{B}$ can be chosen such that every element of $\mathfrak{B}$ is a zero-set. Such compactifications will be called *z-compactifications.*

E. F. Steiner [6] proved that if there is a normal base of closed sets of a compact space X such that every element of the base is a *regular closed set* then X is a Wallman-type compactification of each of its dense subsets.

In this case, we shall say that X is a *regular Wallman compactification* of each of its dense subsets. He also proved that every compact subspace of the real numbers, or every product of compact subsets of real numbers is a regular Wallman compactification of each of its dense subspaces.

Theorem 1. ([4]) *Every (totally) orderable compact space and even every product of orderable compact spaces is regular Wallman-type and, moreover, a z-compactification of each of its dense subsets.*

J. de Groot and J. M. Aarts [1] gave another internal characterization of complete regularity which is a generalization of Frink's theorem and naturally fits between regularity and normality.

To generalize this, we introduce the following notions:

Definition 1. Two subsets A and B of a space X are said to *be screened by a finite family* $\mathfrak{B}$ if $\mathfrak{B}$ covers X and each element of $\mathfrak{B}$ meets at most one of the sets A and B.

We shall say that two subsets A and B of a space X *are screened by the closed (sub) base* $\mathfrak{T}$ of X if A and B are screened by a finite subcollection of $\mathfrak{T}$.

Definition 2. Two subsets A and B of X are said to *be weakly screened by* $\mathfrak{T}$ if there are $A_i \in \mathfrak{T}$, $i = 1, \ldots, n$ and $B_j \in \mathfrak{T}$, $j = 1, \ldots, m$ such that

$$A \subset \bigcup_{i=1}^{n} A_i, \quad B \subset \bigcup_{j=1}^{m} B_j$$

and for every $i = 1, \ldots, n$, $j = 1, \ldots, m$ the subsets A_i and B_j are screened by a finite subcollection of $\mathfrak{T}$.

Theorem 2. *A space X is completely regular if and only if there is a subbase $\mathfrak{T}$ for the closed subsets of X such that:*

(1) (*Weak subbase-regularity.*) *If $S \in \mathfrak{T}$, $x \notin S$, then S and $\{x\}$ are weakly screened by $\mathfrak{T}$.*

(2) (*Weak subbase-normality.*) *Every two disjoint elements of $\mathfrak{T}$ are weakly screened by $\mathfrak{T}$.*

The proofs of these theorems can be found in [4] and [5].

References

[1] *J. M. Aarts and J. de Groot:* Complete regularity as a separation axiom. Canad. J. Math. *21* (1969), 96—105.

[2] *O. Frink:* Compactifications and semi-normal spaces. Amer. J. Math. *86* (1964), 602—607.

[3] *E. Deák und P. Hamburger:* Interne Charakterisation der kompaktifizierbaren Räume. Periodica Math. Hung. (to appear).

[4] *P. Hamburger:* On Wallman-type, regular Wallman-type and *z*-compactifications. Periodica Math. Hung. *1* (1971), 303—309.

[5] *P. Hamburger:* Complete regularity as a separation axiom. Periodica Math. Hung. (to appear).

[6] *E. F. Steiner:* Wallman spaces and compactifications. Fund. Math. *61* (1968), 295—304.

[7] *H. Wallman:* Lattices and topological spaces. Ann. of Math. *39* (1938), 112—126.

MATHEMATICAL INSTITUTE OF THE HUNGARIAN ACADEMY OF SCIENCES

GAME-THEORETICAL APPROACH TO SOME MODIFICATIONS OF GENERALIZED TOPOLOGIES

J. HANÁK

Brno

0. There is an interesting possibility of a use of generalized topologies in the theory of extensive games: In my works, I introduced the so-called SN-games [= simultaneous nondeterministic games; at each nonfinal position of an SN-game all the players play mutually independently (the simultaneousness), knowing the preceding course of play, but their common influence need not determine the next position uniquely (the (local) nondeterminateness)], and I have shown that a certain reduced description (see [5], § 2.30.2 etc.) of SN-games is sufficient for the introduction of analogues of usual game-theoretical notions, for defining certain significant classes of SN-games, and, among other, for proving various strong game-theoretical theorems. (Cf., e.g., [3], [5], [6].) (Such a conception of SN-games has also admitted the introduction of the corresponding "descriptive theory", see [3], § 9, or [5], § 7, and of topological games, see [4], or [5], § 8; of course, various known results concerning games with perfect information etc. are included as corollaries in theorems on SN-games.)

In this communication, we shall show that the game-theoretical interpretation of some partial unary operations (in particular, of those corresponding to some modifications) can be used for investigating their algebraic properties. In contradiction to [3]–[6], this paper contains no principal theorem: the goal is to show typical approaches, and to present a number of particular results obtained by means of them. A very short introduction of necessary auxiliary definitions, comments, and properties is given.

1.1. "$\mathscr{A} := \mathscr{B}$" means "$\mathscr{A}$ is defined to be equal to $\mathscr{B}$". $\operatorname{dom} f$ means the domain of the mapping f. P will be a set, $\exp P := \{A \mid A \subseteq P\}$, $\mathscr{T}(P) := (\exp P)^{\exp P}$; P will be usually fixed, and then we shall write, e.g., only $\mathscr{T}$ instead of $\mathscr{T}(P)$. Let $\mathbf{1}$, $(-\mathbf{1})$, $\underline{X}$ (where $X \subseteq P$) be such that $\mathbf{1}A = A$, $(-\mathbf{1})\,A = P - A$, $\underline{X}A = X$ for any $A \subseteq P$. $\mathscr{T}$ together with the usual composition of mappings forms a monoid (the identity $\mathbf{1}$ is its unit), and $\mathscr{T}$ together with the usual partial ordering $\leqq$ ($u \leqq v$ iff $uA \subseteq vA$ for any $A \subseteq P$) forms a complete lattice. $\leqq$ can be introduced in a somewhat different way, too: let Φ be the mapping for which $\operatorname{dom} \Phi = \mathscr{T}$, $\Phi(u) = \{(x, A) \mid A \subseteq P,\ x \in uA\}$ ($u \in \mathscr{T}$); then Φ is a one-to-one ("canonical") mapping of $\mathscr{T}$ onto $\exp (P \times \exp P)$, but the latter set is a complete lattice with respect to $\subseteq$,

$\cup$, $\cap$, and this complete lattice structure can be transformed by Φ^{-1} back onto $\mathcal{T}$ [thus, we shall write shortly, e.g., $u \subseteq v$, $u \cup v$ instead of $\Phi(u) \subseteq \Phi(v)$, $\Phi^{-1}(\Phi(u) \cup \cup \Phi(v))$, respectively]; it is easy to see that this (transformed) $\subseteq$ is the same as $\leqq$. (Besides, there is the canonical one-to-one mapping of $\mathcal{T}$ onto $(\exp\exp P)^P$.) Cf. [12], and [5], § 2a.

1.2. If $R(.)$, $R_1(.)$, ..., $R_k(.)$ are some (one-variable) propositional functions defined at least on $\mathcal{T}$, we denote $\mathcal{T}_R := \{u \mid u \in \mathcal{T},\ R(u)\}$, $\mathcal{T}_{R_1 \ldots R_k} := \mathcal{T}_{R_1} \cap \cap \ldots \cap \mathcal{T}_{R_k}$. The following propositional functions are often used (we write "R:" instead of "$R(u)$ iff"): R: $u\emptyset = \emptyset$; M: $A \subseteq B \subseteq P \Rightarrow uA \subseteq uB$; E: $\mathbf{1} \subseteq u$; I: $u \subseteq \mathbf{1}$; U: $u^2 = u$; A: $A_1, A_2 \subseteq P \Rightarrow u(A_1 \cup A_2) = uA_1 \cup uA_2$. Now, $\mathcal{T}_{\mathrm{REUA}}$ is the set of (all) *topologies* (on P). Various kinds of *generalized topologies* obtained by replacing the Kuratowski axioms by some weaker ones have been investigated (let us quote several of many papers concerning these problems: [1], [7]–[14]; especially, closure spaces ($\mathcal{T}_{\mathrm{REA}}$, see [2], Sec. 14) and *Čech topological spaces* ($\mathcal{T}_{\mathrm{REM}}$ – in the sense of [1]; of course, $\mathcal{T}_{\mathrm{A}} \subseteq \mathcal{T}_{\mathrm{M}}$) have been often investigated (in particular, the problems of modifying Čech topologies and the properties of the constellations of Čech topologies have been studied circumstantially; see, e.g., [1], [8], [9], [11]), and some topological considerations have been performed even for general elements of $\mathcal{T}$ itself (*Koutský topologies*, or "topologies without axioms", see [7], [13], and a part of [14]).

In this communication, we shall consider another kind of generalized topologies, namely $\mathcal{T}_{\mathrm{RM}}$; they will be called *game topologies* (on P), cf. [5], §§ 2.8, 2.26.3. Under a *type* we shall mean (P, P_0) where $P_0 \subseteq P$ (in [5], § 1.1, $P \neq \emptyset$ was supposed besides); $\mathcal{T}_{\mathrm{RM}}(P, P_0) := \{u \mid u \in \mathcal{T}_{\mathrm{RM}}(P),\ uP = P - P_0\}$ will be the set of game topologies of the type (P, P_0). ([5], § 2.13 etc.)

1.3. Under a *game system* (on P) we shall mean a non-empty system $\mathcal{U} = (u_j)_{j\in J} \in (\mathcal{T}_{\mathrm{RM}})^J$ such that all the u_j have the same type, and $\bigcap_{j\in J} A_j = \emptyset$ implies $\bigcap_{j\in J} u_j A_j = \emptyset$ for every $(A_j)_{j\in J} \in (\exp P)^J$. [For an SN-game $\mathcal{G}$, let P (P_0, J) be the set of its positions (final positions, players, respectively), let (for $j \in J$) $u_j \in \mathcal{T}(P)$ be such that $u_j A = \{x \mid x \in P$, in $\mathcal{G}$, j can guarantee at x that the next position will (exist and) belong to $A\}$ for any $A \subseteq P$, let $Z := P - P_0$; then $\mathcal{U} := (u_j)_{j\in J}$ is a game system, $u_j P = Z$ for each $j \in J$, and Z is the set of nonfinal positions of $\mathcal{G}$. On the other hand, if some $\mathcal{U}$ is a game system, then $\mathcal{U}$ can be obtained in the above mentioned way, to a suitable $\mathcal{G}$. See [5], § 2.29–30, § 2 (47), (26) etc.]

1.4. Under an *operator* we shall mean a partial unary operation in $\mathcal{T}$, i.e., a mapping of a subset of $\mathcal{T}$ into $\mathcal{T}$. We shall need, in particular, these operators: $u \to u^k$ ($k = 0, 1, \ldots$), defined as the kth power in the monoid $\mathcal{T}$; $u \to \tilde{u} := \underline{P - uP} \cup \cup u$; $u \to \bar{u} := (-\mathbf{1}) . u . (-\mathbf{1})$; $u \to u' := \underline{uP} \cap \bar{u}$ (i.e., $u'A = uP - u(P - A)$); $u \to \check{u}$ where $\check{u}A = uA \cup (A - uP)$.

It holds: if $u \in \mathscr{T}_{\mathrm{RM}}(P, P_0)$ (where $P_0 \subseteq P$), then $u' \in \mathscr{T}_{\mathrm{RM}}(P, P_0)$, $u'' = u$; if $u_1 \in \mathscr{T}_{\mathrm{RM}}(P, P_0)$, then u' is the greatest (under $\subseteq$) element of $\{u_2 \mid u_2 \in \mathscr{T}_{\mathrm{RM}}, (u_1, u_2)$ is a game system$\}$. (See [5], § 4 (26)–(28), § 4.8.4 etc.) An SN-game $\mathscr{G}$ is said to be complete iff card $J = 2$ and $(u_{j_2})' = u_{j_1}$ for $\{j_1, j_2\} = J$ (where J, u_j etc. have the meaning given by the remark in Sec. 1.3). Complete games have very significant properties, similarly as their particular case – two-player Bergean games with perfect information. (Cf. [5], §§ 4d, 6c, and [6].)

1.5. For $R_1(.), \ldots, R_k(.)$ (Sec. 1.2), under the *upper* [*lower*] $R_1 \ldots R_k$-*modification* of u ($\in \mathscr{T}$) we mean $v \in \mathscr{T}_{R_1 \ldots R_k}$ such that (i) $u \subseteq v$ [$u \supseteq v$], and (ii) $v \subseteq w$ [$v \supseteq w$] whenever $w \in \mathscr{T}_{R_1 \ldots R_k}$ and $u \subseteq w$ [$u \supseteq w$]; of course, there exists at most one upper [lower] $R_1 \ldots R_k$-modification of u, thus the forming of upper [lower] $R_1 \ldots R_k$-modifications may be considered an operator. (Cf. [2], Sec. 31 B.)

Let $\mathbf{1} \nparallel u \in \mathscr{T}_{\mathrm{M}}$; the well-known construction of the transfinite powers of u can be expressed in such a way: $u^0 := \mathbf{1}$, $u^\xi := \lim_{0 \leqq \eta < \xi} u \,.\, u^\eta$ ($\xi > 0$ is an ordinal number); we put formally $u^\infty := u^\xi$ for (any) ξ such that $u^\xi = u^{\xi+1}$ (∞ is not an ordinal number). It is known that u^∞ is the upper [lower] U-*modification* of u if $u \in \mathscr{T}_{\mathrm{EM}}$ [$u \in \mathscr{T}_{\mathrm{IM}}$]. For each $u \in \mathscr{T}_{\mathrm{M}}$, $\mathbf{1} \cup u$ [$\mathbf{1} \cap u$] belongs to $\mathscr{T}_{\mathrm{EM}}$ [$\mathscr{T}_{\mathrm{IM}}$], and $u^\triangle := (\mathbf{1} \cup u)^\infty$ [$u_\nabla := (\mathbf{1} \cap u)^\infty$] is the *upper* UEM-*modification* [*lower* UIM-*modification*] *of* u. The modifications $^\triangle$, $_\nabla$, and also the operators $u \to u_\triangle := (\tilde{u})_\nabla$, $u \to u^\nabla := (\tilde{u})^\triangle$ (where $u \in \mathscr{T}_{\mathrm{M}}$; of course, then $\tilde{u} \in \mathscr{T}_{\mathrm{M}}$) are important in game considerations.

For $u \in \mathscr{T}_{\mathrm{M}}$ it holds: $\overline{u^\triangle} = (\bar{u})_\nabla$, $\overline{u_\nabla} = (\bar{u})^\triangle$, $u_\triangle = \overline{(u')^\triangle}$, $u^\nabla = \overline{(u')_\nabla}$; if, moreover, $u \in \mathscr{T}_{\mathrm{RM}}$, then $u^\triangle = \overline{(u')_\triangle}$, $u_\nabla = \overline{(u')^\nabla}$. ([5], § 5.18.)

2.1. To a given type (P, P_0) we introduce: $Z := P - P_0$, $\mathbf{Z} := \bigcup_{0 \leqq l < \omega_0} Z^{\{0, \ldots, l\}}$, $\boldsymbol{S} := ((\exp P) - \{\emptyset\})^{\mathbf{Z}}$, and

$$\mathbf{P} := \bigcup_{0 \leqq l \leqq \omega_0} \{\mathbf{x} = (x_k \mid 0 \leqq k < 1 + l) \mid x_k \in Z \text{ if } 0 \leqq k < l,\ x_k \in P_0 \text{ if } k = l\}\,,$$

where ω_0 is the first infinite ordinal number. For $\mathbf{z} = (z_0, \ldots, z_l) \in \mathbf{Z}$ we define $\kappa(\mathbf{z}) := z_l$, $l(\mathbf{z}) := l$. For $\mathbf{x} = (x_k \mid 0 \leqq k < 1 + l) \in \mathbf{P}$ we write shortly $\mathbf{x} = (x_k)$ and define $l(\mathbf{x}) := l$. For $\boldsymbol{\sigma} \in \boldsymbol{S}$, $x \in P$ we put

$$\mathrm{s}(\boldsymbol{\sigma}) := \{\mathbf{x} = (x_k) \mid \mathbf{x} \in \mathbf{P},\ x_{k+1} \in \boldsymbol{\sigma}(x_0, \ldots, x_k) \text{ if } 0 \leqq k < l(\mathbf{x})\}\,,$$
$$\mathrm{s}(x, \boldsymbol{\sigma}) := \{\mathbf{x} = (x_k) \mid \mathbf{x} \in \mathrm{s}(\boldsymbol{\sigma}),\ x_0 = x\} \quad (\neq \emptyset)\,.$$

2.2. Let $u \in \mathscr{T}_{\mathrm{RM}}(P, P_0)$ in this section. We put $\boldsymbol{S}(u) := \{\boldsymbol{\sigma} \mid \boldsymbol{\sigma} \in (\exp P)^{\mathbf{Z}}, \kappa(\mathbf{z}) \in u(\boldsymbol{\sigma}\mathbf{z})$ for every $\mathbf{z} \in \mathbf{Z}\}$ ($\subseteq \boldsymbol{S}$). Further, let $\boldsymbol{u} \in (\exp P)^{\exp \mathbf{P}}$ be given by $\boldsymbol{u}\mathbf{A} := \{x \mid x \in P,\ \mathrm{s}(x, \boldsymbol{\sigma}) \subseteq \mathbf{A}$ for some $\boldsymbol{\sigma} \in \boldsymbol{S}(u)\}$. There holds (see [5], § 3a): $\boldsymbol{u}\emptyset = \emptyset$; $\mathbf{A} \subseteq \mathbf{B} \subseteq \mathbf{P} \Rightarrow \boldsymbol{u}\mathbf{A} \subseteq \boldsymbol{u}\mathbf{B}$; if $(u_j)_{j \in J}$ is a game system such that $u_j P = Z$ (cf. Sec. 1.3), then $\bigcap_{j\in J} \boldsymbol{A}_j = \emptyset$ implies $\bigcap_{j\in J} \boldsymbol{u}_j\boldsymbol{A}_j = \emptyset$ for any $(\boldsymbol{A}_j)_{j\in J} \in (\exp \boldsymbol{P})^J$ ([5], § 4.18). But if $v := u'$ (cf. Sec. 1.4), it may happen there is $\boldsymbol{A} \subseteq \boldsymbol{P}$ such that $\boldsymbol{v}\boldsymbol{A} = \boldsymbol{u}\boldsymbol{P} - \boldsymbol{u}(\boldsymbol{P} - \boldsymbol{A})$ does not hold ([5], §§ 4.21.5, 4.23–24); this very important fact concerns immediately some connections with axiomatic set theories. [In terms of the game interpretation mentioned in the remarks in Sec. 1.3, $\boldsymbol{P}$ ($\exp \boldsymbol{P}$) is the set of variants (aims) at $\mathscr{G}$, $\boldsymbol{S}(u_j)$ is the set of player j's strategies; $x \in \boldsymbol{u}_j\boldsymbol{A}$ means that j can enforce $\boldsymbol{A}$ from x, etc. Cf. [5], §§ 2c, 3a.]

3.1. In Sec. 3, let (P, P_0) be a type, $u \in \mathscr{T}_{\mathrm{RM}}(P, P_0)$, $v := u'$. For $\boldsymbol{p} \in (\exp \boldsymbol{P})^{\exp \boldsymbol{P}}$ (= the set of *aim-mappings at* (P, P_0)) we define $\bar{\boldsymbol{p}} \in (\exp \boldsymbol{P})^{\exp \boldsymbol{P}}$ by $\bar{\boldsymbol{p}}A = \boldsymbol{P} - \boldsymbol{p}(P - A)$ (cf. Sec. 1.4!). Similarly as in [5] (§ 5.8, or [3], too), we shall denote a general aim-mapping by symbol $\boldsymbol{p}^\varepsilon$ and we shall write $\boldsymbol{p}_\varepsilon := \overline{\boldsymbol{p}^\varepsilon}$ (this symbolism admits various concrete forms, e.g.: $\boldsymbol{p}^\varepsilon = \boldsymbol{p}^\triangle, \boldsymbol{p}_\square, \tilde{\boldsymbol{p}}, \boldsymbol{p}_\varepsilon = \boldsymbol{p}_\triangle, \boldsymbol{p}^\square, \underset{\sim}{\boldsymbol{p}}$, respectively). To such $\boldsymbol{p}^\varepsilon$ we define $u^\varepsilon := \boldsymbol{u} \,.\, \boldsymbol{p}^\varepsilon$; $^\varepsilon$ itself is then considered an operator (dom $^\varepsilon$ = $= \mathscr{T}_{\mathrm{RM}}(P, P_0)$), and $\boldsymbol{p}^\varepsilon$ is said to be the *aim-meaning of* $^\varepsilon$. $\boldsymbol{p}^\varepsilon$ and $^\varepsilon$ are said to be *normal* iff (cf. Sec. 2.2!) $(v^\varepsilon A =) \boldsymbol{v} \,.\, \boldsymbol{p}^\varepsilon A = \boldsymbol{u}\boldsymbol{P} - \boldsymbol{u}(\boldsymbol{P} - \boldsymbol{p}^\varepsilon A)$ $(= P - \boldsymbol{u} \,.\, \boldsymbol{p}_\varepsilon(P - A) =$ $= P - u_\varepsilon(P - A) = \overline{u_\varepsilon}A)$, i.e., iff $(u')^\varepsilon = \overline{u_\varepsilon}$ (identically). In general, always $(u')^\varepsilon \subseteq$ $\subseteq \overline{u_\varepsilon}$, and $^\varepsilon$ is normal iff $_\varepsilon$ is normal.

3.2. Some operators (e.g., $u \to u^0$ $(= \mathbf{1})$) have simple normal aim-meanings. The union (intersection, product) of two operators having normal aim-meanings need not have an aim-meaning. But if $u \to u^\varepsilon$ is an operator having a [normal] aim-meaning, then $u \to \boldsymbol{u} \,.\, u^\varepsilon$, $u \to \breve{u} \,.\, u^\varepsilon$, and $u \to \tilde{u} \,.\, u^\varepsilon$ have [normal] aim-meanings.

3.3. (Cf. [3], §§ 2.8.1, 6.9.4, or [5], part IV.) Let $\boldsymbol{p}^\varepsilon$, $\boldsymbol{p}_\varepsilon$ $(\varepsilon = \triangle, \nabla, \square)$ be such that

$$\boldsymbol{x} \in \boldsymbol{p}^\triangle A \; [\boldsymbol{x} \in \boldsymbol{p}_\triangle A] \Leftrightarrow x_k \in A \text{ for some [each] } k\,,$$

$$\boldsymbol{p}^\nabla A = \boldsymbol{P}_{\mathrm{F}} \cup \boldsymbol{p}^\triangle A\,, \quad \boldsymbol{p}_\nabla A = (\boldsymbol{P} - \boldsymbol{P}_{\mathrm{F}}) \cap \boldsymbol{p}_\triangle A\,,$$

$$\boldsymbol{x} \in \boldsymbol{p}^\square A \Leftrightarrow \text{for each } \; k \; \text{ there exists } \; r \geqq k \; \text{ such that } \; x_r \in A\,,$$

$$\boldsymbol{x} \in \boldsymbol{p}_\square A \Leftrightarrow \text{there exists } \; k \; \text{ such that } \; x_r \in A \; \text{ for each } \; r \geqq k\,,$$

where

$$\boldsymbol{x} = (x_k) \in \boldsymbol{P}\,, \quad A \subseteq P\,, \quad 0 \leqq k\,, \quad r < 1 + l(\boldsymbol{x})\,,$$

$$\boldsymbol{P}_{\mathrm{F}} := \{\boldsymbol{x} \mid \boldsymbol{x} \in \boldsymbol{P},\; l(\boldsymbol{x}) < \omega_0\}\,;$$

it is easy to see that, indeed, $\boldsymbol{p}_\varepsilon = \overline{\boldsymbol{p}^\varepsilon}$ for $\varepsilon = \triangle, \nabla, \square$ (cf. Sec. 3.1).

Now, $\boldsymbol{u}^\varepsilon$ and $\boldsymbol{u}_\varepsilon$ are defined twice for $\varepsilon = \triangle, \nabla$ (and $u \in \mathscr{T}_{\mathrm{RM}}$), namely by Sec. 3.1 and 3.3, and by 1.5; nevertheless, it can be proved that these two definitions yield the same concepts. Moreover, it can be shown, among others, that $\boldsymbol{p}^\varepsilon$, $\boldsymbol{p}_\varepsilon$

($\varepsilon = \triangle, \nabla, \square$) are normal. From the definitions it follows trivially that u^ε, $u_\varepsilon \in \mathscr{T}_{\mathrm{M}}(P)$, $u^\triangle \emptyset = \emptyset$, $u_\triangle P = P$, and $u_\nabla \subseteq u_\triangle \subseteq u_\square \subseteq u^\square \subseteq u^\triangle \subseteq u^\nabla$, $u_\triangle \subseteq \mathbf{1} \subseteq u^\triangle$, $u^\nabla = u^\triangle . (\mathbf{1} \cup \underline{P_0'})$, $u_\nabla = u_\triangle . (\mathbf{1} \cap \underline{Z})$, etc.

3.4. It holds $\breve{u} = \breve{\bar{v}} \in \mathscr{T}_{\mathrm{RM}}(P, \emptyset)$; we shall write (cf. Sec. 3.2!) $u^{\smile\varepsilon} := \breve{u} . u^\varepsilon$, $u_{\smile\varepsilon} := \breve{u} . u_\varepsilon$ (then $u_{\smile\varepsilon} = \overline{(u')^{\smile\varepsilon}}$ if $u_\varepsilon = \overline{(u')^\varepsilon}$).

Clearly, $(\breve{u})^\nabla = (\breve{u})^\triangle$, $(\breve{u})_\nabla = (\breve{u})_\triangle$. The following important equalities hold:

$$(\breve{u})^\varepsilon = u^\varepsilon\,, \quad (\breve{u})_\varepsilon = u_\varepsilon \quad \text{for } \varepsilon = \triangle, \square\,,$$

$$u^\square = (u^{\smile\triangle})_{\smile\triangle}\,, \quad u_\square = (u_{\smile\triangle})^{\smile\triangle}\,.$$

(To derive the propositions presented in this paper, it is natural to use the latter two equalities to prove the normality of $^\square$ and $_\square$; nevertheless, there are other ways, cf. [3], § 6a, or § 7b, or § 9b.)

4.1. Theorem. *Let* (P, P_0) *be a type,* $u \in \mathscr{T}_{\mathrm{RM}}(P, P_0)$. *Then:*

A.

(1) $$u^\nabla . u^\varepsilon = u^\nabla \quad \textit{for} \quad \varepsilon = \triangle, \nabla$$

(2) $$u^\triangle . u^\varepsilon = u^\varepsilon \quad \textit{for} \quad \varepsilon = \triangle, \nabla, \square$$

(3) $$u^\varepsilon . u_\square = u_\square . u_\square = u_\square \quad \textit{for} \quad \varepsilon = \triangle, \square$$

(4) $$u_\varepsilon . u^\triangle . u_\triangle = u^\triangle . u_\triangle \quad \textit{for} \quad \varepsilon = \triangle, \square$$

(5) $$u_\varepsilon . u^\triangle . u_\nabla = u^\triangle . u_\nabla \quad \textit{for} \quad \varepsilon = \nabla, \triangle, \square$$

(6) $$u^\nabla . u^\square . u^\nabla = u^\square . u^\nabla$$

(7) $$u_\nabla . u^\square . u_\nabla = u^\square . u_\nabla$$

(8) $$u_\triangle . u^\nabla . u^\square = u^\nabla . u^\square$$

(9) $$u_\varepsilon . u^\nabla . u_\triangle = u^\nabla . u_\triangle \quad \textit{for} \quad \varepsilon = \triangle, \square$$

(10) $$u_\varepsilon . u^\nabla . u_\nabla = u^\nabla . u_\nabla \quad \textit{for} \quad \varepsilon = \triangle, \square$$

B. *For* $j = 1, \ldots, 10$, *let* (j') *be obtained from* (j) *by the replacement of each* u^δ $[u_\delta]$ *by* u_δ $[u^\delta]$, $\delta = \triangle, \nabla, \square, \varepsilon$. *Then* (j') *holds, too.*

C. *If* (e) *is obtained from the equality in* (j) *or* (j') $(j = 1, 3, 4, 9, 10)$ *by choosing* $\varepsilon := \varepsilon_0 \in \{\triangle, \nabla, \square\}$ *where* ε_0 *does not belong to those written in* (j), *then for suitable* (P, P_0) *and* u *the equality* (e) *does not hold.*

4.2. Part B is based on the normality of the operators under consideration. Counter-examples for proving Part C can be easily presented. The theorem contains assertions which follow immediately from the other ones (by means of the simple properties mentioned in Sec. 3.3: (10) is a corollary of (9); in (3), (4), (5), (9), (10), always only one case is "essential" while the others follow from it and (2); (1) and (2) can be "reduced", too).

4.3. (We suppose the same as in Sec. 3.1; recall that symbol $\boldsymbol{p}^\varepsilon$ admits also $\boldsymbol{p}_\square$ etc. as concrete "values".) For $\mathbf{x} = (x_k) \in \boldsymbol{P}$ and $0 \leq m < 1 + l(\mathbf{x})$, the sequence $\mathbf{x}^{[m]} := (x_{m+k} \mid 0 \leq k < 1 + (l(\mathbf{x}) - m))$ (here $\omega_0 - m = \omega_0$) belongs to $\boldsymbol{P}$ and is said to be a *remainder of* $\mathbf{x}$, while $\mathbf{x}$ is called an *extension of* $\mathbf{x}^{[m]}$. We say that $\boldsymbol{A} \subseteq \boldsymbol{P}$ has the *property* I_0 $[\mathrm{I}^0]$ iff $\boldsymbol{A}$ is closed under forming remainders [extensions]. Clearly, $\boldsymbol{A}$ has I^0 $[\mathrm{I}_0]$ iff $\boldsymbol{P} - \boldsymbol{A}$ has I_0 $[\mathrm{I}^0]$. We say that an aim-mapping $\boldsymbol{p}^\varepsilon$ has *property* $K \in \{\mathrm{I}^0, \mathrm{I}_0\}$ iff $\boldsymbol{p}^\varepsilon A$ has this property for any $\boldsymbol{A} \subseteq \boldsymbol{P}$. Thus, if $\boldsymbol{p}^\varepsilon$ has I^0 $[\mathrm{I}_0]$, then $\boldsymbol{p}_\varepsilon$ has I_0 $[\mathrm{I}^0]$. In particular, $\boldsymbol{p}^\varepsilon$ $[\boldsymbol{p}_\varepsilon]$ has the property I^0 $[\mathrm{I}_0]$ for $\varepsilon = \triangle, \nabla, \square$.

Lemma 1. *Let* $\boldsymbol{p}^\varepsilon$ *either* (1) *be* $\boldsymbol{p}^\triangle$, *or* (2) *have the property* I_0. *Let* $\boldsymbol{\sigma} \in \boldsymbol{S}(u)$, $x \in P$, $A \subseteq P$, $\mathbf{x} = (x_k) \in \mathrm{s}(x, \boldsymbol{\sigma}) \subseteq \boldsymbol{p}^\varepsilon A$, $0 \leq m < 1 + l(\mathbf{x})$. *Then* $x_m \in u^\varepsilon A$ (1) *if* $\{x_0, \ldots, x_{m-1}\} \cap A = \emptyset$, (2) *always* (*respectively*).

Lemma 2. *Let* $\boldsymbol{p}^\varepsilon$ *have the property* I^0 $[\mathrm{I}_0]$. *Then* $u^\triangle . u^\varepsilon = u^\varepsilon$ $[u_\triangle . u^\varepsilon = u^\varepsilon]$.

(Cf. [3], §§ 2.1.2, 2.4.2, 2.6.2, 2 (14)–(15), 6.2, 6.5.1, 6.5.3.)

4.4. Now, assertions (2) and (3) of the theorem follow immediately by means of Sec. 4.3. In the other "essential" equalities, one inclusion follows trivially from $u_\varepsilon \subseteq \mathbf{1} \subseteq u^\varepsilon$ ($\varepsilon = \triangle, \nabla$); the other inclusion can be proved by means of Sec. 4.3 and 3.3.

As an example, let us present the idea of the proof of (8). Let $A \subseteq P$; there exists $\boldsymbol{\sigma}_1 \in \boldsymbol{S}(u)$ $[\boldsymbol{\sigma}_2 \in \boldsymbol{S}(u)]$ such that $\mathrm{s}(x, \boldsymbol{\sigma}_1) \subseteq \boldsymbol{p}^\square A$ $[\mathrm{s}(x, \boldsymbol{\sigma}_2) \subseteq \boldsymbol{p}^\nabla u^\square A]$ for each $x \in u^\square A$ $[x \in u^\nabla u^\square A]$. We choose $\boldsymbol{\sigma} \in \boldsymbol{S}(u)$ in the following way: If $\mathbf{z} = (z_0, \ldots, z_l) \in \mathbf{Z}$, then if $z_l \notin u^\square A$, then $\boldsymbol{\sigma}\mathbf{z} = \boldsymbol{\sigma}_2\mathbf{z}$, if $z_l \in u^\square A$, then $\boldsymbol{\sigma}\mathbf{z} = \boldsymbol{\sigma}_1(z_r, \ldots, z_l)$ where $r = \min\{k \mid 0 \leq k \leq l, \{z_k, \ldots, z_l\} \subseteq u^\square A\}$. Let $x \in u^\nabla u^\square A$, $\mathbf{x} = (x_k) \in \mathrm{s}(x, \boldsymbol{\sigma})$. Using the evident relations $A \cap P_0 \subseteq u^\square A$ and $P_0 \cup u^\square A \subseteq u^\nabla u^\square A$, and the equality $u^\nabla u^\square A = u^\triangle(P_0 \cup u^\square A)$ (Sec. 3.4), and applying both the cases – successively (2) (for $x \in u^\square A$ firstly) and (1) – of Lemma 1, we obtain that $\mathbf{x} \in \boldsymbol{p}_\triangle(u^\nabla u^\square A)$. Hence, $x \in u_\triangle u^\nabla u^\square A$ for any $A \subseteq P$ and $x \in u^\nabla u^\square A$. Consequently, $u^\nabla u^\square A \subseteq u_\triangle u^\nabla u^\square A \subseteq u^\nabla u^\square A$.

5. In this paper, only a minor part of the results on SN-games ([3], [5] etc.) was used; especially, the min-max results (which were proved in versions being in several ways more general, by various approaches and in connection with the investigation of the game-theoretical meaning of extreme properties of certain operators or their generalizations) have been reduced to the normality of some operators having aim-meanings, graphs of SN-games have not been used, etc. [Further, game-theoretical considerations can be performed even for $\mathcal{T}_M$; I have studied the latter possibility (cf. [6]), using some new operators (e.g., $u \to u^*$, $u^*A = u\emptyset \div u(P - A) \div uP$ ($\div$ is the symmetric difference)).]

References

[1] *E. Čech:* Topologické prostory. Časopis Pěst. Mat. Fys. *66* (1937), 225—264.

[2] *E. Čech:* Topological spaces. (Revised by *Z. Frolík* and *M. Katětov.*) Academia, Prague, 1966.

[3] *J. Hanák:* Simultaneous nondeterministic games. Mathem. Theory of Polit. Decisions, Research Memorandum No. *3*, J. E. Purkyně University, Brno. (December 1967. Cyclostyled, 76 pp.)

[4] *J. Hanák:* Topological SN-games. Mathem. Theory of Polit. Decisions, Research Memorandum No. *7*, J. E. Purkyně University, Brno. (April 1968. Cyclostyled, 20 pp.)

[5] *J. Hanák:* Simultaneous nondeterministic games (I)—(IV). Arch. Math.; I: *5* (1969), 29—60; II: *6* (1970), 115—144; III: *7* (1971), 123—144; IV: to appear.

[6] *J. Hanák:* Equilibrium points in some SN-games. (To appear.)

[7] *K. Koutský:* Určenost topologických prostorů pomocí úplných systémů okolí bodů. Publ. Fac. Sci. Univ. Masaryk *374* (1956), 153—163.

[8] *K. Koutský, V. Polák and M. Sekanina:* On the commutativity of modification. Spisy Přírod. Fak. Univ. Brno *454* (1964), 275—292.

[9] *J. Lihová:* About the first and the second constellations belonging to the topology *u*. Acta Fac. Rerum Natur. Univ. Comenian. mimoriadne číslo 1971, 57—62.

[10] *Z. P. Mamuzić:* Note sur les espaces de voisinages (*V*) et les ordres semi-topogènes. General Topology and its Relations to Modern Analysis and Algebra, II. (Proc. Second Prague Topological Sympos., 1966). Academia, Prague, 1967, 246—247.

[11] *V. Polák, N. Poláková and M. Sekanina:* Remarks to inner constellation in topological spaces. Spisy Přírod. Fak. Univ. Brno *507* (1969), 317—332.

[12] *J. Schmidt:* Symmetric approach to the fundamental notions of general topology. General Topology and its Relations to Modern Analysis and Algebra, II. (Proc. Second Prague Topological Sympos., 1966). Academia, Prague, 1967, 308—319.

[13] *M. Sekanina:* Úplné systémy okolí množin v obecných topologických prostorech. Publ. Fac. Sci. Univ. Masaryk *374* (1956), 185—192.

[14] *М. Секанина:* Системы топологий на данном множестве. Чехослов. матем. журнал *15* (*90*) (1965), 9—29.

J. E. PURKYNĚ UNIVERSITY, BRNO

REMARKS ON DIMENSIONS OF MAPPINGS

J. HEJCMAN

Praha

In the dimension theory, besides the dimension of spaces, the dimension of mappings is often examined. If X, Y are topological spaces, $f : X \to Y$ a continuous mapping, we put $\dim f = \sup \{\dim f^{-1}[y] \mid y \in Y\}$. Similarly for uniform spaces, in addition to the uniform dimension Δd of spaces (see [2]), a uniform dimension of mappings can be defined. If $(X, \mathscr{U})$, $(Y, \mathscr{V})$ are uniform spaces, $f : X \to Y$ a uniformly continuous mapping, then $\Delta d\, f \leqq n$ means that for each U in $\mathscr{U}$ there exist V in $\mathscr{V}$ and W in $\mathscr{U}$ such that for any V-small subset M of Y there exists a W-cover $\mathscr{K}$ of $f^{-1}[M]$ consisting of U-small sets and such that each point of $f^{-1}[M]$ belongs to at most $n + 1$ sets of $\mathscr{K}$. Some results are stated in [1], let us mention here three properties only.

(a) If f maps a uniform space X onto a one-point space then $\Delta d\, f = \Delta d\, X$ (therefore the same symbol Δd is used).

(b) If g is the restriction of f onto a dense subspace then $\Delta d\, g = \Delta d\, f$.

(c) If $f : X \to Y$, then $\Delta d\, X \leqq \Delta d\, Y + \Delta d\, f$.

If X, Y are normal (T_1) topological spaces, we may consider the spaces endowed with some uniformities such that every continuous mapping $f : X \to Y$ becomes uniformly continuous and search for a connection between $\dim f$ and $\Delta d\, f$. This also enables us to derive some results on dim from the properties of Δd. We have the following theorems.

Theorem 1. *Let X be a normal space, Y a paracompact space, $f : X \to Y$ a closed continuous mapping. If both spaces X and Y are endowed with the fine uniformity, then* $\Delta d\, f = \dim f$.

Theorem 2. *Let X be a normal space, Y a paracompact space, $f : X \to Y$ a closed continuous mapping. Suppose Y is compact or* $\dim Y$ *is finite. If both spaces X and Y are endowed with the Čech uniformity, then* $\Delta d\, f = \dim f$.

Using Theorem 1 or 2, the equality of Δd and dim for both spaces and the above property (c), we immediately obtain the following version of Hurewicz theorem: *If X is a normal space, Y a paracompact space, $f : X \to Y$ a closed continuous mapping, then* $\dim X \leqq \dim Y + \dim f$. This result was also obtained by Pasynkov

[3]. His proof is essentially based on the same theorem for both X, Y paracompact which was proved by Skljarenko by means of the theory of sheaves.

Let X, Y be spaces endowed with the Čech uniformity, $f: X \to Y$ a continuous mapping, βf the extension of f onto the Čech-Stone compactifications, which are the completions of the spaces X and Y. Then the above property (b) and Theorem 1 or 2 imply $\Delta d\, f = \Delta d\, \beta f = \dim \beta f$. Thus Theorem 2 also concerns the equality of $\dim f$ and $\dim \beta f$. The additional assumption on the space Y in Theorem 2 cannot be omitted, which can be shown by an example. In this example, we also construct a closed continuous mapping $f: X \to Y$ with $\dim f = 0$ (moreover with finite pre-images of points), but $\dim \beta f > 0$; the spaces X, Y are metric, locally compact and σ-compact. The construction essentially depends on the following lemma.

Lemma. *Let $G_1, \ldots, G_n$ be open sets which cover the n-dimensional cube I^n. Then at least one set G_j contains a component which joins two opposite faces of the cube I^n.*

A detailed paper containing the proofs of all assertions is intended for publication in Czechoslovak Mathematical Journal.

References

[1] *J. Hejcman:* Uniform dimension of mappings. General Topology and its Relations to Modern Analysis and Algebra, II. (Proc. Second Prague Topological Sympos., 1966.) Academia, Prague, 1967, 182—183.

[2] *J. R. Isbell:* Uniform spaces. Providence, 1964.

[3] *Б. А. Пасынков:* О формуле В. Гуревича. Вестник Москов. Унив. Сер. 1 Мат. Мех. *20*, No 4 (1965), 3—5.

INSTITUTE OF MATHEMATICS OF THE CZECHOSLOVAK ACADEMY OF SCIENCES, PRAHA

ON DIFFICULTIES IN EMBEDDING LATTICE-ORDERED INTEGRAL DOMAINS IN LATTICE-ORDERED FIELDS

M. HENRIKSEN

Claremont

A *lattice ordered ring* (or *l-ring*) $A = A(+, \cdot, \vee, \wedge)$ is an abstract algebra closed under four binary operations $+, \cdot, \vee, \wedge$ such that $A(+, \cdot)$ is a ring, $A(\vee, \wedge)$ is a lattice, and if 0 is the identity element of $A(+)$, then

$$a, b \geqq 0 \quad \text{imply that} \quad a + b \geqq 0 \quad \text{and} \quad a \cdot b \geqq 0 .$$

As usual, we say that $a \geqq 0$ if $a \vee 0 = a$, and $a \geqq b$ if $(a - b) \geqq 0$. Moreover, we let $|a| = a \vee (-a)$.

Totally ordered rings and lattice-ordered rings of real-valued functions have long been studied, but the first systematic attack on the structure of l-rings as abstract algebras is due to Birkhoff and Pierce [2]. While they made progress with finite-dimensional l-algebras over totally ordered fields, they got a reasonable structure theory only for subdirect sums of totally ordered rings (which they call *f-rings*). For information about f-rings, see the paper by D. G. Johnson [6], the book by L. Fuchs [5], and for the most thorough summary of the literature on l-rings, see the doctoral dissertation of S. A. Steinberg [7].

The kernel of a homomorphism of one l-ring into another is called an *l-ideal*; I is an l-ideal if it is a ring ideal and $|b| \leqq |a|$, and a in I imply that b is in I. An l-ideal P is said to be *prime* if $|a|A|b|$ contained in P implies a or b is in P. If we let $M(A)$ denote the intersection of all the prime l-ideals of A, then J. E. Diem shows in [4] that $M(A)$ contains all the positive nilpotent elements of A and that $A/M(A)$ is a subdirect sum of l-rings such that

$$a > 0 \quad \text{and} \quad b > 0 \quad \text{imply that} \quad a \cdot b > 0 .$$

He calls such an l-ring by the name *l-domain* and notes that it may have proper divisors of zero, even if we assume also that $a^2 \geqq 0$ for all a in A.

This motivates us to seek answers to the following questions:

(1) *Is it always possible to embedd a (commutative) lattice-ordered integral domain A in a lattice-ordered field (preserving the lattice-order of A).*

By capitalizing on work of Conrad and Dauns in [3], Steinberg exhibits a way of embedding a lattice-ordered integral domain A in lattice-ordered field of formal power series under severe restrictions on the structure of the lattice-ordered group $A(+, \vee, \wedge)$. The general question remains open even if one assumes that $a^2 \geqq 0$ for every a in A.

In [3], Conrad and Dauns ask:

(2) *Under what conditions can the field $Q(A)$ of formal quotients of a lattice-ordered integral domain A be made into a lattice-ordered ring (preserving the lattice-order of A).*

To see that there is a difference between these two questions, we give a non-trivial example.

The ring $R[x]$ of real polynomials becomes an l-ring if we let

$$\sum_{k=1}^{n} a_k x^k \geqq 0 \quad \text{providing} \quad a_k \geqq 0 \quad \text{for} \quad k = 0, 1, \ldots, n .$$

Now, $R[x]$ is a sub-l-ring of the lattice-ordered field $R((x))$ of formal Laurent series

$$\sum_{k=-m}^{\infty} a_k x^k$$

ordered by saying that such a series is non-negative if each of its coefficients is non-negative.

But the field of quotients of elements of $R[x]$ is not a sublattice of $R((x))$.

For if α is any real number, then

$$\sum_{k=0}^{\infty} (\cos k\alpha)\, x^k = \frac{1}{2}\left[\frac{1}{x - e^{i\alpha}} + \frac{1}{x - e^{-i\alpha}}\right]$$

is in $R[x]$, while it can be shown that its absolute value $\sum_{k=0}^{\infty} |\cos k\alpha|\, x^k$ does not represent a rational function unless α is a rational multiple of π. (This result may be derived from a general result in [1], but a direct computational proof was obtained by my colleague S. Busenberg.)

Note that this latter result depends essentially on the fact that infinitely many coefficients of the power series in question are irrational, so it seems natural to ask:

(3) *If Q is the field of rational numbers, is the field $Q(x)$ of rational functions, ordered as above, a sublattice of the field $Q((x))$ of formal Laurent series? This is, if*

$$\sum_{k=-m}^{\infty} a_k x^k$$

represents a rational function (in some neighborhood of the origin) where each a_k

is a rational number, does

$$\sum_{k=-m}^{\infty} |a_k| \, x^k$$

represent a rational function? Indeed, what if each a_k is an integer?

A similar question is posed by Steinberg in [7].

I have one small contribution to make towards an answer to (1). Note that the embedding given above of $R[x]$ into $R((x))$ is *order-convex* in the sense that if $|b| \leqq |a|$ and b is in $R[x]$, so is a.

I can show that:

Suppose A is a lattice-ordered integral domain such that $a^2 \geqq 0$ for every a in A, and that A has no nonzero proper l-ideals. Then, if A is an order-convex sub-l-ring of a lattice-ordered field F, then $A = F$.

If we order $R[x]$ by letting

$$\sum_{k=0}^{n} a_k x^k > 0 \quad \text{if} \quad a_n > 0 \quad \text{and} \quad n \geqq 2\,,$$

and by letting

$$a_0 + a_1 x \geqq 0 \quad \text{if} \quad a_0 \geqq 0 \quad \text{and} \quad a_1 \geqq 0\,,$$

we get an example of a lattice-ordered integral domain that cannot be embedded as an order-convex sub-*l*-ring of a lattice-ordered field.

Generalizing a question posed by Conrad and Dauns [3] for *l*-fields, we may also ask:

(4) *Under what conditions on an l-ring can its order be extended to a linear order?*

I am able to show that *the order on any lattice-ordered integral domain A in which $a^2 \geqq 0$ for every a in A can be extended to a linear order*, but the general question remains open.

References

[1] *Benali Benzaghou:* Algébres de Hademand. Bull. Soc. Math. France *98* (1970), 209—252.
[2] *G. Birkhoff and R. S. Pierce:* Lattice-ordered rings. An. Acad. Brasil. Ci. *28* (1956), 41—69.
[3] *P. Conrad and J. Dauns:* An embedding theorem for lattice-ordered fields. Pacific J. Math. *30* (1969), 385—398.
[4] *J. E. Diem:* A radical for lattice-ordered rings. Pacific J. Math. *25* (1968), 71—82.
[5] *L. Fuchs:* Partially ordered algebraic systems. Pergammon Press, London, 1963.
[6] *D. G. Johnson:* A structure theory for a class of lattice-ordered rings. Acta Math. *104* (1960), 163—215.
[7] *S. A. Steinberg:* Lattice-ordered rings and modules. Thesis, University of Illinois, Urbana, Illinois, 1970.

HARVEY MUDD COLLEGE, CLAREMONT, CALIFORNIA

A GENERALIZATION OF PERFECT MAPS

H. HERRLICH

Bielefeld

This is a preliminary version of work that was done in cooperation with S. P. Franklin and D. Pumplün. Details, applications, and examples will appear elsewhere.

A. Perfect Maps. Assume all spaces to be completely regular.

Proposition 1. *For each continuous function $f : X \to Y$ the following conditions are equivalent*:

(1) *f is perfect,*

(2) *for each space Z the function $f \times 1_Z$ is perfect,*

(3) *each pullback of f is perfect,*

(4) *for any continuous function g and any dense embedding e the equality $f = g \cdot e$ implies that e is a homeomorphism,*

(5) *for any continuous function g and any dense compact-extendable map e the equality $f = g \cdot e$ implies that e is a homeomorphism,*

(6) *for any commutative diagram*

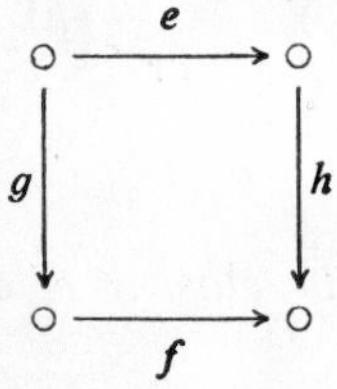

with e being dense and compact-extendable there exists a unique l such that the diagram

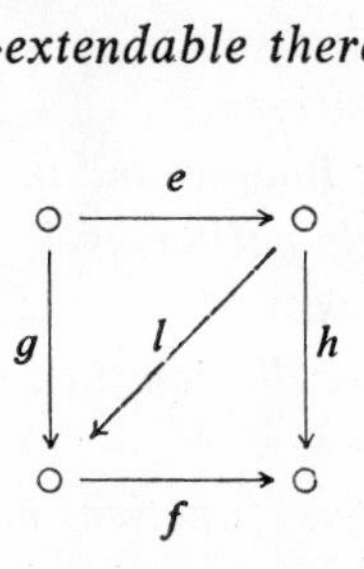

commutes,

(7) *any commutative diagram*

$$\begin{array}{ccc} \circ & \xrightarrow{e} & \circ \\ {\scriptstyle f}\downarrow & & \downarrow{\scriptstyle g} \\ \circ & \xrightarrow[m]{} & \circ \end{array}$$

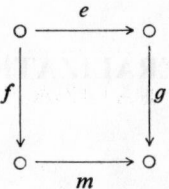

for which e is a dense embedding and m is an embedding must be a pullback,

(8) *any commutative diagram*

$$\begin{array}{ccc} \circ & \xrightarrow{e} & \circ \\ {\scriptstyle f}\downarrow & & \downarrow{\scriptstyle g} \\ \circ & \xrightarrow[m]{} & \circ \end{array}$$

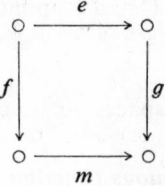

for which e and m are dense and compact-extendable must be a pullback,

(9) *if* $\beta_X : X \to \beta X$ *and* $\beta_Y : Y \to \beta Y$ *denote the Čech-Stone compactification of X and Y, respectively, then the diagram*

$$\begin{array}{ccc} X & \xrightarrow{\beta_X} & \beta X \\ {\scriptstyle f}\downarrow & & \downarrow{\scriptstyle \beta f} \\ Y & \xrightarrow[\beta_Y]{} & \beta Y \end{array}$$

is a pullback.

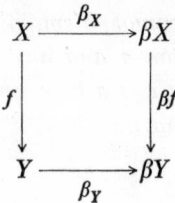

Proposition 2 (well-known). *The class* $\mathscr{P}$ *of all perfect maps has the following properties*

(1) $\mathscr{P}$ *contains all closed embeddings,*
(2) $\mathscr{P}$ *is closed under composition,*
(3) $\mathscr{P}$ *is closed under inverse images and more generally under pullbacks,*
(4) $\mathscr{P}$ *is closed under multiple pullbacks,*
(5) $\mathscr{P}$ *is closed under products,*
(6) $\mathscr{P}$ *is closed under left cancellation, i.e.,* $f \,.\, g \in \mathscr{P}$ *implies* $g \in \mathscr{P}$.

Proposition 3. *Each continuous function has an essentially unique (dense and compact-extendable, perfect) – factorization.*

B. Generalizations. Let $\mathscr{C}$ be a category.

Definition 1. Let $\mathscr{K}$ be a class of objects in $\mathscr{C}$ (resp. a full subcategory of $\mathscr{C}$). A morphism $f : X \to Y$ is called *$\mathscr{K}$-extendable* provided for each $K \in \mathscr{K}$ and each $g : X \to K$ there exists a $\bar{g} : Y \to K$ with $g = \bar{g} \,.\, f$.

Denote the class of $\mathscr{K}$-extendable epimorphisms by $E(\mathscr{K})$.

Proposition 4. (1) *$E(\mathscr{K})$ contains all isomorphisms of $\mathscr{C}$.*

(2) *$E(\mathscr{K})$ is closed under composition.*

(3) *$E(\mathscr{K})$ is closed under pushouts.*

(4) *$E(\mathscr{K})$ is left-cancellative with respect to epis, i.e. $f \,.\, g \in E(\mathscr{K})$ and g epi implies $g \in E(\mathscr{K})$.*

If $\mathscr{C}$ is complete, cocomplete, well-powered, and cowell-powered then:

(5) *For each $\mathscr{C}$-object X there exists a morphism $e_X : X \to \tilde{X}$ in $E(\mathscr{K})$ which is characterized by the fact that for each $e : X \to Y$ in $E(\mathscr{K})$ there exists some f with $e_X = f \,.\, e$.*

(6) *The full subcategory $\mathscr{K}$ of $\mathscr{C}$ consisting of those $\mathscr{C}$-objects X for which each $e \in E(\mathscr{K})$ is $\{X\}$-extendable is the epireflective hull of $\mathscr{K}$ in $\mathscr{C}$.*

(7) *A class $\mathscr{E}$ of epimorphisms in $\mathscr{C}$ is of the form $E(\mathscr{K})$ for some $\mathscr{K}$ iff $\mathscr{E}$ satisfies the conditions corresponding to (1)–(5).*

Definition 2. a) Let $\mathscr{E}$ be a class of epimorphisms in $\mathscr{C}$. Denote by $M(\mathscr{E})$ the class of all morphisms f such that whenever

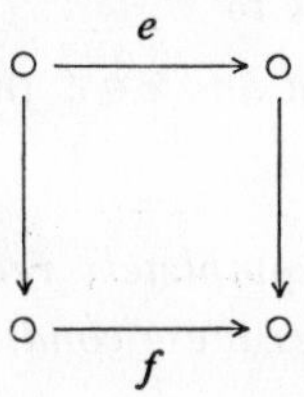

commutes and $e \in \mathscr{E}$ then there exists a morphism l such that

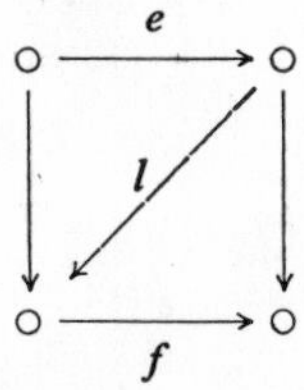

commutes.

b) Let $\mathscr{K}$ be a class of objects in $\mathscr{C}$. Then $P(\mathscr{K}) = M(E(\mathscr{K}))$ is called the class of *$\mathscr{K}$-perfect morphisms.*

Proposition 5. (1) *$M(\mathscr{E})$ contains all strong monomorphisms.*

(2) *$M(\mathscr{E})$ is closed under composition.*
(3) *$M(\mathscr{E})$ is closed under pullbacks.*
(4) *$M(\mathscr{E})$ is closed under multiple pullbacks.*
(5) *$M(\mathscr{E})$ is closed under products.*
(6) *$M(\mathscr{E})$ is closed under left-cancellation.*

Theorem. *Let $\mathscr{C}$ be complete, cocomplete, well-powered, and cowell-powered. Let $\mathscr{K}$ be a class of $\mathscr{C}$-objects, $\tilde{\mathscr{K}}$ be its epireflective hull in $\mathscr{C}$, $\mathscr{E} = E(\mathscr{K})$, and $\mathscr{P} = M(\mathscr{E}) = P(\mathscr{K})$. Then:*

(I) *For each $\mathscr{C}$-object A the following conditions are equivalent:*

(1) $A \in \tilde{\mathscr{K}}$,
(2) *if T is terminal then the unique $\mathscr{C}$-morphism $A \to T$ belongs to $\mathscr{P}$,*
(3) *each morphism with domain A belongs to $\mathscr{P}$,*
(4) *for each $\mathscr{C}$-object B the projection $A \times B \to B$ belongs to $\mathscr{P}$.*

(II) *For each $\mathscr{C}$-morphism $f : A \to B$ the following conditions are equivalent:*

(1) $f \in \mathscr{P}$,
(2) *for each $\mathscr{C}$-object C the morphism $f \times 1_C$ belongs to $\mathscr{P}$,*
(3) *each pullback of f belongs to $\mathscr{P}$,*
(4) *for any $\mathscr{C}$-morphism g and any $e \in \mathscr{E}$ the equality $f = g \,.\, e$ implies that e is an isomorphism.*

In case $\mathscr{C}$ is the category of completely regular spaces and continuous maps and $\mathscr{K}$ contains all compact spaces these conditions are equivalent to

(5) *any commutative diagram*

$$\begin{array}{ccc} \circ & \xrightarrow{e} & \circ \\ {\scriptstyle f}\downarrow & & \downarrow \\ \circ & \xrightarrow[e']{} & \circ \end{array}$$

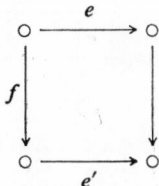

with $\{e, e'\} \subset E$ is a pullback,

(6) *if* $e_A : A \to \gamma A$ *and* $e_B : B \to \gamma B$ *denote the* $\tilde{\mathscr{K}}$*-reflections of* A *resp.* B *then the diagram*

$$\begin{array}{ccc} A & \xrightarrow{e_A} & \gamma A \\ {\scriptstyle f}\downarrow & & \downarrow{\scriptstyle \gamma f} \\ B & \xrightarrow[e_B]{} & \gamma B \end{array}$$

is a pullback.

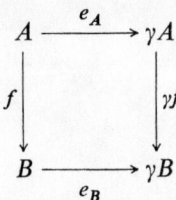

(III) *For each epimorphism* f *of* $\mathscr{C}$ *the following conditions are equivalent*:

(1) $f \in \mathscr{E}$,

(2) $M(\mathscr{E}) \subset M(\{f\})$.

(IV) *Each* $\mathscr{C}$*-morphism has an essentially unique* $(\mathscr{E}, \mathscr{P})$*-factorization.*

Problem. Characterize those classes $\mathscr{P}$ of $\mathscr{C}$-morphisms for which there exists a class $\mathscr{K}$ of $\mathscr{C}$-objects with $\mathscr{P} = P(\mathscr{K})$.

HARMONIC ANALYSIS AND TOPOLOGY

E. HEWITT

Seattle

The mathematical discipline that goes by the name of abstract harmonic analysis has close points of contact with set-theoretic topology, in a variety of directions.

First let us say briefly what harmonic analysis is. The classical theories of Fourier series and Fourier integrals are a highly developed branch of analysis. They have attracted the best mathematical minds for three centuries and may be regarded as part of the tradition of the human race. Given an integrable function f on $[-\pi, \pi[$, we form its Fourier coefficients

$$\hat{f}(k) = \frac{1}{2\pi} \int_{-\pi}^{\pi} f(t) \exp(-ikt)\, dt, \quad k = 0, \pm 1, \pm 2, \ldots,$$

and we may ask if the Fourier coefficients determine the function, how to reconstruct f from its Fourier coefficients, what relations exist between the size of f and the size of its Fourier coefficients. Or we can look at closed subspaces of $\mathfrak{L}_p([-\pi, \pi[)$ $(1 \leqq p \leqq \infty)$ invariant under translation modulo 2π and attempt to classify all of these subspaces, perhaps in terms of the Fourier coefficients. Similarly for $f \in \mathfrak{L}_1(-\infty, \infty)$, we can form the Fourier transform

$$\hat{f}(y) = (2\pi)^{-1/2} \int_{-\infty}^{\infty} f(x) \exp(-ixy)\, dx$$

and ask precisely the same questions.

In abstract harmonic analysis, one replaces the group $\mathbf{T} = [-\pi, \pi[$ (a compact Abelian group under addition modulo 2π and the usual topology) and $\mathbf{R}$ (a locally compact, noncompact, Abelian group under addition and the usual topology) by a locally compact group G. In studying the structure of such groups, it is plain that the topology must play a decisive role. Thus even while one takes the first step into harmonic analysis, one inescapably draws upon topology.

A locally compact group G admits an (essentially) unique Haar measure, which at first is constructed as a translation-invariant nonnegative linear functional L on the space $\mathfrak{C}_{00}(G)$ of all continuous complex-valued functions on G having compact supports. The shortest proof of the existence of L uses Tihonov's theorem, although it can be avoided by an ingenious construction due to Henri Cartan. Using L to construct a Borel measure λ on G for which $L(f) = \int_G f(t)\,d\lambda(t)$ for all $f \in \mathfrak{C}_{00}(G)$, we can construct function spaces $\mathfrak{L}_p(G, \lambda)$ for $0 < p \leqq \infty$, and proceed to study these function spaces. We can also consider the space $\mathfrak{C}_0(G)$ consisting of all continuous complex-valued functions on G that are arbitrarily small in absolute value outside of compact sets (the completion of $\mathfrak{C}_{00}(G)$ in the uniform metric) and construct its conjugate space $\mathbf{M}(G)$, the space of all complex-valued, countably additive, regular, Borel measures on G. Abstract harmonic analysis is primarily the study of the structure of the spaces $\mathfrak{L}_p(G)$ and of $\mathbf{M}(G)$ and of entities which arise in studying them.

For the sake of staying within reasonable boundaries, we will suppose henceforward that our groups G are either compact or Abelian and locally compact. For locally compact Abelian groups, the functions $t \mapsto \exp(ikt)$ on $\mathbf{T}$ and $x \mapsto \exp(ixy)$ on $\mathbf{R}$ are analogized by the continuous characters of G, which are the continuous complex functions χ of absolute value 1 on G such that $\chi(xy) = \chi(x)\chi(y)$ for all $x, y \in G$. The set $\mathbf{X}$ of all continuous characters is an Abelian group under pointwise multiplication, and topologized by the compact-open topology, it is also locally compact. The celebrated duality theorem of Pontryagin and van Kampen asserts that the character group of $\mathbf{X}$ is G; every continuous character of $\mathbf{X}$ has the form $\chi \mapsto \chi(a)$ for some fixed $a \in G$, and the topology of G as characters of $\mathbf{X}$ is its original topology. The *Fourier transform* of $f \in \mathfrak{L}_1(G)$ is the function $\hat{f}$ on $\mathbf{X}$ such that

$$\hat{f}(\chi) = \int_G f(t)\,\overline{\chi(t)}\,d\lambda(t) \quad \text{for all} \quad \chi \in \mathbf{X}.$$

Given a measure $\mu \in \mathbf{M}(G)$, its *Fourier-Stieltjes transform* is the function $\hat{\mu}$ on $\mathbf{X}$ such that

$$\hat{\mu}(\chi) = \int_G \overline{\chi(t)}\,d\mu(t) \quad \text{for all} \quad \chi \in \mathbf{X}.$$

For compact non-Abelian groups G, the role of the character group is taken over by irreducible unitary representations of G. Given a complex Hilbert space H, we denote by $\mathfrak{B}(H)$ the algebra of all bounded linear operators on H and by $\mathfrak{U}(H)$ the group of unitary operators on H. Now consider a homomorphism $U\colon x \mapsto U_x$ of G into $\mathfrak{U}(H)$ such that $U_e = I$. If there are no proper closed subspaces J of H such that $U_x(J) \subset J$ for all $x \in G$, U is called *irreducible*. If the map $x \mapsto \langle U_x\xi, \eta\rangle$ is continuous for all $\xi, \eta \in H$, then U is called *continuous*. All continuous unitary irreducible representations of a compact group G are finite-dimensional. Given two representations U and U' of G, with representation spaces H and H', we say that U and U' are *equivalent* if there is a linear isometry A of H onto H' such that $AU_xA^{-1} = U'_x$ for all $x \in G$. Let Σ denote the set of all equivalence classes of continuous unitary irreducible representations of G. That is, an element σ of Σ consists of a representation $U^{(\sigma)}$ and all representations equivalent to $U^{(\sigma)}$. We call Σ the *dual object of G.* To define Fourier transforms we need also a *conjugation D* on our Hilbert space H; i.e., an additive map D of H onto itself such that $D(\alpha\xi) = \bar{\alpha}\, D(\xi)$ for all $\alpha \in \mathbf{C}$ and $\xi \in H$. (Such conjugations all have the form $D(\Sigma\alpha_\iota\zeta_\iota) = \Sigma\bar{\alpha}_\iota\zeta_\iota$ for some orthonormal basis in H.) Now for every $\sigma \in \Sigma$, choose some $U^{(\sigma)} \in \sigma$ with representation space H_σ and a conjugation D_σ on H_σ. Let d_σ be $\dim(H_\sigma)$. For $f \in \mathfrak{L}_1(G)$, we define the *Fourier transform* $\hat{f}$ as the $\mathfrak{B}(H_\sigma)$-valued function on Σ such that

$$\langle \hat{f}(\sigma)\, \xi, \eta\rangle = \int_G \langle D_\sigma U_x^{(\sigma)} D_\sigma \xi, \eta\rangle\, f(x)\, d\lambda(x)$$

for all $\sigma \in \Sigma$ and $\xi, \eta \in H_\sigma$. For $\mu \in \mathbf{M}(G)$, the *Fourier-Stieltjes* transform $\hat{\mu}$ is defined similarly:

$$\langle \hat{\mu}(\sigma)\, \xi, \eta\rangle = \int_G \langle D_\sigma U_x^{(\sigma)} D_\sigma \xi, \eta\rangle\, d\mu(x)\,.$$

Plainly $\hat{f}(\sigma)$ and $\hat{\mu}(\sigma)$ depend upon the particular choice of $U^{(\sigma)}$ and D_σ, but all such choices yield unitarily equivalent operators, and for all known norms of transforms and all presently studied structural properties of $\mathfrak{L}_p(G)$ and $\mathbf{M}(G)$, this is sufficient.

We also introduce some norms into the space of Fourier transforms. Given a finite-dimensional Hilbert space H of dimension d, and $A \in \mathfrak{B}(H)$, write the positive-definite operator AA^* as $\sum_{j=1}^{d} a_j P_j$, where the P_j's are 1-dimensional projections. For $1 \leqq p < \infty$, define the *von Neumann norm* $\|A\|_{\Phi_p}$ as $(\sum_{j=1}^{d} a_j^{p/2})^{1/p}$ and $\|A\|_{\Phi_\infty}$ as $\max_{1\leqq j\leqq d} a_j^{1/2}$. The norm $\|A\|_{\Phi_\infty}$ is the operator norm, and $\|A\|_{\Phi_2}$ is Frobenius's trace norm $[\operatorname{tr}(A^*A)]^{1/2}$. Define

$$\|\hat{f}\|_p = (\sum_{\sigma\in\Sigma} d_\sigma \|\hat{f}(\sigma)\|_{\Phi_p}^{p})^{1/p} \quad (1 \leqq p < \infty)$$

and

$$\|\hat{f}\|_\infty = \sup_{\sigma\in\Sigma} \|\hat{f}(\sigma)\|_{\Phi_\infty}\,.$$

The Weyl-Peter theorem asserts that if $f \in \mathfrak{L}_2(G)$, then $\|f\|_2 = \|\hat{f}\|_2$, and the Riemann-Lebesgue lemma that $\|\hat{f}(\sigma)\|_{\Phi_\infty}$ is arbitrarily small off of a finite set in Σ.

The function spaces $\mathfrak{L}_p(G)$ $(0 < p \leqq \infty)$ admit isometries defined by translation. For $a \in G$ and any function f on G, define ${}_af$ and f_a by ${}_af(x) = f(ax)$ and $f_a(x) = f(xa)$. A space $\mathfrak{S}$ of functions is called (left, right) invariant if $f \in \mathfrak{S}$ implies $({}_af, f_a) \in \mathfrak{S}$ for all $a \in G$. A main problem of harmonic analysis is the classification of all closed invariant subspaces of the spaces $\mathfrak{L}_p(G)$. For compact groups G and $1 \leqq p \leqq \infty$, this classification is complete. For $1 \leqq p < \infty$, every two-sided invariant subspace of $\mathfrak{L}_p(G)$ is defined by the vanishing of $\hat{f}(\sigma)$ for all σ in a certain subset of Σ. Slight additional complications arise for $p = \infty$ and for one-sided invariant subspaces.

For noncompact Abelian groups G, the invariant subspaces are completely known only for $\mathfrak{L}_2(G)$. For $0 < p < 1$, nothing is known for any infinite group G.

The space of measures $\mathbf{M}(G)$ is a Banach algebra under *convolution:*

$$\mu * \nu(E) = \int_G \mu(Et^{-1})\, d\nu(t) = \int_G \nu(t^{-1}E)\, d\mu(t)$$

for all Borel subsets E of G. Of course $\mathfrak{L}_1(G)$ is also a Banach algebra under convolution:

$$f * g(x) = \int_G f(xy)\, g(y^{-1})\, dy\,.$$

Closed ideals in $\mathfrak{L}_1(G)$ are exactly the closed invariant subspaces of $\mathfrak{L}_1(G)$ (left, right, and two-sided, respectively), so the ideal structure of $\mathfrak{L}_1(G)$ is coextensive with its invariant subspace structure.

The algebra $\mathbf{M}(G)$ is very little understood, in spite of considerable recent progress. For locally compact Abelian G, one has the Gel'fand theory of its maximal ideals, and its compact structure space $\mathscr{S}$. However, the topology of $\mathscr{S}$ is formidably complicated, as has been long known, and there seems no hope of unravelling its puzzles completely in the near future. See for example Yu. A. Šreĭder [15] and Hewitt and Kakutani [5], [6]. Further progress in understanding $\mathbf{M}(G)$ may depend upon brilliant applications of functional analysis like those made by J. L. Taylor [16], [17], [18], and N. Th. Varopoulos [19], [20], [21], (the latter to be sure in part for other purposes). In the author's opinion, however, decisive progress can be expected only through refined study of the arithmetic and analytic properties of certain perfect subsets of G.

In a brief survey it is plainly impossible to give a complete picture of the role of topology in harmonic analysis. We will therefore limit ourselves to two examples. The first is the theory of *lacunary sets,* where the theory of topological linear spaces is applied. The basic idea of a lacunary set is this. If f is a function whose Fourier transform vanishes except on a "thin" or "lacunary" set (of characters in the Abelian case, or of representations in the non-Abelian case), then f can be expected to have some extraordinary properties. Ancestors of many such theorems are two theorems of S. Sidon [13], [14]. A *Hadamard set* $\{n_k\}_{k=1}^{\infty}$ of positive integers is defined by the property that for some constant $q > 1$, we have

$$n_{k+1}/n_k \geqq q \quad \text{for all} \quad k\,.$$

Sidon proved that if a real function f is in $\mathfrak{L}_\infty([-\pi, \pi[)$ (or merely in $\mathfrak{L}_1([-\pi, \pi[)$) and if $\hat{f}(n) = 0$ except for $n = \pm n_k$, where the n_k's are in a Hadamard set, then $\sum_{n\in\mathbf{Z}} |\hat{f}(n)| < \infty$ (or merely $\sum_{n\in\mathbf{Z}} |\hat{f}(n)|^2 < \infty$).

We make these theorems the bases of definitions.

Definition. Let G be any compact group, with dual object Σ. A subset $\mathbf{P}$ of Σ is called a *Sidon set* if for every function $f \in \mathfrak{C}(G)$ such that $\hat{f}(\sigma) = 0$ for all σ in in $\Sigma \setminus \mathbf{P}$, we have $\|\hat{f}\|_1 < \infty$.

Definition. For $1 < p < \infty$, a subset $\mathbf{P}$ of Σ is called a Λ_p *set* if for every function $f \in \mathfrak{L}_1(G)$ such that $\hat{f}(\sigma) = 0$ for all $\sigma \in \Sigma \setminus \mathbf{P}$, we have $f \in \mathfrak{L}_p(G)$.

Both of these definitions are abstractions from Sidon's theorems. Now we can list a large number of properties equivalent to each of the above properties.

Theorem. *Let* $\mathbf{P}$ *be a subset of* Σ. *The following assertions are equivalent*:

(i) $\mathbf{P}$ *is a Sidon set*;

(ii) *for every operator function* $(E_\sigma)_{\sigma \in \mathbf{P}}$ *with* $E_\sigma \in \mathfrak{B}(H_\sigma)$ *such that* $\sup_{\sigma \in \mathbf{P}} \|E_\sigma\|_{\Phi_\infty} < \infty$, *there is a measure* $\mu \in \mathbf{M}(G)$ *such that* $\hat{\mu}(\sigma) = E_\sigma$ *for all* $\sigma \in \mathbf{P}$;

(iii) *for every* $(E_\sigma)_{\sigma \in \mathbf{P}}$ *as in* (ii) *such that* $\|E_\sigma\|_{\Phi_\infty}$ *is arbitrarily small outside of finite sets, there is a function* $f \in \mathfrak{L}_1(G)$ *such that* $\hat{f}(\sigma) = E_\sigma$ *for all* $\sigma \in \Sigma$;

(iv) *if* $f \in \mathfrak{L}_\infty(G)$ *and* $\hat{f}(\sigma) = 0$ *for all* $\sigma \in \Sigma \setminus \mathbf{P}$, *then* $\|\hat{f}\|_1 < \infty$;

(v) *there is a constant* $\varkappa$ *such that* $\|\hat{f}\|_1 \leqq \varkappa \|f\|_\infty$ *for all* $f \in \mathfrak{L}_\infty(G)$ *such that* $\hat{f}(\sigma) = 0$ *for* $\sigma \in \Sigma \setminus \mathbf{P}$;

(vi) *the same as* (v) *for continuous* f;

(vii) *the same as* (v) *for* f *such that* $\hat{f}(\sigma) \neq 0$ *only for a finite set of* $\sigma \in \Sigma$;

(viii) *for every unitary operator-valued function* $U_\sigma \in \mathfrak{U}(H_\sigma)$ *defined for* $\sigma \in \mathbf{P}$, *there is a* $\mu \in \mathbf{M}(G)$ *such that* $\sup_{\sigma \in \mathbf{P}} \|U_\sigma - \hat{\mu}(\sigma)\|_{\Phi_\infty} < 1$.

Obviously the proof of such a string of equivalences is long. However at bottom it is merely an exercise in functional analysis. The open mapping theorem and the Hahn-Banach theorem are all that one really needs. Such applications of functional analysis occur repeatedly in harmonic analysis.

A similar theorem applies to Λ_p sets: they may be characterized in a variety of ways using the techniques of functional analysis. It can be shown that every Sidon set is a Λ_p set for all $p > 1$. For infinite Abelian G, the character group always contains a set $\mathbf{P}$ that is a Λ_p set for all $p > 1$ but is not a Sidon set, as was recently proved by R. E. Edwards, K. A. Ross, and the author [3].

Our second example applies the Čech-Stone compactification, or if you like, the theory of ultrafilters, to a problem in harmonic analysis. Let us consider a compact non-Abelian group G. Closed ideals in the algebra $\mathfrak{L}_1(G)$, as noted above, are completely known. In particular, the maximal closed two-sided ideals in $\mathfrak{L}_1(G)$ all have the form $\mathfrak{J}_\tau = \{f \in \mathfrak{L}_1(G) : \hat{f}(\tau) = 0\}$ for some fixed $\tau \in \Sigma$. The simple quotient algebra $\mathfrak{L}_1(G)/\mathfrak{J}_\tau$ is easily seen to be isomorphic to $\mathfrak{B}(H_\tau)$ and so is finite-dimensional. Now consider the measure algebra $\mathbf{M}(G)$. Does the same phenomenon persist in this large algebra? That is, if $\mathbf{I}$ is a maximal two-sided ideal in $\mathbf{M}(G)$, is the simple algebra $\mathbf{M}(G)/\mathbf{I}$ finite-dimensional? Of course this question is trivial if G is Abelian. For Abelian G, $\mathbf{M}(G)$ is commutative, and the Gel'fand-Mazur theorem shows that all algebras $\mathbf{M}(G)/\mathbf{I}$ are isomorphic with $\mathbf{C}$. Negative answers for some G are obtained, however, with the use of ultrafilters. The construction is joint work of Dusa McDuff and the author [7].

Let us sketch the construction. Consider an infinite index set I and a countably infinite dissection $\mathscr{D} = \{A_k\}_{k=1}^{\infty}$ of I into subsets, some possibly void. For each $\iota \in A_k$ consider a Hilbert space H_ι of dimension k, and form the product algebra $\mathbf{P}_{\iota \in I} \mathfrak{B}(H_\iota)$. In this algebra, single out the elements $E = (E_\iota)_{\iota \in I}$ such that $\|E\| = \sup_{\iota \in I} \|E_\iota\|_{\Phi_\infty} < \infty$, and call this Banach algebra $\mathfrak{E}(I, \mathscr{D})$. The algebra $\mathfrak{E}(I, \mathscr{D})$ is a noncommutative analogue of the algebra $\mathfrak{C}_b(I)$ consisting of all bounded complex-valued functions on I, to which of course it reduces if $I = A_1$. All of the maximal two-sided ideals in $\mathfrak{E}(I, \mathscr{D})$ can be constructed by using the Čech-Stone compactification βI of I, where we regard I as a discrete space. To do this, we associate a certain real-valued function with every $E = (E_\iota)_{\iota \in I}$ in $\mathfrak{E}(I, \mathscr{D})$. Define $\psi_E(\iota)$ as $k^{-1/2}\|E_\iota\|_{\Phi_2}$ for all $\iota \in A_k$. It is trivial that $0 \leqq \psi_E(\iota) \leqq \|E_\iota\|_{\Phi_\infty} \leqq \|E\|_\infty < \infty$, so that ψ_E is a bounded real-valued function on I. By the defining property of βI, ψ_E admits a continuous extension over βI, which we will continue to call ψ_E. For every point $x \in \beta I$, we define

$$\mathfrak{J}_x = \{E \in \mathfrak{E}(I, \mathscr{D}) : \psi_E(x) = 0\}.$$

Then $\mathfrak{J}_x$ is a maximal two-sided ideal in $\mathfrak{E}(I, \mathscr{D})$, and every maximal two-sided ideal in $\mathfrak{E}(I, \mathscr{D})$ is $\mathfrak{J}_x$ for some $x \in \beta I$. This is a special case of a theorem of F. B. Wright [22]. The theorem of Pospíšil [10] shows that there are exp (exp (card (I))) such ideals: plainly if $x \neq x'$, then $\mathfrak{J}_x \neq \mathfrak{J}_{x'}$.

As is well known from the work of P. S. Aleksandrov and others, the points of βI are in one-to-one correspondence with the ultrafilters on I. In fact, we may regard βI as the set of all ultrafilters on I. Let x stand for an ultrafilter on I. If some set A_k belongs to x, then the quotient algebra $\mathfrak{E}(I, \mathscr{D})/\mathfrak{J}_x$ is isomorphic with $\mathfrak{B}(H)$ where H is a k-dimensional Hilbert space. On the other hand, if x contains no set A_k, then $\mathfrak{E}(I, \mathscr{D})/\mathfrak{J}_x$ is infinite-dimensional and is in fact nonseparable. Thus the algebras $\mathfrak{E}(I, \mathscr{D})$ admit infinite-dimensional simple homomorphs provided that the dimensions of the spaces H_ι are unbounded.

Given a metric space Y and a subset S of Y, we say that S is *scattered* if $\inf\{\varrho(x, y) : x, y \in S, x \neq y\} > 0$. Using Frolík's construction of sums of ultrafilters [4], we can construct an x for which $\mathfrak{E}(I, \mathscr{D})/\mathfrak{J}_x$ contains a very large scattered set. The precise result is the following. Let $\mathfrak{m}$ be any infinite cardinal, and I a set of cardinal number $\mathfrak{m}$. Then there is a decomposition $\mathscr{D} = \{A_k\}_{k=1}^{\infty}$ of I and an ultrafilter x in I such that the quotient algebra $\mathfrak{E}(I, \mathscr{D})/\mathfrak{J}_x$ contains a scattered set of cardinal number exp $(\mathfrak{m})$.

To obtain results like this for measure algebras $\mathbf{M}(G)$, we use Sidon sets. Suppose for example that G is a product $\mathbf{P}_{\iota \in I} \mathfrak{U}(H_\iota)$, where each finite-dimensional unitary group is given its usual [compact] topology. For $U = (U_\iota) \in G$, and a fixed $\Theta \in I$, the map $U \mapsto U_\Theta$ is a continuous irreducible unitary representation of G. Let π_Θ be the element of Σ containing this representation. It can be shown that the set $\{\pi_\iota : \iota \in I\}$ is a Sidon set in Σ. Thus the mapping $\mu \mapsto (\hat{\mu}(\iota))_{\iota \in I}$ of $\mathbf{M}(G)$ into $\mathfrak{C}(I, \mathscr{D})$ is an *on*to mapping, and all of the previous paragraph applies to $\mathbf{M}(G)$. The mapping

$$\mathbf{M}(G) \mapsto \mathfrak{C}(I, \mathscr{D}) \mapsto \mathfrak{C}(I, \mathscr{D})/\mathfrak{I}_x$$

is a homomorphism, and for x as above the image is not only infinite-dimensional but contains very large scattered sets. By a closer study of the structure of ultrafilters, it might well be possible to distinguish yet more refinements in simple homomorphs of $\mathbf{M}(G)$.

In conclusion we mention some of the literature on harmonic analysis. The classical treatises of Zygmund [23] and Bari [1] deal exhaustively with the groups $\mathbf{T}$, $\mathbf{Z}$, and $\mathbf{R}$, and in less detail with $\mathbf{T}^m$ and $\mathbf{R}^m$ ($m = 2, 3, \ldots$). They do not have the group-theoretic point of view. R. E. Edwards [2] has written a monograph on Fourier series from the group-theoretic viewpoint, which serves as a fine bridge between the classical and abstract treatment of the subject. A similar attitude is adopted in the book [9] of Katznelson, although his treatment is terser and more technical than Edwards's. A sophisticated and relatively short exposition of analysis on groups has been given by Rudin [12]. Reiter [11] pursues a number of special topics, also mostly on groups. The author and Kenneth A. Ross [8] have written a two-volume monograph on abstract harmonic analysis exclusively from the group-theoretic point of view. A study of either Rudin or Hewitt and Ross should probably be preceded by a close reading of Edwards, if one is not to lose sight of the woods for the multitude of trees.

References

[1] *N. K. Bari (Н. К. Бари):* Trigonometric series (Тригонометрические ряды). Gos. Izdat. Fiz.-mat. Lit., Moskva, 1961. English translation: A treatise on trigonometric series. The Macmillan Co., New York, N. Y., 1964.

[2] *R. E. Edwards:* Fourier series: a modern introduction. Holt, Rinehart, and Winston, Inc., New York, N. Y., 1965.

[3] *R. E. Edwards, E. Hewitt and K. A. Ross:* Lacunarity for compact groups, I. Indiana Univ. Math. J. *21* (1972), 787—806.

[4] *Z. Frolík:* Sums of ultrafilters. Bull. Amer. Math. Soc. *73* (1965), 87—91.

[5] *E. Hewitt and S. Kakutani:* A class of multiplicative functionals on the measure algebra of a locally compact Abelian group. Illinois J. Math. *4* (1960), 553—574.

[6] *E. Hewitt and S. Kakutani:* Some multiplicative linear functions on M(*G*). Ann. of Math. *79* (2) (1964), 489—505.

[7] *E. Hewitt and D. McDuff:* Некоторые патологические максимальные идеалы в алгебрах операторов и алгебрах мер на группах. Mat. Sb. *83* (*125*) (1970), 527—546.

[8] *E. Hewitt and K. A. Ross:* Abstract harmonic analysis. Springer-Verlag, Heidelberg; 1963, 1970.

[9] *Y. Katznelson:* An introduction to harmonic analysis. John Wiley & Sons, New York, N. Y., 1968.

[10] *B. Pospíšil*: Remark on bicompact spaces. Ann. of Math. *38* (2) (1937), 845—846.

[11] *H. J. Reiter:* Classical harmonic analysis and locally compact groups. Oxford Mathematical Monographs. Oxford University Press, Oxford, England, 1968.

[12] *W. Rudin:* Fourier analysis on groups. Interscience Publishers, New York, N. Y., 1962.

[13] *S. Sidon:* Ein Satz über die absolute Konvergenz von Fourierreihen, in denen sehr viele Glieder fehlen. Math. Ann. *97* (1926—1927), 418—419.

[14] *S. Sidon:* Ein Satz über trigonometrische Polynome mit Lücken und seine Anwendungen in der Theorie der Fourier-Reihen. J. Reine Angew. Math. *163* (1930), 251—252.

[15] *Yu. A. Šreĭder* (*Ю. А. Шрейдер*)*:* Construction of maximal ideals in rings of measures with convolution (Строение максимальных идеалов в кольцах мер со сверткой). Mat. Sb. *27* (*69*) (1950), 297—318.

[16] *J. L. Taylor:* The structure of convolution measure algebras. Trans. Amer. Math. Soc. *119* (1965), 150—166.

[17] *J. L. Taylor:* Convolution measure algebras with group maximal ideal spaces. Trans. Amer. Math. Soc. *128* (1967), 257—263.

[18] *J. L. Taylor:* Noncommutative convolution measure algebras. Pacific J. Math. *31* (1969), 809—826.

[19] *N. Th. Varopoulos:* A direct decomposition of the measure algebra of a locally compact Abelian group. Ann. Inst. Fourier (Grenoble) *16* (1966), 121—143.

[20] *N. Th. Varopoulos:* Tensor algebras and harmonic analysis. Acta Math. *119* (1967), 51—112.

[21] *N. Th. Varopoulos:* Groups of continuous functions in harmonic analysis. Acta Math. *125* (1970), 109—154.

[22] *F. B. Wright:* A reduction for algebras of finite type. Ann. of Math. *60* (2) (1954), 560—570.

[23] *A. Zygmund:* Trigonometric series. Cambridge University Press, Cambridge, England, 1959. Russian translation: Тригонометрические ряды. Izdatel'stvo Mir, Moskva, 1965.

A CLASS OF CONNECTED SPACES WITH MANY RAMIFICATIONS

J. L. HURSCH and A. VERBEEK

Amsterdam

In this note we study the following class RAM of spaces, which emerged as counterexamples in the characterization of linearly orderable and weakly linearly orderable, connected spaces.

Definitions. $X \in$ RAM if X is a connected topological space which has a point x_0 such that every connected subset of X containing x_0 is closed.

Following the terminology of [1] we say that a space X is weakly linearly orderable iff there exists a linear order on X for which the order topology is weaker (coarser) than the given topology. If both topologies coincide X is said to be linearly orderable. Moreover we say that X has property (H) iff every connected subset has at most two non-cutpoints.

In [1] it was proved that for a connected T_2-space X, weak linear orderability is equivalent to (H) + each point separates in at most two components (X is linearly orderable iff moreover X is locally connected). It was asked whether property (H) suffices. Now it is easily seen from properties 2, 7 and 8 below that each $X \in$ RAM has property (H) but can never be weakly linearly orderable.

For $X \in$ RAM we have the following properties:

1. First notice that every connected space contains at least one nonclosed connected subset (except spaces with ≤ 1 point).

2. The x_0 from the definition of $X \in$ RAM is unique. For every $x \in X \setminus \{x_0\}$ the subspace $X \setminus \{x\}$ has infinitely many components.

3. X is not separable metrizable (nor is any open subset).

4. For each $x \in X$ there exists an open subset $O \subset X$ such that $x \in O \in$ RAM and x is the "x_0 of O".

5. X is nowhere locally compact.

6. X is nowhere locally connected.

7. RAM $\neq \emptyset$, even $\exists Y \in$ RAM Y is T_2 and Y has countably many points.

8. Each connected subset of X has at most one non-cutpoint.

We do not even know whether no $X \in$ RAM is metrizable, although we conjecture that for each $X \in$ RAM each continuous real-valued function is constant.

Properties 1 and 2 are easily verified, while 3 follows from the following theorem which can essentially be found in [2], p. 75, K 4.

Theorem. *If X is an m-separable connected T_1-space then the number of points $x \in X$ for which $X \setminus \{x\}$ has at least three components does not exceed m.*

The properties 4–8 are proved by exploring the relation R on X defined by: xRy iff $x = x_0$ or x separates x_0 from y. This turns out to be a partial ordering, which is related to the topology in a rather unusual way.

References

[1] *H. Kok:* On conditions equivalent to the orderability of a connected space. Nieuw Arch. Wisk. *18* (1970), 250–270.

[2] *L. F. McAuley:* On decomposition of continua into aposyndetic continua. Trans. Amer. Math. Soc. *81* (1956), 74–91.

SIMPLE CATEGORIES OF TOPOLOGICAL SPACES

M. HUŠEK

Praha

This contribution deals with productive and closed-hereditary classes of uniformizable Hausdorff spaces (= spaces), i.e., [16], with the epireflective subcategories of the category of all spaces. Such classes can be described as the classes $\mathscr{K}(\mathscr{E})$ of all $\mathscr{E}$-compact spaces for suitable classes $\mathscr{E}$ of spaces; here $\mathscr{E}$-compact means that the space can be embedded as a closed subspace into a product of spaces from $\mathscr{E}$ (see [8]). Clearly, if a class $\mathscr{C}$ is productive and closed-hereditary then $\mathscr{C} = \mathscr{K}(\mathscr{C})$ but the problem is to find a nice and small class $\mathscr{E}$ such that $\mathscr{C} = \mathscr{K}(\mathscr{E})$. If $\mathscr{E}$ may be found to be a set or, which is the same, to consist of a single space E, then $\mathscr{C}$ is called simple – [8]. The simple class $\mathscr{K}(E)$ generated by a space E and called E-compact spaces was introduced and studied in [17], [2].

Up today there is no known characterization of simple classes among all the productive and closed-hereditary classes. Here we want to present two comments concerning the problem: the first one describes a nice class of non-simple categories, and the second one shows a connection of the problem with set-theory.

We need some more concepts: $\mathscr{E}$-regular spaces are the homeomorphs of subspaces of products of spaces from $\mathscr{E}$ (hence if $\mathscr{E}$ is productive then $\mathscr{E}$-regular spaces are just homeomorphs of subspaces of members from $\mathscr{E}$). Relatively measurable cardinals are infinite cardinals α admitting non-trivial two valued measures which are β-additive for all $\beta < \alpha$; we denote by $\{\alpha_\xi\}$ the class of all relatively measurable cardinals arranged into the increasing transfinite sequence together with a symbol α_η following all the cardinals if all the relatively measurable cardinals form a set with the index-set $\{\xi \mid \xi < \eta\}$ (hence $\alpha_0 = \omega_0$ and an uncountable relatively measurable cardinal α_ξ is the first cardinal admitting a non-trivial two-valued measure which is α_ζ-additive for all $\zeta < \xi$). The term "nonmeasurable cardinal" is used in the sense of [6], i.e. any cardinal smaller than α_1. The symbol α^+ means the successor of cardinal α. For more details on $\mathscr{E}$-compact and $\mathscr{E}$-regular spaces or epireflective subcategories or relatively measurable cardinals we refer the reader to [8] or [9] or [15], respectively, where one can find further references.

The proofs of the following three assertions can be found in [13].

Theorem 1. *Let $\mathscr{C}$ be a productive class of spaces containing all the discrete spaces of cardinalities α_ζ, $\zeta < \xi$ for a given ordinal $\xi > 0$. If $\mathscr{C}$ is preserved by* *perfect maps onto $\mathscr{C}$-regular spaces, then $\mathscr{C}$ is not a part of $\mathscr{K}(E)$ for any space E of cardinality smaller than α_ξ.*

Corollary. *Let $\mathscr{C}$ be a productive and closed-hereditary class of spaces containing a space which is not countably compact. If $\mathscr{C}$ is preserved by perfect maps onto $\mathscr{C}$-regular spaces, then $\mathscr{C}$ is a part of no $\mathscr{K}(E)$ for any space E of nonmeasurable cardinality.*

Corollary. *If E is a space of nonmeasurable cardinality which is not countably compact, then the class $\mathscr{K}(E)$ is not preserved by perfect maps onto E-regular spaces.*

In the assertions stated above, the perfect maps (i.e. closed continuous maps with compact preimages of points) may be replaced by at most 2 to 1 closed continuous maps (at most 2 to 1 means that preimages of points contain at most two points).

The proof of Theorem 1 is based on the proof of a theorem in [5]; in fact, for nonmeasurable cardinals, Theorem 1 is a reformulation of a theorem from [5] without requiring regularity of certain cardinals.

It is clear that all the productive and closed-hereditary classes $\mathscr{C}$ of compact spaces are preserved by perfect maps onto $\mathscr{C}$-regular spaces without any regard to the simplicity of $\mathscr{C}$. So there remains a problem what is the situation if $\mathscr{C}$ is a productive and closed-hereditary class of countably compact spaces containing a space which is not compact. We believe the solution of this problem can be then generalized to other relatively measurable cardinals. Another problem is to find a general way how to find generators for simple classes; in most cases (e.g. [11], [12]) they were found by chance.

The class $\mathscr{S}$ of all spaces having a complete uniformity is productive and closed-hereditary. Is this class simple? Realize that $\mathscr{S}$ contains the class of all compact spaces which is somehow connected with the first relatively measurable cardinal $\alpha_0 = \omega_0$, and $\mathscr{S}$ also contains the bigger class of all realcompact spaces which is connected with α_1. Defining other bigger and bigger classes which are connected with other relatively measurable cardinals we obtain a monotone transfinite sequence of classes the union of which is the $\mathscr{S}$. Now it is sufficient to prove that each such class is simple and we receive the following

Theorem 2. *The class $\mathscr{S}$ of all spaces having complete uniformities is simple if and only if the class of all relatively measurable cardinals is a set.*

For the sake of completeness we shall describe here generators of the classes mentioned above. All the details and proofs may be found in [12]. Denote by $S(\alpha)$ the metrizable hedgehog with α prickles, i.e. α copies of the closed unit interval $[0, 1]$ sewed together in the point 0; denote $S_0 = S(1) = [0, 1]$ and for an ordinal

$\xi > 0$, $S_\xi = \Pi\{S(\alpha_\zeta) \mid \zeta < \xi\}$, where α_ζ are relatively measurable cardinals (see the paragraph before Theorem 1) – we need not take here all the ζ's smaller than ξ but only a cofinal part so that we may also define $S_{\xi+1} = S(\alpha_\xi)$. The covering character of a uniformity (see [14]) is the first cardinal α such that the uniform space contains no uniformly discrete subspaces of cardinality α; a space is said to be pseudo-α-compact if the covering character of its fine uniformity is at most α – [3], [14].

The following theorem may be regarded as a generalization of a Shirota's theorem from [18].

Theorem 3. *Let P be a space and* $\alpha_\zeta \in \{\alpha_\xi\}$. *Then the following conditions* (1)–(4) *are equivalent to each other; if* $\zeta > 0$ *then* (5) *is equivalent to* (1)–(4).

(1) *P is* S_ζ*-compact.*

(2) *P has a complete uniformity and no closed discrete (or closed discrete* C^**-embedded or closed discrete C-embedded, respectively) subspace of cardinality* α_ζ.

(3) *P has a complete uniformity and no uniformly discrete subspace of cardinality* α_ζ *(i.e., P has a complete uniformity and is pseudo-*α_ζ*-compact).*

(4) *P has a complete uniformity with covering character at most* α_ζ.

(5) *P has a complete uniformity with covering character at most* $\sup\{\alpha_\eta^+ \mid \eta<\zeta\}$.

We restate Theorem 3 once more for the special case $\zeta = 1$ with additional characterizations by means of Herrlich's α-compact spaces [7] and of van der Slot's α-ultracompact spaces [19] (for the definitions see parts (g), (h), (i) of the next theorem). For $\zeta = 0$ we would obtain well-known characterizations of compact spaces.

Theorem 4. *For a space P, the following conditions are equivalent:*

(a) *P is realcompact.*

(b) *P can be embedded as a closed subspace into a power of the metrizable hedgehog* $S(\omega_0)$.

(c) *P has a complete uniformity and each of its closed discrete (or closed discrete* C^**-embedded or closed discrete C-embedded, respectively) subspaces is of nonmeasurable cardinality.*

(d) *P has a complete uniformity and is pseudo-*α_1*-compact.*

(e) *P has a complete uniformity with covering character at most* α_1.

(f) *P has a complete uniformity with covering character at most* ω_1.

(g) *P has a complete uniformity and is* α_1*-compact (i.e., each maximal filter of zero-sets in P with nonmeasurable intersection property is fixed).*

(h) *P has a complete uniformity and is ω_1-ultracompact (i.e., each ultrafilter with countable intersection property for closed sets converges).*

(i) *P has a complete uniformity and is α_1-ultracompact (i.e., each ultrafilter with nonmeasurable intersection property for closed sets converges).*

Using the foregoing assertions for the classes $\mathscr{C}_\alpha$ or $\mathscr{U}_\alpha$ of all α-compact or α-ultracompact spaces, respectively, we obtain the following interesting results:

(α) If α is not a relatively measurable cardinal, then $\mathscr{C}_\alpha$ is not preserved by perfect maps.

(β) If α is not a relatively measurable cardinal, then $\mathscr{C}_\alpha$ is a proper subclass of $\mathscr{U}_\alpha$ (a partial solution of Problem 8 from [10]).

(γ) For any uncountable cardinal α and any nonmeasurable cardinal β there is an α-ultracompact zerodimensional space which is not β-compact.

(δ) α-compactness and β-ultracompactness coincide on spaces with complete uniformities for α, β having the same first greater relatively measurable cardinal.

There remains an open problem what is the connection between α-compactness and α-ultracompactness for uncountable relatively measurable α; under the condition S posed on relatively measurable cardinals in [1] these concepts coincide so that the problem seems to be in a close connection with properties of relatively measurable cardinals. Any way, for every α, α-ultracompact spaces are just uniformizable perfect images of α-compact spaces (see [4]).

References

[1] *W. W. Comfort and S. Negrepontis:* Some topological properties associated with measurable cardinals. Fund. Math. *69* (1970), 191—205.

[2] *R. Engelking and S. Mrówka:* On *E*-compact spaces. Bull. Acad. Polon. Sci. Sér. Sci. Math. Astronom. Phys. *6* (1958), 429—436.

[3] *Z. Frolík:* Generalizations of compact and Lindelöf spaces. Czechoslovak Math. J. *9* (1959), 172—217.

[4] *Z. Frolík:* Prime filters with countable intersection property. (To appear.)

[5] *Z. Frolík and S. Mrówka:* Perfect images of *R*- and *N*-compact spaces. Bull. Acad. Polon. Sci. Sér. Sci. Math. Astronom. Phys. *19* (1971), 369—371.

[6] *L. Gillman and M. Jerison:* Rings of continuous functions. Van Nostrand, Princeton, 1960.

[7] *H. Herrlich:* Fortsetzbarkeit stetiger Abbildungen und Kompaktheitsgrad topologischer Räume. Math. Z. *96* (1967), 64—72.

[8] *H. Herrlich:* $\mathscr{E}$-kompakte Räume. Math. Z. *96* (1967), 228—255.

[9] *H. Herlich:* Topologische Reflexionen und Coreflexionen. Lecture Notes *78*, Springer, Berlin, 1968.

[10] *H. Herrlich:* Categorical topology. General Topology and its Applications *1* (1971), 1—15.

[11] *M. Hušek:* The class of *k*-compact spaces is simple. Math. Z. *110* (1969), 123—126.

[12] *M. Hušek:* Topological spaces with complete uniformities. Reports of Math. Centrum at Amsterdam *3* (1971), 16 pp.

[13] *M. Hušek:* Perfect images of *E*-compact spaces. Bull. Acad. Polon. Sci. Sér. Sci. Math. Astronom. Phys. *20* (1) (1972), 41—45.

[14] *J. R. Isbell:* Uniform spaces. Math. Surveys *12*, Amer. Math. Soc., Providence, R. I., 1964.

[15] *H. J. Keisler and A. Tarski:* From accessible to inaccessible cardinals. Fund. Math. *53* (1964), 225—308.

[16] *J. F. Kennison:* Reflective functors in general topology and elsewhere. Trans. Amer. Math. Soc. *118* (1965), 303—315.

[17] *S. Mrówka:* A property of Hewitt extension vX of topological spaces. Bull. Acad. Polon. Sci. Sér. Sci. Math. Astronom. Phys. *6* (1958), 95—96.

[18] *T. Shirota:* A class of topological spaces. Osaka J. Math. *4* (1952), 23—40.

[19] *J. van der Slot:* Some properties related to compactness. Thesis, Math. Centrum, Amsterdam, 1968.

CHARLES UNIVERSITY, PRAHA

CONTINUOUS MAPPINGS OF EXTENSIONS OF A TOPOLOGICAL SPACE

V. M. IVANOVA and A. A. IVANOV

Leningrad

1. An extension X' of a topological space X is said to be a true extension of X if the system $\{\bar{F}^{X'} \mid F \subset X\}$ is a basis of closed sets of the space X' and for any point $x \in X' \setminus X$ the set $\{x\}$ is a closed set of X'. Here we shall consider only true compact extensions of topological spaces. The statements will be formulated in terms of contiguity relations, the latter being a relation between elements of finite systems of closed sets of X with the following properties:

C 1: If every element of a system α contains a certain element of a system β and $\sigma(\beta)$ takes place (σ holds for the system β) then $\sigma(\alpha)$ takes place.

C 2: If $\sigma(\alpha \vee \beta)$ takes place then either $\sigma(\alpha)$ or $\sigma(\beta)$ or both take place ($\alpha \vee \beta = \{A \cup B \mid A \in \alpha, B \in \beta\}$).

C 3: If a system α contains the empty set then $\sigma(\alpha)$ does not take place ($\bar{\sigma}(\alpha)$ takes place).

C 4: If the intersection of all elements of a system α is not empty then $\sigma(\alpha)$ takes place.

If a contiguity relation σ holds for every finite subsystem of a system γ of closed sets of X, the system γ being finite or infinite, then the system γ is said to be a σ-contiguity system. One can consider maximal σ-contiguity systems and distinguish among them the disappearing systems, that is the systems with the empty intersection.

Let $\hat{X}$ be the set of all maximal disappearing σ-contiguity systems and we put $\hat{F} = \{\gamma \mid F \in \gamma, \gamma \in \hat{X}\}$ for every closed set F of X. Now denote $X \cup \hat{X}$ by σX and consider the topological structure on the set σX induced by the closed basis $\{F \cup \hat{F}\}$. The topological space σX is found to be a true compact extension of the space X and, moreover, every true compact extension of the space X is equivalent to an extension σX for a proper contiguity relation σ on X. That is why in the theory of true compact extensions one may consider only such extensions σX. At this point we should like to note [1] that for any finite system α of closed sets of X, $\sigma(\alpha)$ takes place iff $\bigcap_{F \in \alpha} \bar{F}^{\sigma X} \neq \emptyset$. This property defines σ-extensions up to the extension equivalence.

We shall say that a contiguity relation σ is weaker than a contiguity relation σ' and write $\sigma \leqq \sigma'$ if any σ'-contiguity system is a σ-contiguity system. If $f : \sigma'X \to \sigma X$ is a natural continuous mapping then $\sigma \leqq \sigma'$. Thus the condition $\sigma \leqq \sigma'$ is a necessary condition for some natural mapping $f : \sigma'X \to \sigma X$ to exist. Moreover, if σX

and $\sigma'X$ are Hausdorff extensions then this condition is sufficient but in general it is not so. The subject of our paper is to find conditions for the existence of such natural mappings.

2. It is possible to establish some sufficient conditions for the existence of a natural mapping $f : \sigma'X \to \sigma X$. But we need some additional notions for this purpose.

A system $H_1, H_2, \ldots, H_q$ of closed sets of X is said to be a σ-majorant of a system $F_1, F_2, \ldots, F_p$ if, given any system $F'_1, F'_2, \ldots, F'_n$ of closed sets of X, $\sigma(F'_1, \ldots, F'_n, F_1, \ldots, F_p)$ implies $\sigma(F'_1, \ldots, F'_n, H_1, \ldots, H_q)$.

We shall say that a contiguity relation σ' satisfies the condition C_σ if $\sigma'(F_1, F_2, \ldots, F_n)$ and $\bar{\sigma}(F, F_1, \ldots, F_n)$ imply the existence of a σ'-majorant H of the system $F_1, F_2, \ldots, F_n$ such that $\sigma(F, H)$ does not take place.

We shall say that a contiguity relation σ satisfies the condition $C^0_{\sigma'\sigma}$ if $\bar{\sigma}(F_1, F_2, \ldots, F_n)$ implies the existence of a system $H_1, H_2, \ldots, H_n$ such that $F_i \subset H_i$, $\bar{\sigma}'(H_1, H_2, \ldots, H_n)$ and $\bar{\sigma}'(H_1, \ldots, H_{j-1}, F, H_{j+1}, \ldots, H_n)$ imply $\bar{\sigma}(F_1, \ldots, F_{j-1}, F, F_{j+1}, \ldots, F_n)$ for any closed set F.

Theorem 1. *Let a contiguity relation σ' on X satisfy the conditions $C_{\sigma'}$ and C_σ and a contiguity relation σ satisfy the condition $C^0_{\sigma'\sigma}$. Then there exists a natural continuous mapping $f : \sigma'X \to \sigma X$.*

To prove this theorem, we show, at first, some auxiliary statements assuming that the conditions of Theorem 1 are fulfilled.

Lemma 1.1. *Let γ' be a maximal σ'-contiguity system and $F \bar{\in} \gamma'$. Then there exists $H \in \gamma'$ such that $\sigma'(H, F)$ does not take place.*

Proof. Let $F \bar{\in} \gamma'$. Then there exist $F_1, F_2, \ldots, F_n$ belonging to γ' such that $\sigma'(F, F_1, \ldots, F_n)$ does not take place and therefore according to $C_{\sigma'}$ there exists a σ'-majorant H of the system $F_1, F_2, \ldots, F_n$ such that $\sigma'(H, F)$ does not take place. Since $H \in \gamma'$, Lemma 1.1 is proved.

Lemma 1.2. *Let γ' be a maximal σ'-contiguity system, $\gamma = \{F \mid \sigma(F, F'_1, \ldots, F'_m)$ for any $F'_1, F'_2, \ldots, F'_m$ belonging to $\gamma'\}$. Then γ is the only maximal σ-contiguity system containing γ'.*

Proof. It is enough to show that γ is a σ-contiguity system since any σ-contiguity system containing the system γ' is within γ. For this it is necessary to show that for any $F'_1, F'_2, \ldots, F'_m$ belonging to γ', $\sigma(F_i, F'_1, \ldots, F'_m)$, $i = 1, 2, \ldots, n$, implies $\sigma(F_1, \ldots, F_n, F'_1, \ldots, F'_m)$.

For $n = 1$ this statement is trivial. Assuming the statement to be true for n let us prove it for $n + 1$.

Let $F_1, F_2, \ldots, F_{n+1}$ be closed sets such that $\sigma(F_i, F'_1, \ldots, F'_m)$ takes place for any $F'_1, F'_2, \ldots, F'_m$ belonging to γ', $i = 1, 2, \ldots, n + 1$, and $\sigma(F_1, \ldots, F_{n+1}, F'_1, \ldots, F'_m)$ does not take place for some $F'_1, F'_2, \ldots, F'_m$. Then there exists $H_1, \ldots, H_{n+1}, H'_1, \ldots$

..., H'_m with the properties ensured by the condition $C^0_{\sigma'\sigma}$. In particular, $\sigma'(H_1, \ldots, H_{n+1}, H'_1, \ldots, H'_m)$ does not take place. Then one of the sets $H_1, H_2, \ldots, H_{n+1}$ does not belong to γ'. Let H_1 be such a set. By Lemma 1.1 there exists $F'_0 \in \gamma'$ such that $\sigma'(H_1, F'_0)$ does not take place and therefore $\sigma'(H_1, \ldots, H_n, F'_0, H'_1, \ldots, H'_m)$ does not take place and due to $C^0_{\sigma'\sigma}$, $\sigma(F_1, \ldots, F_n, F'_0, F'_1, \ldots, F'_m)$ does not take place. The contradiction proves Lemma 1.2.

Lemma 1.3. *Let γ' be a maximal σ'-contiguity system, $\gamma = \{F \mid \sigma(F, F')$ for any $F' \in \gamma'\}$. Then γ is the only maximal σ-contiguity system containing γ'.*

Proof. Using Lemma 1.2 it is enough to show that if $\sigma(F, F')$ takes place for any $F' \in \gamma'$ and a set F then $\sigma(F, F'_1, \ldots, F'_m)$ takes place for F and any $F'_1, F'_2, \ldots, F'_m$ belonging to γ'. Suppose that it is not the case, that is, $\sigma(F, F'_1, \ldots, F'_m)$ does not take place for some $F'_1, F'_2, \ldots, F'_m$ belonging to γ'. Then due to C_σ there exists a σ'-majorant H of the system $F'_1, F'_2, \ldots, F'_m$ such that $\sigma(F, H)$ does not take place. Since $H \in \gamma'$, the contradiction proves Lemma 1.3.

Now we prove Theorem 1. Let $f : \sigma'X \to \sigma X$ with $f(x) = x$ for $x \in X$, $f(\gamma') = \gamma$ for $\gamma' \subset \gamma$ and $\gamma' \in \sigma'X \setminus X$. By Lemma 1.2 the mapping f is defined uniquely and it is enough to show f to be continuous, that is, to show that for any closed set F_0 of X the set $f^{-1}(\bar{F}_0^{\sigma X})$ is a closed set of $\sigma'X$.

Let $\gamma'_0 \bar{\in} f^{-1}(\bar{F}_0^{\sigma X})$, that is, $\gamma'_0 \subset \gamma_0 \bar{\in} \bar{F}_0^{\sigma X}$. Then $F_0 \bar{\in} \gamma_0$ and by Lemma 1.3 there exists $F_1 \in \gamma'_0$ such that $\sigma(F_0, F_1)$ does not take place. Thus due to $C^0_{\sigma'\sigma}$, there exist H_0, H_1 with the properties ensured by this condition. In particular, $F_0 \subset H_0$, $F_1 \subset H_1$, $\sigma'(H_0, H_1)$ does not take place. Since $F_1 \in \gamma'_0$, $F_1 \subset H_1$ so $H_1 \in \gamma'_0$ and therefore $H_0 \bar{\in} \gamma'_0$, that is, $\gamma'_0 \bar{\in} \bar{H}_0^{\sigma'X}$. On the other hand, if a maximal σ'-contiguity system γ' does not belong to $\bar{H}_0^{\sigma'X}$ then $H_0 \bar{\in} \gamma'$ and by Lemma 1.1. there exists $F \in \gamma'$ such that $\sigma'(H_0, F)$ does not take place. Thus, taking into account the choice of H_0, we see that $\sigma(H_0, F)$ does not take place, therefore $\sigma(F_0, F)$ does not take place. It means that for any $\gamma'_0 \bar{\in} f^{-1}(\bar{F}_0^{\sigma X})$ there exists a closed set $\bar{H}_0^{\sigma'X}$ of $\sigma'X$ containing $f^{-1}(\bar{F}_0^{\sigma X})$ but not containing γ'_0, which again means $f^{-1}(\bar{F}^{\sigma X})$ is closed and Theorem 1 is proved.

3. Consider now another system of conditions for contiguity relations σ', σ ensuring not only the existence of a natural continuous mapping $f : \sigma'X \to \sigma X$ but also some properties of such a mapping.

Denote the set of all subsets of a set X by PX ($PX = \{A \mid A \subset X\}$), according to which we have $P^2X = \{A \mid A \subset PX\}$. Consider a mapping $\mathscr{A} : PX \to P^2X$ associating with every set $A \subset X$ some system $\mathscr{A}(A)$ of subsets of X. We call the mapping $\mathscr{A} : PX \to P^2X$ majorizing if the following conditions are fulfilled

A_1: If $F' \in \mathscr{A}(F)$ then $F \subset F'$.

A_2: If $F' \subset F''$, $F' \in \mathscr{A}(F)$ then $F'' \in \mathscr{A}(F)$.

A_3: $\bigcap_{F' \in \mathscr{A}(F)} F' = F$.

A majorizing $\mathscr{A}$ is a $\sigma'\sigma$-majorizing if $\sigma(F_1, F_2, \ldots, F_n)$ takes place iff for any $\mathscr{A}$-majorants $F'_1, F'_2, \ldots, F'_n$ of sets $F_1, F_2, \ldots, F_n$, $\sigma'(F'_1, F'_2, \ldots, F'_n)$ takes place.

We shall say that a $\sigma'\sigma$-majorizing $\mathscr{A}$ satisfies the condition $C\mathscr{A}$ if for any $\mathscr{A}$-majorant F' of a set F and any H, $\bar{\sigma}'(H, F')$ implies $\bar{\sigma}(H, F)$.

Theorem 2. *Let a contiguity relation σ' on X satisfy the condition $C_{\sigma'}$ and let there exist a $\sigma'\sigma$-majorizing $\mathscr{A}$ satisfying the condition $C\mathscr{A}$. Then there exists a natural continuous mapping $f : \sigma'X \to \sigma X$ (onto σX).*

To prove this theorem we show at first some auxiliary statements assuming that the conditions of Theorem 2 are fulfilled.

Lemma 2.1. *Let γ' be a maximal σ'-contiguity system. If $F \bar{\in} \gamma'$ then there exists $F' \in \gamma'$ such that $\sigma'(F', F)$ does not take place.*

Proof. The proof is quite analogous to that of Lemma 1.1.

Lemma 2.2. *Let γ' be a maximal σ'-contiguity system and $\gamma = \{F \mid \sigma(F, F')$ for any $F' \in \gamma'\}$. Then γ is the only maximal σ-contiguity system containing γ'.*

Proof. It is sufficient to show that γ is a σ-contiguity system. If $\sigma(F_1, F_2, \ldots, F_n)$ does not take place then $\sigma'(H_1, H_2, \ldots, H_n)$ does not take place for some $\mathscr{A}$-majorants of the sets $F_1, F_2, \ldots, F_n$. Then at least one of $H_1, H_2, \ldots, H_n$ does not belong to γ', for example H_1. Now note that by Lemma 2.1 there exists $F' \in \gamma'$ such that $\sigma'(H_1, F')$ does not take place. Therefore $\sigma(F_1, F')$ does not take place and $F_1 \bar{\in} \gamma$. Thus $F_i \in \gamma$, $i = 1, 2, \ldots, n$, implies $\sigma(F_1, F_2, \ldots, F_n)$. This proves Lemma 2.2.

Lemma 2.3. *Let γ be a maximal disappearing σ-contiguity system. Then there exists a maximal disappearing σ'-contiguity system γ' such that $\gamma' \subset \gamma$.*

Proof. Consider the system $\mathscr{A}(\gamma) = \{F' \mid F' \in \mathscr{A}(F), F \in \gamma\}$ of closed sets of X. If $F'_1, F'_2, \ldots, F'_n$ belong to $\mathscr{A}(\gamma)$ then these sets are the $\mathscr{A}$-majorants of some $F_1, F_2, \ldots, F_n$ belonging to γ and as $\sigma(F_1, F_2, \ldots, F_n)$ takes place, $\sigma'(F'_1, F'_2, \ldots, F'_n)$ takes place, too. Thus the system $\mathscr{A}(\gamma)$ is the σ'-contiguity system, $\mathscr{A}(\gamma)$ being a disappearing system due to A_3. Let γ' be a maximal σ'-contiguity system containing $\mathscr{A}(\gamma)$, $F' \in \gamma'$, $F_1, F_2, \ldots, F_n$ belong to γ and $H', H_1, \ldots, H_n$ be $\mathscr{A}$-majorants of sets $F', F_1, \ldots, F_n$, respectively. Since $H_1, H_2, \ldots, H_n$ belong to $\mathscr{A}(\gamma)$, $\sigma'(F', H_1, \ldots, H_n)$ takes place and all the more $\sigma'(H', H_1, \ldots, H_n)$ takes place. It implies $\sigma(F', F_1, \ldots, F_n)$ by arbitrary choice of $\mathscr{A}$-majorants. This shows that $F' \in \gamma'$ implies $F' \in \gamma$, which proves Lemma 2.3.

Now we prove Theorem 2. The mapping $f : \sigma'X \to \sigma X$ is defined in the usual way, f being the mapping of $\sigma'X$ onto σX by Lemma 2.3. The continuity of this mapping is proved in the same way as in Theorem 1, the only difference being in that $C\mathscr{A}$ is used instead of $C^0_{\sigma'\sigma}$.

4. If $f : \sigma'X \to \sigma X$ is a natural continuous mapping, F is a closed set of X then in a general case $f(\bar{F}^{\sigma'X}) \neq \bar{F}^{\sigma X}$ even if the conditions of Theorem 2 hold. The following example shows this:

Let $X = A \cup B$, $A = \{(0, y) \mid 1 < y \leqq 2\}$, $B = \{(x, 0) \mid 1 < x \leqq 2\}$. The sets A, B are both open and closed subsets of X, the topological structure on B is standard, the topological structure on A is induced by the basis of closed sets consisting of finite unions of sets of the form $A_a = \{(0, y) \mid 1 < y \leqq a\}$ and finite sets. Define the contiguity relation σ on X setting $\sigma(F_1, F_2, \ldots, F_n)$ to take place iff at least one of the following conditions is fulfilled

1. $\bigcap_{i=1}^{n} F_i \neq \emptyset$.

2. $F_i \supset A$, $i = 1, 2, \ldots, m$, and with $i > m$ the point $(1, 0)$ is a limit point of $\bigcap F_i$ in the usual topology.

The majorizing $\mathscr{A}$ is defined as follows:

$\mathscr{A}(F) = \{F' \mid F \subset F'\}$ for $F \subset A$, $F \neq A$;
$\mathscr{A}(F) = \{F' \mid F' \cap A \neq \emptyset\}$ for $F \subset B$, $(1, 0)$ is the limit point of F;
$\mathscr{A}(A) = \{A \cup B_a\}$, $B_a \supset \{(x, 0) \mid 1 < x \leqq a\}$;
$\mathscr{A}(F) = \{F' \mid F \subset F'\}$ for $F \subset B$, $(1, 0)$ is not the limit point of F.

It is easy to see that the majorizing $\mathscr{A}$ is an $\omega\sigma$-majorizing ($C\mathscr{A}$ being satisfied) but $f(\bar{A}^{\omega X}) \neq \bar{A}^{\sigma X}$.

In connection with the above discussion consider a system of conditions for contiguity relations σ', σ (stronger than in 3) namely instead of the condition $C\mathscr{A}$ consider the following condition

$C\mathscr{A}^*$: For any $\mathscr{A}$-majorants $F_1', F_2', \ldots, F_n'$ of sets $F_1, F_2, \ldots, F_n$ and any set F, $\bar{\sigma}'(F, F_1', \ldots, F_n')$ implies $\bar{\sigma}(F, F_1, \ldots, F_n)$.

Theorem 3. *Let a contiguity relation σ' on X satisfy the condition $C_{\sigma'}$ and let there exist a $\sigma'\sigma$-majorizing $\mathscr{A}$ which satisfies the condition $C\mathscr{A}^*$. Then there exists a natural continuous mapping $f : \sigma'X \to \sigma X$ for which $f(\bar{F}^{\sigma'X}) = \bar{F}^{\sigma X}$ for any $F \subset X$.*

Proof. The existence and continuity of the natural mapping $f : \sigma'X \to \sigma X$ results from Theorem 2. Thus, the only thing we need is to prove the equality $f(\bar{F}^{\sigma'X}) = \bar{F}^{\sigma X}$ for any $F \subset X$. Let $\gamma \in \bar{F}^{\sigma X}$, $F \in \gamma$, $\mathscr{A}(\gamma)$ be the system consisting of $\mathscr{A}$-majorants of the sets belonging to γ. Then the system $\{F\} \cup \mathscr{A}(\gamma)$ is the σ'-contiguity system due to $C\mathscr{A}^*$ and is a part of a maximal σ'-contiguity system γ'. But it was noticed in the proof of Lemma 2.3 that $\gamma' \subset \gamma$, $F \in \gamma'$. Therefore $\gamma' \in \bar{F}^{\sigma'X}$, $\gamma = f(\gamma')$, $\gamma \in f(\bar{F}^{\sigma'X})$. Hence $\bar{F}^{\sigma X} \subset f(\bar{F}^{\sigma'X})$. Since the inverse inclusion follows from the continuity of f, Theorem 3 is proved.

We shall say that a contiguity relation σ satisfies the condition C_σ^* if $\sigma'(F_1, F_2, \ldots, F_n)$ and $\bar{\sigma}(F_1', \ldots, F_m', F_1, \ldots, F_n)$ imply the existence of a σ'-majorant H of the system $F_1, F_2, \ldots, F_n$ such that $\sigma(F_1', \ldots, F_m', H)$ does not take place.

Theorem 4. *Let a contiguity relation σ on X satisfy the condition $C_{\sigma'}^*$ and let there exist a natural continuous mapping $f : \sigma'X \to \sigma X$ such that $f(\bar{F}^{\sigma'X}) = \bar{F}^{\sigma X}$*

for any $F \subset X$. Then there exists a $\sigma'\sigma$-majorizing $\mathscr{A}$ which satisfies the condition $C\mathscr{A}^*$.

Proof. Put $\mathscr{A}(F) = \{F' \mid F \subset F', f^{-1}(\bar{F}^{\sigma X}) \subset \bar{F}'^{\sigma' X}\}$ for any $F \subset X$. It is easy to show $F \to \mathscr{A}(F)$ to be majorizing. Let $F'_1, F'_2, \ldots, F'_n$ be $\mathscr{A}$-majorants of sets $F_1, F_2, \ldots, F_n$. If $\sigma(F_1, F_2, \ldots, F_n)$ takes place, that is $\bigcap_{i=1}^{n} \bar{F}_i^{\sigma X} \neq \emptyset$, then $\bigcap_{i=1}^{n} \bar{F}_i'^{\sigma' X} \supset \bigcap_{i=1}^{n} f^{-1}(\bar{F}_i^{\sigma X}) \neq \emptyset$. Thus $\sigma'(F'_1, F'_2, \ldots, F'_n)$ takes place. Now let $\sigma'(F'_1, F'_2, \ldots, F'_n)$ take place for any $\mathscr{A}$-majorants of sets $F_1, F_2, \ldots, F_n$, correspondingly. Consider a system $\mathscr{A}(F_1, F_2, \ldots, F_n)$ of $\mathscr{A}$-majorants of sets $F_1, F_2, \ldots, F_n$. It is easy to show by $C^*_{\sigma'}$ that $\mathscr{A}(F_1, F_2, \ldots, F_n)$ is a σ'-contiguity system, thus the intersection of closures of its elements in $\sigma' X$ is not empty. This intersection is equal to $\bigcap_{i=1}^{n} f^{-1}(\bar{F}_i^{\sigma X})$, therefore $\bigcap_{i=1}^{n} \bar{F}_i^{\sigma X} \neq \emptyset$, that is, $\sigma(F_1, F_2, \ldots, F_n)$ takes place. Thus, $\mathscr{A}$ is a $\sigma'\sigma$-majorizing.

Let $F_1, F_2, \ldots, F_n$ be closed sets of X, $H_1, H_2, \ldots, H_n$ their $\mathscr{A}$-majorants, $\sigma'(F, H_1, \ldots, H_n)$ not taking place, that is $\bar{F}^{\sigma' X} \cap (\bigcap_{i=1}^{n} \bar{H}_i^{\sigma' X}) = \emptyset$. Then $\bar{F}^{\sigma' X} \cap f^{-1}(\bigcap_{i=1}^{n} \bar{F}_i^{\sigma X}) = \emptyset$ and, finally, as $f(\bar{F}^{\sigma' X}) = \bar{F}^{\sigma X}$, we obtain $\bar{F}^{\sigma X} \cap (\bigcap_{i=1}^{n} \bar{F}_i^{\sigma X}) = \emptyset$, that is, $\sigma(F, F_1, \ldots, F_n)$ does not take place. This proves $C\mathscr{A}^*$ and Theorem 4, as well.

Corollary 1. *Let a contiguity relation σ' on X satisfy the condition $C^*_{\sigma'}$. Then there exists a natural continuous mapping $f : \sigma' X \to \sigma X$ with $f(\bar{F}^{\sigma' X}) = \bar{F}^{\sigma X}$ for every $F \subset X$ iff there exists a $\sigma'\sigma$-majorizing $\mathscr{A}$ satisfying the condition $C\mathscr{A}^*$.*

This statement is simplified in the case of $\sigma' = \omega$, that is the contiguity relation under intersection ($\omega(F_1, F_2, \ldots, F_n)$ takes place iff $\bigcap_{i=1}^{n} F_i \neq \emptyset$).

Corollary 2. *There exists a natural continuous mapping $f : \omega X \to \sigma X$ with $f(\bar{F}^{\omega X}) = \bar{F}^{\sigma X}$ for every $F \subset X$ iff there exists an $\omega\sigma$-majorizing $\mathscr{A}$ satisfying the condition $C\mathscr{A}^*$.*

This statement gives us a description of the class, say $\mathscr{A}^*$, of natural continuous images of Wallman extension ωX which contains the so-called $\omega\alpha$-extensions [2]. Note that for any $\omega\alpha$-extension X' of a space X there exists a natural both closed and continuous mapping $f : \omega X \to X'$, while the corresponding natural mapping for any extension of the class $\mathscr{A}^*$ satisfies a weaker condition.

References

[1] *В. М. Иванова и А. А. Иванов:* Пространства смежности и бикомпактные расширения топологических пространств. Изв. АН СССР сер. матем. *23* (4) (1959), 613—634.

[2] *П. К. Осматеску:* $\omega\alpha$-расширения. Вестник ЛГУ. сер. VI, математика, механика (6) (1963), 45—54.

DISCONNECTED BOUNDED *PL* MANIFOLDS IN EUCLIDEAN SPACES

I. IVANŠIĆ
Zagreb

In this paper we consider the following problem: Under what conditions one can extend a given embedding $g: \partial M \to E^q$ to an embedidng of the whole M, where M is disconnected bounded PL (piecewise linear) manifold of dimension n and all embeddings are assumed to be piecewise linear. Under a disconnected bounded PL manifold we understand a compact PL manifold such that each component of M has a nonempty boundary. We say that M unknots rel ∂M in E^q, if every two extensions are isotopic keeping the boundary fixed.

Proposition 1. *Every PL embedding* $g: \partial M \to E^{2n+1}$ *extends to a PL embedding of M. Furthermore, M unknots* rel ∂M *in* E^{2n+1}.

Proposition 2. *Every PL embedding* $g: \partial M \to E^{2n}$ *extends to a PL embedding of M. M in general knots* rel ∂M.

Denote by $M_1 \cup M_2 = M$ a disjoint union of two manifolds.

Theorem 1. *Let* $M = M_1 \cup M_2$ *be a closed orientable PL n-manifold in* E^{2n+1}, $n \geqq 2$. *Then the linking number* $L(M_1, M_2)$ *classifies, up to an ambient isotopy, the embeddings of M into* E^{2n+1}.

Theorem 2. *Let* $M = M_1 \cup M_2$ *be a compact bounded orientable PL n-manifold,* $n \geqq 3$, *such that each* ∂M_i $(i = 1, 2)$ *is nonempty and connected, and* $g: \partial M = \partial M_1 \cup \partial M_2 \to E^{2n-1}$ *is a PL embedding. Then g extends to a PL embedding of M if and only if the linking number* $L(g(\partial M_1), g(\partial M_2)) = 0$.

Theorem 3. *Let* $M = M_1 \cup M_2$ *be a compact bounded PL n-manifold. Assume that* M_1 *is* $(k+1)$*-connected and* M_2 *is k-connected,* $2k + 2 < n$, $k \leqq n - 4$. *Let* $g: \partial M \to E^{2n-k-1}$ *be a PL embedding. If the embeddings* $g_1 = g|\partial M_1 : \partial M_1 \to E^{2n-k-1} - g(\partial M_2)$ *and* $g_2 = g|\partial M_2 : \partial M_2 \to E^{2n-k-1} - g(\partial M_1)$ *are inessential (homotopic to a constant map), then g extends to a PL embedding of M.*

Using the same technique as in the proofs of the above statements one can prove results about unlinking in Euclidean spaces. We say that two polyhedra X

and Y in E^q are geometrically unlinked if there is a q-ball which contains one of them and does not intersect the other.

Proposition 3. *Let M_1 and M_2 be compact bounded PL n-manifolds in E^{2n}. Then M_1 and M_2 are geometrically unlinked.*

Theorem 4. *Let M_1 and M_2 be two compact closed PL n-manifolds in E^{2n+1}, $n \geqq 2$. M_1 and M_2 are geometrically unlinked in E^{2n+1} if and only if the given embedding $M_1 \cup M_2 \to E^{2n+1}$ extends to an embedding of a cone $C(M_1 \cup M_2) \to$ $\to E^{2n+1}$.*

TOPOLOGICAL REPRESENTATIONS OF MEASURABLE SPACES

J. E. JAYNE

Pisa

1. Measurable spaces

1.1. A measurable space is a pair $(X, \mathscr{B})$, where X is a set and $\mathscr{B}$ is a σ-algebra of subsets of X. By a topological representation of a measurable space $(X, \mathscr{B})$ we mean a completely regular Hausdorff topology τ on X such that $\mathscr{B}$ coincides with the family of Baire sets generated by τ (i.e. the smallest family of subsets of X containing the zero sets of τ-continuous real valued functions and closed under countable unions and countable intersections).

All proofs are omitted, and the references (which would accompany the proofs) are restricted. However, we point out that the work of Z. Frolík, much of which is surveyed in [1], is central to several of our demonstrations.

1.2. Notation. Let $(X, \mathscr{B})$ be a measurable space. We denote by $B(X, \mathscr{B})$ the space of measurable functions from $(X, \mathscr{B})$ into $(\mathbf{R}, \mathscr{B}_0)$, where $\mathbf{R}$ denotes the real line and $\mathscr{B}_0$ its family of Baire sets (equivalently, Borel sets). $B^*(X, \mathscr{B})$ denotes the subspace of bounded functions in $B(X, \mathscr{B})$.

If $(X, \mathscr{B})$ is topologically representable, then both $B(X, \mathscr{B})$ and $B^*(X, \mathscr{B})$ form lattice-ordered rings under the pointwise algebraic and lattice operations.

1.3. Theorem. *If $(X, \mathscr{B})$ and $(Y, \mathscr{F})$ both admit realcompact representations, then the following are equivalent:*

1) *$(X, \mathscr{B})$ is measurably isomorphic to $(Y, \mathscr{F})$,*
2) *$B(X, \mathscr{B})$ is ring isomorphic to $B(Y, \mathscr{F})$,*
3) *$B(X, \mathscr{B})$ is lattice isomorphic to $B(Y, \mathscr{F})$,*
4) *$B(X, \mathscr{B})$ is multiplicative semigroup isomorphic to $B(Y, \mathscr{F})$.*

This theorem along with the fact that every Baire function on a completely regular space X has a unique extension to the Hewitt realcompactification of X (due to P. R. Meyer – 1961, unpublished – in the more precise form that a Baire function of class α extends uniquely to one of class α on the realcompactification) allows us to apply the argument given in [4] to obtain the following:

1.4. Theorem. *If $(X, \mathscr{B})$ and $(Y, \mathscr{F})$ are topologically representable spaces, then the following are equivalent:*

1) *$B(X, \mathscr{B})$ is ring isomorphic to $B(Y, \mathscr{F})$,*
2) *$B(X, \mathscr{B})$ is lattice isomorphic to $B(Y, \mathscr{F})$,*
3) *$B(X, \mathscr{B})$ is multiplicative semigroup isomorphic to $B(Y, \mathscr{F})$.*

1.5. Theorem. *If $(X, \mathscr{B})$ and $(Y, \mathscr{F})$ are topologically representable and have the property that for all $x \in X$ and $y \in Y$ we have $\{x\} \in \mathscr{B}$ and $\{y\} \in \mathscr{F}$, then the following are equivalent:*

1) *$(X, \mathscr{B})$ is measurably isomorphic to $(Y, \mathscr{F})$,*

2), 3, 4) *$B(X, \mathscr{B})$ is ring (lattice, multiplicative semigroup) isomorphic to $B(Y, \mathscr{F})$,*

5), 6), 7), 8) *$B^*(X, \mathscr{B})$ is ring (lattice, multiplicative semigroup, Banach space) isomorphic to $B^*(Y, \mathscr{F})$.*

1.6. The results of Theorems 1.3, 1.4, and 1.5 can be demonstrated for the spaces of Baire functions of class α for each countable ordinal α and explicit relations between the various automorphisms of these spaces and the measurable automorphisms of the underlying measurable space can be given (see [6]).

1.7. Let $(X, \mathscr{B})$ be a topologically representable space and let $\hat{X}$ denote the compactification of X (considering X with the weak topology generated by $B^*(X, \mathscr{B})$) determined by $B^*(X, \mathscr{B})$. The points of $\hat{X}$ are in one to one correspondence with the $\mathscr{B}$-ultrafilters on X, each point $p \in \hat{X}$ corresponding to the unique $\mathscr{B}$-ultrafilter U_p on X which converges to it [3]. The $\mathscr{B}$-ultrafilters are in one to one correspondence with the maximal ring ideals in $B(X, \mathscr{B})$ [6]. Thus for each point $p \in \hat{X}$ there corresponds a maximal ring ideal M_p in $B(X, \mathscr{B})$ defined by

$$M_p = \{f \in B(X, \mathscr{B}) : Z(f) \in U_p\},$$

where $Z(f) = \{x : f(x) = 0\}$.

We have the following analog of the Gelfand-Kolmogoroff theorem:

1.8. Theorem. *If $(X, \mathscr{B})$ is topologically representable, then each maximal ring ideal M_p $(p \in \hat{X})$ in $B(X, \mathscr{B})$ is of the form*

$$M_p = \{f \in B(X, \mathscr{B}) : p \in \operatorname{cl}_{\hat{X}} Z(f)\},$$

where $\operatorname{cl}_{\hat{X}} Z(f)$ denotes the closure in $\hat{X}$ of $Z(f)$.

In [6] this theorem is demonstrated for the rings of Baire functions of class α for every countable ordinal α, thus giving a generalization of the Gelfand-Kolmogoroff theorem.

1.9. Theorem. *If $(X, \mathscr{B})$ admits a realcompact representation, then the intersection of all free maximal ring ideals in $B(X, \mathscr{B})$ is the ideal of functions with finite support.*

1.10. Definition. A completely regular Hausdorff space X is called a *ZS-space* if it is obtainable from the zero sets of its Stone-Čech compactification βX by Suslin's operation (A), i.e. of the form

$$\bigcup_{\sigma \in \mathbf{N}^{\mathbf{N}}} \bigcap_{s < \sigma} Z(f_s),$$

where $\mathbf{N} = \{1, 2, \ldots\}$, $s < \sigma$ means that s is a restriction of σ, and the f_s are continuous real valued functions on βX.

The class of ZS-spaces will be denoted $\mathscr{S}$ and its subclass of spaces which are Baire subsets of their Stone-Čech compactifications by $\mathscr{Z}$ (in view of the fact that they are zero sets of Baire functions on βX).

Each space in $\mathscr{S}$ is Lindelöf and therefore realcompact. Thus the preceding theorems apply to representations in this class of spaces.

1.11. Theorem. *If $(X, \mathscr{B})$ admits a representation in $\mathscr{S}$, then the following are equivalent:*

1) *The Baire class hierarchy is not strictly increasing for all countable ordinals for some representation of $(X, \mathscr{B})$ in $\mathscr{S}$,*

2) *The Baire class hierarchy is not strictly increasing for all countable ordinals for every representation of $(X, \mathscr{B})$ in $\mathscr{S}$,*

3) *Every $B \in \mathscr{B}$ is multiplicative Baire class* 1 *for some representation of $(X, \mathscr{B})$ in $\mathscr{S}$,*

4) *Every $B \in \mathscr{B}$ is multiplicative Baire class* 1 *for every representation of $(X, \mathscr{B})$ in $\mathscr{S}$,*

5) *$\mathscr{B}$ is invariant under the operation* (A),

6) *Some representation of $(X, \mathscr{B})$ in $\mathscr{S}$ does not contain a compact perfect subset,*

7) *Every representation of $(X, \mathscr{B})$ in $\mathscr{S}$ does not contain a compact perfect subset,*

8) *$B(X, \mathscr{B})$ is the space of all real valued functions which are continuous in the weak topology generated on X by $B(X, \mathscr{B})$,*

9) *The weak topology generated on X by $B(X, \mathscr{B})$ has the Lindelöf property.*

1.12. Theorem. *If $(X, \mathscr{B})$ admits a metrizable representation in $\mathscr{S}$, then every representation of $(X, \mathscr{B})$ in $\mathscr{S}$ is metrizable. All of these representations are separable absolute analytic topologies (analytic subsets of their metric completions).*

2. Spaces of representations

2.1. We now turn to the analysis of spaces of representations of a fixed measurable space.

Let $(X, \mathscr{B})$ be a measurable space admitting a separable metrizable absolute Borel set representation. Let T denote the space of compact representations of $(X, \mathscr{B})$ (each of which must be metrizable by Theorem 1.12). E. R. Lorch has introduced a topology on T for which the automorphism group G of $(X, \mathscr{B})$ acts naturally as a homeomorphism group of T [8]. He and H. Tong have characterized the discrete points of this topology as the finite dimensional topologies in T and related this to the corresponding uniformities (mentioned above) on G [10].

A variant approach is to fix a class of topological spaces, introduce the equivalence relation of Baire isomorphism, and attempt to describe the resulting equivalence classes. In Lorch's setting above the orbits of the action of G on T form the single equivalence class in the class of all compact (uncountable) metrizable spaces. For the class of separable metrizable absolute Borel sets it is known that cardinality alone determines the Baire isomorphism classes [7]. For the class of all metrizable absolute Borel sets A. H. Stone has given a complete set of topological invariants of Baire isomorphism (see [13] and [14]).

For the class of analytic non-Borel subsets of separable completely metrizable spaces the problem is open. It is known that if there exists a coanalytic subset of the real line of power of the continuum which does not contain a homeomorph of the Cantor set, then there exist two analytic non-Borel subsets of separable completely metrizable spaces which are not Baire isomorphic [11]. The hypothesis of this result is consistent [2] and independent [12] of the axioms of Zermelo-Fraenkel set theory including the axiom of choice. It is implied by Gödel's axiom of constructibility [2] and its negation is implied by the existence of a measurable cardinal [12]. It is also known that any two universal analytic non-Borel subsets of the plane are Baire isomorphic [11].

Since analytic subsets of separable completely metrizable spaces belong to $\mathscr{S}$, this is part of the more extensive problem of describing all of the Baire isomorphism classes in $\mathscr{S}$. Toward this goal we give $\mathscr{S}$ the structure of a category by defining the morphisms between two spaces to be perfect maps (i.e. closed, continuous, and the inverses of points are compact). Note that $\mathscr{Z}$ is then a full subcategory.

2.2. Theorem. 1) *A space in $\mathscr{S}$ is a projective object if and only if it is extremally disconnected.*

2) *Every space in $\mathscr{S}$ has a projective resolution. If a space is in $\mathscr{Z}$, then its projective resolutions are also in $\mathscr{Z}$.*

3) *The Baire isomorphism class of every projective space in $\mathscr{Z}$ contains a compact space.*

4) *Every projective space X in $\mathscr{X}$ has the property that*

$$\bigcup_{\tau \in K(X,\mathscr{B})} C(X_\tau)$$

is uniformly dense in $B^*(X, \mathscr{B})$, where $\mathscr{B}$ is the Baire σ-algebra of X, $K(X, \mathscr{B})$ is the set of compact representations of $(X, \mathscr{B})$, and $C(X_\tau)$ is the space of τ-continuous real valued functions on X.

Part 4) of this theorem for a Borel subset of a separable completely metrizable space is due to E. R. Lorch and H. Tong [9].

A Baire quotient map from a space X into a space Y is one with the property that the inverse of a set in Y is a Baire set in X if and only if the set is a Baire set in Y. It follows from parts 2) and 3) of the above theorem that every space in $\mathscr{X}$ is the Baire quotient image of a compact space.

References

[1] *Z. Frolík:* A survey of the separable descriptive theory of sets and spaces. Czechoslovak Math. J. *20* (1970), 406—467.

[2] *K. Gödel:* The consistency of the axiom of choice and the generalized continuum-hypothesis. Proc. Nat. Acad. Sci. U.S.A. *24* (1938), 556—557.

[3] *H. Gordon:* Compactifications defined by means of generalized ultrafilters. Ann. Mat. Pura Appl. *86* (1970), 15—23.

[4] *M. Henriksen:* On the equivalence of the ring, lattice, and semigroup of continuous functions. Proc. Amer. Math. Soc. *7* (1956), 959—960.

[5] *J. E. Jayne:* Descriptive set theory in compact spaces. Notices Amer. Math. Soc. *17* (1970), 268.

[6] *J. E. Jayne:* Spaces of Baire functions, Baire classes, and Suslin sets. Doctoral dissertation, Columbia University, New York, 1971.

[7] *K. Kuratowski:* Topology, Vol. I. Academic Press, New York, 1966.

[8] *E. R. Lorch:* On compact metric spaces and the group of Baire equivalences. Studia Math. *31* (1968), 243—252.

[9] *E. R. Lorch and H. Tong:* Continuity of Baire functions and order of Baire sets. J. Math. Mech. *16* (1967), 991—996.

[10] *E. R. Lorch and H. Tong:* Caractérisation de certain topologies compactes. Enseignement Math. *15* (1969), 153—157.

[11] *A. Maitra and C. Ryll-Nardzewski:* On the existence of two analytic non-Borel sets which are not isomorphic. Bull. Acad. Polon. Sci. Sér. Sci. Math. Astronom. Phys. *18* (1970), 177—178.

[12] *R. M. Solovay:* On the cardinality of Σ_2^1 sets of reals. Foundations of Mathematics, Symposium Papers Commemorating the Sixtieth Birthday of Kurt Gödel. Springer-Verlag, Berlin, 1969, 58—73.

[13] *A. H. Stone:* On σ-discreteness and Borel isomorphism. Amer. J. Math. *85* (1963), 655—666.

[14] *A. H. Stone:* Non-separable Borel sets II. (To appear.)

SCUOLA NORMALE SUPERIORE, PISA

THE UTILITY OF EMPTY INVERSE LIMITS

F. B. JONES

Riverside

By an inverse system we shall mean a collection of pairs $\{X_\alpha, f_{\alpha\beta}\}$ where α, β, and γ belong to a directed index set A, X_α is a nonempty set and $f_{\alpha\beta}$ is a mapping from X_α to X_β such that $f_{\beta\gamma}[f_{\alpha\beta}] = f_{\alpha\gamma}$ for $\gamma < \beta < \alpha$. In fact, we shall consider only the special case where A is the set of all ordinals less than a given limit ordinal and the maps $f_{\alpha\beta}$ are "onto". The inverse limit of such a system is the subset $\{x_\alpha\}$ of the cartesian product $\times X_\alpha$ $(\alpha \in A)$ such that $x_\beta = f_{\alpha\beta}(x_\alpha)$ for $\beta < \alpha$. Usually one is interested in an inverse system because the inverse limit is nonempty. However, there are situations when the fact that the inverse limit is empty or trivial is of interest.

In studying Souslin spaces an inverse system whose inverse limit is empty arises naturally. In 1935 G. Kurepa called such systems ramifications and studied them at some length in [8] and [9]. In addition to many positive results (about linearly ordered spaces in particular) he posed some beautiful problems, some of which have apparently turned out to be unsolvable. At about the same time, in connection with the normal Moore space metrization problem [4] (and certain other problems [5]), I also constructed inverse systems of this sort (with X_α countable for each $\alpha < \omega_1$) whose inverse limits were empty. What seemed at that time a promising method of constructing a non-metric normal Moore space has also wound up among the logician's models [11].

About 1950 Tukey raised with Henkin the question as to the existence of an inverse system (with onto bonding maps) whose inverse limit was empty. To understand Henkin's construction [2] it will help to see how a very simple construction would yield such a system if we did not require the bonding maps to be onto.

For each positive integer i let H_i be the set of positive integers greater than $i - 1$. Define $f_{21}(n)$ to be n (in H_1) if $n > 1$. And in general, $f_{\alpha\beta}(n) = n$ (in H_β) for $n \geqq \alpha$. This defines an inverse system $\{H_\alpha, f_{\alpha\beta}\}$ for $\beta < \alpha < \omega_0$ whose inverse limit space is empty. The easiest way to picture this system as a tree is in the first quadrant of the xy-plane with H_1 being the (positive) integral points of x-axis; H_2, those on the line $y = 1$, etc., with the branches of the tree being arcs running straight up from H_1 to the highest integral point below the main diagonal (the line $y = x$). Clearly the inverse limit space is empty because no arc (or ray) in the union runs all the way to the top of the plane. So when the tree is $\aleph_0$ wide ($|H_n| = \aleph_0$) and $\aleph_0$ tall, no branch need run all the way to the top of the tree. (Actually to complete the tree one must run the branches from H_1 down to a common trunk but this just amounts to adding H_0 to the system with H_0 degenerate, say $H_0 = \{0\}$ and $f_{\alpha 0}(n) = 0$ for all $n \geqq \alpha$ and $0 < \alpha < \omega_0$.) Obviously the maps are not onto. But a similar thing can be done to define a system $\{X_\alpha, f_{\alpha\beta}\}$ for $\beta < \alpha < \omega_1$ where the bonding maps are all "onto".

Begin by letting X_1 be the ordinal points in $y = 1$ of $X \times Y$ where both X and Y are "long rays" (from 1 to ω_1 including 1 but not ω_1). For each element x in X_1 and countable limit ordinal α, run a ray $L_\alpha(x)$ straight up from x which intersects every horizontal ray $y = \beta$ (where β is an ordinal less than α) but not intersecting $y = \alpha$. Consider all of these rays to be disjoint except that for α and β different countable limit ordinals $L_\alpha(x) \cap L_\beta(x) = x$, i.e., no two of the rays emanating from points of X_1 intersect except when they emanate from the same point and when they do emanate from the same point, they have that point and no other in common.

Now let X_2 be the set of all points of $y = 2$ belonging to rays $L_\alpha(x)$ for x in X_1. So for each x in X_1 and each countable limit ordinal α there is a unique point of $L_\alpha(x)$ in X_2 and f_{21} maps this point to x. Again for x in X_2 and limit ordinal $\alpha < \omega_1$ run a ray straight up from x up to but not intersecting the horizontal line $y = \alpha$ and consider all of these disjoint except their common emanation point in case they emanate from the same point. The set X_3 is the set of all points at the $y = 3$ level which belong to vertical rays thus defined emanating from points of $X_1 \cup X_2$. The definition of the bonding maps is obvious. And in fact, the entire construction is now obvious, X_{ω_0} (for example) being those points of $y = \omega_0$ belonging to those rays running straight up from points of X_β for $\beta < \omega_0$ as already defined. Hence if the inverse limit of the system were non-empty some branch would have to run straight up from some line $y = \alpha$ $(\alpha < \omega_1)$ to (but not intersecting) the line $y = \omega_1$, for if a path changed branches infinitely many times, say at $\alpha_1 < \alpha_2 < \alpha_3 < \ldots$ then such a path would not reach $y = \lim \alpha_i$. But no branch runs straight up to $y = \omega_1$. So the inverse limit is empty.

At this point one can see that even when the bonding maps are "onto" it is not surprising for the inverse limit to be empty when $|X_\alpha| \geqq \aleph$ $(\alpha \in A)$ and $|A| \leqq \aleph$ unless some countable subset of A is cofinal with A. On the other hand, if A is a limit ordinal (i.e., the set of all ordinals less than A) and $\aleph$ is a cardinal such that $|X_\alpha| < < \aleph < |A|$ for $\alpha < A$, the inverse limit is non-empty. The spirit of this condition would be violated if some subsequence of A cofinal with A had cardinality less than $|A|$. Roughly speaking, if the tree is substantially taller than it is wide, then from some level on, the branches run straight on up to the top. See [5] for a more accurate statement of this rather simple fact. So the more interesting cases occur when the height is larger than the width but just barely so. This is the case in an example due to Higman and Stone, which I shall describe next.

If the spaces in the inverse system $\{X_\alpha, f_{\alpha\beta}\}$ are groups or rings, etc., and the mappings are "onto" homomorphisms, the inverse limit is a group or ring, etc., in such a natural way that its natural projection onto X_α will be an "onto" homomorphism. Zelinsky raised the question as to the necessity of the projection being "onto" and Higman and Stone constructed an example where the cardinality of the inverse limit is too small for the projection be "onto" (in fact, contained only one element). Basic to their construction (for the case of groups, etc.) is the following inverse system whose inverse limit is empty [3].

For each initial closed interval $[1, \alpha]$ of the countable ordinals $(\alpha < \omega_1)$ let G_α denote the collection of all increasing bounded functions from $[1, \alpha]$ to the real numbers. Now for each $\alpha < \omega_1$, X_α is defined by induction to be a countable subset of G_α subject to the following conditions:

(1) if $\beta < \alpha < \omega_1$ and x_α belongs to X_α, there is one and only one element x_β of X_β such that $x_\beta \subset x_\alpha$, and

(2) if $\beta < \alpha < \omega_1$, n is a natural number, and x_β belongs to X_β, then there is an element x_α of X_α such that $x_\beta \subset x_\alpha$ and $x_\beta(\beta) - x_\alpha(\alpha) < 1/2^n$.

Choosing X_1 to be some arbitrary countably infinite subset of G_1 there would be no difficulty in constructing the system if one realizes that for each n and each $\alpha < \omega_1$ there is an increasing function from $[1, \alpha]$ of the ordinals into $[0, 1/2^n]$ of the real numbers. Furthermore, the bonding maps $f_{\alpha\beta}$ are defined so that for x_α in X_α, $f_{\alpha\beta}(x_\alpha) = x_\beta$ in X_β if and only if $x_\beta \subset x_\alpha$. From this property it follows immediately that the inverse limit is empty because there exists no increasing function from $[1, \omega_1)$ to the real numbers (uncountable subsets of the real numbers contain points which are condensation points from both sides).

This construction is quite like an old one of mine (about 1946) in which the branches of the tree were arcs whose lengths were diadic rational numbers. The functions x_α above for $[0 < \beta \leqq \alpha]$ could be chosen so that the arc length from H_0 to its end point in H_β is $x_\alpha(\beta)$ [6].

The use of inverse systems with empty inverse limits in examples is perhaps to be expected. But such systems are quite useful in the proof of certain kinds of theorems. For instance, Roy in proving Arhangelskii's theorem on the cardinality of first countable, Lindelöf spaces S uses an inverse system $\{X_\alpha, f_{\alpha\beta}\}$ where for each $\alpha < \omega_1$, X_α is a closed covering of S with $|X_\alpha| \leqq \mathbf{c}$. While the bonding maps preserve inclusion (i.e., $f_{\alpha\beta}^{-1}(x_\beta)$ is a decomposition of x_β for $\alpha = \beta + 1$ when $f_{\alpha\beta}^{-1}(x_\beta)$ is nondegenerate) and the maps are onto, the tree does not branch at $x_\beta \in X_\beta$ if $|x_\beta| \leqq \mathbf{c}$, i.e., if $f_{\alpha\beta}(x_\alpha) = x_\beta$ and $|x_\beta| \leqq \mathbf{c}$ then $x_\alpha = x_\beta$. So the inverse system itself does not have an empty inverse limit. However the subsystem of those sets x_α such that $|x_\alpha| > \mathbf{c}$ does have an empty inverse limit and it is this fact that is central in the proof.

Perhaps it may be instructive to outline Roy's proof [10], leaving unproved certain set theoretic lemmas which are either known or in any case easy to establish.

Arhangelskii's Theorem [1]. *Let S denote a Hausdorff space in which the First Countability Axiom holds true* $(\chi(S) = \aleph_0)$. *Then if S is Lindelöf* $|S| \leqq \mathbf{c}$.

Let $X_0 = \{S\}$. Let X_1 denote a collection of closed subsets of S covering S such that $|X_1| \leqq \mathbf{c}$. Define f_{10} from X_1 onto X_0 in the obvious way: $f_{10}(x_1) = S$

for all x_1 in X_1. Construct X_1 so that one of its elements x_1^c has the following three properties:

(1) $|x_1^c| \leqq \mathbf{c}$,

(2) if $x_1 \in X_1 - \{x_1^c\}$ then $x_1 \cap x_1^c = \emptyset$ (if $|x_1^c| \leqq \mathbf{c}$, then x_1^c is the intersection of at most $\mathbf{c}$ open subsets of S), and

(3) if $|S| \leqq \mathbf{c}$, $X_1 - \{x_1^c\} = \emptyset$.

In general, given X_β, X_α is defined from X_β in exactly the same way when $\alpha = \beta + 1$ so that:

If $x_\beta \in X_\beta$, $f_{\alpha\beta}^{-1}(x_\beta)$ is a collection of no more than $\mathbf{c}$ closed subsets of x_β covering x_β which contains an element x_α^c such that

(1) $|x_\alpha^c| \leqq \mathbf{c}$,

(2) if $x_\alpha \in f_{\alpha\beta}^{-1}(x_\beta) - \{x_\alpha^c\}$ then $x_\alpha \cap x_\alpha^c = \emptyset$, and

(3) if $|x_\beta| \leqq \mathbf{c}$ then $x_\alpha^c = x_\beta$ and $f_{\alpha\beta}^{-1}(x_\beta) - \{x_\alpha^c\} = \emptyset$.

More specifically, this is how X_α is constructed if β is not a limit ordinal.

When β is a limit ordinal define X_β as follows: If $\{x_\gamma\}$ $(\gamma < \beta)$ is an element of the inverse limit of $\{X_\gamma, f_{\gamma\delta}\}$ then $\bigcap x_\gamma$ $(\gamma < \beta)$ is an element of X_β and conversely. Obviously X_β is a closed cover of S and the definition of $f_{\beta\gamma}$ for $\gamma < \beta$ is the natural one. Furthermore, if $\beta < \omega_1$, $|X_\beta| \leqq \mathbf{c}$.

Returning to the definition of $|X_\alpha|$ when $\alpha = \beta + 1$ and β is a limit ordinal with $x_\beta = f_{\alpha\beta}(x_\alpha^c)$ and $x_\beta = \bigcap x_\gamma$ as above, we require that $x_\alpha^c \cup \{x_\gamma^c \mid f_{\gamma(\gamma-1)}(x_\gamma^c) = f_{\gamma(\gamma-1)}(x_\gamma)\}^*$ (for all $\gamma < \beta$) be closed. (* means union.)

Now suppose that the subsystem of all those elements of X_α for all $\alpha < \omega_1$ whose cardinality exceeds $\mathbf{c}$ has a non-empty inverse limit. Let $\{x_\alpha\}$ be an element of this limit. Then $|x_\alpha| > \mathbf{c}$ for each $\alpha < \omega_1$ and $\{x_\alpha^c\}^*$ for which $f_{\alpha(\alpha-1)}(x_\alpha^c) = f_{\alpha(\alpha-1)}(x_\alpha)$ is closed. But $\{x_\alpha^c\}^*$ is covered by $\{S - x_\alpha\}$ $(\alpha < \omega_1)$ but by no countable subcollection. Since S is Lindelöf this is a contradiction. Hence the subsystem has an empty inverse limit. It follows that the collection of all elements of $\bigcup X_i$ of cardinality $\mathbf{c}$ or less covers S. But $|\bigcup X_i| \leqq \mathbf{c}$; so $|S| \leqq \mathbf{c}$.

When one grasps the salient points of this construction, one sees that it generalizes to cardinals larger than $\aleph_0$. (Juhász [7] has done this for Arhangelskii's argument.)

Theorem (Juhász). *Suppose that S is a Hausdorff space and $\aleph$ is a transfinite cardinal such that $\chi(p) \leqq \aleph$ for each p in S. Then if each open cover of S contains an open subcover of cardinality $\aleph$ or less, $|S| \leqq \exp \aleph$.*

First let us check some elementary lemmas.

Lemma 1. *If p is a limit point of a point set M then p is the limit of a net T of distinct points of M such that $|T| \leqq \aleph$.*

Lemma 2. *If M is a closed subset of S, such that $|M| \leqq \exp \aleph$, then M is the intersection of no more than* $\exp \aleph$ *open sets.*

Proof. For each point p of M let $\mathscr{U}(p)$ denote a topological basis at p such that $|\mathscr{U}(p)| \leqq \aleph$. If U is any open set containing M some subcollection $\mathscr{V}$ of $\bigcup_{p \in M} \mathscr{U}(p)$ covers M such that $U \supset \mathscr{V}^*$. Without loss of generality we may assume that $|\mathscr{V}| \leqq \aleph$. Hence the total number of such collections $\mathscr{V}$ required so that we have at least one for each $U \supset M$ is $(\exp \aleph)^{\aleph}$ and $(\exp \aleph)^{\aleph} = \exp \aleph$.

Lemma 3. *If $\mathscr{M}$ is a collection of disjoint closed point sets M such that $|\mathscr{M}| \leqq \aleph$ and $|M| \leqq \exp \aleph$, then $|\mathrm{Cl}\, \mathscr{M}^* - \mathscr{M}^*| \leqq \exp \aleph$.*

Indication of proof. If $p \in \mathrm{Cl}\, \mathscr{M}^* - \mathscr{M}^*$, then p is the limit of a net T obtained by selecting no more than one point from each element of $\mathscr{M}$. The total number of such nets T is at most $(\exp \aleph)^{\aleph} = \exp \aleph$.

Now for the space in the theorem we construct our inverse system $\{X_\alpha, f_{\alpha\beta}\}$ for all $\alpha, \beta < \omega^+$ where ω^+ is the smallest ordinal such that $|\omega^+| = \aleph^+$ (the smallest cardinal greater than $\aleph$). Suppose that $\alpha = \beta + 1$ and x_β is an element of X_β. Then if $|x_\beta| \leqq \exp \aleph$, $f_{\alpha\beta}^{-1}(x_\beta)$ contains just one element x_α of X_α; otherwise, $f_{\alpha\beta}^{-1}(x_\beta)$ is a collection of closed subsets of x_β covering x_β one of whose elements x_α^c is of cardinality no more than $\exp \aleph$ and is disjoint from all the others. If β is a limit ordinal, $x_\beta = \bigcap f_{\beta\gamma}(x_\beta)$ for all $\gamma < \beta$ and x_α^c contains all of the limit points of $\{x_\gamma^c \mid f_{\alpha(\gamma-1)}(x_\alpha^c) = f_{\beta(\gamma-1)}(x_\beta)\}^*$ which belong to x_β.

Again, as in Roy's proof of Arhangelskii's theorem, the subsystem composed of those elements of $\bigcup X_\alpha$ $(\alpha < \omega^+)$ whose cardinality exceed $\exp \aleph$ has an empty inverse limit. Consequently the subcollection of $\bigcup X_\alpha$ of those of cardinality $\exp \aleph$ or less covers S. Since the cardinality of this collections is $\exp \aleph$ or less, the cardinality of its union and hence of S is also $\exp \aleph$ or less.

References

[1] *A. V. Arhangelskii:* On the cardinality of first countable compacta. Dokl. Akad. Nauk SSSR *187* (1969), 967—978.

[2] *L. Henkin:* A problem on inverse mapping systems. Proc. Amer. Math. Soc. *1* (1950), 224—225.

[3] *G. Higman and A. H. Stone:* On inverse systems with trivial limits. J. London Math. Soc. *29* (1954), 233—236.

[4] *F. B. Jones:* On certain well-ordered monotone collections of sets. J. Elisha Mitchell Scientific Society *69* (1953), 30—34.

[5] *F. B. Jones:* On a property related to separability in metric spaces. J. Elisha Mitchell Scientific Society *70* (1954), 30—33.

[6] *F. B. Jones:* Remarks on the normal Moore space metrization problem. Annals Math. Stud. *60* (1966), 115—119.

[7] *I. Juhász:* Arhangelskii's solution of Alexandroff's problem. Mathematisch Centrum Publication ZW 1969—013, Amsterdam, 1969.

[8] *G. Kurepa:* Ensembles ordonnés et ramifiés. Publications Mathématiques de l'Université de Belgrade *4* (1935), 1—138.

[9] *G. Kurepa:* Ensembles linéaires et une classe de tableaux ramifiés (Tableaux ramifiés de M. Aronszajn). Publications Mathématiques de l'Université de Belgrade *6—7* (1938), 129—160.

[10] *P. Roy:* The cardinality of first countable spaces. Bull. Amer. Math. Soc. (to appear).

[11] *F. D. Tall:* Set-theoretic consistency results and topological theorems concerning the normal Moore space conjecture and related problems. Dissertation, University of Wisconsin, 1969.

UNIVERSITY OF CALIFORNIA, RIVERSIDE, CALIFORNIA

CARDINAL FUNCTIONS ON PRODUCTS

I. JUHÁSZ

Budapest

Cardinal functions, i.e., functions defined on classes of topological spaces and having cardinal numbers as values, can be used to unify a diversity of cardinality problems arising in general topology (cf. [5]). Thus e.g., many problems concerning product spaces have the following general form:

Let us be given a cardinal function φ on a productive class $\mathscr{C}$, and spaces $R_i \in \mathscr{C}$, $i \in I$. Evaluate or estimate $\varphi(R)$, where $R = \bigtimes_{i \in I} R_i$, in terms of the values $\varphi(R_i)$ and $|I|$. (We always assume that none of the spaces R_i is indiscrete.) We shall mention several results of this kind in this lecture.

If φ is one of the functions w (weight), π (π-weight) or χ (character) defined on the class $\mathscr{T}$ of all topological spaces, or the function ψ (pseudo-character) defined on the class $\mathscr{T}_1$ of T_1 spaces, we have the following exact formula:

$$\varphi(R) = |I| \,.\, \sup \{\varphi(R_i) : i \in I\} .$$

A different exact formula can be given for the density function d on the class of spaces containing two disjoint non-empty open sets as follows:

$$\mathrm{d}(R) = \log |I| \,.\, \sup \{\mathrm{d}(R_i) : i \in I\} .$$

Here $\log \alpha = \min \{\beta : 2^\beta \geqq \alpha\}$ and the $\leqq$ holds on the whole $\mathscr{T}$, according to a well-known theorem of E. S. Pondiczery and E. Hewitt (see [9] or [4]).

The case of the cellularity number c is especially interesting, because it is closely connected to undecidable set theoretic problems, such as the Suslin hypothesis. Indeed, G. Kurepa [6] has shown that if this hypothesis fails, i.e., there exists a Suslin continuum X, then we have $\mathrm{c}(X) = \omega$ but $\mathrm{c}(X \times X) = \omega_1$. On the other hand Martin's axiom (see [8], [5], or [2]) implies that $\mathrm{c}(R) = \omega$ if $\mathrm{c}(R_i) = \omega$ for all $i \in I$. We do not know whether it is consistent to assume

$$\mathrm{c}(R) = \sup \{\mathrm{c}(R_i) : i \in I\}$$

on $\mathscr{T}$. However, the following estimate, due to G. Kurepa [7] (see also [3]) is valid on $\mathscr{T}$ without any special set-theoretic assumptions:

$$\mathrm{c}(R) \leqq \sup \{2^{\mathrm{c}(R_i)} : i \in I\} .$$

I do not know whether this formula can be sharpened as follows:

$$c(R) \leqq \sup \{c(R_i)^+ : i \in I\}$$

(here α^+ is the successor cardinal of α).

Concerning the spread function s, defined as the supremum of the cardinalities of discrete subspaces, the following formula has been recently established by A. Hajnal and the present author [1] for the class $\mathcal{T}_2$ of Hausdorff spaces:

$$s(R) \leqq |I| \, . \sup \{2^{s(R_i)} : i \in I\} \, .$$

This settles a conjecture from [5], Chapter 4. The Sorgenfrey line S is known to have $s(S) = \omega$ and $s(S \times S) = 2^\omega$, which shows that this estimate cannot be improved. The proof of this result is quite difficult, requiring the construction of a very complicated ramification system.

We conjecture that a similar formula is valid for the Lindelöf degree $\mathcal{L}$, too ($\mathcal{L}(X) = \min \{\alpha : X$ is α-Lindelöf$\}$), however we cannot even show that the product of two ω-Lindelöf spaces is 2^ω-Lindelöf.

References

[1] *A. Hajnal and I. Juhász:* Discrete subspaces of product spaces. General Topology and its Applications (to appear).

[2] *A. Hajnal and I. Juhász:* A consequence of Martin's axiom. Nederl. Akad. Wetensch. Indag. Math. *33* (1971), 457—463.

[3] *Z. Hedrlin:* An application of Ramsey's theorem to the topological products. Bull. Acad. Polon. Sci. Sér. Sci. Math. Astronom. Phys. *14* (1966), 25—26.

[4] *E. Hewitt:* A remark on density character. Bull. Amer. Math. Soc. *52* (1946), 641—643.

[5] *I. Juhász:* Cardinal functions in topology. Math. Center Tract No. *34*, Amsterdam, 1971.

[6] *G. Kurepa:* La condition de Souslin et une proprieté characteristique des nombres réels. C. R. Acad. Sci. Paris Sér. A—B *231* (1950), 1113—1114.

[7] *G. Kurepa:* The Cartesian multiplication and the cellularity number. Publ. Inst. Math. (Béograd) *2* (1962), 121—139.

[8] *D. A. Martin and R. M. Solovay:* Internal Cohen extensions. Ann. Math. Logic *2* (1970), 143—178.

[9] *E. S. Pondiczery:* Power problems in abstract spaces. Duke Math. J. *11* (1944), 835—837.

ON RIGIDITY AND GROUPS OF HOMEOMORPHISMS

V. KANNAN and M. RAJAGOPALAN

Madurai

If X is an object of a category, we denote by $A(X)$ the group of all automorphisms of X. Thus for example, if X is a topological space, $A(X)$ is the group of all homeomorphisms on X. In this paper, we investigate $A(X)$ for objects in the categories of topological spaces and apply the results to obtain corresponding theorems in the categories of lattices, Boolean algebras, semigroups, graphs, and some other related categories. The groups of homeomorphisms have been investigated by J. de Groot [7] (see also [1], [5], [8] and [9]), the groups of automorphisms of lattices by G. Birkhoff [4] (see also [15]) and the groups of automorphisms of Boolean algebras by M. Katětov [13] and J. de Groot [7] (see also [12], [17]). Here we solve some open problems on this topic, generalize many known results and give different proofs for several results. The following are some of the main theorems of this paper:

Theorem 1. *Every zero-dimensional Hausdorff space is a subspace of a compact rigid zero-dimensional Hausdorff space.*

The above theorem gives a new proof for the solution of Birkhoff's problem 74 [3] and considerably generalizes the results of [12], [14] and [17].

The following theorem supplements the main theorems of J. de Groot [7]. Unlike his proof, which uses the theory of graphs, here we use only the results in topology. J. de Groot's results on graphs are then deduced as corollaries.

Theorem 2. *Let X be any Hausdorff space and G any group. Then there exists a topological space X^* having X as a closed subspace such that $A(X^*)$ is isomorphic to G. Further, X^* can be chosen so that:*

(i) *X^* is connected and locally connected.*

(ii) *Under some mild conditions on X, X and X^* have the same cardinality.*

We have that X^* is sequential if and only if X is. If X is Tychonoff then we can choose X^* to be compact and connected. If X is normal, then X^* can be chosen to be compact, connected, and of the same positive dimension as X.

If $\mathscr{A}$ is the category of all partially ordered topological spaces and if X is an object of $\mathscr{A}$, then the group $A(X)$ is a partially ordered group in a natural way.

Theorem 3. *Given a partially ordered topological space X and a partially ordered group G, we can find a partially ordered space X^* which is an extension of X such that $A(X)$ and G are isomorphic as partially ordered groups.*

Definition. A Hausdorff space X is said to be *chaotic* if any two distinct point x, $y \in X$ have neighbourhoods U_x and U_y, respectively, such that no two non-empty open subspaces of U_x and U_y are homeomorphic.

Theorem 4. (a) *Every Hausdorff k-space is a closed subspace of a chaotic space of the same cardinality.*

(b) *There exists a proper class of completely normal connected locally connected chaotic spaces.*

After writing this paper, we found that E. S. Berney [2] has announced that the real line contains a chaotic subspace, answering in the affirmative parts (a) and (b) of the question in [16]. However, our theorem answers the question completely and the two proofs are entirely different. We are thankful to E. S. Berney for having sent us a preprint of his results.

J. de Groot [7] was the first to construct an example of a Hausdorff space in which every continuous self-map is either identity or constant. H. Herrlich [10] used this result to construct pathological reflections. In [11] he poses the following problem:

Does there exist a proper class of Hausdorff spaces such that any continuous map between any two of them is either constant or identity?

We answer this question in the affirmative. Our example combines another pathology also with this. For, we have

Theorem 5. *Let $\mathcal{T}$ be the category of all topological spaces with continuous maps as morphisms. Then there exists a full subcategory $\mathcal{A}$ of $\mathcal{T}$ such that*

(i) *$\mathcal{A}$ is not small.*

(ii) *Every object in $\mathcal{A}$ is a connected Hausdorff space with a dispersion set of two points.*

(iii) *Every morphism in $\mathcal{A}$ is either constant or identity.*

Generalizing a result of G. Birkhoff [4] we show that

Theorem 6. *Every lattice is a sublattice of a lattice having a given group of automorphisms. If the lattice is distributive, then the extension can also be chosen to be distributive.*

We give a proof via topology to the following theorem of J. de Groot [7] and G. Sabidussi [18].

Theorem 7. *Any group can be realized as the automorphism group of a graph.*

J. de Groot has raised the following question in [7]: What can we say about the cardinality of rigid Boolean algebras? Here we give a complete answer, assuming the generalized continuum hypothesis.

Theorem 8. *Let m be a cardinal. Then there exists a rigid Boolean algebra of cardinality m if and only if m is uncountable, or* $m \leqq 2$.

This theorem improves a result of [14], which answers a weaker question posed in [8].

All rigid Boolean algebras constructed hitherto are, to the best of our knowledge, not σ-complete. M. Katětov [13] asked whether there exist σ-complete rigid Boolean algebras. We answer this in the affirmative.

Theorem 9. *Any Boolean algebra can be embedded in a rigid* σ*-complete Boolean algebra.*

The proofs will appear elsewhere.

References

[1] *R. D. Anderson:* Zero-dimensional compact groups of homeomorphisms. Pacific J. Math. *7* (1957), 797—810.

[2] *E. S. Berney:* A separable metric space which is chaotic. Notices Amer. Math. Soc. *18* (1971), 261—262.

[3] *G. Birkhoff:* Lattice theory. New York, 1945.

[4] *G. Birkhoff:* On groups of automorphisms. Rev. Un. Mat. Argentina *11* (1945), 155—157.

[5] *N. J. Fine and G. E. Schweigert:* On the groups of homeomorphisms of an arc. Ann. of Math. *62* (1955), 237—253.

[6] *J. de Groot and R. J. Wille:* Rigid continua and topological group-pictures. Arch. Math. *9* (1958), 441—446.

[7] *J. de Groot:* Groups represented by homeomorphism groups I. Math. Ann. *138* (1959), 80—102.

[8] *J. de Groot and R. H. Mc Dowell:* Auto-homeomorphism groups of 0-dimensional spaces. Compositio Math. *15* (1963), 203—209.

[9] *J. de Groot and M. A. Maurice:* On the existence of rigid compact ordered spaces. Proc. Amer. Math. Soc. *19* (1958), 844—846.

[10] *H. Herrlich:* On the concept of reflections in general topology. Contributions to Extension Theory of Topological Structures (Proc. Sympos., Berlin, 1967). Deutsch. Verlag Wissensch., Berlin, 1969, 105—114.

[11] *H. Herrlich:* Categorical topology. General Topology and its Applications *1* (1971), 1—15.

[12] *B. Jonsson:* A Boolean algebra without proper automorphisms. Proc. Amer. Math. Soc. *2* (1951), 766—770.

[13] *M. Katětov:* Remarks on Boolean algebras. Colloq. Math. *2* (1951), 229—235.

[14] *F. W. Lozier:* A class of compact rigid 0-dimensional spaces. Canad. J. Math. *21* (1969), 817—821.

[15] *K. D. Magill, Jr.:* The lattice of compactifications of a locally compact space. Proc. London Math. Soc. *18* (1968), 231—244.
[16] *E. D. Nix:* Chaotic spaces. Bull. Amer. Math. Soc. *76* (1970), 975—976.
[17] *L. Rieger:* Some remarks on automorphisms of Boolean algebras. Fund. Math. *38* (1951), 209—216.
[18] *G. Sabidussi:* Graphs with given infinite group. Monatsh. Math. *64* (1960), 64—67.

MADURAI UNIVERSITY, MADURAI

ON DESCRIPTIVE CLASSIFICATION OF FUNCTIONS

M. KATĚTOV

Praha

In this note, an extension of the classic descriptive theory of sets and functions is developed which makes possible, in principle, a classification of all discontinuous functions on any topological space admitting of a one-to-one continuous mapping onto a separable metrizable space. The basic device are filters on countably infinite sets and their types. It turns out e.g. that every Baire class is generated by a certain filter which is described explicitly. It is also possible to define explicitly a filter generating a class containing all Baire functions. Assuming the continuum hypothesis, it can be shown that the class of all Baire functions is generated e.g. by the intersection of two ultrafilters. Various other theorems concerning filters, their types and filter-generated classes are included, though some of them are not directly connected with problems of the descriptive classification.

No proofs are given. Some of the results, and also their proofs, are contained in the author's paper "On descriptive classes of functions" (referred to as DFC here), which is to appear in "Abhandlungen aus Mengenlehre und Topologie, dem Andenken Felix Hausdorffs gewidmet", Greifswald.

1. Preliminaries

1.1. We use the standard terminology and notation with slight modifications. The ordered pair x, y is denoted by $\langle x, y\rangle$. Symbols like $\{x_a \mid a \in A\}$, $\{x \mid f(x) < 0\}$, etc. denote either a set or a family (an indexed set); the meaning will always be clear from the context. If M is a set, we put $\exp M = \{X \mid X \subset M\}$.

"Space" will mean a topological space. As a rule, the same symbol will stand for a space and its underlying set.

1.2. Let a set A be given. Then, for any $\mathcal{M} \subset \exp A$, $\mathcal{M}^*$ denotes the collection of the sets $X \subset A$ intersecting all $M \in \mathcal{M}$.

1.3. If T is a set or a space, then $F(T)$ denotes the set of all real-valued functions on T, and 2^T denotes the set of all families $\{u_t \mid t \in T\}$ with $u_t = 0, 1$. As a rule, $F(T)$ and 2^T are considered as spaces with the topology of Cartesian product. The set of all continuous $f \in F(T)$ is denoted by $C(T)$. The canonical mapping of exp T onto 2^T is denoted by χ_T; if $\mathcal{M} \subset \exp T$, we usualy write $\chi\mathcal{M}$ instead of $\chi_T(\mathcal{M})$.

Remark. As usual, we consider functions on distinct spaces as distinct; thus, $F(T_1) \cap F(T_2) = \emptyset$ whenever T_1, T_2 are distinct spaces (possibly with the same underlying set).

1.4. A non-void collection $\mathcal{F} \subset \exp A$ is called a *quasi-filter* on A if (1) A is countably infinite, (2) $X \in \mathcal{F}$, $Y \in \mathcal{F}$, $X \cap Y \subset Z \subset A$ implies $Z \in \mathcal{F}$, (3) $\emptyset$ non $\in \mathcal{F}$. If, in addition, $\bigcap \mathcal{F} = \emptyset$, then $\mathcal{F}$ is called a *filter*. Thus, we consider filters on countably infinite sets only.

1.5. Convention. Whenever a definition refers to filters, it is tacitly assumed that it is also valid for quasi-filters.

Letters $\mathcal{F}$, $\mathcal{G}$, $\mathcal{U}$, $\mathcal{V}$, possibly with subscripts, will denote filters. The Fréchet filter on N, consisting of all $X \subset \mathsf{N}$ with $\mathsf{N} - X$ finite, will be denoted by $\mathcal{N}$.

1.6. If $\mathcal{F}$ is a filter on A, and $B \in \mathcal{F}^*$, then the collection of all $F \cap B$, $F \in \mathcal{F}$, is a filter on B, which will be called the *trace* of $\mathcal{F}$ on B.

1.7. If $\{X_a \mid a \in A\}$ is a family of sets, we denote by $\sum\{X_a \mid a \in A\}$ the set of all $\langle a, x\rangle$ where $a \in A$, $x \in X_a$. Let $\mathcal{F}$ be a filter on A; for every $a \in A$, let $\mathcal{G}_a$ be a filter on a set B_a. Then the collection of all $\sum\{G_a \mid a \in F\} \cup \sum\{H_a \mid a \in A - F\}$, where $F \in \mathcal{F}$, $G_a \in \mathcal{G}_a$, $H_a \subset B_a$, is a filter on $\sum\{B_a \mid a \in A\}$. It will be denoted by $\mathcal{F}$-$\sum\{\mathcal{G}_a \mid a \in A\}$.

1.8. Let $\mathcal{F}$ and $\mathcal{G}$ be filters on A and B, respectively. Put $\mathcal{G}_a = \mathcal{G}$ for every $a \in A$. The filter $\mathcal{F}$-$\sum\{\mathcal{G}_a\}$ is denoted by $\mathcal{F} . \mathcal{G}$ and called the *product* of filters $\mathcal{F}$ and $\mathcal{G}$. The filter $\mathcal{F} . \mathcal{F}$ is denoted by $\mathcal{F}^2$.

1.9. Let $\mathcal{F}$ and $\mathcal{G}$ be filters on A and B, respectively. Let φ be a single-valued relation with the domain A, ranging in B and such that $\varphi^{-1}(G) \in \mathcal{F}$ whenever $G \in \mathcal{G}$. The triple $\langle\varphi, \mathcal{F}, \mathcal{G}\rangle$ is called a morphism from $\mathcal{F}$ to $\mathcal{G}$. If $\langle\varphi^{-1}, \mathcal{G}, \mathcal{F}\rangle$ is also a morphism, then $\langle\varphi, \mathcal{F}, \mathcal{G}\rangle$ is called an isomorphism.

Convention. The terminology and notation used for mappings will be also applied to morphisms of filters.

1.10. If there exists a morphism from $\mathcal{F}$ to $\mathcal{G}$, we shall write $\mathcal{F} \geqq \mathcal{G}$. Clearly, $\geqq$ is a transitive reflexive relation on the class of all quasi-filters.

1.11. Proposition. *In the class of all quasi-filters endowed with the quasiorder* $\geqq$, *every countable non-void subset has a supremum and an infimum, and every subset of cardinality* $\leqq \exp \aleph_0$ *is bounded.*

1.12. Definition. If there exists an isomorphism from $\mathscr{F}$ to $\mathscr{G}$, we shall write $\mathscr{F} \cong \mathscr{G}$ and say that $\mathscr{F}$ is *isomorphic* to $\mathscr{G}$. If $\mathscr{F} \geqq \mathscr{G}$, $\mathscr{G} \geqq \mathscr{F}$, we shall write $\mathscr{F} \approx \mathscr{G}$ and say that $\mathscr{F}$ is *equivalent* to $\mathscr{G}$. Observe that "equivalent" has been used in [1] and in [2] in the sense of "isomorphic" as just defined.

1.13. Proposition. *Two equivalent ultrafilters are isomorphic.*

See e.g. [2], Proposition 1.15. – Observe that a filter equivalent to an ultrafilter need not be isomorphic to it.

2. Some properties of filters

2.1. If A is a set and $\boldsymbol{P}$ is a property of subsets of the space 2^T, then $\mathscr{M} \subset \exp A$ will be said to possess the property χ-$\boldsymbol{P}$, if $\chi\mathscr{M}$ (see 1.3) has property $\boldsymbol{P}$; the prefix "χ-" will be often omitted. E.g., $\mathscr{M} \subset \exp A$ is called Souslin if $\chi\mathscr{M}$ is a Souslin (= analytic) subset of 2^A.

2.2. Proposition. *A filter is Souslin if it has a base which is a Souslin collection.*

2.3. If A is a countably infinite set, then μ_A denotes the "canonical" measure on 2^A. If $\mathscr{M} \subset \exp A$, then the μ_A-measure of $\chi\mathscr{M}$ will be called the measure of $\mathscr{M}$; and similarly for other related concepts, in accordance with 2.1. – Cf. DCF, 4.3.

2.4. Theorem. *The interior measure of every filter is zero. No ultrafilter is measurable. The intersection of a non-measurable filter (on a set A) and an ultrafilter (on the same set) is not measurable.* – See DCF, 4.4.

3. Types of filters. Filter-limits

3.1. Definition. Two single-valued relations, typ and Typ, are chosen once for all. The domain of both typ and Typ is the class of all quasi-filters. The equality typ $\mathscr{F} =$ = typ $\mathscr{G}$ holds if and only if $\mathscr{F}$ is isomorphic to $\mathscr{G}$; Typ $\mathscr{F}$ = Typ $\mathscr{G}$ holds if and only if $\mathscr{F}$ is equivalent to $\mathscr{G}$. We shall call typ $\mathscr{F}$ and Typ $\mathscr{F}$ the *isomorphism type* and the *equivalence type* of $\mathscr{F}$, respectively. We put typ $\mathscr{F} \geqq$ typ $\mathscr{G}$, Typ $\mathscr{F} \geqq$ $\geqq$ Typ $\mathscr{G}$ if and only if $\mathscr{F} \geqq \mathscr{G}$. If ξ, η are equivalence types, then $\xi > \eta$ means that $\xi \geqq \eta$ and $\xi \neq \eta$.

Observe that isomorphism types of filters are called simply "types of filters" in [2]. It is to be noted that isomorphism types (called simply "types") of ultrafilters have been introduced by Z. Frolík [1].

3.2. Clearly, all those quasi-filters which are not filters are equivalent; their equivalence type will be denoted by et 0 or simply by 0. The equivalence type of $\mathscr{N}$ be denoted by et 1 or simply by 1.

3.3. It is easy to show that the cardinality of the set of isomorphism types, and also that of the set of equivalence types is exp exp $\aleph_0$.

3.4. Theorem. *A countable non-void set of isomorphism types of filters has a supremum and an infimum. A countable set of equivalence types of filters has a unique supremum and a unique infimum.* – See 1.11.

3.5. Theorem. *If M is a set of cardinality $\leqq$ exp $\aleph_0$ consisting of isomorphism types (or equivalence types) of filters, then M is bounded.* – See 1.11.

3.6. Let P be a Hausdorff space. If $\mathscr{F}$ is a quasi-filter on a set A and $\{x_a \mid a \in A\}$ is a family of points of P, then the *$\mathscr{F}$-limit* of $\{x_a\}$, denoted by $\mathscr{F}$-lim $\{x_a\}$, is defined in the usual way: $x = \mathscr{F}\text{-lim}\,\{x_a\}$, if, for any neighborhood V of x, there is a set $F \in \mathscr{F}$ such that $x_a \in V$ whenever $a \in F$.

3.7. If P is a Hausdorff space and $\mathscr{F}$ is a filter, then, for any $S \subset P$, the set of all points $\mathscr{F}$-lim $\{x_a\}$, $x_a \in S$, will be denoted by $\mathscr{F}$-Lim S. If P and S are fixed, we shall sometimes say that $\mathscr{F}$ *generates* the set $\mathscr{F}$-lim S (and also every point $y \in \mathscr{F}\text{-Lim}\,S$).

3.8. If T is a set, $\mathscr{F}$ is a filter on a set A, and $\{f_a \mid a \in A\}$ is a family of functions on T, then the *upper* and the *lower $\mathscr{F}$-limit* of $\{f_a\}$ are defined in the usual way. E.g., the value assumed by the upper $\mathscr{F}$-limit of $\{f_a\}$ at a point $t \in T$ is equal to the g.l.b. of numbers ξ such that $\{a \mid f_a(t) < \xi\}$ belongs to $\mathscr{F}$; values $-\infty$ and ∞ are admitted. The upper and the lower $\mathscr{F}$-limit of $\{f_a\}$ will be denoted by $\mathscr{F}\text{-lim}^+\,\{f_a\}$ and $\mathscr{F}\text{-lim}^-\,\{f_a\}$ respectively.

3.9. If T is a set or a space, $\mathscr{F}$ is a filter (on a set A) and $Y \subset F(T)$, then we denote by $\mathscr{F}\text{-Lim}^+\,Y$ the set of all $f \in F(T)$ such that $f = \mathscr{F}\text{-lim}^+\,\{f_a\}$ for some $f_a \in Y$. The symbol $\mathscr{F}\text{-Lim}^-\,Y$ is defined in an analogous way.

3.10. Proposition. *Let P be a Hausdorff space, $S \subset P$. If $\mathscr{F} \geqq \mathscr{G}$, then $\mathscr{F}\text{-Lim}\,S \supset \mathscr{G}\text{-Lim}\,S$. If $\mathscr{F} \approx \mathscr{G}$, then $\mathscr{F}\text{-Lim}\,S = \mathscr{G}\text{-Lim}\,S$.*

3.11. If P is a Hausdorff space, $S \subset P$, and ξ is an equivalence type, by 3.10, we may put $\xi\text{-Lim}\,S = \mathscr{F}\text{-Lim}\,S$, where Typ $\mathscr{F} = \xi$. If P and S are fixed, we shall say that ξ *generates* the set ξ-Lim S (and also every point $y \in \xi\text{-Lim}\,S$).

3.12. Theorem. *If P is a space admitting a one-to-one continuous mapping onto a separable metrizable space, then every $f \in F(T)$ is in some $\mathscr{F}$-Lim $C(T)$.*

3.13. Remarks. 1) $\mathscr{F} \geqq \mathscr{G}$ does not imply $\mathscr{F}\text{-Lim}^+\,C(T) \supset \mathscr{G}\text{-Lim}^+\,C(T)$. Example: $\mathscr{F}$, $\mathscr{U}$ are non-isomorphic ultrafilters on N, $\mathscr{G} = \mathscr{F} \cap \mathscr{U}$. 2) $\mathscr{F} \approx \mathscr{G}$ does not imply $\mathscr{F}\text{-Lim}^+\,C(T) = \mathscr{G}\text{-Lim}^+\,C(T)$.

3.14. Problems. 1) Does there exist a Hausdorff space P such that $\mathscr{F} \geqq \mathscr{G}$ whenever $\mathscr{F}$-Lim $S \supset \mathscr{G}$-Lim S for every $S \subset P$? 2) Does there exist a Hausdorff space P such that $\mathscr{F} \approx \mathscr{G}$ whenever $\mathscr{F}$-Lim $S = \mathscr{G}$-Lim S for every $S \subset P$?

4. Filter-descriptive classes and types of functions

4.1. We denote by $\mathscr{T}$ the class of all topological spaces. For any filter $\mathscr{F}$, we put $\mathrm{Cl}(\mathscr{F}) = \bigcup\{\mathscr{F}\text{-Lim } C(T) \mid T \in \mathscr{T}\}$, $\mathrm{Cl}^{+}(\mathscr{F}) = \bigcup\{\mathscr{F}\text{-Lim}^{+} C(T) \mid T \in \mathscr{T}\}$, $\mathrm{Cl}^{-}(\mathscr{F}) = \bigcup\{\mathscr{F}\text{-Lim}^{-} C(T) \mid T \in \mathscr{T}\}$. If ξ is an equivalence type, we may, by 3.11, put $\mathrm{Cl}(\xi) = \mathrm{Cl}(\mathscr{F})$ where $\mathrm{Typ}\,\mathscr{F} = \xi$. However, it is not possible to define $\mathrm{Cl}^{+}(\xi)$ in an analogous way; see 3.13.

We shall call $\mathrm{Cl}(\mathscr{F})$ the *bilateral filter-descriptive class* (*descriptive class*, for short) *generated* by $\mathscr{F}$ (or by $\mathrm{Typ}\,\mathscr{F}$); $\mathrm{Cl}^{+}(\mathscr{F})$ and $\mathrm{Cl}^{-}(\mathscr{F})$ will be called the *upper* and the *lower unilateral descriptive class* (abbreviated: *upper* and *lower descriptive class*) *generated by* $\mathscr{F}$.

Examples: $\mathrm{Cl}(0)$ consists of all continuous functions on topological spaces; $\mathrm{Cl}(1) = \mathrm{Cl}(\mathscr{N})$ consists of all functions of the first Baire class.

4.2. Proposition. *If* $\mathscr{F} \geqq \mathscr{G}$, *then* $\mathrm{Cl}(\mathscr{F}) \supset \mathrm{Cl}(\mathscr{G})$. *If* $\mathscr{F} \approx \mathscr{G}$, *then* $\mathrm{Cl}(\mathscr{F}) = \mathrm{Cl}(\mathscr{G})$. – See 3.10.

4.3. Theorem. *If T is a space admitting a one-to-one continuous mapping onto a separable metrizable space, then every real-valued function on T is in some filter-descriptive class.* – See 3.12.

4.4. Proposition. *For every* $n \in \mathrm{N}$, *let* $\mathscr{D}_n$ *be a bilateral filter-descriptive class of functions. Then* $\mathscr{D} = \{f \mid f \in \mathscr{D}_n$ *for every* $n \in \mathrm{N}\}$ *is a bilateral filter-descriptive class; if* $\mathscr{D}_n = \mathrm{Cl}(\mathscr{F}_n)$, *then* $\mathscr{D}$ *is generated by an infimum of filters* $\mathscr{F}_n$ (*or: by the infimum of types* $\mathrm{Typ}\,\mathscr{F}_n$). – See 1.11.

4.5. Proposition. *For every* $n \in \mathrm{N}$, *let* $\mathscr{D}_n$ *be a bilateral filter-descriptive class of functions. Then there exists a smallest bilateral filter-descriptive class containing all* $\mathscr{D}_n$. *If* $\mathscr{D}_n = \mathrm{Cl}(\mathscr{F}_n)$, *then this class is generated by a supremum of filters* $\mathscr{F}_n$. – See 1.11.

4.6. Proposition. *Let M be a set of cardinality* $\leqq \exp \aleph_0$. *For every* $m \in M$, *let* $\mathscr{C}_m$ *be a bilateral filter-descriptive class. Then there exists a filter-descriptive class* $\mathscr{C}$ *such that* $\mathscr{C}_m \subset \mathscr{C}$ *for every* $m \in M$. – See 1.11.

4.7. Let T be a non-void compact metrizable space. Denote by $\mathscr{Y}(T)$ the collection of all countable $Y \subset C(T)$ which are dense in $C(T)$ endowed with the topology of uniform convergence. If $f \in F(T)$, $Y \in \mathscr{Y}(T)$, put $Y_f = Y \cup (f)$ if f is continuous, $Y_f = Y$ if not. Denote by $\Phi(f, Y)$ the filter on Y_f a subbase of which consists of all sets $\{g \mid g \in Y_f, f(t) - \varepsilon < g(t) < f(t) + \varepsilon\}$, where $t \in T$, $\varepsilon > 0$. It is easy to show that, for any Y_1, Y_2 in $\mathscr{Y}(T)$, $\Phi(f, Y_1)$ is isomorphic to $\Phi(f, Y_2)$. We put $\Phi(f) = \operatorname{Typ} \Phi(f, Y)$ where $Y \in \mathscr{Y}(T)$; $\Phi(f)$ will be called the *descriptive type* of f.

4.8. Theorem. *Let T be a non-void compact metrizable space. If $\mathscr{F}$ is a quasi-filter, then a function $f \in F(T)$ belongs to* $\mathrm{Cl}(\mathscr{F})$ *if and only if* $\operatorname{Typ} \mathscr{F} \geqq \Phi(f)$.

In other words, the descriptive type of a function $f \in F(T)$ is the infimum (in fact, the "minimum") of equivalence types of filters generating f.

4.9. Problems. 1) Does $\mathrm{Cl}^+(\mathscr{F}) = \mathrm{Cl}^+(\mathscr{G})$ imply $\mathscr{F} \approx \mathscr{G}$ (or even $\mathscr{F} \cong \mathscr{G}$)? 2) Does $\mathrm{Cl}(\mathscr{F}) = \mathrm{Cl}(\mathscr{G})$ imply $\mathscr{F} \approx \mathscr{G}$? 3) Does $\mathrm{Cl}(\mathscr{F}) \supset \mathrm{Cl}(\mathscr{G})$ imply $\mathscr{F} \geqq \mathscr{G}$? 4) Is there a compact metrizable T, a function $f \in F(T)$ and an ultrafilter $\mathscr{U}$ such that $\Phi(f) \geqq \operatorname{Typ} \mathscr{U}$? 5) If T is a topological space, does there always exist an infimum of all equivalence types of filters generating f? If such an infimum exists, does it generate the function f?

5. Special filters and transforms of filters

5.1. Definition. We put $\mathscr{N}^0 = \{X \mid X \subset \mathsf{N},\ 0 \in X\}$, $D^{(0)} = \mathsf{N}$, $\mathscr{N}^1 = \mathscr{N}$, $D^{(1)} = \mathsf{N}$. Let $\alpha > 1$ be a countable ordinal and suppose that the filters $\mathscr{N}^\xi$ on $D^{(\xi)}$ have been defined for all $\xi < \alpha$. If $\alpha = \beta + 1$, we put $\mathscr{N}^\alpha = \mathscr{N} . \mathscr{N}^\beta$, $D^{(\alpha)} = \mathsf{N} \times D^{(\beta)}$. If α is a limit ordinal, we put $\mathscr{N}^\alpha = \mathscr{O}(\alpha)\text{-}\sum\{\mathscr{N}^\xi \mid 0 < \xi < \alpha\}$ where $\mathscr{O}(\alpha)$ denotes the filter on $\{\xi \mid 0 < \xi < \alpha\}$ with a base consisting of all sets $\{\xi \mid \beta < \xi < \alpha\}$, $\beta < \alpha$. Then $\mathscr{N}^\alpha$ is a filter on the set $D^{(\alpha)} = \sum\{D^{(\xi)} \mid 0 < \xi < \alpha\}$. – See DCF, 2.7.

5.2. The equivalence type of $\mathscr{N}^\alpha$ will be denoted by et α or simply by α.

5.3. Proposition. *If α, β are countable ordinals, then $\mathscr{N}^\alpha . \mathscr{N}^\beta$ is isomorphic to $\mathscr{N}^{\beta+\alpha}$.* – See DCF, 2.8.

5.4. Definition. If A is a set, then $\mathrm{e}A$ will denote the collection of all finite subsets of A. If $\mathscr{F}$ is a filter on A, consider the collection of sets $\{x \mid x \in \mathrm{e}A,\ x \subset F\}$ where $F \in \mathscr{F}$, and $\{x \mid x \in \mathrm{e}A,\ x \cap H \neq \emptyset\}$ where $H \in \mathscr{F}^*$ (see 1.2). Clearly, this collection is a subbase of a filter on $\mathrm{e}A$, which will be denoted by $\mathbf{e}\mathscr{F}$ and called the **e**-transform of $\mathscr{F}$.

5.5. Proposition. *For any filter $\mathscr{F}$, $\mathbf{e}\mathscr{F} \geqq \mathscr{F}$.*

5.6. Proposition. *If $\mathscr{U}$ is an ultrafilter, then $\mathbf{e}\mathscr{U} \approx \mathscr{U}$.*

In more detail: $\mathscr{U}$ is isomorphic to the trace of $\mathbf{e}\mathscr{U}$ on the set $\{(x) \mid x \in A\} \subset \mathrm{e}A$.

5.7. Proposition. $\mathcal{N} \leqq \mathbf{e}\mathcal{N} \leqq \mathcal{N}^2$.

5.8. If A is a set, then wA will denote the set of all finite sequences of elements of A. If $\alpha \in \mathrm{w}A$, $\beta \in \mathrm{w}A$, then $\alpha . \beta$ denotes the "concatenation" of α and β (i.e., if $\alpha = \{a_0, \ldots, a_m\}$, $\beta = \{b_0, \ldots, b_n\}$, then $\alpha . \beta = \{a_0, \ldots, a_m, b_0, \ldots, b_n\}$).

5.9. Definition. A set $V \subset \mathrm{w}A$ will be called *sequentially finite*, if, for any $\alpha \in A^{\mathrm{N}}$, there are only finitely many such segments of α which are also segments of some $\beta \in V$.

5.10. Lemma. *If A is countable non-void, then the collection of complements of sequentially finite sets is a filter on* wA.

5.11. Let $\mathcal{F}$ be a filter on A. For any $\varphi \in \mathcal{F}^{\mathrm{w}A}$, we denote by $G(\varphi)$ the set of all $\alpha \in \mathrm{w}A$ possessing the following property: if $\alpha = \beta . \{c\} . \delta$, then $c \in \varphi(\beta)$. It is easy to show that, for any $\varphi_i \in \mathcal{F}^{\mathrm{w}A}$, $G(\varphi_1) \cap G(\varphi_2) = G(\varphi)$ where $\varphi \in \mathcal{F}^{\mathrm{w}A}$, $\varphi(\alpha) = \varphi_1(\alpha) \cap \varphi_2(\alpha)$ for every $\alpha \in \mathrm{w}A$, and that no $G(\varphi)$ is sequentially finite.

We denote by $\mathbf{w}\mathcal{F}$ the collection of all $H \subset \mathrm{w}A$ such that $H \supset G(\varphi) - V$ for some $\varphi \in \mathcal{F}^{\mathrm{w}A}$ and some sequentially finite V. It is easily seen that $\mathbf{w}\mathcal{F}$, called the **w**-transform of $\mathcal{F}$, is a filter on wA.

5.12. Proposition. *For any filter $\mathcal{F}$, $(\mathbf{w}\mathcal{F})^2 \geqq \mathbf{w}\mathcal{F} \geqq \mathcal{F}$.*

5.13. Theorem. *For any filter $\mathcal{F}$, both $\mathcal{F} . \mathbf{w}\mathcal{F}$ and $(\mathbf{w}\mathcal{F}) . \mathcal{F}$ are isomorphic to $\mathbf{w}\mathcal{F}$.*

5.14. Theorem. *For any filter $\mathcal{F}$, the class* Cl $(\mathbf{w}\mathcal{F})$ *is closed under $\mathcal{F}$-limits (hence also under $\mathcal{N}$-limits, i.e. the usual limits of sequences).*

5.15. Let T be a space. A set $S \subset T$ will be called a *Souslin set* (in T) if it can be obtained by the Souslin operation from a family $\{X_\alpha \mid \alpha \in \mathrm{wN}\}$ where X_α are zero-sets in T, i.e. sets of the form $\{t \mid t \in T, f(t) = 0\}$, $f \in C(T)$. A function $f \in F(T)$ will be called *Souslin* (*co-Souslin*) if all sets $\{t \mid t \in T, f(t) > c\}$, c a real number, are Souslin (co-Souslin).

5.16. Theorem. *The class* Cl $(\mathbf{w}\mathcal{N})$ *contains all Souslin and all co-Souslin functions.*

5.17. Remark. It can be shown that $\chi(\mathbf{w}\mathcal{N}) \subset 2^{\mathrm{wN}}$ belongs to the projective class ***PCA*** (for the definition of this class see e.g. [3]).

5.18. Problems. 1) Does there exists an equivalence type ν such that $1 < \nu < 2$? In a more general way, do there exist, for countable ordinals ξ, equivalence types ν_ξ

with $\xi < \nu_\xi < \xi + 1$? 2) Is there a filter $\mathscr{G}$ with $\mathscr{G} \approx \mathscr{G}^2$? Does $(\mathbf{w}\mathscr{F})^2 \approx \mathbf{w}\mathscr{F}$ hold for some $\mathscr{F}$? For $\mathscr{F} = \mathscr{N}$? For all $\mathscr{F}$? 3) The same questions with $\cong$ instead of $\approx$. 4) Is $\mathbf{w}\mathscr{N}$ measurable (see 2.3)?

6. Unilateral and bilateral descriptive classes

6.1. Theorem. *Let T be a set. Let $D \subset F(T)$ possess the following property: if $f_1 \in D$, $f_2 \in D$, $\varepsilon > 0$, then there is a function $g \in D$ such that $|g(t) - \max(f_1(t), f_2(t)| < \varepsilon$ for all $t \in T$. Then, for any filter $\mathscr{F}$, $\mathscr{F}\text{-Lim}^+ D \subset \mathbf{e}\mathscr{F}\text{-Lim } D$, $\mathscr{F}\text{-Lim}^- D \subset \mathbf{e}\mathscr{F}\text{-Lim } D$.*

6.2. Theorem. *Let T be a topological space. Then, for any filter $\mathscr{F}$, we have $\mathrm{Cl}^+(\mathscr{F}) \subset \mathrm{Cl}(\mathbf{e}\mathscr{F})$, $\mathrm{Cl}^-(\mathscr{F}) \subset \mathrm{Cl}(\mathbf{e}\mathscr{F})$. Thus, every unilateral filter-descriptive class is contained in a bilateral filter-descriptive class.*

7. Baire functions and filter-descriptive classes

7.1. Theorem. *For every countable ordinal α, the filter-descriptive class $\mathrm{Cl}(\mathrm{et}\ \alpha)$ consists of all functions (on topological spaces) of Baire class α.* – See DCF, 2.17.

7.2. Theorem. *Under the continuum hypothesis, there exist ultrafilters $\mathscr{U}$, $\mathscr{V}$ on N such that $\mathscr{U} \cap \mathscr{V}$ generates the class of all Baire functions on topological spaces.* – See DCF, 5.6.

7.3. Theorem. *Under the continuum hypothesis, there exists an ultrafilter $\mathscr{U}$ (on $\mathrm{w}\mathsf{N}$) such that $\mathscr{U} \cap \mathbf{w}\mathscr{N}$ generates the class of all Baire functions on topological spaces.*

Added in proof. The author wishes to point out that filtered sums of filters (see 1.7) and products of filters (see 1.8) have been introduced by G. Grimeisen [4]. A part of Theorem 7.1 is due, in a different setting, to G. Grimeisen [5].

References

[1] *Z. Frolík:* Sums of ultrafilters. Bull. Amer. Math. Soc. *73* (1967), 791.
[2] *M. Katětov:* Products of filters. Comment. Math. Univ. Carolinae *9* (1968), 173–189.
[3] *K. Kuratowski:* Topology, Vol. I. 1966.
[4] *G. Grimeisen:* Gefilterte Summation von Filtern und iterierte Grenzprozesse. I. Math. Ann. *141* (1960), 318–342.
[5] *G. Grimeisen:* Ein Approximationssatz für Bairesche Funktionen. Math. Ann. *146* (1962), 189–194.

INSTITUTE OF MATHEMATICS OF THE CZECHOSLOVAK ACADEMY OF SCIENCES, PRAHA

ON MILD AND WICKED EMBEDDINGS

A. KIRKOR

Warszawa

A subset A of a triangulated space X is said to be *tame* iff there is an auto-homeomorphism of X mapping A onto a polyhedron in X. Otherwise A is said to be *wild* in X [1]. It is well-known that every tame subset of X is a deformation retract of a neighborhood in X (which may be taken arbitrarily small) [4]. However, as simple polyhedra as the arc, the simple closed curve, the disk, the sphere and the ball may be so embedded in some 3-manifolds that they are deformation retracts of no neighborhood [2]. This gives rise to the following definition. A subset A of a topological space X is *mildly embedded* in X (for short – is *mild* in X) iff A is a deformation retract of a neighborhood in X. Otherwise, A is said to be *wickedly embedded* in X (is *wicked* in X). It is clear that a wicked topological polyhedron in a triangulated space must be wild.

Obviously enough, if A and X are AR-spaces, then A is mild in X. But, if A or X is only an ANR-space, the situation is quite different as we have already mentioned. Perhaps the most interesting case is the one where X is the Euclidean n-dimensional space E^n or the n-sphere S^n, and A is an m-dimensional manifold with $1 \leqq m \leqq n$. It has been proved [2] that a topological $(n-1)$-sphere is always mild in S^n and there are some indications to support the conjecture that such is the case of any ANR-set which is homotopy equivalent to S^m with $0 \leqq m < n$. On the other hand, any orientable closed surface of positive genus may be wicked in S^3 [3].

It can be proved that an orientable closed surface is mild in S^3 if it can be homeomorphically approximated by unknotted surfaces in the sense that each of them bounds a cube with handles in every complementary domain.

Thus, one is led to ask the following questions.

1. Let A be an ANR-subset of S^n which is homotopy equivalent to S^m (or simply, $A \underset{\text{top}}{=} S^m$). Is A mild in S^n?

If $m = 1$ and $n = 3$, the answer is yes.

2. Do there exist wicked embeddings of an m-manifold in S^n for $1 < m \leqq n$ and $n > 3$?.

3. Does there exist a sufficient condition for mild embedding of an $(n-1)$-manifold in S^n $(n > 3)$ similar to the one for closed orientable surfaces in S^3?

References

[1] *R. H. Fox and E. Artin:* Some wild cells and spheres in 3-space. Ann. of Math. *49* (1948), 979—990.
[2] *A. Kirkor:* Every topological (n — 1)-sphere is a deformation retract of an open neighborhood in the n-sphere. Bull. Acad. Polon. Sci. Sér. Sci. Math. Astronom. Phys. *17* (1969), 801—807.
[3] *A. Kirkor:* A positional characterization of the 2-sphere. Bull. Acad. Polon. Sci. Sér. Sci. Math. Astronom. Phys. *18* (1970), 437—442.
[4] *E. C. Zeeman:* Seminar on combinatorial topology. Inst. Hautes Études Sci. Publ. Math., 1963.

INSTITUTE OF MATHEMATICS, POLISH ACADEMY OF SCIENCES

SOME MAPPING AND FIXED POINT THEOREMS

J. KOLOMÝ

Praha

1. This remark concerns some mapping and fixed point theorems. Some of these results are related to those of Pochožajev [1], Browder [2], Edelstein [3], Belluce and Kirk [4], Daneš [5] and our paper [6].

Let X, Y be normed linear spaces. In the following we use the symbols "$\rightarrow$", "$\rightharpoonup$" to denote strong and weak convergence, respectively. To fix our terminology we introduce here the following notions. A set $M \subset X$ is said to be (a) weakly closed if for each $u_n \in M$, $u_n \rightharpoonup u_0 \Rightarrow u_0 \in M$; (b) weakly compact if for each $u_n \in M$ there is a subsequence u_{n_k} which is weakly convergent in X. A mapping $F : X \rightarrow Y$ is said to be

(1) weakly continuous if $u_n \in X$, $u \in X$, $u_n \rightharpoonup u \Rightarrow F(u_n) \rightharpoonup F(u)$;
(2) demicontinuous if u_n, $u \in X$, $u_n \rightarrow u \Rightarrow F(u_n) \rightharpoonup F(u)$;
(3) p-positively homogeneous if $F(tu) = t^p F(u)$ for each $u \in X$, $t \geqq 0$, where $p > 0$.

We shall say that a functional φ is quasi-convex on a convex set $M \subset X$ if u, $v \in M$, $\lambda \in [0, 1] \Rightarrow \varphi(\lambda u + (1 - \lambda) v) \leqq \max [\varphi(u), \varphi(v)]$. A functional f is said to be weakly lower-semicontinuous at $u_0 \in X$ if $u_n \in X$, $u_n \rightharpoonup u_0 \Rightarrow f(u_0) \leqq \varliminf_{n \to \infty} f(u_n)$. By $B_\delta(u)$ we denote an open ball of a space X centered at u and with the radius $\delta > 0$.

2. We start with the following

Theorem 1. *Let X be a reflexive Banach space, $F : X \rightarrow X$ a mapping such that for some $\lambda > 0$, u, $v \in X$, $u \neq v \Rightarrow \|u - v - \lambda(F(u) - F(v))\| < \|u - v\|$. If $F(X)$ is weakly closed in X, then $F(X) = X$.*

Theorem 2. *Let X, Y be normed linear spaces, Y reflexive, $F : X \rightarrow Y$ a mapping such that $F(X)$ is weakly closed in Y. Let $H : X \rightarrow Y$ be a p-positively homogeneous map of X onto Y. Suppose that for each $u \in X$ there exist constants α_u, δ_u ($0 \leqq \alpha_u < 1$, $\delta_u > 0$) and a mapping $G_u : X \rightarrow Y$ such that $v \in B_{\delta_u}(u) \Rightarrow \|F(v) - F(u) - G_u(v - u)\| \leqq \alpha_u \|H(v - u)\|$. Assume there is $R > 0$ and $\varepsilon_u \geqq 0$ such that $v \in B_R(0) \Rightarrow \|G_u(v) - H(v)\| \leqq \varepsilon_u \|H(v)\|$, $u \in X$. If $\varepsilon_u + \alpha_u < 1$ for each $u \in X$, then $F(X) = Y$.*

Corollary 1. *Let X, Y be normed linear spaces, Y reflexive, $K : X \to Y$ a linear (i.e., additive and homogeneous) mapping of X onto Y, $F : X \to Y$ a map such that $(K + F)(X)$ is weakly closed. Assume that for each $u \in X$ there are constants α_u, δ_u $(0 \leqq \alpha_u < 1, \delta_u > 0)$ such that $v \in B_{\delta_u}(u) \Rightarrow \|F(v) - F(u)\| \leqq \alpha_u \|K(v - u)\|$. Then $(K + F)(X) = Y$.*

Remark. The conclusions of Theorems 1, 2 remain true if X, Y are normed linear spaces, $F : X \to Y$ is weakly continuous, $F(0) = 0$ and $\{u \in X \mid \|F(u)\| \leqq a\}$ is weakly compact for each $a \geqq 0$. Here we do not assume that $F(X)$ is weakly closed in Y.

Theorem 3. *Let X, Y be normed linear spaces, X reflexive, $F : \overline{B_R(0)} \to Y$ a given mapping, $G : X \to Y$ a suitable p-positively homogeneous mapping of X onto Y so that $u, v \in B_R(0) \Rightarrow \|F(u) - F(v) - G(u - v)\| \leqq \alpha \|G(u - v)\|$, for some $\alpha \in [0, 1)$. Suppose there is a point $u_0 \in B_R(0)$ such that $f(u_0) < \min_{\|u\| = R} f(u)$, where $f(u) = \|F(u)\|$, $u \in \overline{B_R(0)}$. If either* a) *F is weakly continuous on $\overline{B_R(0)}$, or* b) *F is demicontinuous on $\overline{B_R(0)}$ and $f(u)$ is quasi-convex on $\overline{B_R(0)}$, then there exists $u^* \in B_R(0)$ such that $F(u^*) = 0$.*

Theorem 4. *Let X, Y be normed linear spaces, $M \subset X$ an open subset, $F : M \to Y$, $G : X \to Y$ mappings such that $f(u) = \|F(u) + G(u)\|$ is weakly lower-semicontinuous on M and that G is a linear mapping from X onto Y. Suppose that $\{u \in M \mid f(u) \leqq c\}$ is weakly compact and non-void for some $c \geqq 0$. If for each point $u \in M$ there exist constants α_u, δ_u $(0 \leqq \alpha_u < 1, \delta_u > 0)$ so that $B_{\delta_u}(u) \subset M$ and $v \in B_{\delta_u}(u) \Rightarrow \|F(v) - F(u)\| \leqq \alpha_u \|G(v - u)\|$, then there exists a point $u^* \in M$ such that $F(u^*) + G(u^*) = 0$.*

As a simple consequence of Theorem 4 one can obtain a new fixed-point theorem for a class of nonlinear mappings which are called local contractions (compare [3]).

A mapping F defined on an open subset M of a normed linear space X with values in X is said to be a feeble local contraction on M if for each $u \in M$ there are constants α_u, δ_u $(\alpha_u \in [0, 1), \delta_u > 0)$ such that $v \in B_{\delta_u}(u) \subset M \Rightarrow \|F(v) - F(u)\| \leqq \alpha_u \|v - u\|$.

Theorem 5. *Let X be a normed linear space, $M \subset X$ an open subset, $F : M \to X$ a feeble local contraction on M such that $\{u \in M \mid \|u - F(u)\| \leqq c\}$ is weakly compact and nonvoid for some $c \geqq 0$. If either* a) *F is weakly continuous on M, or* b) *M is convex, F is demicontinuous and $\psi(u) = \|u - F(u)\|$ is quasi-convex on M, then there is $u^* \in M$ such that $u^* = F(u^*)$.*

Theorem 6. *Let X be a normed linear space, M a non-void subset of X, $F : M \to M$ such that $u, v \in M$, $u \neq v \Rightarrow \|F(u) - F(v)\| < \|u - v\|$. If either* a) *$X$ is* *reflexive and* $(\mathrm{id} - F)(M)$ *is weakly closed, or* b) $(\mathrm{id} - F)(M)$ *is weakly closed and weakly compact, then there is a unique point* $u^* \in M$ *such that* $F(u^*) = u^*$.

Theorem 7. *Let* X *be a reflexive Banach space,* M *an open subset of* X, $M \neq \emptyset$, $F : M \to X$ *a feeble local contraction on* M. *If* $(\mathrm{id} - F)(M)$ *is weakly closed, then there is a point* $u^* \in M$ *such that* $u^* = F(u^*)$.

Remark. In comparison with Banach's contraction principle we need not assume in Theorem 5 that X is complete, M is closed, and that F is a map of M into M.

References

[1] *S. I. Pochožajev:* Normalnaja razrešimosť nelinejnych uravnenij v ravnomerno vypuklych prostranstvach. Funkcional. Anal. i Priložen. *3* (1969), 80—84.

[2] *F. E. Browder:* On the Fredholm alternative for nonlinear operators. Bull. Amer. Math. Soc. *76* (1970), 993—998.

[3] *M. Edelstein:* An extension of Banach's contraction principle. Proc. Amer. Math. Soc. *12* (1961), 7—10.

[4] *L. P. Belluce and W. A. Kirk:* Some fixed point theorems in metric and Banach spaces. Canad. Math. Bull. *12* (1969), 481—491.

[5] *J. Daneš:* Fixed point theorems, Nemyckii and Uryson operators, and continuity of nonlinear mappings. Comment. Math. Univ. Carolinae *11* (1970), 481—500.

[6] *J. Kolomý:* Some mapping theorems and solvability of nonlinear equations in Banach space. Comment. Math. Univ. Carolinae *12* (1971), 413—418.

CHARLES UNIVERSITY, PRAHA

ON SOME CONVERGENCE CLOSURES GENERATED BY FUNCTIONS

V. KOUTNÍK

Praha

0. In this note we shall consider several convergences defined on a given closure space and investigate relations between corresponding convergence closures and their modifications.

Let L be a set. Let $\mathfrak{L}$ be a set of pairs $(\{x_n\}, x)$ where $x_n \in L$, $n \in N$, and $x \in L$ satisfying the following axioms

($\mathscr{L}_0$) If $(\{x_n\}, x) \in \mathfrak{L}$, $(\{x_n\}, y) \in \mathfrak{L}$ then $x = y$.

($\mathscr{L}_1$) If $x_n = x$, $n \in N$, then $(\{x_n\}, x) \in \mathfrak{L}$.

($\mathscr{L}_2$) If $(\{x_n\}, x) \in \mathfrak{L}$ and $\{n_i\}$ is any subsequence of $\{n\}$, then $(\{x_{n_i}\}, x) \in \mathfrak{L}$.

Then $\mathfrak{L}$ is called a *convergence* on L. For each $A \subset L$ let $\lambda A = \{x \mid x \in L, \exists \{x_n\}, x_n \in A, n \in N, \ni (\{x_n\}, x) \in \mathfrak{L}\}$. Then (L, λ) is a T_1-closure space. It is denoted by $(L, \mathfrak{L}, \lambda)$ and called a *convergence space* [4]. Note that in general $\lambda^2 A \neq \lambda A$ and hence a convergence space may not be a topological space. To each convergence $\mathfrak{L}$ there corresponds convergence $\mathfrak{L}^*$ inducing the same convergence closure and satisfying the Urysohn axiom

($\mathscr{L}_3$) If each subsequence $\{x_{n_i}\}$ of a sequence $\{x_n\}$ contains a subsequence $\{x_{n_{i_j}}\}$ converging to a point x, then the sequence $\{x_n\}$ itself converges to x.

Let $(L, \mathfrak{L}, \lambda)$ be a convergence space. The finest topology coarser than λ is called a *topological modification of* λ and denoted by λ^{ω_1}. Recall that a convergence space $(L, \mathfrak{L}, \lambda)$ is called a *Fréchet space* if $\lambda^{\omega_1} = \lambda$, i.e., if (L, λ) is a topological space. A topological space (P, u) is called a *sequential space* if there exists a convergence closure μ for P such that $\mu^{\omega_1} = u$, i.e., if u is a topological modification of some convergence closure. If (P, u) is a closure space we shall denote by $C(u)$ [1]) the set of all continuous real-valued functions on (P, u).

A convergence space $(L, \mathfrak{L}, \lambda)$ is called *sequentially regular* [4] if for each point $x \in L$ and each sequence $\{x_n\}$ of points of L such that $x \in L - \lambda \bigcup_{n=1}^{\infty} (x_n)$ there is a function $f \in C(\lambda)$ such that $\{f(x_n)\}$ does not converge to $f(x)$.

---

[1]) We write simply $C(u)$ instead of $C((P, u))$ because we shall consider different closures for the same given set P.

Let $(L, \mathfrak{L}, \lambda)$ be a convergence space such that $C(\lambda)$ separates the points of L. The finest sequentially regular convergence closure for L coarser than λ is called a *sequentially regular modification of* λ and denoted by $\hat{\lambda}$ [3]. The finest completely regular topology for L coarser than λ is called a *completely regular modification of* λ and denoted by $\tilde{\lambda}$ [3]. The following relations hold

$$\lambda < \hat{\lambda} < \hat{\lambda}^{\omega_1} < \tilde{\lambda},$$
$$\lambda < \lambda^{\omega_1} < \tilde{\lambda},$$
$$\tilde{\hat{\lambda}} = \tilde{\lambda}.$$

1. Let (P, u) be a closure space and suppose that $C(u)$ separates the points of P. Clearly (P, u) is a separated space.

Consider the following convergences on P:

$\mathfrak{P}$: $(\{x_n\}, x) \in \mathfrak{P}$ if for each neighborhood U of x we have $x_n \in U$ for nearly all $n \in N$,

$\mathfrak{P}_{C(u)}$: $(\{x_n\}, x) \in \mathfrak{P}_{C(u)}$ if $\{f(x_n)\}$ converges to $f(x)$ for each $f \in C(u)$.

Denote π and $\pi_{C(u)}$ the corresponding convergence closures. $\mathfrak{P}$ is the usual convergence on P. The convergence space $(P, \mathfrak{P}, \pi)$ was called a *convergence space associated with* (P, u) in [3] and π is called a *sequential modification of* u in [1]. The convergence $\mathfrak{P}_{C(u)}$ was introduced by J. Novák in [5] who pointed out that there are interesting relations between the closures u, π, $\pi_{C(u)}$ and their modifications. Let us define still another convergence

$\mathfrak{P}_{C(\pi)}$: $(\{x_n\}, x) \in \mathfrak{P}_{C(\pi)}$ if $\{f(x_n)\}$ converges to $f(x)$ for each $f \in C(\pi)$.

Denote $\pi_{C(\pi)}$ the corresponding convergence closure. We shall show in Example 1 that generally $C(\pi) \neq C(u)$.

Lemma 1. $\pi < \pi_{C(\pi)} < \pi_{C(u)}$.

Proof. Let $(\{x_n\}, x) \in \mathfrak{P}$. By definition of $C(\pi)$ and $\mathfrak{P}_{C(\pi)}$ we have $(\{x_n\}, x) \in \mathfrak{P}_{C(\pi)}$. Since $\pi < u$ we have $C(u) \subset C(\pi)$ and hence $(\{x_n\}, x) \in \mathfrak{P}_{C(u)}$.

Proposition 1. *The convergence spaces* $(P, \mathfrak{P}_{C(\pi)}, \pi_{C(\pi)})$ *and* $(P, \mathfrak{P}_{C(u)}, \pi_{C(u)})$ *are sequentially regular.*

Proof. The assertion follows immediately from the definition of sequential regularity and from definitions of $\mathfrak{P}_{C(\pi)}$ and $\mathfrak{P}_{C(u)}$.

In view of Corollary 3 in [3] we have

Corollary 1. *The following are equivalent*

(a) $\pi_{C(\pi)} = \pi_{C(u)}$.
(b) $\pi^{\omega_1}_{C(\pi)} = \pi^{\omega_1}_{C(u)}$.
(c) $\tilde{\pi}_{C(\pi)} = \tilde{\pi}_{C(u)}$.

The question arises whether $\pi_{C(\pi)} = \pi_{C(u)}$ does not always hold. The following example shows that we may have $\pi_{C(\pi)} \neq \pi_{C(u)}$.

Example 1. Let $P = [0, 1]$. For $x \neq 0$ let $U_n = P \cap (x - 1/n,\ x + 1/n)$ be the usual local base at x and let the local base at 0 consist of sets $U_{n,S} = \{0\} \cup \cup ((0, 1/n) - S)$ where $n \in N$ and $|S| \leq \aleph_0$. Denote u the corresponding topology. Clearly $(P, \mathfrak{P}_{C(u)}, \pi_{C(u)})$ is the interval $[0, 1]$ with the usual topology. On the other hand 0 is π-isolated and therefore also $\pi_{C(\pi)}$-isolated. Hence $\pi_{C(\pi)} \neq \pi_{C(u)}$. Note that both spaces $(P, \mathfrak{P}_{C(\pi)}, \pi_{C(\pi)})$ and $(P, \mathfrak{P}_{C(u)}, \pi_{C(u)})$ are Fréchet spaces.

Proposition 2. *If* $C(\pi) = C(u)$ *then* $\pi_{C(\pi)} = \pi_{C(u)}$.

Proof. $C(\pi) = C(u)$ implies that $\mathfrak{P}_{C(\pi)} = \mathfrak{P}_{C(u)}$.

Corollary 2. *If* (P, u) *is a convergence or a sequential space then* $\pi_{C(\pi)} = \pi_{C(u)}$.

The condition in Proposition 2 is not necessary as the following example shows.

Example 2. Let $P = \bigcup_{m=1}^{\infty} \bigcup_{n=1}^{\infty} (x_{mn}) \cup (x_0)$. The points x_{mn} are isolated and the local base at x_0 consists of sets $U_{k,r} = \bigcup_{m=k}^{\infty} \bigcup_{n=r(m)}^{\infty} (x_{mn}) \cup (x_0)$ where $k \in N$ and r is any mapping of N into itself. Denote u the corresponding topology. The space (P, u) is normal. The spaces $(P, \mathfrak{P}, \pi)$, $(P, \mathfrak{P}_{C(\pi)}, \pi_{C(\pi)})$, and $(P, \mathfrak{P}_{C(u)}, \pi_{C(u)})$ are all discrete and hence $\pi_{C(\pi)} = \pi_{C(u)}$ while $C(\pi) \neq C(u)$.

Proposition 3. *If* $(\{x_n\}, x) \in \mathfrak{P}$ *whenever* $\{f(x_n)\}$ *converges to* $f(x)$ *for each* $f \in C(u)$ *then* $\pi_{C(\pi)} = \pi_{C(u)}$.

Proof. The condition implies $\mathfrak{P}_{C(u)} \subset \mathfrak{P}$ and the assertion follows by Lemma 1.

Corollary 3. *If* (P, u) *is completely regular then* $\pi_{C(\pi)} = \pi_{C(u)}$.

Again the condition in Proposition 3 is not necessary as the following example shows.

Example 3. Let $P = [0, 1]$. For $x \neq 0$ let $U_n = P \cap (x - 1/n,\ x + 1/n)$ be the usual local base at x and let the local base at 0 consist of sets $V_n = [0, 1/n) - \bigcup_{m=1}^{\infty} (1/m)$. Denote u the corresponding topology. We have $(\{1/m\}, 0) \notin \mathfrak{P}$ while $\{f(1/m)\}$ converges to $f(0)$ for each $f \in C(u)$. However, (P, u) is a Fréchet space and hence $\pi_{C(\pi)} = \pi_{C(u)}$ by Corollary 2.

Problem 1. What is the necessary and sufficient condition for the equality $\pi_{C(\pi)} = \pi_{C(u)}$?

2. Now let us characterize the convergence closures $\pi_{C(\pi)}$ and $\pi_{C(u)}$.

Theorem 1. $\pi_{C(\pi)} = \hat{\pi}$.

Proof. By Lemma 1 and Proposition 1 $\pi_{C(\pi)}$ is a sequentially regular convergence closure coarser than π. On the other hand let λ be a sequentially regular convergence closure for P coarser than π. To complete the proof we must show that $\pi_{C(\pi)} < \lambda$. Let $A \subset P$ and $x \in \pi_{C(\pi)}A$. Then there is a sequence $\{x_n\}$ of points of A which $\mathfrak{P}_{C(\pi)}$-converges to x. Hence $\{f(x_n)\}$ converges to $f(x)$ for each $f \in C(\pi)$. Since $\pi < \lambda$ it follows that $\{g(x_n)\}$ converges to $g(x)$ for each $g \in C(\lambda)$. Because λ is sequentially regular $\{x_n\}$ $\mathfrak{L}^*$-converges to x by Lemma 2 in [3]. Therefore $x \in \lambda A$.

Let (P, u) be a closure space and let w be the weak topology for P [2]. We shall denote by $(P, \mathfrak{P}_w, \pi_w)$ the convergence space associated with (P, w), i.e., π_w is the sequential modification of w.

Theorem 2. $\pi_{C(u)} = \pi_w$.

Proof. If $(\{x_n\}, x) \in \mathfrak{P}_{C(u)}$ then $\{f(x_n)\}$ converges to $f(x)$ for each $f \in C(u)$ and hence for each w-neighborhood U of x we have $x_n \in U$ for nearly all $n \in N$. Therefore $(\{x_n\}, x) \in \mathfrak{P}_w$. On the other hand if $(\{y_n\}, y) \in \mathfrak{P}_w$ then clearly $\{f(y_n)\}$ converges to $f(y)$ for each $f \in C(u)$ so that $(\{y_n\}, y) \in \mathfrak{P}_{C(u)}$.

Since (P, w) is a completely regular space it follows that if v is any of the closures u, π, $\pi_{C(\pi)}$, $\pi_{C(u)}$ or their modifications then we have $\pi < v < w$.

3. Finally let us consider the relations between the convergence closures π, $\pi_{C(\pi)}$, $\pi_{C(u)}$ and the closure u.

If (P, u) is just a closure space then the only statement we can make is the obvious $\pi < u$.

If (P, u) is a topological space then clearly $\pi < \pi^{\omega_1} < u$. However, both $\pi_{C(u)} \lneqq u$ (Example 2) and $u \lneqq \pi_{C(u)}$ (Example 3) may occur. The same holds for $\pi_{C(\pi)}$.

Finally, if (P, u) is a completely regular space, then $\tilde{\pi}_{C(u)} < u$. It follows from Lemma 6 and Theorem 6 in [3] that $\tilde{\pi}_{C(u)} = u$ if and only if $C(\pi) = C(u)$.

References

[1] *E. Čech:* Topological spaces. Academia, Praha, 1966.

[2] *L. Gillman and M. Jerison:* Rings of continuous functions. D. Van Nostrand Company, Inc., Princeton, New Jersey, 1960.

[3] *V. Koutník:* On sequentially regular convergence spaces. Czechoslovak Math. J. *17 (92)* (1967), 232–247.

[4] *J. Novák:* On convergence spaces and their sequential envelopes. Czechoslovak Math. J. *15 (90)* (1965), 74–100.

[5] *J. Novák:* On some topologies defined by a class of real-valued functions. General Topology and its Applications *1* (1971), 247–251.

INSTITUTE OF MATHEMATICS OF THE CZECHOSLOVAK ACADEMY OF SCIENCES, PRAHA

ON A CONVERGENCE PROPERTY OF SET ALGEBRAS

P. KRATOCHVÍL

Praha

Let X be a nomenpty set, 2^X the algebra of all subsets of the set X and λ the *convergence closure operator* on 2^X. We recall its definition. For each $\mathfrak{M} \subset 2^X$,

$$\lambda\mathfrak{M} = \{A; A \in 2^X \text{ and there is a sequence of sets } A_n \in \mathfrak{M} \text{ such that}$$

$$A = \lim_{n\to\infty} A_n = \bigcup_{k=1}^{\infty} \bigcap_{n=k}^{\infty} A_n = \bigcap_{k=1}^{\infty} \bigcup_{n=k}^{\infty} A_n\} .$$

A *power* of λ is defined by the transfinite induction: $\lambda^0\mathfrak{M} = \mathfrak{M}$, $\lambda^\alpha\mathfrak{M} = \bigcup_{\beta<\alpha} \lambda(\lambda^\beta\mathfrak{M})$ for an ordinal $\alpha \neq 0$ and $\mathfrak{M} \subset 2^X$.

Let a set algebra $\mathfrak{A}$, $\mathfrak{A} \subset 2^X$, be given. It has been noticed (see [2]) that $\lambda^\alpha\mathfrak{A}$ is also a set algebra for an arbitrary ordinal α and $\lambda^{\omega_1}\mathfrak{A}$ (ω_1 = the first uncountable ordinal) is equal to the σ-algebra generated by $\mathfrak{A}$. An easy completion of the well-known statement (see [1], Chap. 1, Exercise 13) claims:

(1) The image $\mathrm{P}[\lambda^{\omega_1}\mathfrak{A}] = \{\mathrm{P}A; A \in \lambda^{\omega_1}\mathfrak{A}\}$ is a closed subset of the real line for each probability measure P.

J. Novák has raised the problem to find the least ordinal α such that $\mathrm{P}[\lambda^\alpha\mathfrak{A}]$ is always closed. The answer is given by

Theorem. *The number* $\alpha = 2$ *is the least ordinal such that* $\mathrm{P}[\lambda^\alpha\mathfrak{A}]$ *is closed.*

Proof. 1) $\mathrm{P}[\lambda^2\mathfrak{A}]$ is closed. Let a real number a belong to the closure of $\mathrm{P}[\lambda^2\mathfrak{A}]$. From (1) it follows that there is $B \in \lambda^{\omega_1}\mathfrak{A}$ such that $\mathrm{P}B = a$. The definition of the outer measure implies the existence of sets $B_{ni} \in \mathfrak{A}$, $n = 1, 2, \ldots, i = 1, 2, \ldots,$ such that $a \leq \mathrm{P}(\bigcup_{i=1}^{\infty} B_{ni}) \leq a + 1/n$ and $B \subset \bigcup_{i=1}^{\infty} B_{ni} \in \lambda\mathfrak{A}$ for each $n = 1, 2, \ldots$. Hence $a = \mathrm{P}(\bigcap_{n=1}^{\infty} \bigcup_{i=1}^{\infty} B_{ni}) \in \mathrm{P}[\lambda^2\mathfrak{A}]$.

2) The image $\mathrm{P}[\lambda\mathfrak{A}]$ need not be closed as the following example shows.

Let R denote the real line, $\mathfrak{R}$ the algebra generated by semiclosed intervals of the form $\langle a, b)$, $a, b \in R$.

Lemma. *If* $A_n \in \mathfrak{R}$, $n = 1, 2, \ldots, A = \lim_{n\to\infty} A_n$, *then there is a set* Y, $\emptyset \neq Y \in \mathfrak{R}$ *such that* $Y \subset A$ *or* $Y \cap A = \emptyset$.

Proof. We denote $B_i = R + A_i + A_{i+1}$, $C_i = \bigcap_{n=i}^{\infty} B_n$, $i = 1, 2, \ldots,$ where $+$ denotes the symmetric difference. Evidently $B_i \in \mathfrak{R}$ and

$$\bigcup_{n=1}^{\infty} C_n = R\,. \tag{2}$$

Now two cases are possible:

1) There is a natural number n_0 such that $C_{n_0} = R$. Then $B_{n_0} = B_{n_0+1} = \ldots = R$ and $A_{n_0} = A_{n_0+1} = \ldots = A$. At least one of the sets $Y = A_{n_0}$ or $Y = R - A_{n_0}$ possesses the declared property.

2) All the sets $C_n \neq R$. Then there is $B_k \neq R$. We choose a compact non-degenerate interval $T_1 \subset R - B_k$. Suppose that compact non-degenerate intervals $T_1 \supset T_2 \supset \ldots \supset T_n$ have been constructed in such a way that $T_i \cap C_i = \emptyset$, $i = 1, 2, \ldots, n$. Denote $T_n^* = T_n - \{r\}$, where r is the right end point of T_n. Now two cases are possible:

a) $T_n^* \subset C_{n+1}$. Then $A_{n+1} \cap T_n^* = A_{n+2} \cap T_n^* = \ldots = A \cap T_n^*$ (otherwise there would be $k > n$ and a point $x \in (A_k + A_{k+1}) \cap T_n^* \subset (R - B_k) \cap C_{n+1} = \emptyset$). Hence the set Y from the Lemma can be found by using a nonempty measurable subset of $T_n^* \cap A$ or $T_n^* - A$.

b) $T_n^* \not\subset C_{n+1}$. Then there is a point $x \in T_n^*$, $x \notin C_{n+1}$, i.e. there is $j \geqq n + 1$ such that $x \notin B_j$. We pick out a non-degenerate compact interval $T_{n+1} \subset T_n^* - B_j$.

In the case a) we have the set Y as desired in the Lemma. If the case a) does not occur, then we have a non-increasing sequence of non-degenerate compact intervals T_n. The intersection of it is disjoint with $\bigcup_{n=1}^{\infty} C_n$ and nonempty. We get a contradiction with (2).

Example. Let $Q = \{q_1, q_2, \ldots\}$ be the set of all rational numbers, $s_n = q_n + \sqrt{2}$, $S = \{s_1, s_2, \ldots\}$. A probability P on $\lambda^2\mathfrak{R}$ is defined by the relations $\mathrm{P}(\{q_n\}) = 2/3^{2n-1}$, $\mathrm{P}(\{s_n\}) = 2/3^{2n}$, $\mathrm{P}(R - Q - S) = 0$. It is easy to see that sets $A_n \in \mathfrak{R}$ can be chosen in such a way that $q_i \in A_n$ and $s_i \notin A_n$ for $i = 1, 2, \ldots, n$. Then $(3/4)(1 - 1/9^n) = \sum_{i=1}^{n} \mathrm{P}q_i \leqq \mathrm{P}A_n \leqq 1 - \sum_{i=1}^{n} \mathrm{P}s_i = 3/4 + 1/(4 \cdot 9^n)$ and hence $\lim_{n\to\infty} \mathrm{P}A_n = 3/4$. Now, let A be any set of $\lambda^2\mathfrak{R}$ such that $\mathrm{P}A = 3/4$. Then the uniqueness of the ternary expansion $3/4 = \sum_{i=1}^{\infty} 2/3^{2i-1} = \sum_{i=1}^{\infty} \mathrm{P}q_i = \mathrm{P}(Q)$ implies $Q \subset A$ and $A \cap S = \emptyset$. Then $A \notin \lambda\mathfrak{R}$ as follows from the Lemma and from the fact that Q and S are dense subsets of R. It follows that $3/4$ is a point of the closure of the P-image of $\mathfrak{R}$ but there is no element $A \in \lambda\mathfrak{R}$ such that $\mathrm{P}A = 3/4$.

Remark. Part 1) of the proof of the Theorem can be proved without using an outer measure. The outer measure can be replaced by Marczewski's characteristic function of a sequence of sets (see [3]). Problems and importance of elimination of the notion of an outer measure from measure theory are treated in [2].

References

[1] *M. Loève:* Probability theory. 2nd edition, Van Nostrand, Princeton, 1960.
[2] *J. Novák:* Über die eindeutigen stetigen Erweiterungen stetiger Funktionen. Czechoslovak Math. J. *8* (1958), 344—355.
[3] *E. Szpilrajn:* The characteristic function of a sequence of sets and some of its applications. Fund. Math. *31* (1938), 207—223.

INSTITUTE OF MATHEMATICS OF THE CZECHOSLOVAK ACADEMY OF SCIENCES, PRAHA

ON CONDENSATION NUMBERS

P. B. KRIKELIS

Athens

In this note we consider the weight (=character) of a topological space at a point and some other cardinal numbers which describe the "condensation" of points at a given point.

Notation. The cardinal number of a set B is denoted by card B. We denote by k a fixed sufficiently large cardinal, K denotes the set of cardinals $\leqq k$, and $\hat{K}$ denotes the Kurepa completion of K; l will denote an element of $\hat{K}$. The letter E denotes a topological space, α a point of E, V a neighbourhood of α in E, B a subset of E, $\bar{B}$ the closure of B. The weight of E at α is denoted by $w(\alpha)$.

$P(\alpha, l)$ will be an abbreviation for "if $\alpha \notin B$ and card $B \leqq l$, then $\alpha \notin \bar{B}$". If α is given, $l(\alpha)$ will denote any l for which $P(\alpha, l)$ holds. We can consider the following elements of $\hat{K}$: $\sup l(\alpha)$, $\inf_{\alpha \in E} \sup l(\alpha)$, $\sup_{\alpha \in E} \sup l(\alpha)$. In particular, if $\sup l(\alpha)$ is constant for $\alpha \in E$, then E is "homogeneous" in a certain sense. If not, then the equality $\sup l(\alpha_1) = \sup l(\alpha_2)$ defines a partition of E which may be worth consideration.

Clearly, if $0 \leqq l \leqq \inf_{\alpha \in E} \sup l(\alpha)$, then $P(\alpha, l)$ holds for all $\alpha \in E$.

Proposition. *For every non-isolated* $\alpha \in E$, $w(\alpha) > \sup l(\alpha)$.

This follows at once from the Axiom of Choice; the inequality is strict, due to considering $\hat{K}$ instead of K.

Let $(D, \geqq)$ be a directed set and let $A \subset D$. Let $\mathscr{T}$ denote the collection of all maximal totally ordered subsets T of D such that for any $a \in A$ there is an element $d \in T$ with $a \geqq d$. In the following we assume that A is such that $\bigcup \mathscr{T} = D$.

For every $T \in \mathscr{T}$ let μ_T denote the least cardinality of a cofinal subset of T. The set of all numbers μ_T will be denoted by $M(D, A)$. It can be shown that the supremum (in $\hat{K}$) of the set $M(D, A)$ is a dimension in the sense of [3], [4]. This supremum will be denoted by dep (D, A) and called the depth of D with respect to A; the least cardinal $\geqq$ dep (D, A) will be denoted by dep* (D, A). The least cardinality of a cofinal subset of D will be called the weight of D and denoted by $w(D)$, and the cardinality of $\mathscr{T}$ will be denoted by br (D, A) and called the breadth of D with respect to A.

Clearly, $w(D) \leqq \text{dep}^* (D, A) \, \text{br} (D, A)$. Hence, $\text{br} (D, A) = w(D)$ whenever $\text{dep}^* (D, A) < w(D)$.

We denote by D_α the collection of all neighbourhoods of α (in E) ordered by inclusion; A_α will denote the collection of all neighbourhoods of α of the form $E - (x)$, $x \in E$. We shall call dep (D_α, A_α) the depth of E at α.

Problem. Does there exist, for any non-void set M of infinite cardinal numbers, a normal space E and a point $\alpha \in E$ such that $M(D_\alpha, A_\alpha) = M$?

Proposition. *If card B is less than the depth of E at* α, *then* $\alpha \notin \bar{B} - B$. *In particular, if* dep* $(D_\alpha, A_\alpha) = w(\alpha)$, *then* $\alpha \in \bar{B} - B$ *implies* $w(\alpha) \leqq$ card B.

Remarks. 1) The condition dep* $(D_\alpha, A_\alpha) = w(\alpha)$ is more general than the assumption of the existence of a totally ordered basis of neighbourhoods of α. — 2) Using the cardinal $\inf_{\alpha \in E} \sup l(\alpha)$ we obtain a classification of topological spaces which starts with the T_1-spaces (namely, E is a T_1-space iff $\inf_{\alpha \in E} \sup l(\alpha) \geqq 1$). — 3) $\inf_{\alpha \in E} \sup l(\alpha)$ and $\sup_{\alpha \in E} \sup l(\alpha)$ are "dimension functions" in the sense of [3]. — 4) Analogous questions can be investigated in generalized topological spaces; cf. also [1]. — 5) There are certain relations between notions introduced in this note, and U-sefs considered in [2]. — 6) In the author's opinion, the term "condensation number at α" is preferable to "weight".

References

[1] *E. Čech et B. Pospíšil:* I. Sur les espaces compacts II. Sur les caractères des points dans les espaces $\mathscr{L}$. Publ. Fac. Sci. Univ. Masaryk *258* (1938), 1—14.

[2] *P. B. Krikelis:* Certain relations between generalized topology and universal algebra. Proc. Internat. Sympos. on Topology and its Applications (Herceg-Novi, 1968). Savez Društava Mat. Fiz. i Astronom., Belgrade, 1969, 233—238.

[3] *S. P. Zervos:* Une notion abstraite de dimension. C. R. Acad. Sci. Paris Sér. A—B *261* (1965), 859—862.

[4] *S. P. Zervos:* Une définition générale de la dimension. Séminaire Delange-Pisot-Poitou, 1965/66, No. 9.

VERY UNLATTICELIKE ORDERED SPACES

E. H. KRONHEIMER

London

All partial orderings $<$ ("*strictly precedes*") are to satisfy $p < q < p \Rightarrow p = q$; the converse implication may, but need not, hold – indeed a point which strictly precedes itself will be called *singular*. The reflexive relation $\leqq$ ("*precedes*") is associated with $<$ in the usual way ($p \leqq q$ iff $p < q$ or $p = q$); and we write

$$L(q) = \{x \mid x < q\}, \quad L[Q] = \bigcup_{q \in Q} L(q).$$

A non-void subset D of a partially ordered set is *directed* (resp. *strictly directed*) if, given two points in D, there exists a point in D succeeding (resp. strictly succeeding) both; D is a *strict ideal* if it is strictly directed and contains all the predecessors of each of its points. Call $L[Q]$ the set *generated* by Q: then every strictly directed set generates a strict ideal, and every strict ideal generates itself. Any set of the form $L(q)$ is called a *principal strict ideal.*

Borrowing a term from Michael [2], we call a partially ordered set a *cushion* if it satisfies any of the following equivalent conditions:

(a) *Every point has a strict predecessor*; *and, whenever* $p_1, p_2 < r$, *there exists* q *satisfying* $p_1, p_2 < q < r$.

(b) *Every directed subset generates a strict ideal.*

(c) *Every principal strict ideal is a strict ideal.*

We equip each cushion with the topology determined by the base $\{\langle p, q]\}_{p<q}$, where $\langle p, q] = \{x \mid p < x \leqq q\}$. The singular points of a cushion are then its isolated points, and it is Hausdorff if and only if $p = q$ whenever $L(p) = L(q)$. A subset S of the cushion X is called a *subcushion* of X if S (with the restricted ordering) is itself a cushion whose topology coincides with its topology as a subspace of X. A function f between cushions is called a *cushion map* if it is continuous and $p < q$ implies $f(p) < f(q)$.

Example 1. Let E be a normal T_1-space. (Somewhat weaker separation axioms are in fact sufficient.) Let cE denote the collection of its open subsets other than $\emptyset$ and E, ordered by putting $U < V$ iff $U^- \subset V$. Then cE is a Hausdorff cushion which is non-singular if and only if E is connected. If E, F are normal T_1-spaces and $\theta : E \to F$ is a closed continuous surjection, then $c\theta : cF \to cE$, where $c\theta(W) = \theta^{-1}[W]$, is a cushion map.

Example 2. Let $\mathbf{R}^-$ denote the non-positive real numbers and M be a pseudometric space. Let kM denote the set $M \times \mathbf{R}^-$, ordered by putting $(m_1, r_1) < (m_2, r_2)$ iff $d(m_1, m_2) < r_2 - r_1$. Then kM is a non-singular cushion which is Hausdorff if and only if M is Hausdorff (i.e., metric). If M, N are pseudometric spaces and the function $\varphi : M \to N$ satisfies $d(\varphi(m_1), \varphi(m_2)) < \lambda\, d(m_1, m_2)$ for some fixed positive number λ, then the function $k_\lambda \varphi : kM \to kN$, where $k_\lambda\, \varphi(m, r) = (\varphi(m), \lambda r)$, is a cushion map.

A cushion in which every strict ideal is principal is called *complete*. Complete cushions have some pleasant properties. Let us, for instance, say that a net $(s_d)_{d \in D}$ (on the directed set D) in a partially ordered space is *increasing* if $d \leqq e$ implies $s_d \leqq s_e$. Then a cushion X is complete (resp. Hausdorff) if and only if every increasing net in X has at least (resp. at most) one limit point. Again: every closed subcushion of a complete cushion is complete, and every complete subcushion of a Hausdorff cushion is closed. The results we shall establish here are two more specific ones.

Theorem 1. *The cushion cE is complete if and only if the topological space E is compact.*

Proof. Assume first that E is compact, and let $\mathscr{P}$ be an ideal in cE; then $P^* = \bigcup_{P \in \mathscr{P}} P$ is open and non-void. Suppose Q is a non-void open set such that $Q^- \subset P^*$. Since $\mathscr{P}$ covers Q^-, so does some finite subcollection $\{P_1, \ldots, P_m\}$ of $\mathscr{P}$. Since $\mathscr{P}$ is directed, some member of $\mathscr{P}$ contains all the P_i and hence contains Q^-. It follows that $Q \in \mathscr{P}$. Since $E \notin \mathscr{P}$, this argument (with $Q = E$) shows that $P^* \neq E$; so $P^* \in cE$. It also shows that $L(P^*) \subset \mathscr{P}$. On the other hand, if $P \in \mathscr{P}$, then $P < P' \in \mathscr{P}$ for some P', so that $P^- \subset P' \subset P^*$: therefore $P \in L(P^*)$. It follows that $\mathscr{P} = L(P^*)$, and hence that cE is complete.

To prove the converse, assume E has an open cover $\mathscr{U}$ with no finite refinement. Let $\mathscr{V}$ denote the collection of all non-void sets expressible as finite unions of members of $L[\mathscr{U}]$. Then $\mathscr{V}$ is a directed subset of cE which fails to generate a principal strict ideal; for if $L[\mathscr{V}] = L(W)$, where $W \in cE$, then (since $\mathscr{V}$ is actually strictly directed) each member of $\mathscr{V}$ is a subset of W, contradicting the fact that $\mathscr{V}$ covers E.

Theorem 2. *The cushion kM is complete if and only if the pseudometric space M is complete.*

Proof. Suppose M is a complete pseudometric space and P is a strict ideal in kM. Let

$$r^* = \sup \{r \mid (x, r) \in P \text{ for some } x \in M\},$$

and choose $x_0, x_1, \dots$ in M such that $(x_n, r^* - 2^{-n}) \in P$ for each n. Then (x_n) is a Cauchy sequence; and $P = L(x^*, r^*)$, where x^* is a limit of (x_n).

Conversely, suppose that kM is a complete cushion and (y_n) is a Cauchy sequence. Define

$$s_n = -2 \sup_{k \geq 0} d(y_n, y_{n+k}) .$$

The set $\{(y_0, s_0), (y_1, s_1), \dots\}$ is directed and therefore generates a strict ideal: call this $L(q)$. Then q must be of the form $(y, 0)$, and y must be a limit of (y_n).

If X is a complete Hausdorff cushion and $f: X \to X$ a cushion map satisfying $a < f(a)$ for some a, then f has a fixed point. (To construct it, put $a_0 = a$, $a_{n+1} = f(a_n)$. If the directed set $\{a_0, a_1, \dots\}$ generates $L(b)$, then $f(b) = b$.) Theorem 2 shows that this result includes the Banach fixed-point theorem: for if $\varphi: M \to M$ satisfies $d(\varphi(m_1), \varphi(m_2)) < \lambda\, d(m_1, m_2)$, where $\lambda < 1$, and if m is any point of M, then (m, r) strictly precedes $k_\lambda\, \varphi(m, r)$ in the cushion kM for all sufficiently large $-r$; and if $k_\lambda \varphi$ has a fixed point, so has φ. (It may be noted that, working with reflexive orderings, one obtains, instead of propositions about the (complete) cushion kM, closely analogous propositions about (Dedekind σ-complete) ordered sets. See [1].)

References

[1] *R. DeMarr:* Partially ordered spaces and metric spaces. Amer. Math. Monthly *72* (1965), 628–631.

[2] *E. Michael:* Yet another note on paracompact spaces. Proc. Amer. Math. Soc. *10* (1959), 309–314.

BIRKBECK COLLEGE, UNIVERSITY OF LONDON

ON A MECHANISM OF CHOOSING MORPHISMS IN CONCRETE CATEGORIES[1])

L. KUČERA and A. PULTR

Praha

It was observed (see, e.g., [2], [3], [4], [5]) that many concrete categories (i.e. categories with fixed forgetful functors) may be viewed upon as follows: a functor $F : \mathbf{Set} \to \mathbf{Set}$ is given, the objects are some couples (X, r) where X is a set and r a subset of $F(X)$, and the morphisms from (X, r) into (Y, s) are those mappings for which $F(f)(r) \subset s$ (or $F(f)(s) \subset r$, if F is contravariant). Thus, e.g., the category of topological spaces and their continuous mappings may be obtained using the contravariant power set functor P^- for F, the category of uniform spaces and uniformly continuous mappings may be obtained using the functor $P^- \circ Q$, where Q sends X into $X \times X$, etc.

In other words, it is often the case that a concrete category $(\mathfrak{K}, U)$ is realizable in an $S(F)$ (to recall the definitions from [2] and [4]: $S(F)$ is the category of all (X, r) with X sets and $r \subset F(X)$ for objects, morphisms from (X, r) into (Y, s) are triples $((X, r), f, (Y, s))$ with mappings $f : X \to Y$ satisfying $F(f)(r) \subset s$ or $F(f)(s) \subset r$ according to the variance of F; $S(F)$ is considered as a concrete category endowed by the forgetful functor sending (X, r) to X and $((X, r), f, (Y, s))$ to f. A concrete category $(\mathfrak{K}, U)$ is said to be realizable in a concrete category $(\mathfrak{L}, V)$ if there is a full embedding $\Phi : \mathfrak{K} \to \mathfrak{L}$ with $V \circ \Phi = U$). The aim of this note is to present a necessary and sufficient condition for realizability in an $S(F)$. The proofs will be outlined very roughly. In detail, they will appear in a longer forthcoming paper.

Obviously, the following two conditions on $(\mathfrak{K}, U)$ are necessary for realizability in an $S(F)$:

(J) If $\alpha : a \to b$ is an isomorphism in $\mathfrak{K}$ and if $U(\alpha) = \mathrm{id}_{U(a)}$, then $\alpha = \mathrm{id}_a$.

(S) For every set X, the class $\{a \mid U(a) = X\}$ is a set.

On the other hand, it is also very easy to show on an ad hoc example that they are not sufficient. We have, however, the following

Theorem 1. *If* $(\mathfrak{K}, U)$ *satisfies* (J), (S) *and*

(R) *for every morphism* α *there are morphisms* β *and* γ *such that* $U(\beta)$ *is one-to-one,* $U(\gamma)$ *onto and* $\alpha = \beta \circ \gamma$,
then there is a covariant F *such that* $(\mathfrak{K}, U)$ *is realizable in* $S(F)$.

[1]) Preliminary communication.

To prove this, we show first that if $(\mathfrak{K}, U)$ satisfies (J), (S) and (R), it is realizable in an $(\mathfrak{L}, V)$ satisfying (J), (S) and

(P) for every morphism α and every two mappings f, g such that $U(\alpha) = f \circ g$ there are morphisms β and γ with $\alpha = \beta \circ \gamma$, $U(\beta) = f$ and $U(\gamma) = g$.

Then, we show that if $(\mathfrak{K}, U)$ satisfies (J), (S) and (P), it is realizable in an $S(F)$ with F constructed as follows:

$$F(X) \quad = \{A \mid A \subset \{a \mid U(a) = X\} \,\&\, ((a \in A \,\&\, \exists \alpha : a' \to a, U(\alpha) = \mathrm{id}) \Rightarrow a' \in A)\},$$
$$F(f)(A) = \{b \mid U(b) = Y, \exists a \in A, \exists \varphi : b \to a, f = U(\varphi)\} \text{ for } f : X \to Y.$$

Of course, the condition (R) is not necessary. To be able to formulate the necessary and sufficient condition mentioned above, let us now give the following

Definition. Let $(\mathfrak{K}, U)$ be a concrete category, a an object of $\mathfrak{K}$ and $m : X \to U(a)$ a one-to-one mapping. Denote by $\mathscr{S}(m)$ the class of all mappings $f : Y \to X$ such that there is a morphism $\alpha : b \to a$ in $\mathfrak{K}$ with $U(\alpha) = m \circ f$. If $\alpha : a \to b$ is a morphism of $\mathfrak{K}$, a U-image of α is any $\mathscr{S}(m)$ such that there is an onto mapping p with $U(\alpha) = m \circ p$. Two morphisms are said to be equivalent if they have a common U-image. We say that $(\mathfrak{K}, U)$ satisfies (E), if

(E) There is a class M of morphisms of $\mathfrak{K}$ such that

(1) for every morphism α there is a $\beta \in M$ equivalent to α,
(2) for every cardinal $\mathfrak{a}$, $\{\alpha \mid \alpha \in M, \text{card range } \alpha \leqq \mathfrak{a}\}$ is a set.

It is not difficult to see that every $S(F)$ has (E) (one can take the class of all embeddings of naturally induced subobjects into objects $(\mathfrak{a}, r)$ for M), and that full concrete subcategories of concrete categories inherit (E). Thus, we have

Statement. (E) *is a necessary condition for* $(\mathfrak{K}, U)$ *to be realizable in an* $S(F)$.

Further, the following lemma holds

Lemma. *If* $(\mathfrak{K}, U)$ *has* (J) *and* (E), *it is realizable in an* $(\mathfrak{L}, V)$ *with* (J), (R) *and* (S).

Consequently, we obtain

Theorem 2. $(\mathfrak{K}, U)$ *is realizable in an* $S(F)$ *iff it has the properies* (J) *and* (E).

Of the two theorems, of course, Theorem 1 is more applicable, since the properties there are very easy to check. As a corollary we obtain, e.g., that every category of topological spaces in which the morphisms are such that every homeomorphism is an isomorphism, and which satisfies (R) is realizable in an $S(F)$. More generally, this holds for any category of structures in the sense of Bourbaki ([1]) satisfying (R).

References

[1] *N. Bourbaki:* Éléments de Mathématique. Livre I (Théorie des ensembles). Hermann & Cie, Paris.

[2] *Z. Hedrlín and A. Pultr:* On categorial embeddings of topological structures into algebraic. Comment. Math. Univ. Carolinae 7 (3) (1966), 377–400.

[3] *Z. Hedrlín, A. Pultr and V. Trnková:* Concerning a categorial approach to topological and algebraic theories. General Topology and its Relations to Modern Analysis and Algebra, II (Proc. Second Prague Topological Sympos., 1966). Academia, Prague, 1967, 176–181.

[4] *A. Pultr:* On selecting of morphisms among all mappings between underlying sets of objects in concrete categories and realisations of these. Comment. Math. Univ. Carolinae *8* (1) (1967), 53–83.

[5] *A. Pultr and V. Trnková:* On realization and boundability of concrete categories in which the morphisms are choiced by local conditions. Comment. Math. Univ. Carolinae *8* (4) (1967), 651–662.

CHARLES UNIVERSITY, PRAHA

THE CATEGORY OF COMPACT HAUSDORFF SPACES IS NOT ALGEBRAIC IF THERE ARE TOO MANY MEASURABLE CARDINALS[1])

L. KUČERA and A. PULTR

Praha

At the 1st Prague Topological Symposium in 1961, J. R. Isbell, who proved before ([5]) that if there is no measurable cardinal the category **Comp** of compact Hausdorff spaces and their continuous mappings is algebraic (i.e., fully embeddable into a category of algebras), put a question whether this holds for the category of all topological spaces. It was positively answered by Hedrlín and Pultr in [2] (communicated at the 2nd Prague Topological Symposium, see also [3]) under a weaker assumption

(M) There is a cardinal $\mathfrak{a}$ such that every ultrafilter closed under intersection of $\mathfrak{a}$ elements is trivial.

In the quoted paper and in further ones (see, e.g., [6]) it was, moreover, proved that every concretizable category which was "constructive" in some sense was algebraic. Since, on the other hand, there was no example of a non-algebraic concretizable category known, the conjecture arose that every concretizable category was algebraic. And this is really the case, by a result of Hedrlín and Kučera proved in 1969 (still unpublished) – under the assumption (M).

In this note, we want to communicate that under non (M), however, the situation is entirely different. Namely, e.g. the categories $\mathbf{Set}^{op}$ and **Comp** become non-algebraic. Full proofs of these facts will be given in a longer forthcoming paper. Here we will sketch them roughly.

Theorem 1. *Under* non (M), $\mathbf{Set}^{op}$ *is not algebraic.*

The proof of this is based on the following

Lemma. *If there exists an $\mathfrak{a}$-additive non-trivial two-valued measure on a set X and if $F : \mathbf{Set} \to \mathbf{Set}$ is a contravariant faithful functor such that* card $F(1) \leqq \mathfrak{a}$, *then there is a mapping $\mu : F(X) \to F(1)$ such that* (1) *for every $\xi : 1 \to X$, $\mu \neq F(\xi)$, and* (2) *if* card $A \leqq \mathfrak{a}$, *then for every $\alpha : A \to X$ there is a $\xi : 1 \to X$ with $\mu \circ \alpha = F(\xi) \circ \alpha$.*

We cannot go here into the proof of this statement. To elucidate what it says, let us only point out that the measure itself has the required properties if F is the

[1]) Preliminary communication.

contravariant power set functor. Let us show now how the theorem follows: Suppose $\mathbf{Set}^{\mathrm{op}}$ is algebraic. Then, by [1], there is a full embedding $\Phi : \mathbf{Set}^{\mathrm{op}} \to \mathfrak{R}$ where $\mathfrak{R}$ is the category of sets with binary relations and relation preserving mappings. Denote by U the natural forgetful functor $\mathfrak{R} \to \mathbf{Set}$. Put $F = U \circ \Phi$. If non (M) holds, there is a mapping $\mu : F(X) \to F(1)$ with the properties described above. By (2), μ carries a morphism from $\Phi(X)$ into $\Phi(1)$ and, by (1), this morphism is not in the image of Φ, in a contradiction with the assumption that Φ is full.

Theorem 2. *Under* non (M), **Comp** *is not algebraic.*

The proof goes, roughly, as follows: If a concrete category $(\mathfrak{R}, U)$ is fully embeddable into a category of algebras in such a way that the underlying set of the algebra corresponding to an object always contains the original underlying set and that the homomorphisms corresponding to morphisms are extensions of their original underlying mappings, it is not difficult to prove that then a category obtained from $(\mathfrak{K}, U)$ endowing the objects by relational systems on their underlying sets, and taking only those morphisms the underlying mappings of which preserve them, is again algebraic.

Using Pontrjagin duality we can prove that $\mathbf{Set}^{\mathrm{op}}$ is fully embeddable into the category thus obtained from the category of compact abelian groups adding two unary relations. Consequently, it is fully embeddable into the category obtained from **Comp** by adding one ternary and two binary relations. Thus, **Comp** endowed by the natural forgetful functor cannot be, by Theorem 1, embedded into a category of algebras in the way described above. But, by Theorem 1.8 in [7], it cannot be then fully embedded into a category of algebras at all.

Let us remark that Theorem 2 in a way completes the answer to the mentioned Isbell's problem on the category of topological spaces. Namely, we have

Corollary. *The following three statements are equivalent*:

(a) (M).

(b) **Comp** *is algebraic.*

(c) *The category of topological spaces and their continuous mappings is algebraic.*

As an immediate consequence of Theorems 1 and 2 we obtain the following

Corollary. *Under* non (M) *the following categories are non-algebraic*:

the category of complete Boolean algebras and their complete homomorphisms,
the category of topological spaces and their closed continuous mappings,
the category of uniform spaces and their uniformly continuous mappings.

Several further categories (topological spaces with open or open continuous mappings, sets with mappings onto etc.) can be shown to be non-algebraic under non (M) using more complicated considerations. In contrast with the situation with (M), under non (M) the property to be algebraic becomes a very special one. In fact, we do not know at the moment an example of a nice category without a small left adequate (see [4], [5]) which would be algebraic under non (M).

References

[1] *Z. Hedrlín and A. Pultr:* On full embeddings of categories of algebras. Illinois J. Math. *10* (3) (1966), 392—406.

[2] *Z. Hedrlín and A. Pultr:* On categorial embeddings of topological structures into algebraic. Comment. Math. Univ. Carolinae *7* (3) (1966), 377—400.

[3] *Z. Hedrlín, A. Pultr and V. Trnková:* Concerning a categorial approach to topological and algebraic theories. General Topology and its Relations to Modern Analysis and Algebra, II (Proc. Second Prague Topological Sympos., 1966). Academia, Prague, 1967, 176—181.

[4] *J. R. Isbell:* Adequate subcategories. Illinois J. Math. *4* (1960), 541—552.

[5] *J. R. Isbell:* Subjects, adequacy, completeness and categories of algebras. Rozprawy matematyczne *36*, Warzawa, 1964.

[6] *A. Pultr:* Limits of functors and realisations of categories. Comment. Math. Univ. Carolinae *8* (4) (1967), 663—682.

[7] *A. Pultr:* On full embeddings of concrete categories with respect to forgetful functors. Comment. Math. Univ. Carolinae *9* (2) (1968), 281—305.

CHARLES UNIVERSITY, PRAHA

A GENERAL APPROACH TO THE THEORY OF SET-VALUED MAPPINGS

K. KURATOWSKI

Warszawa

Summary. Given an arbitrary set Y, a countably additive lattice $\boldsymbol{L}$ of subsets of Y, and a metric separable space X, we consider set-valued mappings $F : Y \to \mathscr{C}(X)$ (where $\mathscr{C}(X)$ is the space of all compact subsets of X) satisfying either condition (3) or (6) below. This is a far going generalization of upper resp. lower semi-continuous mappings (case where $\boldsymbol{L}$ is the lattice of all open subsets of Y). Other important applications are obtained by substituting to $\boldsymbol{L}$ the lattice of all Borel sets of additive class α, the lattice of measurable sets, of projective sets etc.

I. Introduction

1. Definitions. Let Y be a set of arbitrary elements and $\boldsymbol{L}$ a countably additive lattice (i.e. closed under countable unions) of subsets of Y containing as members the empty set and the set Y.

Denote $(-\boldsymbol{L}) = \{E : (Y - E) \in \boldsymbol{L}\}$.

Denote by A_σ (resp. A_δ) the lattice generated by the family of sets A and closed under countable unions (resp. intersections).

Let Z be a topological space. A mapping $f : Y \to Z$ will be called an *$\boldsymbol{L}$-mapping*, briefly $f \in \boldsymbol{L}^0$, if

$$f^{-1}(G) \in \boldsymbol{L} \text{ whenever } G \text{ is open in } Z\,; \tag{1}$$

equivalently: if

$$f^{-1}(K) \in (-\boldsymbol{L}) \text{ whenever } K \text{ is closed in } Z \tag{2}$$

(compare [4], Chapter IX).

2. Examples. In the following examples Y is assumed to be a topological space.

1. Let $\boldsymbol{L}$ be the lattice $\boldsymbol{G}$ of all *open* subsets of Y. Then the mapping $f : Y \to Z$ is an $\boldsymbol{L}$-mapping iff f is continuous.

2. Let $\boldsymbol{L}$ be the lattice of all *Borel* subsets of Y. Then f is an $\boldsymbol{L}$-mapping iff it is B-measurable (a Baire mapping).

3. Let $\boldsymbol{L}$ be the lattice of *Borel subsets of Y of additive class* $\alpha < \Omega$ (recall that the families: of open sets, of F_σ-sets, of $G_{\delta\sigma}$-sets etc. are additive Borel classes; the families of closed sets, of G_δ-sets, of $F_{\sigma\delta}$-sets etc. are multiplicative Borel classes). Here an $\boldsymbol{L}$-mapping means B-measurable (Baire mapping) of class α. (For an outline of a theory of set-valued B-measurable mappings, see [9].)

4. Let Y be a Polish space (= complete separable) and let $\boldsymbol{L}$ denote its *nth projective class*; recall that the 0-projective class is the class of all Borel sets, the first projective class (the class of Souslin sets or A-sets) is composed of continuous images of sets of class 0, the second projective class (the class of CA-sets) is composed of the complements to the sets of the first projective class and so on; in general, the projective class $2n + 1$ consists of continuous images of sets of class $2n$ and the $2n$-class consists of complements of sets of class $2n - 1$.

Note that the projective classes are closed under countable union and countable intersection. If Z has a countable open base, then, if $f : Y \to Z$ is an $\boldsymbol{L}$-mapping, i.e. if $f^{-1}(G)$ is of projective class n whenever G is open, then $f^{-1}(G)$ is also of class $n + 1$.

In particular, if $n = 1$, f is B-measurable (because by a known theorem of Souslin, a set which is simultaneously A and CA is a Borel set).

5. Let $\boldsymbol{L}$ be the lattice of *measurable sets* (more generally: a σ-algebra of subsets of Y). Then an $\boldsymbol{L}$-mapping means $\boldsymbol{L}$-measurable mapping.

II. Compact-valued mappings

We are going to consider *set-valued* mappings (called also *multifunctions*) $F : Y \to \mathscr{C}(X)$, where Y is an arbitrary set, X a normal space with a countable open base (equivalently: a metric separable space) and $\mathscr{C}(X)$ the space of all compact subsets of X. Thus $F : Y \to \mathscr{C}(X)$ means that for each $y \in Y$, $F(y)$ is a compact subset of X (of course, if X is compact, then $\mathscr{C}(X) = 2^X$, space of all closed subsets of X, and $F(y)$ is a closed subset of X).

The space $\mathscr{C}(X)$ is endowed with the Vietoris topology, which means (compare [11]) that the collection of all sets which are either of the form

(i) $\{F : F \subset G\}$ or
(ii) $\{F : F \cap G \neq \emptyset\}$,

where F is compact and G open in X, is an *open subbase* of $\mathscr{C}(X)$. Therefore, the collection of all sets of the form

(iii) $\{F : F \subset G) \cap \{F : F \cap G_1 \neq \emptyset\} \cap \ldots \cap \{F : F \cap G_n \neq \emptyset\}$

is an open base of $\mathscr{C}(X)$. Finally, since X has a *countable base*, so does $\mathscr{C}(X)$, and hence

(iv) *every open subset of* $\mathscr{C}(X)$ *is a countable union of sets of the form* (iii).

3. Definition. Given a lattice $\boldsymbol{L}$ (like in § 1), the mapping $F: Y \to \mathscr{C}(X)$ will be called of class $\boldsymbol{L}^+$ or briefly, $F \in \boldsymbol{L}^+$, if $F^{-1}(\mathscr{C}(G)) \in \boldsymbol{L}$, i.e., if

$$(3) \qquad \{y : F(y) \subset G\} \in \boldsymbol{L} \text{ whenever } G \text{ is open in } X\,,$$

equivalently, if

$$(4) \qquad \{y : F(y) \cap K = \emptyset\} \in \boldsymbol{L} \text{ whenever } K \text{ is closed in } X\,.$$

Symmetrically, $F \in \boldsymbol{L}_-$, if $F^{-1}(\mathscr{C}(K)) \in (-\boldsymbol{L})$, i.e. if

$$(5) \qquad \{y : F(y) \subset K\} \in (-\boldsymbol{L}) \text{ whenever } K \text{ is closed in } X\,,$$

equivalently, if

$$(6) \qquad \{y : F(y) \cap G = \emptyset\} \in (-\boldsymbol{L}) \text{ whenever } G \text{ is open in } X\,.$$

Let us note that, according to (1), $F \in \boldsymbol{L}^0$ iff $F^{-1}(G) \in \boldsymbol{L}$ for each G open in 2^X.

4. Examples and Remarks. 1. Let Y be a topological space, X compact and $\boldsymbol{L}$ the lattice of all open subsets of Y. Then $F \in \boldsymbol{L}^+$ means that F is upper semi-continuous and $F \in \boldsymbol{L}_-$ means that F is lower semi-continuous.

2. Let Y be a metric space, X compact and $\boldsymbol{L}$ the additive Borel class $\alpha < \Omega$. Then $F \in \boldsymbol{L}^+$ means that the sets $F^{-1}(2^G)$ are of additive class α. Similarly $F \in \boldsymbol{L}_-$ means that the sets $F^{-1}(2^K)$ are of multiplicative class α.

(Instead of $F \in \boldsymbol{L}^+$ ($F \in \boldsymbol{L}_-$) we also say that F is of Baire class α^+ (class α_-).)

5. Elementary properties of classes $\boldsymbol{L}^+$ and $\boldsymbol{L}_-$.

Theorem 1. *If F is constant, say $F(y) = K_0$ for each $y \in Y$, then $F \in \boldsymbol{L}^0$.*

Because $F^{-1}(\boldsymbol{G}) = Y$ or $\emptyset$ according to whether $K_0 \in \boldsymbol{G}$ or $K_0 \notin \boldsymbol{G}$. In both cases $F^{-1}(\boldsymbol{G}) \in \boldsymbol{L}$.

Theorem 2. *Let $f: Y \to X$ and $F(y) = \{f(y)\}$ (i.e., f is point-valued). If $F \in \boldsymbol{L}^+ \cup \boldsymbol{L}_-$, then f is an $\boldsymbol{L}$-mapping.*

This follows easily from the formula

$$\{y : F(y) \subset A\} = \{y : f(y) \in A\} = f^{-1}(A)$$

and from (3) and (1), resp. from (5) and (2) (substituting $A = G$ or $A = K$, resp.).

Theorem 3. $\boldsymbol{L}^0 \subset \boldsymbol{L}^+ \cap \boldsymbol{L}_-$. (Here $\boldsymbol{L}$ is not assumed to be countably additive.)

This follows from (1) and (3), resp. from (2) and (4), because $\mathscr{C}(G)$ is open, and $\mathscr{C}(K)$ is closed in $\mathscr{C}(X)$.

Theorem 4. $L^+ \cap L_- \subset L^0$. *Hence* $L^0 = L^+ \cap L_-$.

Proof. Let G be an open subset of $\mathscr{C}(X)$. Let $F \in L^+ \cap L_-$. We have to show that $F^{-1}(G) \in L$. Now by (iv), $F^{-1}(G)$ is a countable union of sets of the form

$$\{y : F(y) \subset G\} \cap \{y : F(y) \cap G_1 \neq \emptyset\} \cap \ldots \cap \{y : F(y) \cap G_n \neq \emptyset\} \tag{7}$$

and by (3) and (6) each of the factors of (7) is a member of L, and therefore the set (7) belongs to L. Hence $F^{-1}(G) \in L$.

Theorem 5. $L^+ \subset ((-L)_\sigma)_-$ *and* $L_- \subset ((-L)_\sigma)^+$.

Proof. 1. Let $F \in L^+$ and let K be closed in X. We have to show that $\{y : F(y) \subset K\} \in -(-L)_\sigma$, i.e. that $\{y : F(y) \subset K\} \in L_\delta$. Put $K = G_1 \cap G_2 \cap \ldots$ where G_n is open. Then

$$\{y : F(y) \subset K\} = \bigcap_n \{y : F(y) \subset G_n\} \tag{8}$$

and the proof is completed because $\{y : F(y) \subset G_n\} \in L$.

2. Let $F \in L_-$ and let G be open in X. We have to show that $\{y : F(y) \subset G\} \in (-L)_\sigma$. Put $G = K_1 \cup K_2 \cup \ldots$ where K_n is closed and $K_n \subset \operatorname{Int}(K_{n+1})$. If $F(y) \subset G$, then – by compactness of $F(y)$ – there is n such that $F(y) \subset K_n$. Thus

$$\{y : F(y) \subset G\} = \bigcup_n \{y : F(y) \subset K_n\} \tag{8'}$$

and the proof is completed because $\{y : F(y) \subset K_n\} \in (-L)$.

Corollary 5′. *If* $L = -L$ *(i.e. if* L *is a* σ*-algebra), then* $L^+ = L_-$.

Corollary 5″. *If* $L \subset (-L)_\sigma$, *then* $(L^+ \cup L_-) \subset ((-L)_\sigma)^0$.

This follows from Theorems 4 and 5 and the formula

$$L^+ \subset ((-L)_\sigma)^+ \quad \text{and} \quad L_- \subset ((-L)_\sigma)_-$$

which is an obvious consequence of $L \subset (-L)_\sigma$.

Remark. The assumption $L \subset (-L)_\sigma$ is satisfied if, for example, L denotes the α additive Borel class. So, in particular, it follows from Corollary 5″ that semi-continuous compact-valued mappings are of the first Baire class.

6. Operations on classes L^+ and L_-.

Theorem 1. L^+ *and* L_- *are closed under the operation of finite union.*

In other terms, if $F_j \in L^+$ for $j = 0, 1$, and $F = F_0 \cup F_1$, then $F \in L^+$. Similarly, if $F_j \in L_-$, then $F \in L_-$.

Here $F = F_0 \cup F_1$ means that $F(y) = F_0(y) \cup F_1(y)$ for each $y \in Y$. (A similar meaning has $F = F_0 \cap F_1$.) The theorem follows immediately from the formula (compare [8], p. 20(2)):

$$F^{-1}(\mathscr{C}(A)) = F_0^{-1}(\mathscr{C}(A)) \cap F_1^{-1}(\mathscr{C}(A)),$$

where $A \subset X$ is open, respectively closed.

Theorem 2. $\boldsymbol{L}^+$ *is closed under countable intersections.*

In other terms, if $F_n \in \boldsymbol{L}^+$ for $n = 1, 2, \ldots$, then $(\bigcap_n F_n) \in \boldsymbol{L}^+$.

The proof, completely similar to that of Theorem 8 of [9], is based on the following two general valid formulas (compare [8], p. 179(2)). Let $F = F_0 \cap F_1$. Then

$$F^{-1}(\mathscr{C}(G)) = \bigcup_{i,j}[F_0^{-1}(\mathscr{C}(G \cup R_i)) \cap F_1^{-1}(\mathscr{C}(G \cup R_j))],$$

where $R_i \cap R_j = \emptyset$, $R_1, R_2, \ldots$ being an open base of X closed under finite unions, and

$$F^{-1}(\mathscr{C}(G)) = \bigcup_n F_n^{-1}(\mathscr{C}(G)),$$

where $F = F_0 \cap F_1 \cap \ldots$ and $F_0 \supset F_1 \supset \ldots$.

Theorem 3. $\boldsymbol{L}_-$ *is closed under the operation of the closure of a countable union.*

More precisely, if $F = \overline{F_1 \cup F_2 \cup \ldots}$ and $F_n \in \boldsymbol{L}_-$ for $n = 1, 2, \ldots$, then $F \in \boldsymbol{L}_-$, provided $F(y)$ is compact for each $y \in Y$.

This follows from the formula (compare [8], p. 164 (iii)):

$$F^{-1}(\mathscr{C}(A)) = \bigcap_n F_n^{-1}(\mathscr{C}(A)).$$

Theorem 4. *If* $F_0 \in \boldsymbol{L}_-$ *and* $F_1 \in \boldsymbol{L}^+$, *then* $\overline{F_0 - F_1} \in \boldsymbol{L}_-$.

In particular, $\overline{X - F_1} \in \boldsymbol{L}_-$.

This follows from the lemma which we are now going to prove (compare also [8], p. 181(2)):

Lemma. *Let* $F_0 : Y \to 2^X$, $F_1 : Y \to 2^X$ *and* $F = \overline{F_0 - F_1}$. *Let* K *be closed in* X *and let* $R_1, R_2, \ldots$ *be the sequence of members of an open base of* X *such that* $K \cap \bar{R}_i = \emptyset$. *Then*

(9) $$F^{-1}(2^K) = \bigcap_i \{Y - [F_1^{-1}(2^{X-\bar{R}_i}) - F_0^{-1}(2^{X-R_i})]\},$$

equivalently

(9′) $$Y - F^{-1}(2^K) = \bigcup_i \{F_1^{-1}(2^{X-\bar{R}_i}) - F_0^{-1}(2^{X-R_i})\}$$

or

$$(9'') \quad [F_0(y) - F_1(y) \not\subset K] \equiv \exists i : [F_1(y) \cap \bar{R}_i = \emptyset \text{ and } F_0(y) \cap R_i \neq \emptyset] .$$

Proof. 1. Let $F_0(y) - F_1(y) \not\subset K$, i.e., $F_0(y) \not\subset K \cup F_1(y)$. So let $x \in F_0(y)$ and $x \notin K \cup F_1(y)$. By the regularity of X, there exists a member of the base, we may call it R_i, such that

$$x \in R_i \quad \text{and} \quad \bar{R}_i \cap (K \cup F_1(y)) = \emptyset .$$

It follows that $F_1(y) \cap \bar{R}_i = \emptyset$ and $F_0(y) \cap R_i \neq \emptyset$, because $x \in F_0(y) \cap R_i$. Thus y satisfies the right member of (9″).

2. On the other hand, if y satisfies the right member of (9″) and $K \cap \bar{R}_i = \emptyset$, then $(F_0(y) - F_1(y)) \cap R_i \neq \emptyset$, because $(F_0(y) - F_1(y)) \cap R_i = (F_0(y) \cap R_i) - (F_1(y) \cap R_i) \neq \emptyset$.

Since $R_i \subset X - K$, it follows that $(F_0(y) - F_1(y)) \cap (X - K) \neq \emptyset$, which means that the left member of (9″) is fulfilled.

Remark. As seen, we don't require in our Lemma that the values of the mappings F_0 and F_1 be *compact*; they are only assumed to be *closed* in the (metric) space X.

The same remark applies, of course, to Theorem 4.

Theorem 5. $\boldsymbol{L}^+$ *and* $\boldsymbol{L}_-$ *are closed under the operation of limit*

$$(10) \qquad F = \operatorname{Lim} F_n ,$$

the convergence being uniform.

More precisely: if $F_n \in \boldsymbol{L}^+$ (resp. $F_n \in \boldsymbol{L}_-$) for $n = 1, 2, \ldots$, then $F \in \boldsymbol{L}^+$ (resp. $F \in \boldsymbol{L}_-$), provided $F(y)$ is compact for each $y \in Y$.

Theorem 5 is a direct consequence of the following lemma which we are going to prove.

Lemma. *Assume that the mappings* $F_n : Y \to \mathscr{C}(X)$, *where* $n = 1, 2, \ldots$ *and where* X *is metric, satisfy the condition* (10); *in other terms, there is a sequence* $m_1 < < m_2 < \ldots$ *such that*

$$(11) \qquad \operatorname{dist}(F(y), F_j(y)) < 1/n \text{ for each } y \in Y \text{ and } j > m_n .$$

Then, we have, for each open G and closed K,

$$(12) \qquad \{y : F(y) \subset G) = \bigcup_n \bigcup_{j > m_n} \{y : F_j(y) \subset G_n)$$

$$(13) \qquad \{y : F(y) \subset K\} = \bigcap_n \bigcap_{j > m_n} \{y : F_j(y) \subset Q_n\} ,$$

where

$$G_n = \{x : \varrho(x, X - G) > 1/n\} \quad \text{and} \quad Q_n = \{x : \varrho(x, K) \leqq 1/n\},$$

and thus

$$(14) \qquad G = G_1 \cup G_2 \cup \ldots, \quad G_1 \subset G_2 \subset \ldots$$

and

$$(15) \qquad K = Q_1 \cap Q_2 \cap \ldots, \quad Q_1 \supset Q_2 \supset \ldots.$$

Proof of (12). 1. Let $F(y) \subset G$. Since $F(y)$ is compact, there is by (14) an n such that $F(y) \subset G_n$. We have to show that there is $j > m_n$ such that $F_j(y) \subset G_n$. Suppose that the contrary is true, i.e., that $F_j(y) - G_n \neq \emptyset$ for each $j > m_n$; but then $\operatorname{Lim}_{j\to\infty} F_j(y) - G_n \neq \emptyset$, i.e. $F(y) \not\subset G_n$, which is a contradiction.

2. Let $F(y) - G \neq \emptyset$. Let $p \in F(y) - G$. By (11) there is, for each n and $j > m_n$, a point $p_j \in F_j(y)$ such that $|p_j - p| \leqq 1/n$, hence $\varrho(p_j, X - G) \leqq 1/n$, i.e., $p_j \notin G_n$. Thus $F_j(y) \not\subset G_n$.

Proof of (13). 1. Let $F(y) \subset K$. Suppose that contrary to (13), there are n and $j > m_n$ such that $F_j(y) \not\subset Q_n$. Let $q_j \in F_j(y) - Q_n$, i.e., $\varrho(q_j, K) > 1/n$ and consequently $\varrho(q_j, F(y)) > 1/n$. Therefore $\operatorname{dist}(F_j(y), F(y)) > 1/n$, contrary to (11).

2. Let $F(y) - K \neq \emptyset$. Let $p \in F(y) - K$. Therefore there is n such that $\varrho(p, K) > > 1/n$, i.e. $p \notin Q_n$. Suppose that for each $j > m_n$, we have $F_j(y) \subset Q_n$. Then $\operatorname{Lim}_{j\to\infty} F_j(y) \subset Q_n$, i.e., $F(y) \subset Q_n$, which is a contradiction.

Remark. A similar formula to (13) is known for point-valued mappings (see [4], p. 268, and [8], p. 386).

Theorem 6. *Let X_n be compact and let $X = X_0 \times X_1 \times \ldots$. If each mapping $F_n : Y \to 2^{X_n}$ belongs to the class $\boldsymbol{L}^+$ (respectively to the class $\boldsymbol{L}_-$) for $n = 0, 1, \ldots$, then so does their Cartesian product $F = F_0 \times F_1 \times \ldots$.*

The proof is completely analogous to the proof of Theorem 12 of [9].

7. Selection problems. In this section we assume that L is a σ-algebra and that X is a Polish (complete separable) space (2^X is endowed with the Vietoris topology).

According to Theorem and Corollary 1 of [10], the following Selection Theorem is true.

Theorem. *If $F : Y \to 2^X$ is either $\boldsymbol{L}^+$ or $\boldsymbol{L}_-$, there exists a selector $f : Y \to X$ (i.e., $f(y) \in F(y)$) which is an $\boldsymbol{L}$-mapping.*

Corollary. *For each* L*-measurable mapping* $F : Y \to 2^X$ *there exists an* L*-measurable selector* $f : Y \to X$.

For further applications (also to the optimal control theory) see e.g. [5], [6], [7], and [12].

III. Case where Y is a topological space

8. Relations to continuous, closed, open mapping etc.

The proof of the following theorem is immediate.

Theorem 1. *Let us assume that all open subsets of* Y *are members of the lattice* L. *Then each continuous mapping* $F : Y \to \mathscr{C}(X)$ *is an* L*-mapping.*

More precisely: if F is upper (lower) semi-continuous, then $F \in L^+ (F \in L_-)$.

Corollary. *If* $Y = \mathscr{C}(X)$, *then the identity,* $F(K) = K$, *is an* L*-mapping.*

Theorem 2. *Let* $f : X \to Y$ *be continuous and onto and let* $f^{-1}(y)$ *be compact for each* $y \in Y$. *Then the mapping* $f^{-1} : Y \to \mathscr{C}(X)$ *satisfies the equivalences*:

$(f^{-1} \in L^+) \equiv (f(K) \in (-L)$ *for each closed* K *in* $X)$,

$(f^{-1} \in L_-) \equiv (f(G) \in L$ *for each open* G *in* $X)$.

This follows by virtue of the general valid formula (see [8], p. 14(3)):

$$f(A) = \{y : A \cap f^{-1}(y) \neq \emptyset\}$$

in which one has to substitute for A either K or G (compare (4) and (6)).

Remark. In the case where L is the lattice of all open subsets of Y, our theorem states that f^{-1} is upper semi-continuous iff f is a closed mapping; f^{-1} is lower semi-continuous iff f is an open mapping (compare [8], p. 177).

9. The graph of the relation $x \in F(y)$. The set

$$J = \{\langle x, y \rangle : x \in F(y)\}$$

is the graph under consideration.

Theorem. *Let* M *be a lattice of subsets of* $X \times Y$, *closed under countable unions and such that*

$$(G \text{ open in } X \text{ and } A \in L) \Rightarrow (G \times A) \in M\,.$$

Let $F \in L^+$, *then* $[(X \times Y) - J] \in M$.

Proof. Let $G_1, G_2, \ldots$ be an open base of X. Then: $x \notin F(y)$ iff there is n such that $x \in G_n$ and $F(y) \cap \bar{G}_n \neq \emptyset$, i.e.,

$$(X \times Y) - J = \bigcup_n \{\langle x, y\rangle : (x \in G_n)(F(y) \cap \bar{G}_n = \emptyset)\} =$$

$$= \bigcup_n [G_n \times \{y : F(y) \cap \bar{G}_n = \emptyset\}].$$

Since $F \in L^+$, we have, by (4), $\{y : F(y) \cap \bar{G}_n = \emptyset\} \in L$; this completes the proof.

Corollary 1. *If F is upper semi-continuous, then J is closed in $X \times Y$.*

Here L denotes the lattice of all open subsets of Y.

Corollary 2. *If F is of Baire class α^+, then J is of Borel multiplicative class α in $X \times Y$.*

Here L denotes the lattice of Borel subsets of additive class α of Y.

Corollary 3. *If L is any projective class (in Y), then J belongs to the projective class $(-L)$ in $X \times Y$.*

Corollary 4. *If F is L-measurable, so is J.*

For the sake of simplicity we put here $X = Y =$ interval.

Remark. In the case where $F(y)$ reduces to a single point, $f(y)$, our Theorem implies some well known statements about the graph of the mapping f

$$J_0 = \{\langle x, y\rangle : x = f(y)\}$$

(see e.g. [8], p. 384 and [3], Section 3).

Let us note that the converse to Corollary 2 is not true: the set J_0 can be G_δ without f being of class 1.

IV. Final remarks

It seems useful in many cases to consider the set Φ of all mappings $F : Y \to \mathscr{C}(X)$ as a *metric space*. The distance between two members F_0 and F_1 of Φ is defined following the regular procedure (compare [8], p. 218). Namely – assuming that X is bounded – we put

$$\text{Dist}(F_0, F_1) = \sup \text{dist}[F_0(y), F_1(y)] \quad \text{where} \quad y \in Y$$

and where "dist" means the Hausdorff distance of sets.

In view of this definition, convergence in the space Φ means the uniform convergence. Thus Theorem 5 of § 6 can be restated as follows.

The sets $\mathbf{L}^+$ *and* $\mathbf{L}_-$ *are closed in the space* Φ.

The space Φ – although not separable, in general – has a number of interesting properties. For example, if X is complete, then so is $\mathscr{C}(X)$ and consequently (see [8], p. 408) also Φ is complete.

Furthermore, the set of semi-continuous compact-valued mappings can be shown to be non-dense in the set of Baire 1st class mappings (under suitable assumptions on X and Y).

An analogous statement is true also for Baire mappings of arbitrary class α.

Let us add that "joint semi-continuity" holds under the above defined topology of the space Φ.

The proofs of these statements and of further properties of the space Φ will appear elsewhere.

References

[1] *C. Castaing:* Quelques problèmes de mesurabilité liés à la théorie de la commande. C. R. Acad. Sci. Paris Sér. A—B *262* (1966), 409.

[2] *R. Engelking:* Quelques remarques concernant les opérations sur les fonctions semi-continues dans les espaces topologiques. Bull. Acad. Polon. Sci. Sér. Sci. Math. Astronom. Phys. *11* (1963), 719—725.

[3] *Z. Frolik:* A survey of separable descriptive theory of sets and spaces. Czechoslovak Math. J. *20* (1970), 406—467.

[4] *F. Hausdorff:* Mengenlehre. W. de Gruyter, 1927.

[5] *C. J. Himmelberg and F. S. Van Vleck:* Some selections theorems for measurable functions. Canad. J. Math. *21* (1969), 394—399.

[6] *C. J. Himmelberg and F. S. Van Vleck:* Selection and implicit function theorems for multifunctions with Souslin graph. Bull. Acad. Polon. Sci. Sér. Sci. Math. Astronom. Phys. *19* (1971), 911—916.

[7] *C. J. Himmelberg, M. Q. Jacobs and F. S. Van Vleck:* Measurable multifunctions, selectors and Filippov's implicit functions lemma. J. Math. Anal. Appl. *25* (1969), 276—284.

[8] *K. Kuratowski:* Topology, Vol. I. Acad. Press — PWN, 1966.

[9] *K. Kuratowski:* On set-valued *B*-measurable mappings and a theorem of Hausdorff. Felix-Hausdorff Gedenkband (in print).

[10] *K. Kuratowski and C. Ryll-Nardzewski:* A general theorem on selectors. Bull. Acad. Polon. Sci. Sér. Sci. Math. Astronom. Phys. *13* (1965), 397—403.

[11] *E. Michael:* Topologies on spaces of subsets. Trans. Amer. Math. Soc. *71* (1951), 152—182.

[12] *C. Olech:* Existence theorems for optimal problems with vector-valued cost functions. Trans. Amer. Math. Soc. *136* (1969), 159—180.

INSTITUTE OF MATHEMATICS, POLISH ACADEMY OF SCIENCES

FACTORIALS AND THE GENERAL CONTINUUM HYPOTHESIS

D. KUREPA

Beograd

1. Number n!

1.1. Definition. For any (cardinal or ordinal) number n we define the *right factorial* $n!$ as the cardinality $\mathrm{k}S!$ of the set $S!$ of all permutations ($\equiv$ one-to-one mappings of S onto S) of S, S being any set having $\mathrm{k}n$ ($\equiv$ card n) members; we put $\mathrm{k}n = n$ for every cardinality n.

1.2. Right factorial hypothesis.

RFH $n! = 2^{\mathrm{k}n}$ for every transfinite (cardinal or ordinal) number n.

1.3. Theorem. Z $\Rightarrow$ RFH, (Z *denoting the choice axiom*).

(See [1], [1[a]] Th. 2.2; in particular we proved that for transfinite cardinalities $n^2 = n \Rightarrow n! = 2^n$; therefore [since $\aleph_\alpha^2 = \aleph_\alpha$] RFH holds for transfinite ordinal numbers.)

1.4. Problem. Does RFH $\Rightarrow$ Z?

2. Left factorials

2.1. In [4] we defined the left factorial $!n$ by

$$!n = \sum m!\,, \quad 0 \leqq m < n \tag{2.1}$$

for any (cardinal or ordinal, finite or transfinite) number n.

2.2. We also proved ([4] Th. 6.2. ((0000)) that the GCH (general continuum hypothesis) implies the following *left factorial hypothesis*

LFH $!n = n$ for every transfinite cardinality n.

2.3. On the other side, GCH $\Rightarrow$ Z (Sierpiński [5]). In other words

$$\mathrm{GCH} \Rightarrow \mathrm{LFH} \wedge \mathrm{Z}\,. \tag{2.2}$$

2.4. *The converse of* (2.2) *also holds.*

Proof. In the opposite case, there would be a pair of transfinite cardinalities (n, r) such that

$$n < r < 2^n . \tag{2.3}$$

Now, $n < r$ implies

$$n! \leqq !r \tag{2.4}$$

because $n!$ is a summand for $!r$. On the other hand

$$Z \Rightarrow n! = 2^n \tag{2.5}$$

for any transfinite cardinality n (Theorem 1.3), and (2.4) would yield $2^n \leqq !r =$ $=$ (by LFH) $= r$, i.e., $2^n \leqq r$, contradicting (2.3).

Consequently, one has the following

2.5. Theorem. GCH $\Leftrightarrow$ LFH $\wedge$ Z.

2.6. Problem. Does LFH $\Rightarrow$ GCH?

In connection with 2.5 and 2.6 one has the following

2.7. Theorem. GCH $\Leftrightarrow 2^{\underline{n}} = n$ *identically for transfinite cardinalities* n (cf. Tarski [6], L. 9a), p. 194), *where, by definition,* $2^{\underline{n}} = \sum_{m<n} 2^m$ (see Tarski [6] Def 4).

References

[1] *D. Kurepa:* O faktorijelama konačnih i beskonačnih brojeva. Rad. Jugosl. Akad. Znan. Umjet. *296* (1953), 105–122.

[1a] *D. Kurepa:* Über die Faktoriellen der endlichen und unendlichen Zahlen. Bull. Internat. Acad. Yougoslave Sci. *12* (1954), 51–64.

[2] *D. Kurepa:* Über das Auswahlaxiom. Math. Ann. *126* (1953), 381–384.

[3] *D. Kurepa:* Sull'ipotesi del continuo. Rendiconti del Seminario Matematico, Torino *18* (1958–1959), 11–20.

[4] *D. Kurepa:* Factorials of cardinal numbers and trees. Glasnik mat.-fiz. astr. *19* (1964), 7–21.

[5] *W. Sierpiński:* L'hypothèse généralisée du continu et l'axiome de choix. Fund. Math. *34* (1947), 1–5.

[6] *A. Tarski:* Sur les classes d'ensembles closes par rapport à certaines opérations élémentaires. Fund. Math. *16* (1930), 181–304.

SEMIGROUPS AND NEAR-RINGS OF CONTINUOUS FUNCTIONS

K. D. MAGILL, JR.

Buffalo

1. Introduction

Let X and G be topological spaces and let $\mathscr{S}(X, G)$ denote the family of all continuous functions from X into G. It has long been recognized that if G has an algebraic structure with which the topological structure is compatible, then one can provide $\mathscr{S}(X, G)$ with an algebraic structure be defining pointwise operations. However, even in the absence of any algebraic structure on G one can, in a natural way, provide $\mathscr{S}(X, G)$ with an algebraic structure. In fact, each continuous function α from G into X induces an associative binary operation on $\mathscr{S}(X, G)$. Specifically, one can define the product fg of any two functions f and g of $\mathscr{S}(X, G)$ by $fg = f \circ \alpha \circ g$, that is, fg is just the composition of the functions f, α and g. We will denote the resulting semigroup by $\mathscr{S}(X, G, \alpha)$. Such semigroups were introduced and first investigated in [2] and [3]. However, it was assumed in the latter papers that α mapped G onto X. We will not generally make that assumption here.

In Section 2 of this paper, a result is proved which gives the form of an isomorphism between two semigroups $\mathscr{S}(X, G, \alpha)$ and $\mathscr{S}(Y, H, \beta)$. In Section 3, we take G to be an additive topological group. This allows us to define point-wise addition on the continuous functions from X into G and the result, with multiplication defined as before, is a near-ring which we denote by $\mathscr{N}(X, G, \alpha)$. If $G = X$ and α is the identity map, then $\mathscr{N}(X, G, \alpha)$ becomes the near-ring of all continuous selfmaps of G under point-wise addition and ordinary composition. In this case, we use the simpler notation $\mathscr{N}(G)$. The isomorphism theorem for semigroups has an analogue for near-rings which is given in Section 3 and this result is then applied to get further results in the case when G is the additive topological group of one of the N-dimensional real number spaces.

2. Semigroups of continuous functions

The following result has not appeared before although most of the basic techniques needed to prove it were actually developed in [2] and [3]. We will make use of various results in those papers. In the statement of the theorem, $\mathscr{R}(\alpha)$ and $\mathscr{R}(\beta)$ denote the ranges of the functions α and β respectively.

Theorem 2.1. *Let α and β be nonconstant continuous functions from G and H into completely regular Hausdorff spaces X and Y respectively. Suppose that each of the subspaces $\mathscr{R}(\alpha)$ and $\mathscr{R}(\beta)$ contains a compact subspace with nonempty interior and suppose also that both G and H are connected, locally arcwise connected metric spaces. Then for each isomorphism φ from $\mathscr{S}(X, G, \alpha)$ onto $\mathscr{S}(Y, H, \beta)$ there exists a unique homeomorphism h from $\mathscr{R}(\alpha)$ onto $\mathscr{R}(\beta)$ and a unique homeomorphism t from G onto H such that the following diagram commutes for each $f \in \mathscr{S}(X, G, \alpha)$.*

$$\begin{array}{ccccc} \mathscr{R}(\alpha) & \xrightarrow{\;f\;} & G & \xrightarrow{\;\alpha\;} & \mathscr{R}(\alpha) \\ \downarrow h & & \downarrow t & & \downarrow h \\ \mathscr{R}(\beta) & \xrightarrow{\;\varphi(f)\;} & H & \xrightarrow{\;\beta\;} & \mathscr{R}(\beta) \end{array}$$

Proof. The existence and uniqueness of the bijections h and t and the fact that the diagram commutes all follow immediately from Theorem (2.3) of [2, p. 83]. We must show that h and t are, in fact, homeomorphisms and we consider h first. For each $p \in X$ and $f \in \mathscr{S}(X, G, \alpha)$, let

$$A(p, f) = \{x \in X : \alpha(f(x)) = p\} .$$

Similarly, for $q \in Y$ and $g \in \mathscr{S}(Y, H, \beta)$, let

$$B(q, g) = \{y \in Y : \beta(g(y)) = q\} .$$

Using the fact that the diagram commutes, one shows with some minor calculations that

$$h[\mathscr{R}(\alpha) \cap A(p, f)] = \mathscr{R}(\beta) \cap B(h(p), \varphi(f))$$

and also that

$$h^{-1}[\mathscr{R}(\beta) \cap B(q, g)] = \mathscr{R}(\alpha) \cap A(h^{-1}(q), \varphi^{-1}(g)) .$$

Therefore, in order to conclude that h is a homeomorphism, it is sufficient to show that

$$\{A(p, f) : p \in X \text{ and } f \in \mathscr{S}(X, G, \alpha)\}$$

is a basis for the closed subsets of X. Since α is nonconstant, we may choose two distinct points $a, b \in \mathscr{R}(\alpha)$ and then choose any two points $v, w \in G$ such that $\alpha(v) = a$ and $\alpha(w) = b$. Since G is both connected and locally arcwise connected, it must also be arcwise connected so we let k be any homeomorphism from the closed unit interval I into G such that $k(0) = v$ and $k(1) = w$. Now let W be any closed subset of X. Since X is completely regular and Hausdorff, there exists, for each $z \in X - W$, a continuous function f_z from X into I such that $f_z(z) = 0$ and $f_z(x) = 1$ for $x \in W$. Now let $k_z = k \circ f_z$. Then $k_z \in \mathscr{S}(X, G, \alpha)$ and one readily shows that

$$W = \bigcap \{A(b, k_z) : z \in X - W\}.$$

It follows from all this that h is a homeomorphism.

Now we show that t is a homeomorphism. Since both G and H are k-spaces, it will be sufficient to show that $t(K)$ is compact for each compact subset K of G and that $t^{-1}(K)$ is compact for each compact subset K of H. In fact, it will be sufficient to show the former since the latter follows in the same manner. So let K be a compact subset of G. We will verify the existence of a continuous function k from X into G such that

$$K \subset k(\mathscr{R}(\alpha)) \tag{2.1.1}$$

and

$$\mathscr{R}(\alpha) \cap k^{-1}(K) \text{ is compact}. \tag{2.1.2}$$

We will first dispose of the case where $K = G$. Then G is a Peano continuum and since α is nonconstant, $\mathscr{R}(\alpha)$ contains two distinct points a and b. Let f be any continuous function from X into the closed unit interval I such that $f(a) = 0$ and $f(b) = 1$. Then let g be any continuous mapping from I *onto* G and let $k = g \circ f$. Since $\mathscr{R}(\alpha)$ is connected, it follows that (2.1.1) is satisfied and (2.1.2) is satisfied since $\mathscr{R}(\alpha)$ is compact.

Now we consider the case where $K \neq G$ and we choose $a \in G - K$. By Theorem 5 of [1, p. 253], there exists a Peano continuum K^* such that

$$K \cup \{a\} \subset K^* \subset G.$$

By hypothesis, there is a point $b \in \mathscr{R}(\alpha)$, an open subset A of $\mathscr{R}(\alpha)$ and a compact subset W such that

$$b \in A \subset W \subset \mathscr{R}(\alpha).$$

Since $\mathscr{R}(\alpha)$ is a connected space with more than one point, it follows that there exists a point $c \in A - \{b\}$. Let $B = A - \{c\}$ and let B^* be an open subset of X such that $B = B^* \cap \mathscr{R}(\alpha)$. Now let f be any continuous function from X into I such that $f(b) = 0$ and $f(x) = 1$ for $x \in X - B^*$. Since K^* is a Peano continuum, there exists a continuous function g from I *onto* K^* such that $g(1) = a$. Then $k = g \circ f$ belongs to $\mathscr{S}(X, G, \alpha)$ and since $f(b) = 0$ and $f(c) = 1$ and $\mathscr{R}(\alpha)$ is connected, it readily follows that (2.1.1) holds. Furthermore one can verify that $\mathscr{R}(\alpha) \cap k^{-1}(K) \subset W$ which implies that (2.1.2) also holds.

Now we are in a position to show that $t(K)$ is compact. Because h is a homeomorphism, if follows from (2.1.2) that $h(\mathscr{R}(\alpha) \cap k^{-1}(K))$ is compact. Consequently,

$\varphi(k)\,[h(\mathscr{R}(\alpha) \cap k^{-1}(K))]$ is compact, but it follows from (2.1.1) (and the fact that the diagram commutes) that this latter set is just $t(K)$. Since t^{-1} behaves in a similar manner, it follows that t is a homeomorphism.

3. Near-rings of continuous functions

The near-ring analogue of Theorem 2.1 follows very quickly. The only additional thing one must do is show that t is, in this case, also a group isomorphism. For $a \in G$, let $\langle a \rangle$ denote the constant function which maps all of X into the point a. Then $\varphi\langle a \rangle = \langle t(a) \rangle$ for all $a \in G$ and for any $a, b \in G$ we have

$$\begin{aligned} \langle t(a + b) \rangle &= \varphi\langle a + b \rangle = \varphi(\langle a \rangle + \langle b \rangle) = \\ &= \varphi\langle a \rangle + \varphi\langle b \rangle = \langle t(a) \rangle + \langle t(b) \rangle = \langle t(a) + t(b) \rangle \end{aligned}$$

which implies that $t(a + b) = t(a) + t(b)$. Thus, t is a group isomorphism and we have the following

Theorem 3.1. *Let G and H be connected, locally arcwise connected metrizable topological groups and let X and Y be completely regular Hausdorff spaces. Let α and β be nonconstant continuous functions from G into X and H into Y, respectively, such that both $\mathscr{R}(\alpha)$ and $\mathscr{R}(\beta)$ contain compact subspaces with nonempty interiors. Then for each isomorphism φ from the near-ring $\mathscr{N}(X, G, \alpha)$ onto the near-ring $\mathscr{N}(Y, H, \beta)$, there exists a unique homeomorphism h from $\mathscr{R}(\alpha)$ onto $\mathscr{R}(\beta)$ and a unique topological isomorphism t from the group G onto the group H such that the following diagram commutes for each $f \in \mathscr{N}(X, G, \alpha)$.*

$$\begin{array}{ccccc} \mathscr{R}(\alpha) & \xrightarrow{\ f\ } & G & \xrightarrow{\ \alpha\ } & \mathscr{R}(\alpha) \\ \Big\downarrow h & & \Big\downarrow t & & \Big\downarrow h \\ \mathscr{R}(\beta) & \xrightarrow{\ \varphi(f)\ } & H & \xrightarrow{\ \beta\ } & \mathscr{R}(\beta) \end{array}$$

Now let R^N denote the additive topological group of the N-dimensional real number space. We use the latter theorem to get information about the automorphisms of the near-rings $\mathscr{N}(X, R^N, \alpha)$. We will find, among other things, that the existence of a certain type of automorphism on $\mathscr{N}(X, R^N, \alpha)$ has a considerable effect on the behavior of the function α.

Theorem 3.2. *Let X be a completely regular Hausdorff space and let α be a quotient map from R^N into X which is injective on some neighborhood of zero. Suppose also that $\mathscr{R}(\alpha)$ contains a compact subspace with nonempty interior.*

Then for each automorphism φ of the near-ring $\mathcal{N}(X, R^N, \alpha)$ there exists a unique homeomorphism h from $\mathcal{R}(\alpha)$ onto $\mathcal{R}(\alpha)$ and a unique linear automorphism t of the vector space R^N such that the following diagram commutes for each $f \in \mathcal{N}(X, R^N, \alpha)$.

$$\begin{array}{ccccc} \mathcal{R}(\alpha) & \xrightarrow{\;f\;} & R^N & \xrightarrow{\;\alpha\;} & \mathcal{R}(\alpha) \\ \big\downarrow h & & \big\downarrow t & & \big\downarrow h \\ \mathcal{R}(\alpha) & \xrightarrow{\;\varphi(f)\;} & R^N & \xrightarrow{\;\alpha\;} & \mathcal{R}(\alpha) \end{array}$$

Moreover, if $\max \{ \sum_{j=1}^{N} |a_{ij}| \}_{i=1}^{N} < 1$ *where* (a_{ij}) *is the matrix of* t *with respect to the canonical basis, then* α *is a homeomorphism. If, in addition to this,* $\mathcal{R}(\alpha) = X$, *then* $\mathcal{N}(X, R^N, \alpha)$ *is isomorphic to* $\mathcal{N}(R^N)$ *and its automorphism group is isomorphic to* $GL(N, R)$, *the full linear group of all real* $N \times N$ *nonsingular matrices.*

Proof. Let φ be an automorphism of $\mathcal{N}(X, R^N, \alpha)$. According to the previous theorem, there exists a unique homeomorphism h and a unique topological group isomorphism t such that the diagram above commutes. Since t is additive, it readily follows that $t(rx) = rt(x)$ for every rational number r and since t is continuous, it follows from this that $t(ax) = at(x)$ for every real number a. Thus, t is a linear automorphism of the vector space R^N.

Now let $M = \max \{ \sum_{j=1}^{N} |a_{ij}| \}_{i=1}^{N}$ and suppose that $M < 1$. We must show that α is a homeomorphism. In view of the fact that it is a quotient map, it is sufficient to show that it is injective so we assume that $\alpha(v) = \alpha(w)$ and we show that $v = w$. First, we take the norm of an element $x = (x_1, x_2, \ldots, x_N) \in R^N$ to be $\max \{|x_i|\}_{i=1}^{N}$. Then, if $\|x\| \leqq 1$, it readily follows that

$$\|t(x)\| = \max \{| \sum_{j=1}^{N} x_j a_{ij}|\}_{i=1}^{N} \leqq M .$$

Thus, $\|t\| < 1$ where $\|t\|$ denotes the norm of the operator t.

Next, let φ^n denote the composition of φ with itself n times. One readily shows that the unique homeomorphism associated with φ^n is h^n and that the unique linear automorphism associated with φ^n is t^n. Since the corresponding diagram commutes, it follows that

$$\alpha(t^n(v)) = h^n(\alpha(v)) = h^n(\alpha(w)) = \alpha(t^n(w)) .$$

However, $\|t^n(v)\| \leqq \|t\|^n \|v\|$ and $\|t^n(w)\| \leqq \|t\|^n \|w\|$ and since $\lim \|t\|^n = 0$, we can choose n so large that both $t^n(v)$ and $t^n(w)$ belong to the neighborhood on which α is injective. Consequently, for such an n, we have $t^n(v) = t^n(w)$ and since t^n is injective, it follows that $v = w$. Thus, α is a homeomorphism. If, in addition to this, $\mathcal{R}(\alpha) = X$, one easily verifies that the mapping which sends $f \in \mathcal{N}(X, R^N, \alpha)$ into $f \circ \alpha$ is an isomorphism from $\mathcal{N}(X, R^N, \alpha)$ onto $\mathcal{N}(R^N)$. To complete the proof of the theorem, we need only verify that the automorphism group of $\mathcal{N}(R^N)$ is isomorphic to $GL(N, R)$. As a matter of fact, it follows from our previous considerations that for each automorphism θ of $\mathcal{N}(R^N)$ there exists a unique linear automorphism s such that $\theta(f) = s \circ f \circ s^{-1}$ for each $f \in \mathcal{N}(R^N)$. One can easily verify that the mapping which sends θ into the matrix of s is an isomorphism from the automorphism group of $\mathcal{N}(R^N)$ onto the full linear group $GL(N, R)$.

References

[1] *K. Kuratowski:* Topology, Vol. II. Academic Press, New York, 1968.
[2] *K. D. Magill, Jr.:* Semigroup structures for families of functions, I; some homomorphism theorems. J. Austral. Math. Soc. 7 (1967), 81–94.
[3] *K. D. Magill, Jr.:* Semigroup structures for families of functions, II; continuous functions. J. Austral. Math. Soc. 7 (1967), 95–107.

STATE UNIVERSITY OF NEW YORK AT BUFFALO, NEW YORK

THE SPACE OF BOUNDED MAPS INTO A BANACH SPACE

E. MAKAI, JR.

Budapest

Let D be a real B-space for which

(1) D is strictly convex,

(2) for every $v \in D^*$, $\|v\| = 1$, and $0 < \delta < 1$ there is a number γ such that the set

$$\{w, w \in D, \|w\| \leqq 1, v(w) = 1 - \delta\}$$

contains a u for which

$$\{w, w \in D, \|w - u\| \leqq \gamma, v(w) = 1 - \delta\} \subset \{w, w \in D, \|w\| \leqq 1, v(w) = 1 - \delta\},$$

and for a fixed v $\gamma/\delta \to \infty$ if $\delta \to 0$,

(3) D has no proper subspace isometrically isomorphic to D,

(4) D is not finite dimensional.

Let X_j, $j = 1, 2$ be realcompact spaces. $C^*(X_j, D)$ denotes the B-space of the bounded continuous functions from X_j to D. For any linear isometry ψ of $C^*(X_1, D)$ onto $C^*(X_2, D)$ there exist a homeomorphism $\varphi : X_2 \to X_1$ and a continuous map A from X_1 to the isometrical isomorphies of D to itself (these taken in the strong operator topology) such that $(\psi f)(x_2) = A(\varphi(x_2)) \,.\, f(\varphi(x_2))$.

Let X_j be compact, let D have property (2). A similar statement holds (with $A(x_1) \equiv$ identity) for the pairs $(i_j, C_w^*(X_j, D))$ where $C_w^*(X_j, D)$ denotes the B-space of bounded weakly continuous functions from X_j to D, $i_j : D \to C_w^*(X_j, D)$, $(i_j d)(x_j) = d$ for every $x_j \in X_j$.

For further development S-compact spaces should be considered, where S is the unit sphere of a B-space (of measurable cardinality), or the unit sphere of a (non-reflexive) B-space with the weak topology.

References

[1] *M. Jerison:* The space of bounded maps into a Banach space. Ann. of Math. *52* (1950), 309—327.

[2] *E. Makai, Jr.:* The space of bounded maps into a Banach space. Publ. Math. Debrecen (to appear).

MATHEMATICAL INSTITUTE OF THE HUNGARIAN ACADEMY OF SCIENCES, BUDAPEST

A SURVEY OF THE SHAPE THEORY OF COMPACTA

S. MARDEŠIĆ

Zagreb

1. The notion of shape

The notion of shape has been introduced by K. Borsuk [2], [3] as a modification of the notion of homotopy type. The idea was to take into account the global properties of compacta and neglect the local ones. This allows an effective application of the theory to compacta as opposed to homotopy theory which only works well on spaces with nice local properties like the ANR's.

In shape theory of compacta one introduces and studies a new category whose objects are compact spaces and whose morphisms, called shape maps, are modifications of classes of homotopic maps. Two compacta are said to be of the same shape if they are isomorphic objects in this category. Moreover, there is a covariant functor S from the category of compact spaces and maps into the shape category. It keeps objects fixed and assigns to every map f a shape map said to be generated by f. In contrast to the case of homotopy theory, not every shape map is generated by a map.

The shape category and the functor S have the following essential features (W. Holsztyński [34]).

(i) S factors through the homotopy category, i.e., homotopic maps generate the same shape map and consequently spaces of the same homotopy type are of the same shape.

(ii) If Y is an ANR, then every shape map into Y is generated by a map unique up to homotopy and therefore on ANR's shape coincides with homotopy type.

(iii) The functor S is continuous with respect to inverse limits.

As an example one can consider the "Polish circle" (the closure of the graph of $\sin 1/x$, $x \in (0, 1]$, closed by an arc) which is easily obtained as an inverse limit of circles with bonding maps of degree 1. After applying the functor S and condition (iii) it is clear that the "Polish circle" is of the shape of a circle although their homotopy types differ.

Shape can be viewed as a "Čech homotopy type" and its relationship to homotopy type is analogous to the relationship of Čech homology to singular homology.

In order to define shapes Borsuk [2] considers compacta X embedded in the Hilbert cube $I^\infty = \Pi I_i$, $I_i = [-1, 1]$, $i \in N$. Instead of mappings $f: X \to Y$ he considers fundamental sequences $\boldsymbol{f}: X \to Y$ defined as sequences of maps $f_n: I^\infty \to I^\infty$ with the property that for every neighborhood V of Y in I^∞ there exist a neighborhood U of X in I^∞ and an integer $n_0 \in N$ such that for $n, n' \geqq n_0$ the restrictions $f_n \mid U$ and $f_{n'} \mid U$ are homotopic in V. Consequently, X is not mapped into Y itself, but into its neighborhoods in I^∞, which can always be chosen with nice local properties.

The composition $\boldsymbol{h} = \boldsymbol{gf}: X \to Z$ of fundamental sequences $\boldsymbol{f}: X \to Y$ and $\boldsymbol{g}: Y \to Z$ is by definition the fundamental sequence consisting of the maps $h_n = g_n f_n: I^\infty \to I^\infty$. The identity fundamental sequence $\boldsymbol{1}_X: X \to X$ consists of a sequence of identity maps $1_{I^\infty}: I^\infty \to I^\infty$.

Two fundamental sequences $\boldsymbol{f}, \boldsymbol{f}': X \to Y$ are considered to be homotopic, $\boldsymbol{f} \simeq \boldsymbol{f}'$, provided for every neighborhood V of Y in I^∞ there exist a neighborhood U of X in I^∞ and an integer $n_0 \in N$ such that for $n \geqq n_0$, $f_n \mid U$ and $f'_n \mid U$ are homotopic in V. It readily follows that $\boldsymbol{f} \simeq \boldsymbol{f}': X \to Y$ and $\boldsymbol{g} \simeq \boldsymbol{g}': Y \to Z$ implies $\boldsymbol{gf} \simeq \boldsymbol{g}'\boldsymbol{f}': X \to Z$.

According to Borsuk, Y is said to fundamentally dominate X provided there exist fundamental sequences $\boldsymbol{f}: X \to Y$ and $\boldsymbol{g}: Y \to X$ such that $\boldsymbol{gf} \simeq \boldsymbol{1}_X$; if also $\boldsymbol{fg} \simeq \boldsymbol{1}_Y$, then X and Y are said to be of the same shape, $\text{Sh}\, X = \text{Sh}\, Y$. In the case of fundamental domination one writes $\text{Sh}\, X \leqq \text{Sh}\, Y$. Equality of shape is an equivalence relation.

To every map $f: X \to Y$ Borsuk assigns the fundamental sequence $\boldsymbol{f}: X \to Y$ consisting of any sequence $f_1, f_1, \ldots$ of extensions $f_1: I^\infty \to I^\infty$ of f. Homotopy of maps implies homotopy of the assigned fundamental sequences. Therefore, the equality of homotopy types implies the equality of shapes.

The notion of shape was also defined for pairs of compacta (X, A) and in particular for pointed compacta [2]. Recently Borsuk has shown in [11] (also see [17]) that the role of the Hilbert cube I^∞ in founding shape theory can also be taken by any compact AR M and the resulting notion of shape is independent of M. The notion of shape was also defined for metric non-compact spaces by Borsuk [11] and by R. H. Fox [26]. For an account of this aspect of shape theory see [18].

Čech cohomology and homology groups are important shape invariants. Borsuk has proved that two continua in the plane E^2 are of the same shape if and only if their first Betti numbers coincide. Thus, there are $\aleph_0$ different shapes of planar continua, as their representatives one can take a single point (trivial shape), a bouquet of n circles, $n \in N$, and the infinite bouquet of circles. A. Trybulec has proved that every 1-dimensional Peano continuum is of the shape of a planar continuum. S. Godlewski [27] has shown that there are $2^{\aleph_0}$ different shapes among continua in E^3; in fact that many different shapes can be found among solenoids. A complete shape classification of n-sphere-like continua was given in [44] and of projective plane-like continua in [53]. Borsuk has shown that there are $2^{\aleph_0}$ different shapes among compacta in E^2. In fact, in [44] it is shown that in E^1 there are at least $\aleph_1$ different shapes of compacta.

2. ANR-system approach to shapes

In [44] the author and J. Segal have developed shape theory on the basis of the notion of an ANR-system. In this approach shapes are defined for arbitrary Hausdorff compact spaces. We shall describe here only the case of metric compacta, where ANR-systems can be replaced by ANR-sequences.

An ANR-sequence $\boldsymbol{X} = \{X_i, p_{ii'}\}$ is an inverse sequence of compact metric ANR's X_i and maps $p_{ii'} : X_{i'} \to X_i$, $i \leqq i'$, $i, i' \in N$. A map of ANR-sequences $\boldsymbol{f} : \boldsymbol{X} \to \boldsymbol{Y} = \{Y_j, q_{jj'}\}$ consists of an increasing function $f : N \to N$ and a sequence of maps $f_j : X_{f(j)} \to Y_j$ such that the following diagram commutes up to homotopy

$$\begin{array}{ccc} X_{f(j)} & \xleftarrow{p_{f(j)f(j')}} & X_{f(j')} \\ \downarrow{\scriptstyle f_j} & & \downarrow{\scriptstyle f_{j'}} \\ Y_j & \xleftarrow[q_{jj'}]{} & Y_{j'} \end{array}$$

The composition $\boldsymbol{h} = \boldsymbol{gf}$ of maps of sequences $\boldsymbol{f} : \boldsymbol{X} \to \boldsymbol{Y}$, $\boldsymbol{g} : \boldsymbol{Y} \to \boldsymbol{Z} = \{Z_k, r_{kk'}\}$ consists of the map $h = fg : N \to N$ and of the sequence of maps $h_k = g_k f_{g(k)} : X_{fg(k)} \to Z_k$. The identity map of sequences $\boldsymbol{1} : \boldsymbol{X} \to \boldsymbol{X}$ consists of the identity $1 : N \to N$ and of the sequence of identity maps $1_{X_i} : X_i \to X_i$. Two maps of sequences $\boldsymbol{f}, \boldsymbol{f}' : \boldsymbol{X} \to \boldsymbol{Y}$ are considered to be homotopic, $\boldsymbol{f} \simeq \boldsymbol{f}'$, provided for every $j \in N$ there exists an integer $i \geqq f(j), f'(j)$, such that

$$f_j p_{f(j)i} \simeq f'_j p_{f'(j)i} .$$

It is well-known that every metric compactum X admits an inverse sequence $\boldsymbol{X}$ of ANR's (even of polyhedra) such that $X = \text{Inv lim } \boldsymbol{X}$; every such ANR-sequence $\boldsymbol{X}$ is said to be associated with X. If $\boldsymbol{X}$ and $\boldsymbol{Y}$ are associated with X and Y respectively, then a map of sequences $\boldsymbol{f} : \boldsymbol{X} \to \boldsymbol{Y}$ is said to be associated with a map $f : X \to Y$ provided for every $j \in N$ the following diagram commutes up to homotopy:

$$\begin{array}{ccc} X_{f(j)} & \xleftarrow{p_{f(j)}} & X \\ \downarrow{\scriptstyle f_j} & & \downarrow{\scriptstyle f} \\ Y_j & \xleftarrow[q_j]{} & Y \end{array}$$

Here $p_i : X \to X_i$ and $q_j : Y \to Y_j$ denote the respective projections.

An essential step in this approach to shapes consists in the result [44] that every map $f : X \to Y$ admits an associated map of sequences $\boldsymbol{f} : \boldsymbol{X} \to \boldsymbol{Y}$ for any choice of $\boldsymbol{X}$ and $\boldsymbol{Y}$; $\boldsymbol{f}$ is determined uniquely up to homotopy. Moreover, if $f \simeq f'$, then

$\mathbf{f} \simeq \mathbf{f}'$. In particular, if $\mathbf{X}$ and $\mathbf{X}'$ are two ANR-sequences associated with X, then there are maps of sequences $\mathbf{i} : \mathbf{X} \to \mathbf{X}'$ associated with the identity map $1_X : X \to X$, determined uniquely up to homotopy. This enables one to compare maps of sequences $\mathbf{f} : \mathbf{X} \to \mathbf{Y}$ and $\mathbf{f}' : \mathbf{X}' \to \mathbf{Y}'$ between different ANR-sequences associated with X and Y respectively. $\mathbf{f}$ and $\mathbf{f}'$ are now said to be homotopic, $\mathbf{f} \simeq \mathbf{f}'$, provided the following diagram commutes up to homotopy

$$\begin{array}{ccc} \mathbf{X} & \xrightarrow{\ \mathbf{f}\ } & \mathbf{Y} \\ {\scriptstyle \mathbf{i}}\downarrow & & \downarrow{\scriptstyle \mathbf{j}} \\ \mathbf{X}' & \xrightarrow[\ \mathbf{f}'\ ]{} & \mathbf{Y}' \end{array} ;$$

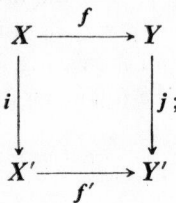

here $\mathbf{j}$ is any map of sequences associated with $1_Y : Y \to Y$. The homotopy relation classifies all the maps of ANR-sequences $\mathbf{X}$ associated with X to ANR-sequences $\mathbf{Y}$ associated with Y. The corresponding classes are called shape maps from X to Y and are denoted by $\mathbf{f} : X \to Y$ (using the notation of a representative $\mathbf{f} : \mathbf{X} \to \mathbf{Y}$). Shape maps correspond to homotopy classes of fundamental sequences in the Borsuk approach. Compact metric spaces and shape maps form the shape category. The functor S from the category of metric compacta to the shape category keeps the objects fixed and sends every $f : X \to Y$ into the shape map whose representative $\mathbf{f} : \mathbf{X} \to \mathbf{Y}$ is any map of sequences associated with f [42].

The value of this approach stems mainly from the fact that in studying the shape of X any ANR-sequence expansion $\mathbf{X}$ of X can be used. In many cases the space X itself is defined by means of an ANR-sequence, e.g., in the case of solenoids which are defined by an inverse sequence of circles. If X is already an ANR, one can use the sequence where all $X_i = X$ and all $p_{ii'} = 1_X$. Then one concludes immediately that maps into such sequences are associated with maps of spaces and so (ii) is satisfied. This approach is also more categorical and therefore allows generalizations to some abstract categories [34].

The equivalence of the ANR-sequence approach to shapes and the Borsuk approach was established in [45]. It is easy to construct a decreasing sequence of closed neighborhoods X_i of a compactum $X \subset I^\infty$ in such a way that the intersection yields X and each X_i is an ANR. Therefore, such a sequence can be considered as a special ANR-sequence associated with X. By suitably restricting the members of a fundamental sequence, one can easily produce a map of such ANR-sequences $\mathbf{f} : \mathbf{X} \to \mathbf{Y}$. To assign to every map of such ANR-sequences a fundamental sequence is more delicate and requires a repeated application of the Borsuk extension theorem (see [45]).

3. The method of infinite-dimensional manifolds

Recently an important and somewhat unexpected connection between shape and infinite-dimensional manifolds has been established. As far as we know the first application of these methods to homotopy is due to K. Borsuk [12], who has used a homeomorphism extension theorem of V. Klee. This approach has been further expanded by D. W. Henderson [33]. The final result in this direction is the following characterization theorem due to T. A. Chapman [21]: Two compacta X, Y contained in the pseudointerior s of the Hilbert cube I^∞, $s = \prod_{i \in N} I_i^\circ$, $I_i^\circ = (-1, 1)$, are of the same shape if and only if their complements $I^\infty \setminus X$ and $I^\infty \setminus Y$ are homeomorphic. It is of interest to point out that this theorem converts the problem of shape, which is essentially a homotopy problem, to a homeomorphism problem. Thus, shape theory can also be considered as a part of general topology. Chapman's result looks less surprising if one recalls that in ∞-dimensional manifolds homotopy and homeomorphism problems often are equivalent. Chapman's work is based on recent deep results on Hilbert cube manifolds due to several authors. It uses specially results by R. D. Anderson, T. A. Chapman, D. W. Henderson, R. M. Schori, J. E. West and R. Y. T. Wong.

Chapman has also succeeded in obtaining an analogous characterization theorem for finite-dimensional compacta [24]. A compactum $X \subset E^n$ is said to be stably embedded in E^n provided X lies in a Euclidean subspace of E^n of codimension $\geqq 2(\dim X) + 1$. A theorem of V. Klee insures that any two stable embeddings of X in E^n are equivalent by a space homeomorphism. Now Chapman's result can be stated as follows: For every integer $m > 0$ there exists an integer $n_1 > 0$ such that whenever $X, Y \subset E^n$ are stably embedded compacta of dimension $\leqq m$ and $n \geqq n_1$, then $\mathrm{Sh}\, X = \mathrm{Sh}\, Y$ if and only if the complements $E^n \setminus X$ and $E^n \setminus Y$ are homeomorphic. $n_1 = 10m + 17$ suffices. The same results holds with E^n replaced by the sphere S^n. The general structure of the proof of this result follows the one for the infinite-dimensional case, but the techniques are ones from piecewise linear topology.

4. Shapes of some quotient spaces

In this section we discuss some specific results concerning the shape of some quotient spaces. Borsuk has proved in [12] that for homeomorphic compacta A, B contained in the n-cell I^n the quotients I^n/A and I^n/B are of the same homotopy type and therefore of the same shape. D. W. Henderson [33] has obtained the same conclusion under the weaker hypothesis that A and B are of the same homotopy type. Using ANR-sequences the author has proved that for $A, B \subset I^n$ and $\mathrm{Sh}\, A = \mathrm{Sh}\, B$, the pairs (I^n, A) and (I^n, B) are of the same shape, which implies that $\mathrm{Sh}\,(I^n/A) = \mathrm{Sh}\,(I^n/B)$. Improving this result Borsuk has shown in [16] that the same theorem holds with I^n replaced by any AR. Finally, using his characterization of shape, Chapman has proved the following: if X is any compactum, A and B closed subsets of X of the same shape and there exists a closed subset $Y \subset X$ of trivial shape such that $A \cup B \subset Y$, then $\mathrm{Sh}(X/A) = \mathrm{Sh}(X/B)$ [23].

Using Chapman's characterization of shape R. D. Anderson [1] has recently proved a "Vietoris theorem" for shapes: Let X and Y be metric compacta and $f: X \to Y$ a continuous surjection with the property that $\mathrm{Sh}(f^{-1}(y))$ is trivial for every $y \in Y$. If in addition the set $Y' \subset Y$ of all points with nondegenerate $f^{-1}(y)$ is of finite dimension, then f is a shape equivalence and $\mathrm{Sh}\, X = \mathrm{Sh}\, Y$.

5. Movable compacta

Borsuk has introduced in [5] an interesting shape invariant property of compacta called movability. $X \subset I^\infty$ is said to be movable provided for every neighborhood U of X in I^∞ there exists a neighborhood $V \subset U$ of X in I^∞ such that for any neighborhood W of X in I^∞, V can be deformed into W within U. If X fundamentally dominates Y and is movable, then so is Y [5]. Every planar compactum is movable. A product of a finite or a countable collection of movable compacta is again movable [5]. The suspension of a movable compactum is movable [10].

In [46] the author and J. Segal have redefined movability in terms of ANR-sequences. X is movable provided every ANR-sequence $\mathbf{X}$ associated with X has the following property: for every $i \in N$ there exists an $i' \in N$, $i \leqq i'$, such that for every $i'' \in N$, $i \leqq i''$, there exists a map $r^{i'i''}: X_{i'} \to X_{i''}$ satisfying the homotopy relation $p_{ii''}r^{i'i''} \simeq p_{ii'}$. If one of the associated sequences $\mathbf{X}$ has this property then so does everyone. Using this property the author and J. Segal [46] have exhibited a 1-dimensional continuum X which is acyclic (all Čech homology and cohomology groups vanish) but fails to be movable. Nevertheless, its suspension is movable and even of trivial shape [41].

The author has proved [39] that every n-dimensional LC^{n-1} continuum is movable. The proof was simplified by R. Overton and J. Segal in [51]. Borsuk [6] has exhibited a 2-dimensional locally connected (LC^0) continuum which fails to be movable, which shows that the dimension in the above theorem cannot be improved.

K. Kuperberg [37] has shown that for movable compacta a form of the Hurewicz theorem holds. One uses the Čech homology groups and the shape groups $\pi_n(X, x_0)$, which are defined as the inverse limit of the inverse sequence $\{\pi_n(X_i, x_{i0}), p_{ii'\#}\}$, where $(\mathbf{X}, \mathbf{x}_0) = \{(X_i, x_{i0}), p_{ii'}\}$ is an ANR-sequence associated with (X, x_0).

The definition of movability given in [46] and applied to pairs yields the notion of a movable pair of compacta. R. Overton has shown that Čech homology with integer coefficients is exact on movable pairs of metric compacta [50].

Recently M. Moszyńska has introduced uniformly movable compacta, a subclass of movable compacta, as metric compacta X such that every ANR-sequence $\mathbf{X}$

associated with X has the properties as in the definition of movability and in addition $p_{i''i'''}r^{i'i'''} \simeq r^{i'i''}$ whenever $i \leqq i'' \leqq i'''$, i.e., the maps $r^{i'i''}: X_{i'} \to X_{i''}$, $i \leqq i''$, form a map of sequences $\boldsymbol{r}^{i'}: X_{i'} \to \boldsymbol{X}^i = \{X_{i''}, p_{i''i'''}, i'' \geqq i\}$. Uniform movability is also preserved under fundamental domination and is therefore a shape invariant too. Using the notion of uniform movability of pointed compacta Moszyńska has proved the following form of the Whitehead theorem: If (X, x_0) and (Y, y_0) are finite-dimensional uniformly movable pointed compacta and $\boldsymbol{f}: (X, x_0) \to (Y, y_0)$ is a shape map which induces isomorphisms of the shape groups $\pi_n(X, x_0) \to \pi_n(Y, y_0)$ for all $n \in N$, then $\boldsymbol{f}$ is a shape equivalence.

6. Retraction in shape theory

In [4] Borsuk has defined and studied fundamental or shape retraction. In terms of ANR-sequences the definition assumes this form: Let X and Y be metric compacta, $X \subset Y$. A shape retraction of Y to X is a shape map (in the sense of § 2) $\boldsymbol{r}: Y \to X$ such that $\boldsymbol{ri} = \boldsymbol{1}_X$, where $\boldsymbol{i}: X \to Y$ is the shape map generated by the inclusion $i: X \to Y$; if such an $\boldsymbol{r}$ exists, then X is said to be a shape retract of Y. Notice that this definition means that for any two ANR-sequences $\boldsymbol{X}$ and $\boldsymbol{Y}$ associated with X and Y respectively, there exists a map of sequences $\boldsymbol{r}: \boldsymbol{Y} \to \boldsymbol{X}$ such that $\boldsymbol{ri} \simeq \boldsymbol{1}_X$. Thus, shape retraction generalizes the notion of homotopy retraction. The techniques of [45] and a theorem of H. Patkowska [52] enable one to show that this definition of a shape retract is equivalent to the original Borsuk definition. Let us also mention that M. Moszyńska has defined the notion of retraction at the level of ANR-sequences [47], which is closer to Borsuk's definition. However, after passing to space, it yields the same notion as described above.

Once shape retracts are defined, it is clear how to define absolute shape retracts ASR (called by Borsuk fundamental absolute retracts FAR) and absolute neighborhood shape retracts ANSR (called by Borsuk FANR) [4], [42].

Borsuk has characterized ASR's as compacta having the shape of a point (trivial shape) (see [9] or [42]). The author has characterized ASR's also as compacta obtainable as the intersection of a decreasing sequence of Hilbert cubes [43]. The same result has been obtained by T. A. Chapman [21] using ∞-dimensional manifolds techniques; the author's proof is based on elementary techniques from *PL*-topology. R. C. Lacher [38] and the author [43] have independently characterized finite-dimensional ASR's as cellular sets in some Euclidean space.

Many results known for AR's carry over to ASR's. For example, if $X_1, X_2, X_0 = X_1 \cap X_2$ are ASR's, then so is $X = X_1 \cup X_2$ [4]. If X and X_0 are ASR's, then so are X_1 and X_2 [22].

ANSR's form an interesting shape invariant class of compacta. Every ANR is an ANSR and ANSR's are movable and uniformly movable. Borsuk has characterized metric ANSR's by means of a property called strong movability [9].

The property of being an ANSR is preserved under fundamental domination ([7] or [42]). Therefore, shape retracts of ANR's are ANSR's. The converse also holds, i.e., every ANSR is a shape retract of some ANR ([4] or [42]).

M. Moszyńska has considered the notion of a shape deformation retract X of Y and has characterized it by the property that the shape map $\boldsymbol{i}: X \to Y$ generated by the inclusion is a shape equivalence. She has also introduced the notion of the mapping cylinder $C_{\boldsymbol{f}}$ of a shape map $\boldsymbol{f}$ and has proved a "Fox theorem" for shapes: A shape map $\boldsymbol{f}: X \to Y$ is a shape equivalence if and only if X is a shape deformation retract of the mapping cylinder $C_{\boldsymbol{f}}$. This result is used in the proof of her Whitehead theorem. She has also proved that $\mathrm{Sh}\, X = \mathrm{Sh}\, Y$ if and only if there exists a compactum Z such that both X and Y are shape deformation retracts of Z [42].

7. The shape dimension

K. Borsuk has also defined and studied the shape or fundamental dimension $\mathrm{Sd}\, X$ as the minimum of $\dim Y$, where Y runs through all compacta which fundamentally dominate X [3]. For example, every cell has fundamental dimension 0. Clearly, $\mathrm{Sd}\, X \leqq \dim X$. Holsztyński has shown that one obtains the same value for $\mathrm{Sd}\, X$ by letting Y run through all compacta having the same shape as X.

It is well-known that the nth cohomotopy group $\pi^n(X)$ is defined if $\dim X < 2n - 1$. Godlewski shows in [30] that $\pi^n(X)$ is defined even if $\mathrm{Sd}\, X < 2n - 1$ and that it is a shape invariant [27].

Recently Borsuk and Holsztyński [20] have proved a "Hopf classification theorem" for shapes: Let X be a compactum with $\mathrm{Sd}\, X \leqq n$ and let Y be any compactum fundamentally dominated by the n-sphere S^n. Two shape maps $\boldsymbol{f}, \boldsymbol{g}: X \to Y$ coincide (i.e., the corresponding maps of sequences are homotopic) if and only if they induce the same homomorphism of homology groups, $\boldsymbol{f}_* = \boldsymbol{g}_*$. From here one can easily deduce that S^n fundamentally dominates only its own shape and the trivial shape.

References

[1] *R. D. Anderson:* Point-like decompositions of the Hilbert cube (in preparation).

[2] *K. Borsuk:* Concerning homotopy properties of compacta. Fund. Math. *62* (1968), 223—254.

[3] *K. Borsuk:* Concerning the notion of the shape of compacta. Proc. Internat. Sympos. on Topology and its Applications (Herceg-Novi ,1968). Savez društava Mat. Fiz. i Astronom., Belgrade, 1969, 98—104.

[4] *K. Borsuk:* Fundamental retracts and extensions of fundamental sequences. Fund. Math. *64* (1969), 55—85.

[5] *K. Borsuk:* On movable compacta. Fund. Math. *66* (1969), 137—146.

[6] *K. Borsuk:* On a locally connected non-movable continuum. Bull. Acad. Polon. Sci. Sér. Sci. Math. Astronom. Phys. *17* (1969), 425—430.

[7] *K. Borsuk:* On the shape of FANR-sets. Bull. Acad. Polon. Sci. Sér. Sci. Math. Astronom. Phys. *17* (1969), 529—532.

[8] *K. Borsuk:* Some remarks concerning the shape of pointed compacta. Fund. Math. *67* (1970), 221—240.

[9] *K. Borsuk:* A note on the theory of shape of compacta. Fund. Math. *67* (1970), 265—278.

[10] *K. Borsuk:* On the shape of the suspension. Colloq. Math. *21* (1970), 247—252.

[11] *K. Borsuk:* On the concept of shape for metrizable spaces. Bull. Acad. Polon. Sci. Sér. Sci. Math. Astronom. Phys. *18* (1970), 127—132.

[12] *K. Borsuk:* On the homotopy types of some decomposition spaces. Bull. Acad. Polon. Sci. Sér. Sci. Math. Astronom. Phys. *18* (1970), 235—239.

[13] *K. Borsuk:* On homomorphisms of homology groups induced by fundamental sequences. Bull. Acad. Polon. Sci. Sér. Sci. Math. Astronom. Phys. *18* (1970), 353—358.

[14] *K. Borsuk:* A note on the shape of quasi-homeomorphic compacta. Comment. Math. Prace Mat. *14* (1970), 25—34.

[15] *K. Borsuk:* Some remarks concerning the shape of pointed compacta. Fund. Math. *67* (1970), 221—240.

[16] *K. Borsuk:* Remark on a theorem of S. Mardešić. Bull. Acad. Polon. Sci. Sér. Sci. Math. Astronom. Phys. *19* (1971), 475—483.

[17] *K. Borsuk:* Theory of shape. Lecture Notes Series No. *28*. Matematisk Inst., Aarhus Univ., 1971, 145 pp.

[18] *K. Borsuk:* Some remarks concerning the theory of shape in arbitrary metrizable spaces. These Proceedings.

[19] *K. Borsuk and W. Holsztyński:* Concerning the ordering of shapes of compacta. Fund. Math. *68* (1970), 107—115.

[20] *K. Borsuk and W. Holsztyński:* Hopf classification theorem in the shape theory. Bull. Acad. Polon. Sci. Sér. Sci. Math. Astronom. Phys. *19* (1971), 387—390.

[21] *T. A. Chapman:* On some applications of infinite-dimensional manifolds to the theory of shape. Fund. Math. (to appear).

[22] *T. A. Chapman:* Some results on shapes and fundamental absolute retracts. Bull. Acad. Polon. Sci. Sér. Sci. Math. Astronom. Phys. *20* (1972), 37—40.

[23] *T. A. Chapman:* Shapes of some decomposition spaces. Bull. Acad. Polon. Sci. Sér. Sci. Math. Astronom. Phys. (to appear).

[24] *T. A. Chapman:* Shapes of finite-dimensional compacta. Mathematisch Centrum Amsterdam, 1971, mimeographed, 28 pp.

[25] *T. A. Chapman:* Characterizing shapes of compacta. Bull. Amer. Math. Soc. (to appear).

[26] *R. H. Fox:* On shape. Fund. Math. *74* (1972), 47—71.

[27] *S. Godlewski:* Homomorphisms of cohomotopy groups induced by fundamental classes. Bull. Acad. Polon. Sci. Sér. Sci. Math. Astronom. Phys. *17* (1969), 277—283.

[28] *S. Godlewski:* On shapes of solenoids. Bull. Acad. Polon. Sci. Sér. Sci. Math. Astronom. Phys. *17* (1969), 623—627.

[29] *S. Godlewski:* Solenoids of comparable shapes are homeomorphic. Bull. Acad. Polon. Sci. Sér. Sci. Math. Astronom. Phys. *18* (1970), 565—566.

[30] *S. Godlewski:* Homotopy dependence of fundamental sequences, relative fundamental equivalence of sets and a generalization of cohomotopy groups. Fund. Math. *69* (1970), 63—91.

[31] *S. Godlewski:* Cohomotopy groups and shape in the sense of Fox. Fund. Math. *75* (1972), 175—185.

[32] *S. Godlewski and W. Holsztyński:* Some remarks concerning Borsuk's theory of shape. Bull. Acad. Polon. Sci. Sér. Sci. Math. Astronom. Phys. *17* (1969), 373—376.

[33] *D. W. Henderson:* Applications of infinite-dimensional manifolds to quotient spaces of complete ANR's. Bull. Acad. Polon. Sci. Sér. Sci. Math. Astronom. Phys. *19* (1971), 747—753.

[34] *W. Holsztyński:* An extension and axiomatic characterization of Borsuk's theory of shape. Fund. Math. *70* (1971), 157—168.
[35] *D. M. Hyman:* On decreasing sequences of compact absolute retracts. Fund. Math. *64* (1969), 91—97.
[36] *D. M. Hyman:* A remark on Fox's paper on shape. Fund. Math. *75* (1972), 205—208.
[37] *K. Kuperberg:* An isomorphism theorem of Hurewicz type in the Borsuk's theory of shape. (To appear.)
[38] *R. C. Lacher:* Cell-like spaces. Proc. Amer. Math. Soc. *20* (1969), 598—602.
[39] *S. Mardešić:* n-dimensional LC^{n-1} compacta are movable. Bull. Acad. Polon. Sci. Sér. Sci. Math. Astronom. Phys. *19* (1971), 505—509.
[40] *S. Mardešić:* On the shape of the quotient space S^n/A. Bull. Acad. Polon. Sci. Sér. Sci. Math. Astronom. Phys. *19* (1971), 623—629.
[41] *S. Mardešić:* A non-movable compactum with movable suspension. Bull. Acad. Polon. Sci. Sér. Sci. Math. Astronom. Phys. *19* (1971), 1101—1103.
[42] *S. Mardešić:* Retracts in shape theory. Glasnik Mat. Ser. III. *6* (*26*) (1971), 153—163.
[43] *S. Mardešić:* Decreasing sequences of cubes and compacta of trivial shape. General Topology and its Applications *2* (1972), 17—23.
[44] *S. Mardešić and J. Segal:* Shapes of compacta and ANR-systems. Fund. Math. *72* (1971), 41—59.
[45] *S. Mardešić and J. Segal:* Equivalence of the Borsuk and the ANR-system approach to shapes. Fund. Math. *72* (1971), 61—68.
[46] *S. Mardešić and J. Segal:* Movable compacta and ANR-systems. Bull. Acad. Polon. Sci. Sér. Sci. Math. Astronom. Phys. *18* (1970), 649—654.
[47] *M. Moszyńska:* On shape and fundamental deformation retracts I, II. Fund. Math.; I: *75* (1972), 145—167; II: to appear.
[48] *M. Moszyńska:* Uniformly movable compact spaces and their algebraic properties. (To appear.)
[49] *M. Moszyńska:* Various approaches to the fundamental groups.
[50] *R. Overton:* Čech homology for movable compacta. (To appear.)
[51] *R. Overton and J. Segal:* A new construction of movable compacta. Glasnik Mat. Ser. III. *6* (*26*) (1971), 361—363.
[52] *H. Patkowska:* A homotopy extension theorem for fundamental sequences. Fund. Math. *64* (1969), 87—89.
[53] *J. Segal:* Shape classification of projective plane-like continua. Glasnik Mat. Ser. III. *6* (*26*) (1971), 365—371.
[54] *R. Geoghegan and R. Summerhill:* Concerning the shapes of finite-dimensional compacta. (To appear.)
[55] *S. Godlewski and S. Nowak:* On two notions of shape. Bull. Acad. Polon. Sci. Sér. Sci. Math. Astronom. Phys. *20* (1972), 387—393.
[56] *D. Handel and J. Segal:* Shape classification of (projective m-space)-like continua. (To appear.)
[57] *D. Handel and J. Segal:* Finite shape classifications. (To appear.)
[58] *W. Holsztyński:* Continuity of Borsuk's shape functor. Bull. Acad. Polon. Sci. Sér. Sci. Math. Astronom. Phys. *19* (1971), 1105—1108.
[59] *S. Mardešić:* Shapes for topological spaces. (To appear.)
[60] *M. Moszyńska:* The Whitehead theorem in the theory of shapes. (To appear.)
[61] *T. J. Sanders:* Shape groups and products. Mimeographed. Univ. of Oklahoma, 1972.
[62] *R. B. Sher:* Realizing cell-like maps in Euclidean spaces. General Topology and its Applications (to appear).

ON TOTAL ORDERINGS IN TOPOLOGY

P. R. MEYER

New York

In an orderable topological space the following cardinal invariants coincide: local neighborhood character, pseudocharacter, sequential character, and Fréchet character (definitions in § 2). This paper considers the extent to which this result can be extended to classes of spaces satisfying some related total order conditions (list in § 1) which have appeared in the literature. The implications which hold between these conditions (there are two more in § 3) and examples to show their independence are given.

1. Some total order conditions from the literature

We begin by listing the order conditions to be considered. Unfortunately the terminology in the literature is not uniform. No attempt is made here to give a complete list of references; those given are either the earliest known to the author or those which are most useful for the present work.

1.1. Definition. Let (X, t) be a topological space.

1) t is *orderable* if there is a total ordering of X for which t is the order topology.

2) t is *locally orderable* if each point has an open neighborhood on which the relative topology is orderable [10].

3) and 4) Assume t is T_1. If $A \subset X$, the "chain net closure" of A consists of all limits of chain nets in A. (A chain net is a net whose directed set is totally ordered.) If the chain net closure operator coincides with the t-closure operator, t is called a *strong chain net topology*. If iteration of the chain net closure operator yields the t-closure operator, t is called a *chain net topology*, [4] and [5].

5) t is called a *nested neighborhood topology* if it is T_1 and each point has a local neighborhood base that is totally ordered by inclusion [2].

6) t is called a *generalized orderable topology* (GO topology) if there is a total ordering for X such that: t is larger (finer) than the order topology, and each point has a t-neighborhood base consisting of (possibly degenerate) order intervals [3].

7) t is called *weakly orderable* if there is a total ordering for X for which order-open rays are topologically open.

We will sometimes find it convenient to refer to these conditions by their initial letters. Two other related conditions, which do not seem to have been studied, are given in § 3.

1.2. Remark. A locally orderable space is T_1 but need not be T_2; in fact sequential limits need not be unique. (The space consisting of a sequence of isolated points converging to two distinct limits is locally orderable.) Thus all spaces considered in this paper are at least T_1, because T_2 is implied by each of the other conditions for which no separation axiom was assumed.

1.3. Theorem. *The above conditions are related as follows*: O → LO → SCN → → CN; O → GO → WO; GO → SCN; NN → SCN. *No other (non-trivial) implications are valid, even for T_2 spaces.*

Proof. The implications are all easy and/or known. Examples 1 through 5 justify the assertion that there are no other implications.

Some terminology and notation will be useful for the examples. **W** (**W***) denotes the set of all ordinals $<\omega_1$ ($\leqq\omega_1$). **N** is the set of positive integers; $\mathbf{N}^* = \mathbf{N} \cup \{\omega_0\}$. If X and Y are totally ordered sets, there are two total orderings on $X \times Y$ which will be used. The order topology induced by ordering by first (last) differences will be called the *left* (*right*) *lexicographic order topology*.

Example 1. The subset of $\mathbf{W}^* \times \mathbf{N}^*$ consisting of the top edge and the right edge, $(\{\omega_1\} \times \mathbf{N}^*) \cup (\mathbf{W}^* \times \{\omega_0\})$, is orderable but not NN.

Example 2. A circle is locally orderable and NN but not WO. (Note however that this topology can be obtained as the intersection of two orderable topologies.)

Example 3. The Sorgenfrey line is a GO space and NN but not locally orderable. (Proof as in [7].)

Example 4. We construct a WO space that is not a chain net space. Let $X = (\mathbf{W} \times \mathbf{N}) \cup \{p\}$, where $p = (\omega_1, \omega_0)$. Let t_1 (t_2) be the left (right) lexicographic order topology on the product with p as last point, and put $t = t_1 \vee t_2$ (join in the lattice of all topologies on X). The join of orderable topologies is a fortiori WO. The fact that t is not CN can be seen by observing that t-neighborhoods of p are the same as relative product neighborhoods from $\mathbf{W}^* \times \mathbf{N}^*$.

2. On extending cardinal invariant properties of orderable spaces

For a topological space (X, t) there are four cardinal invariants to be considered here. For simplicity we consider only infinite cardinals. The *local neighborhood character* χX is the least cardinal $\mathfrak{m}$ such that each point has a local neighborhood base of cardinality not greater than $\mathfrak{m}$. The *pseudocharacter* ψX is the least $\mathfrak{m}$ such that each point can be expressed as the intersection of at most $\mathfrak{m}$ open sets.

The other invariants arise from two interpretations of the question: what is the smallest $\mathfrak{m}$ such that one can recover t from the t-convergent $\mathfrak{m}$-nets? (An $\mathfrak{m}$-*net* is a net whose directed set has cardinality $\leqq \mathfrak{m}$. The cardinality of the directed set is also called the *size* of the net.) The $\mathfrak{m}$-closure of a subset A consists of all limits of t-convergent $\mathfrak{m}$-nets in A. The least $\mathfrak{m}$ for which the $\mathfrak{m}$-closure operator is the t-closure operator is called the *Fréchet character* of X, ϕX. The least $\mathfrak{m}$ for which iteration of the $\mathfrak{m}$-closure operator yields the t-closure operator is called the *sequential character*, σX [10]. The spaces for which $\sigma X = \aleph_0$ ($\phi X = \aleph_0$) are the sequential (Fréchet) spaces.

In general $\sigma X \leqq \phi X \leqq \chi X$ and $\psi X \leqq \chi X$, and all of the inequalities can be strict. However, for orderable spaces all four invariants coincide. To what extent does this result extend to spaces satisfying the total order conditions in § 1? The complete answer:

2.1. Theorem. (i) *If* (X, t) *is any one of the following*:

GO *space,*

nested neighborhood space,

locally orderable space,

then $\sigma X = \phi X = \chi X = \psi X$.

(ii) *If* (X, t) *is a strong chain net space then* $\sigma X = \phi X$. *In other words, if the t-closure operator can be obtained in one step using totally ordered directed sets of minimal size, then allowing iteration will not enable one to use smaller directed sets, totally ordered or otherwise.*

(iii) *No other such equalities hold in any of the classes of spaces in* § 1.

Proof. Cases of (i) were proved in [9] and [10]; the other proofs are similar. Examples 5 through 8 justify (iii).

Part of the next Corollary was obtained independently by C. Aull [1].

2.2. Corollary. *The notions of sequential space, Fréchet space and first countable space are equivalent in the following classes of spaces*: GO *spaces, locally orderable spaces, and nested neighborhood spaces.*

Proof. This is the countable case of (i).

We now give the examples to show that no other cardinal equalities are possible. We begin by showing that three well known examples can be obtained as enlargements (more open sets) of the same order topology. Let $M = (\mathbf{N} \times \mathbf{N}) \cup \{p\}$

and $u =$ the order topology on M from the following ordering: left lexicographic order on $\mathbf{N} \times \mathbf{N}$ with p as last element.

Example 5. Let t_1 be the sequential topology on M generated by the following u-convergent sequences: $((k, 1)) \to p$ and $((k, n)) \underset{n}{\to} (k+1, 1)$ for each k (i.e., the bottom row converges and each column converges). Then $t_1 \supset u$ and (M, t_1) is a WO space which is CN but not SCN and for which $\psi = \sigma = \aleph_0$, $\sigma < \phi$, and $\psi < \chi$.

Example 6. Let t_2 be the Fréchet topology on M generated by the u-convergent rows: for each n $((k, n)) \underset{k}{\to} p$. (M, t_2) is a WO and SCN space for which $\psi = \sigma = \phi = \aleph_0$, but $\phi < \chi$.

Example 7. Let t_3 be the topology on M in which all points except p are isolated and a set containing p is open if and only if it contains all but finitely many points in all but finitely many columns. Thus $t_3 \supset u$ and (M, t_3) is a WO space in which $\aleph_0 = \psi < \chi$ and $\psi < \sigma = \phi$.

Example 8. The one point compactification of an uncountable discrete space provides an example of a SCN space in which $\aleph_0 = \sigma = \phi < \chi = \psi$.

The next theorem describes product spaces which satisfy cardinal equalities. It is a slight extension of [10, Theorem 3.2]. The method of proof is the same.

2.3. Theorem. *Let X be the product of the family $\{X_i : i \in I\}$, where each X_i is a T_1 space with at least two points.*

(i) *If $|I| \geqq \chi X_i$ for each i, then $\sigma X = \phi X = \chi X = \psi X$.*

(ii) *If for each i there is a j such that $\psi X_j = \chi X_j \geqq \chi X_i$ then $\psi X = \chi X$. Corresponding statements with ψ replaced be either σ or ϕ are also valid.*

3. Lattice operations on orderable topologies.

There are two other order conditions for a topological space (X, t) which are related to those in § 1:

(I) t can be expressed in the form $t_1 \cap t_2$ where each t_i is an orderable topology.

(J) t can be expressed as the join (in the lattice of all topologies on X) of two orderable topologies.

3.1. Proposition. *The lattice conditions are related to the conditions in* § 1 *as follows*: O → (I) → CN *and* O → (J) → WO.

There are two examples which I do not have: a GO space which is not (I) [1]), and a GO space which is not (J). Aside from these two open problems, no other implications among the nine conditions are possible; examples 4, 7, 9, and 10 complete the justification. The implications in Theorem 1.3 and Proposition 3.1 are summarized in Figure 1.

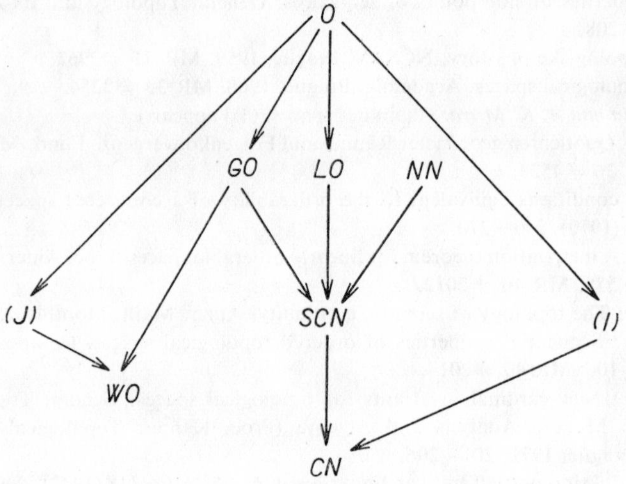

Figure 1.

Example 9. The cross topology on $\mathbf{R} \times \mathbf{R}$ of J. Novák [11] (or see [8]) is the intersection of the left and right lexicographic order topologies. ($\mathbf{R}$ denotes the real line.) This is known to be not a strong chain net topology and can be shown to be not WO (by a connectedness argument [6]).

Example 10. The example in Remark 1.2 satisfies (I), but, if the sequence converges to three points instead of two, we have an example of a LO and NN space which is not (I).

Note added in proof.

The results in § 2 on equality of cardinal functions can be extended to include tightness. More precisely, tightness of X can be added to the equalities in the conclusions of 2.1 Theorem (i) and (ii) and 2.3 Theorem (i). A corresponding statement of 2.3 Theorem (ii) in which tightness of $X = \chi X$ is also valid.

The tightness of X is the least $\mathfrak{m}$ with the property: If A is a subset of X and x is a limit point of A, then x is a limit point of B for some subset B of A with $|B| \leqq \mathfrak{m}$.

[1]) Example 11. A GO space which is not (I) can be obtained by enlarging the usual topology for the real line so as to make the irrationals discrete.

In general tightness of $X \leqq \sigma X$. For more on the relationship of tightness to the other cardinal functions, see the paper of A. V. Arhangelskij in these proceedings.

References

[1] *C. Aull:* Properties of side points of sequences. General Topology and its Applications *1* (1971), 201–208.
[2] *E. Čech:* Topologické prostory. NČSAV, Prague, 1959. MR 21 #2962.
[3] *E. Čech:* Topological spaces. Academia, Prague, 1966. MR 35 #2254.
[4] *S. P. Franklin and A. K. Misra:* Chain net spaces. (To appear.)
[5] *H. Herrlich:* Quotienten geordneter Räume und Folgenkonvergenz. Fund. Math. *61* (1967), 79–81. MR 36 #4528.
[6] *H. Kok:* On conditions equivalent to the orderability of a connected space. Nieuw Arch. Wisk. *18* (3) (1970), 250–270.
[7] *D. J. Lutzer:* A metrization theorem for linearly orderable spaces. Proc. Amer. Math. Soc. *22* (1969), 557–558. MR 40 #2012.
[8] *P. R. Meyer:* The topology of separate continuity. Amer. Math. Monthly *75* (1968), 1128.
[9] *P. R. Meyer:* Sequential properties of ordered topological spaces. Compositio Math. *21* (1969), 102–106. MR 40 #2014.
[10] *P. R. Meyer:* New cardinal invariants for topological spaces. General Topology and its Relations to Modern Analysis and Algebra (Proc. Kanpur Topological Conf., 1968). Academia, Prague, 1971, 201–206.
[11] *J. Novák:* Induktion partiell stetiger Funktionen. Math. Ann. *118* (1942), 449–461. MR 6, 164.

LEHMAN COLLEGE OF THE CITY UNIVERSITY OF NEW YORK, NEW YORK

ON TWO THEOREMS OF V. V. FILIPPOV

E. MICHAEL

Seattle

1. Introduction. The following two beautiful theorems were recently proved by V. V. Filippov [3, Theorems 1.1 and 2.1].

Theorem 1.1. (Filippov). *If $f : X \to Y$ is a bi-quotient s-map, and if X has a point-countable base, so does Y.*

Theorem 1.2. (Filippov). *If $f : X \to Y$ is a quotient s-map, if X has a point-countable base, and if Y is a Hausdorff space of pointwise countable type, then f is bi-quotient.*

The purpose of this note is to briefly outline a proof for Theorem 1.1 which is somewhat shorter and simpler than Filippov's (for details, see [1]), and to indicate how Theorem 1.2 can be strengthened in two directions (for details, see [4, Theorem 9.5]).

Let us briefly explain our terminology. All maps are continuous and onto. A map $f : X \to Y$ is *bi-quotient* [1] [2] if, whenever $y \in Y$ and $\mathscr{U}$ is a cover of $f^{-1}(y)$ by open subsets of X, then $y \in (\bigcup f(\mathscr{V}))^0$ for some finite $\mathscr{V} \subset \mathscr{U}$. (We use A^0 to denote the interior of A.) A map $f : X \to Y$ is an *s-map* if $f^{-1}(y)$ has a countable base for every $y \in Y$. A space Y is of *pointwise countable type* if every $y \in Y$ is contained in a compact subset K of Y of countable character in Y (i.e., there is a countable base for the neighborhoods of K in Y).

It should be remarked that, in Theorem 1.1, the cardinal $\aleph_0$ (which appears in the definition of "s-map" and "point-countable") can be replaced by any other infinite cardinal. This was also observed by Filippov.

2. A new proof for Theorem 1.1. Let $\mathscr{B}$ be a point-countable base for X, and let $\mathscr{P} = f(\mathscr{B})$. Let $\Phi = \{\mathscr{F} \subset \mathscr{P} : \mathscr{F} \text{ finite}\}$. For each $\mathscr{F} \in \Phi$, let

$$\mathscr{M}(\mathscr{F}) = \{P \in \mathscr{P} : P \subset (\bigcup\mathscr{F})^0,\ P \not\subset (\bigcup\mathscr{E})^0 \text{ if } \mathscr{E} \subsetneqq \mathscr{F}\},$$
$$V(\mathscr{F}) = (\bigcup(\mathscr{M}(\mathscr{F})))^0 .$$

Let $\mathscr{V} = \{V(\mathscr{F}) : \mathscr{F} \in \Phi\}$. Then $\mathscr{V}$ is the required point-countable base for Y.

The verification that $\mathscr{V}$ is a base for Y is fairly routine, but the proof that it is point-countable requires some work. For details, see [1].

3. A strengthening of Theorem 1.2. Consider the following property of a space Y.

(*) If (F_n) is a decreasing sequence of subsets of Y with a common accumulation point y, then there exist closed (in Y) subsets $A_n \subset F_n$ whose union $\bigcup_{n=1}^{\infty} A_n$ is not closed in Y.

Property (*) is a useful hypothesis in a number of theorems (see [4, section 9]). Every T_1-space of pointwise countable type has property (*), but not conversely.

According to [5], a space Y is *determined by countable subsets* if a subset A of Y is closed in Y whenever $\bar{C} \subset A$ for every countable $C \subset A$. Clearly, every sequential space has this property.

We can now state the following theorem, which is easily seen to imply Theorem 1.2. The proof is given in [4, Theorem 9.5].

Theorem 3.1. *Let $f : X \to Y$ be a quotient map. Suppose that Y is a Hausdorff space satisfying (*), that X or Y is determined by countable subsets, and that $(f^{-1}(E))^-$ is Lindelöf for every countable $E \subset Y$. Then f is bi-quotient.*

References

[1] *D. Burke and E. Michael:* On a theorem of V. V. Filippov. Israel J. Math. (to appear).

[2] *E. Michael:* Bi-quotient maps and the cartesian products of quotient maps. Ann. Inst. Fourier (Grenoble) *18* (2) (1968), 287—302.

[3] *V. V. Filippov:* Quotient spaces and multiplicity of a base. Mat. Sb. *80* (1969), 521—532. (=Math. USSR-Sb. *9* (1969), 487—496.)

[4] *E. Michael:* A quintuple quotient quest. General Topology and its Applications (to appear).

[5] *R. C. Moore and G. S. Mrowka:* Topologies determined by countable objects. Notices Amer. Math. Soc. *11* (1964), 554.

ON A METHOD WHICH LEADS TO EXTREMALLY DISCONNECTED COVERS

J. MIODUSZEWSKI

Katowice

Recently Hager [4] has found a simple method (using Kuratowski-Zorn Lemma) to get for each compact Hausdorff space the associated extremally disconnected cover (the construction due originally to Gleason [3] and simplified by Rainwater [8]). The aim of this note is to show that this method may be applied in order to get extremally disconnected covers for arbitrary Hausdorff spaces.

Throughout this text all the spaces are assumed (this is in fact not always necessary) to be Hausdorff and all the maps are assumed to be continuous.

A map $p: Y \xrightarrow{\text{onto}} X$ will be said to be *irreducible* iff for each closed subset F of Y we have $\mathrm{Cl}\, p(F) \neq X$ whenever $F \neq Y$; in the case of compact Hausdorff spaces this notion coincides with the usual one. Irreducible maps $Y \to X$ will be said also to be *covers* of X, and in the case when Y is extremally disconnected, *extremally disconnected covers* of X.

It was shown by L. Rudolf and the author in [7] that

1. *If $p: Y \to X$ is irreducible, then for each V regularly open in Y there exists a U regularly open in X, such that* $V = \mathrm{Int}\, p^{-1}(\mathrm{Cl}\, U) = \mathrm{Int}\, \mathrm{Cl}\, p^{-1}(U)$, *and this determines U for a given V.*

The above lemma established a one-to-one correspondence, $p_*: U \to \mathrm{Int}\, p^{-1}(\mathrm{Cl}\, U)$, between the families of regularly open subsets of X and Y, which preserves the Boolean operations. It is easy to check that if u is an ultrafilter in X (we consider only filters consisting of regularly open subsets), then $p_* u = \{p_*(U) : U \in u\}$ is an ultrafilter in Y. Thus the underlying set of Y may be regarded as a subset of the family $R(X)$ of all ultrafilters in X, and consequently, every class of non-homeomorphic covers of X is a set. Let $\mathrm{Cov}\, X$ be a fixed set of covers of X such that each cover of X is homeomorphic to one of $\mathrm{Cov}\, X$.

The following cancellation law may be found in [7] (see [8] for the compact case):

2. *If $p: Y \to X$ is irreducible and $h: Y \to Y$ is such that $p \circ h = p$, then h is the identity.*

This leads to a corollary that we get a partial ordering in Cov X if we set $p \prec q$ for covers $p: Y \to X$ and $q: Z \to X$ whenever there exists $h: Z \to Y$ such that $p \circ h = q$.

Let R, $R \subset R(X)$, be a set of ultrafilters in X such that for each x in X there exists a u in R such that $x \in \bigcap\{\mathrm{Cl}\, U : U \in u\}$; in particular, the whole $R(X)$ satisfies this condition. A cover $p: Y \to X$ will be said to be an *R-cover* if for each u in R there exists a y in Y such that $y \in \bigcap\{\mathrm{Cl}\, p_*(U) : U \in u\}$; then $p(y)$ is the only point in $\bigcap\{\mathrm{Cl}\, U : U \in u\}$. Let $\mathrm{Cov}_R\, X$ be the set of all R-covers of X.

Let us note an easy fact (cf. [3]):

3. *If V is regularly open in Y, then the projection $(Y - V) + \mathrm{Cl}\, V \to Y$ is an R-cover for any R, $R \subset R(Y)$*; here "+" stands for the free union. The projection is not a homeomorphism unless V is closed-open. This leads to the conclusion that the maximal elements in $\mathrm{Cov}_R\, X$ are extremally disconnected covers. To examine the set $\mathrm{Cov}_R\, X$ with respect to maximal elements, note first that

4. *If $p: Y \to X$ and $q: Z \to X$ are R-covers, $h: Z \to Y$ is such that $p \circ h = q$ and $u \in R$, then $h(z) = y$, whenever $\{y\} = \bigcap\{\mathrm{Cl}\, p_*(U) : U \in u\}$ and $\{z\} = \bigcap\{\mathrm{Cl}\, q_*(U) : U \in u\}$.*

The proof consists in the calculation which makes use of Assertion 1, that $h^{-1}(y) = \bigcap\{\mathrm{Cl}\, q_*(U) : U \in u\}$ (cf. [7], p. 29, for a similar calculation in a more general situation).

Now we shall prove

5. *The set $\mathrm{Cov}_R\, X$ is closed with respect to inverse limits of directed systems.*

Proof. Let P be a directed system of R-covers of X. Let Z be the inverse limit space of the system and let $p_X: Z \to X$ be the projection.

The projection p_X is onto. To see this, let $x \in X$. Let u in R be such that $\{x\} = \bigcap\{\mathrm{Cl}\, U : U \in u\}$. If Y is a space of the system P, then let z_Y be the only point in $\bigcap\{\mathrm{Cl}\, (p_X^Y)_* (U) : U \in u\}$, where $p_X^Y: Y \to X$ is a map from the system. The point z whose coordinates are z_Y just defined, Y running over the spaces of the system, is an element of Z, in virtue of Assertion 4. Clearly, $p_X(z) = x$.

The projection p_X is irreducible. In fact, let F be closed in Z and such that $F \neq Z$. Consider $W = Z - F$, open and non-empty. Since the system P is directed, there exists a space Y in the system and a non-empty open subset V of Y such that $p_Y^{-1}(V) \subset W$. Thus we have $\mathrm{Cl}\, p_Y(F) \subset Y - V$, and consequently $\mathrm{Cl}\, p_Y(F) \neq Y$. From the irreducibility of p_X^Y it follows that $\mathrm{Cl}\, p_X(F) \neq X$.

The projection p_X is an R-cover. In fact, let $u \in R$ and let us take an x in X such that $\{x\} = \bigcap\{\mathrm{Cl}\, U : U \in u\}$. Let z in Z be taken for x and for u as at the beginning of the proof, so that $p_X(z) = y$. For this z we get $z \in \bigcap\{\mathrm{Cl}\, (p_X)_* (U) : U \in u\}$.

Now, using the Kuratowski-Zorn Lemma, we get the maximal elements in $\mathrm{Cov}_R X$.

Note that the maximal elements in $\mathrm{Cov}_R X$ are maximal in the whole $\mathrm{Cov}\, X$ if we neglect a difference in topologies modulo regularly open subsets (i.e., if we regard topologies as equivalent if the families of regularly open sets in these topologies are equal). To see this, let $p : E \to X$ be a maximal element in $\mathrm{Cov}_R X$. The space E is extremally disconnected, in virtue of the remark following Assertion 3. Then the announced fact follows from the assertion below (Błaszczyk [1]; for more usual version, see Gleason [3] and Flachsmeyer [2]):

6. *If E is extremally disconnected then each irreducible map onto E is a bijection which preserves regularly open subsets.*

There are many extremally disconnected covers for a given X, e.g. depending on the choice of R, $R \subset R(X)$. A partial ordering in the set E Cov X, of extremally disconnected covers of X, is introduced if we set $p \leqq q$ iff there exists an h such that $q \circ h = p$ (notice the change of the order of composition in comparison to the ordering in the whole Cov X). The map h which realizes the inequality is an "embedding" onto a dense subset, if we neglect the difference of topologies modulo regularly open subsets. We get the greatest extremally disconnected cover of X, if we take for R the whole $R(X)$; this is the Iliadis extremally disconnected cover as defined in [7], p. 31 (see also [6]).

References

[1] *A. Błaszczyk:* A characterization of extremally disconnected spaces by means of irreducible maps. Notes in Mathematics, Silesian University, Katowice (to appear).

[2] *J. Flachsmeyer:* Topologische Projektivräume. Math. Nachr. *26* (1963), 355–365.

[3] *A. M. Gleason:* Projective topological spaces. Illinois J. Math. *2* (1958), 482–489.

[4] *A. W. Hager:* The projective resolution of a compact space. Proc. Amer. Math. Soc. *28* (1971), 262–266.

[5] *С. Илиадис:* Абсолюты хаусдорфовых пространств. ДАН СССР *149* (1963), 22—25.

[6] *J. Mioduszewski and L. Rudolf: H*-closed spaces and projectiveness. Contributions to Extension Theory of Topological Structures (Proc. Sympos., Berlin, 1967). Deutsch. Verlag Wissensch., Berlin, 1969, 153–156.

[7] *J. Mioduszewski and L. Rudolf: H*-closed and extremally disconnected Hausdorff spaces. Dissertationes Math. *66*, 1969.

[8] *J. Rainwater:* A note on projective resolutions. Proc. Amer. Math. Soc. *10* (1959), 734–735.

SILESIAN UNIVERSITY, KATOWICE

HOMOTOPICAL STRUCTURE OF LINEAR GROUPS OF BANACH SPACES

B. MITJAGIN

Moskva

Some geometrical conditions are given for a Banach space X which implicate contractibility of the linear group $\mathrm{GL}(X)$, i.e., the group of all automorphisms of X with the topology induced by the norm

$$\|A\| = \sup\{\|Ax\| : \|x\| \leqq 1\}.$$

Definition 1. A Banach space X is *weakly infinitely decomposed* (WID) if

a) there exists a total system of disjoint projections $\{P_k, k \geqq 0\}$, i.e., $P_kP_i = P_iP_k = 0$, $\forall i \neq k$, and $(P_kx = 0, \forall k) \Rightarrow x = 0$;

b) all images P_kX, $k \geqq 0$, are isomorphic to X, more exactly, there exist isomorphisms $T_k : P_kX \rightrightarrows X$, $\forall k \geqq 0$;

c) X is isomorphic to its Cartesian square $X \times X$;

d) there exist bounded operators (left and right shifts) $S, S' : X \to X$ such that $T_kP_kSx = T_{k+1}P_{k+1}x$, $\forall k \geqq 0$ and $T_kP_kS'x = T_{k-1}P_{k-1}x$, $\forall k \geqq 1$, $P_0S'x = 0$.

e) for any $B : X \to X$ there exists an operator $\tilde{B} : X \to X$ such that $T_kP_k\tilde{B}x = BT_kP_kx$, $\forall k \geqq 0$, $\forall x \in X$, i.e., the diagonal representation of $L(X)$ is continuous.

Definition 2. A Banach space X has the property of *smallness of operator blocks* (SOB) if for any compact family $B = (b)$ of operators in X and $\varepsilon > 0$ there exist projections Q_1 and Q_2 such that $Q_1Q_2 = Q_2Q_1 = 0$, its images Q_iX, $i = 1, 2$, are isomorphic to X, and $\|Q_1bQ_2\| < \varepsilon$, $\forall b \in B$.

The pointing out of these conditions by the author of [1] is based on Kuiper's, 1965, and Neubauer's, 1967, constructions, and the further generalization of their results is the following

Theorem ([1], § 2). *Let a Banach space X have properties* WID *and* SOB. *Then* $\mathrm{GL}(X)$ *is contractible to* 1_X.

This statement has been used for a proof of contractibility of $\mathrm{GL}(X)$ for particular Banach spaces, namely, a) $L^p[0, 1]$, $1 < p < \infty$ (C. McCarthy and the author, [1], § 5); b) $C^k(M)$, $k \geqq 1$, M is a differentiable manifold ([1], § 4); c) $L^1[0, 1]$ (I. Edelstein, E. Semenov and the author [3], [1], § 4); d) $C(K)$ for a wide class of compacts (the same authors [3], Theorem 1).

Speaking in a more detailed way, the group GL $(C(K))$ is contractible if K is one of the following compact Hausdorff spaces: 1) an uncountable compact metric space; 2) an infinite compact topological group; 3) an infinite product of non-one-point compact metric spaces; 4) the Stone space of an infinite homogeneous measure algebra; 5) βN – the Stone-Čech compactification of integers.

Nevertheless there exist ([2], § 9) such compacts K that GL $(C(K))$ is not contractible. More precisely, let K_1 be a compact of ordinals less than or equal to ω_1, the first uncountable ordinal, with the interval topology; then $\mathrm{GL}(C_R(K_1)) \simeq Z_2$ and $\mathrm{GL}(C_c(K_1)) \simeq S^1$. More generally, if $K_n =$ the union of n copies of K_1 then $\mathrm{GL}(C_R(K_n)) \simeq O(n)$ and $\mathrm{GL}(C_c(K_n)) \simeq U(n)$.

The James spaces also give homotopically-non-trivial linear groups. The above is a brief resume of [1], [2], [3].

References

[1] *Б. С. Митягин:* Гомотопическая структура линейной группы Банахова пространства. Успехи математических наук *25* (5) (1970), 63—106.

[2] *Б. С. Митягин и И. С. Эдельштейн:* Гомотопический тип линейных групп двух классов Банаховых пространств. Функциональный Анализ и его приложения *4* (3) (1970), 61—72.

[3] *I. Edelstein, B. Mitjagin and E. Semenov:* The linear groups of C and L_1 are contractible. Bull. Acad. Polon. Sci. Sér. Sci. Math. Astronom. Phys. *18* (1) (1970), 27—33.

A CONTRIBUTION TO THE THEORY OF MODULAR SPACES

J. MUSIELAK and A. WASZAK

Poznań

1. In this paper we introduce and investigate some modular spaces and connections between these spaces. In the first part the definition of a modular and a pseudomodular and of a modular space are given. Next, some examples of modular spaces depending on a parameter are given. In the second part of this paper a property of these spaces and connections between them are considered.

1.1. Let a real linear space X be given and let ϱ be a functional defined on X with values $-\infty < \varrho(x) \leqq +\infty$. This functional will be called a *pseudomodular*, if it satisfies the following conditions:

$$\varrho(0) = 0,$$
$$\varrho(-x) = \varrho(x),$$
$$\varrho(\alpha x + \beta y) \leqq \varrho(x) + \varrho(y) \text{ for every } \alpha, \beta \geqq 0,\ \alpha + \beta = 1.$$

If ϱ satisfies the condition

$$\varrho(x) = 0 \text{ if and only if } x = 0$$

instead of condition one, then ϱ is called a *modular*. It is easily seen that if ϱ is a pseudomodular on X, then we have always $\varrho(x) \geqq 0$. Now, we define the *modular space*

$$X_\varrho = \{x : \varrho(\lambda x) \to 0 \text{ as } \lambda \to 0,\ x \in X\}.$$

It is quite obvious that defining modulars in different manners, we obtain various modular spaces (see [2]).

1.2. Let X be a real linear space, and let Ξ be an abstract set. Let $\mathfrak{X}$ be a σ-algebra of subsets of the set Ξ, and let m be a nonnegative measure on $\mathfrak{X}$. We consider an extended real-valued function ϱ defined on $\Xi \times X$, satisfying the following conditions:

1. $\varrho(\xi, x)$ is a pseudomodular in X for almost every $\xi \in \Xi$,
2. if $\varrho(\xi, x) = 0$ for almost every $\xi \in \Xi$, then $x = 0$,
3. $\varrho(\xi, x)$ is measurable in Ξ for every $x \in X$.

By means of this function ϱ we define the following functionals in X:

$$\varrho(x) = \int_{\Xi} p(\xi) \frac{\varrho(\xi, x)}{1 + \varrho(\xi, x)} \, dm ,$$

where $p(\xi)$ is measurable, $0 < p(\xi) < \infty$, $\int_{\Xi} p(\xi) \, dm = 1$,

$$\varrho_0(x) = \operatorname*{supess}_{\xi} \varrho(\xi, x) , \quad \varrho_u(x) = \sup_{\xi} \varrho(\xi, x) .$$

Moreover, let $\mathfrak{M} = \{m_\eta\}$, $\eta \in \mathfrak{Y}$, be a family of nonnegative measures on $\mathfrak{X}$, where $\mathfrak{Y}$ is a set of indices. Then we define

$$\varrho_{\sigma(\mathfrak{M})}(x) = \sup_{\eta} \int_{\Xi} \varrho(\xi, x) \, dm_\eta .$$

In particular, if m_η are absolutely continuous with respect to m, then

$$\varrho_{\sigma(\mathfrak{M})}(x) = \sup_{\eta} \int_{\Xi} a(\xi, \eta) \, \varrho(\xi, x) \, dm ,$$

where the kernel $a(\xi, \eta) \geqq 0$ is measurable in Ξ for every $\eta \in \mathfrak{Y}$. Two special cases will be of importance. In the first one with $a(\xi, \eta) \equiv 1$ we shall write ϱ_s in place of $\varrho_{\sigma(\mathfrak{M})}$, i.e.,

$$\varrho_s(x) = \int_{\Xi} \varrho(\xi, x) \, dm .$$

The second one is obtained taking $\Xi = \langle 0, \infty)$, $\mathfrak{Y} = \langle \eta^*, \infty)$, where $\eta^* > 0$, m is the Lebesgue measure in Ξ, and

$$a(\xi, \eta) = \begin{cases} 1/\eta & \text{for} \quad \xi \leqq \eta , \\ 0 & \text{for} \quad \xi > \eta . \end{cases}$$

Then we shall write ϱ_σ in place of $\varrho_{\sigma(\mathfrak{M})}$, i.e.,

$$\varrho_\sigma(x) = \sup_{\eta \geqq \eta^*} \frac{1}{\eta} \int_{\eta^*}^{\eta} \varrho(\xi, x) \, d\xi .$$

It is easily verified that ϱ, ϱ_0, ϱ_s and ϱ_σ are modulars in X. If $\varrho(\xi, x)$ is a pseudomodular in X for every $\xi \in \Xi$, then ϱ_u is also a modular, and $\varrho_{\sigma(\mathfrak{M})}$ is in general a pseudomodular in X. The respective modular spaces will be denoted by X_ϱ, X_{ϱ_0}, X_{ϱ_s}, X_{ϱ_σ}, X_{ϱ_u}, and $X_{\varrho_{\sigma(\mathfrak{M})}}$. Let us remark, that taking Ξ to be the set of positive integers, $\mathfrak{X}$ the σ-algebra of all subsets of the set Ξ, and mB the number of elements of the set $B \subset \Xi$, $p(\xi) = (\frac{1}{2})^\xi$, then X_ϱ and $X_{\varrho_0} = X_{\varrho_u}$ are countably modulared space and uniformly countably modulared space, respectively (see [1]). Taking also $\mathfrak{Y}$ to be the set of positive integers and defining m_n, $n \in \mathfrak{Y}$, by means of a matrix $A = (a_{ni})$, $a_{ni} \geqq 0$, i.e., $m_n(i) = a_{ni}$, we obtain the space $X_{\varrho_{\sigma(A)}}$ defined in [3].

2. In this section we shall investigate some properties and connections between the above introduced spaces without any further assumptions on X. It is easily observed that

2.1. *We have* $X_{\varrho_u} \subset X_{\varrho_0} \subset X_{\varrho}$.

2.2. *If* $m\Xi < \infty$, *then* $X_{\varrho_0} \subset X_{\varrho_s}$.

This follows from the inequality $\varrho_s(x) \leqq m\Xi \,.\, \varrho_0(x)$.

2.3. *If* Ξ *consists of a countable number of pairwise disjoint atoms* $A_1, A_2, \ldots$ *with respect to the measure m, and* $\inf_k mA_k > 0$, *then* $X_{\varrho_s} \subset X_{\varrho_u}$.

This is obtained from the inequality $\varrho_s(x) \geqq \inf_k mA_k \,.\, \varrho_0(x)$.

2.4. $X_{\varrho_s} \subset X_{\varrho}$.

To prove this inclusion, let us write $A_n = \{\xi : p(\xi) > n\}$. Then $mA_n \to 0$ as $n \to \infty$. Let us choose an $\varepsilon > 0$. Then there exists an integer n such that $\int_{A_n} p(\xi)\, dm < \frac{1}{2}\varepsilon$, and so

$$\varrho(x) < \tfrac{1}{2}\varepsilon + \int_{\Xi \setminus A_n} p(\xi) \,.\, \varrho(\xi, x)\, dm \leqq \tfrac{1}{2}\varepsilon + n \,.\, \varrho_s(x)\,.$$

Let $x \in X_{\varrho_s}$, then $\varrho_s(\lambda x) \to 0$ as $\lambda \to 0$, and so there exists a $\lambda_\varepsilon > 0$ such that $\varrho_s(\lambda x) < \varepsilon/(2n)$ for $0 < \lambda < \lambda_\varepsilon$. Hence $\varrho(\lambda x) < \varepsilon$ for $0 < \lambda < \lambda_\varepsilon$, and thus $x \in X_\varrho$.

2.5. *An element* $x \in X$ *belongs to* X_ϱ, *if and only if,* $\varrho(\xi, \lambda x) \to 0$ *as* $\lambda \to 0$ *almost everywhere in* Ξ.

Proof. Let $\lambda_k \downarrow 0$ and let us denote

$$h_k(\xi) = p(\xi)\,\frac{\varrho(\xi, \lambda_k x)}{1 + \varrho(\xi, \lambda_k x)}\,.$$

Now, let $\varrho(\xi, \lambda x) \to 0$ as $\lambda \to 0$ almost everywhere in Ξ. Then $h_k(\xi) \leqq p(\xi) \,.\, \varrho(\xi, \lambda_k x) \to 0$ as $k \to \infty$ and $h_k(\xi) \leqq p(\xi)$. Hence, by Lebesgue dominated convergence theorem, $\int_\Xi h_k(\xi)\, dm \to 0$, i.e., $\varrho(\lambda_k x) \to 0$. Thus $x \in X_\varrho$.

Conversely, let $x \in X_\varrho$, then $\int_\Xi h_k(\xi)\, dm \to 0$ and so $h_k(\xi) \to 0$ in measure m. By the well-known Riesz theorem, $h_{k_i}(\xi) \to 0$ almost everywhere in Ξ, where $\{k_i\}$ is a subsequence of indices. Hence $\varrho(\xi, \lambda_{k_i} x) \to 0$ as $i \to \infty$ almost everywhere in Ξ. Since $\varrho(\xi, \lambda x)$ is a nondecreasing function of $\lambda > 0$, it follows $\varrho(\xi, \lambda x) \to 0$ as $\lambda \to 0$ almost everywhere in Ξ.

From 2.5 it follows immediately that

2.6. *If* $\varrho(\xi, \lambda x) \to 0$ *as* $\lambda \to 0$ *in measure m, then* $x \in X_\varrho$.

The converse statement is true under an additional assumption, namely

2.7. *If the measure m is absolutely continuous with respect to the measure* $nA = \int_A p(\xi)\,dm$, $A \in \mathfrak{X}$, *and* $x \in X_\varrho$, *then* $\varrho(\xi, \lambda x) \to 0$ *as* $\lambda \to 0$ *in measure m.*

Proof. Since $x \in X_\varrho$, by 2.5 we get $\varrho(\xi, \lambda x) \to 0$ as $\lambda \to 0$ almost everywhere with respect to measure m. But the measure n is absolutely continuous with respect to m, and so $\varrho(\xi, \lambda x) \to 0$ as $\lambda \to 0$ almost everywhere with respect to measure n. Since the measure n is finite, this implies convergence $\varrho(\xi, \lambda x) \to 0$ as $\lambda \to 0$ in measure n. Since m is absolutely continuous with respect to n, this implies $\varrho(\xi, \lambda x) \to 0$ as $\lambda \to 0$ in measure m.

Let us remark that the assumption of absolute continuity of m with respect to n in 2.7 cannot be omitted in general. For example, taking Ξ as the set of positive integers, mB as the number of elements of the set $B \subset \Xi$, and $p(\xi) = (\frac{1}{2})^\xi$, the condition $\varrho(\xi, \lambda x) \to 0$ as $\lambda \to 0$ in measure m is equivalent to the condition $x \in X_{\varrho_0}$, and not to the condition $x \in X_\varrho$.

Now, we proceed to investigation of the space $X_{\varrho_{\sigma(\mathfrak{M})}}$; it is convenient to assume absolute continuity of the measures m_η with respect to m, i.e., the modular $\varrho_{\sigma(\mathfrak{M})}(x)$ is defined by means of a kernel $a(\xi, \eta) \geqq 0$ (see 1).

2.8. *Suppose that a sequence of sets* $A_k \in \mathfrak{X}$, $k = 1, 2, \ldots$, *a sequence of indices* $\{\eta_k\}$ *and a sequence of numbers* $\{M_k\}$ *are given such that* $\bigcup_{k=1}^{\infty} A_k = \Xi$ *and* $a(\xi, \eta_k) > M_k$ *for every* $\xi \in A_k$. *Then* $X_{\varrho_{\sigma(\mathfrak{M})}} \subset X_\varrho$.

Proof. Let $x \in X_{\varrho_{\sigma(\mathfrak{M})}}$, then $\int_\Xi a(\xi, \eta) \,.\, \varrho(\xi, \lambda x)\,dm \to 0$ as $\lambda \to 0$ uniformly in $\mathfrak{Y}$. Taking $\lambda_i \downarrow 0$ and choosing an $\varepsilon > 0$ we have

$$M_k \int_{A_k} \varrho(\xi, \lambda_i x)\,dm \leqq \int_{A_k} a(\xi, \eta_k) \,.\, \varrho(\xi, \lambda_i x)\,dm < \varepsilon$$

for any k and for i sufficiently large. Hence $\varrho(\xi, \lambda_i x) \to 0$ in measure in the set A_k. Since $\varrho(\xi, \lambda x)$ is a nondecreasing function of $\lambda > 0$, the well-known Riesz theorem implies $\varrho(\xi, \lambda_i x) \to 0$ as $i \to \infty$ almost everywhere in A_k. Thus $\varrho(\xi, \lambda x) \to 0$ as $\lambda \to 0$ almost everywhere in Ξ, and so according to 2.5, $x \in X_\varrho$.

2.9. *If* $\sup_\eta \int_\Xi a(\xi, \eta)\,dm < \infty$, *then* $X_{\varrho_0} \subset X_{\varrho_{\sigma(\mathfrak{M})}}$.

This result follows from the inequality

$$\varrho_{\sigma(\mathfrak{M})}(x) \leqq \sup_\eta \int_\Xi a(\xi, \eta)\,dm \,.\, \varrho_0(x)\,.$$

2.10. *If* $\sup_\eta \operatorname{supess}_\xi a(\xi, \eta) < \infty$, *then* $X_{\varrho_s} \subset X_{\varrho_{\sigma(\mathfrak{M})}}$.

This follows from the inequality

$$\varrho_{\sigma(\mathfrak{M})}(x) \leqq \sup_\eta \operatorname{supess}_\xi a(\xi, \eta) \, . \, \varrho_s(x) \, .$$

Remark. Let us note that the results obtained in this paper are generalizations of some results of [3], taking the set of natural numbers as Ξ, the σ-algebra of all subsets of Ξ as $\mathfrak{X}$, the measure mB defined as the number of elements of the set $B \subset \Xi$, and $p(\xi) = (\frac{1}{2})^\xi$.

References

[1] *J. Albrycht and J. Musielak:* Countably modulared spaces. Studia Math. *31* (1968), 331—337.
[2] *J. Musielak and W. Orlicz:* On modular space. Studia Math. *18* (1959), 49—65.
[3] *J. Musielak and A. Waszak:* Some new countably modulared spaces. Comment. Math. Prace Mat. *15* (1971), 209—215.

INSTITUTE OF MATHEMATICS, A. MICKIEWICZ UNIVERSITY, POZNAŃ

A SURVEY OF THE THEORY OF GENERALIZED METRIC SPACES

J. NAGATA

Pittsburgh

1. Introduction

In the present lecture a survey will be given on the remarkable development of theory of generalized metric spaces which has taken place mainly since 1966, the year of the second Prague Symposium. There are various surveys, [5], [51], [42], [57], [10], [25] on this aspect of general topology, so efforts will be made to avoid too much overlap with those works. (Note that all spaces in this lecture are at least T_1 though most definitions as well as some theorems may be valid for spaces without T_1; all paracompact spaces are Hausdorff, and all single-valued maps (=mappings) are continuous. As for general terminology and symbols in general topology, see [46].)

Our main concern are generalized metric spaces of the following two types.

Type 1. M, p and related spaces.

Type 2. M_i, σ and related spaces.

Some spaces discussed here may be classified to neither of the two types but only related to them to various extents. The importance of spaces of type 1 largely relies on the fact that they are general enough to generalize both metric spaces and compact spaces and still concrete enough to allow beautiful extensions of theorems from those classical spaces. The importance of spaces of type 2 relies on the fact that they have, in various aspects, nicer properties than metric spaces do. For example, if a space is the sum of countably many closed sets which are σ-spaces as subspaces, then it is a σ-space while the same is not true for metrizable spaces. Accordingly some popular spaces (e.g. some infinite complices with weak topology) are spaces of this type though they are not metrizable. Thus our knowledge of topological properties of complices, for example, is not satisfactory without a full study of type 2 spaces. (Paracompactness seems too general to well represent properties of many important spaces which are "nearly metrizable".)

2. Basic properties

Since M, p, σ and M_i $(i = 1, 2, 3)$-*spaces* are becoming quite popular (originated by K. Morita, A. Arhangelskii, A. Okuyama and J. Ceder, respectively), no definition of them will be given in this lecture. The reader is referred to [40], [5] and [10] for their definitions. The above basic concepts are accompanied by many generalizations and modifications which are less popular. M^* and $w\Delta$, for example, generalize M-space.

Definition 1 ([9], [29]). X is a $w\Delta$- (M^*-) *space* if it has a sequence $\{\mathcal{U}_i \mid i = 1, 2, \ldots\}$ of open covers (locally finite closed covers) satisfying

(M) if $x_i \in S(x, \mathcal{U}_i)$, $i = 1, 2, \ldots$ for a fixed point x, then the point sequence $\{x_i \mid i = 1, 2, \ldots\}$ has a cluster point.

M^*-space is especially interesting as a useful supplement to M-space because the former is often easier to handle than the latter while *M and M^* coincide for every normal space* as proved by T. Ishii [30]. Semi-stratifiable is an interesting concept to generalize stratifiable ($=M_3$-) space.

Definition 2 ([15]). X is a *semi-stratifiable space* if to every open set U of X we can assign a sequence $\{U_i \mid i = 1, 2, \ldots\}$ of closed sets such that

(i) $U = \bigcup_{i=1}^{\infty} U_i$,

(ii) if $U \subset V$ for open sets U and V, then $U_i \subset V_i$, $i = 1, 2, \ldots$.

When a new class $\mathcal{C}$ of spaces is established, we are interested in properties of $\mathcal{C}$ like those in the following.

(1) If $X \in \mathcal{C}$ and $X' \subset X$, then $X' \in \mathcal{C}$.

(2) If $X_i \in \mathcal{C}$, $i = 1, 2, \ldots$, then $\prod_{i=1}^{\infty} X_i \in \mathcal{C}$.

(3) If $X \in \mathcal{C}$, and there is a closed map from X onto Y, then $Y \in \mathcal{C}$.

(4) If $X = \bigcup_{i=1}^{\infty} X_i$ for closed sets $X_i \in \mathcal{C}$, $i = 1, 2, \ldots$, then $X \in \mathcal{C}$.

(5) If X is dominated by a closed cover $\{X_\alpha \mid \alpha \in A\}$ with $X_\alpha \in \mathcal{C}$, then $X \in \mathcal{C}$.

In this respect spaces of type 1 usually show poor quality. M-spaces, for example, satisfy none of (1)–(5); especially, as shown by T. Ishiwata [32], *the product of two Tychonoff M-spaces need not be M. The image of an M-space by a perfect map is not necessarily M* as proved by K. Morita [41]. Paracompact M-spaces satisfy only (2). (See [42].) On the contrary, as suggested in the previous section, spaces of type 2 usually show good records. σ-spaces, for example, have all of the five properties though regularity for (3) and normality for (5) are required (see [57]). Stratifiable ($=M_3$-) spaces have all but (4) (see [10]). As for (4), R. Heath [24] proved that *there is a regular space with countably many points which is not stratifiable.* Relations between the classes of generalized metric spaces and some classical spaces are partially given in the following diagram.

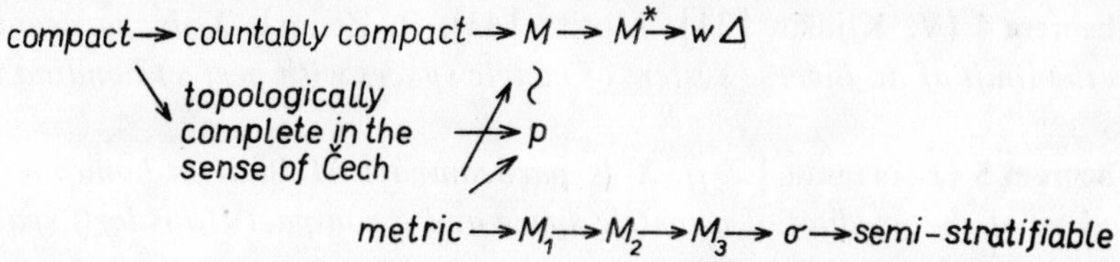

(We need Hausdorff axiom for the implication compact → topologically complete; M and p coincide if the space is paracompact though they differ in general case.) Most of the relations are easy to prove, but the implication $M_3 \to \sigma$ had been a major open question until *R. Heath* [23] *recently proved it.* On the other hand "$M_3 \to M_2$? $M_2 \to M_1$?" still remains as a major open question as it was posed by J. Ceder ten years ago. C. Borges recently proved in an unpublished paper that if each point of every stratifiable space has a σ-closure preserving nbd (local) base, then M_3 is M_1. But it is unknown if every stratifiable space has such nbd basis. In this connection we ask: *Is the image of a metric space by a closed map M_1 (or M_2)?*, which might be a little easier than Ceder's problem if we could expect affirmative answers at all. (Note that the closed image of a metric space is stratifiable. It has been announced quite recently that this problem has been solved affirmatively by F. Slaughter.) There are several remarkable relations which do not appear in the above diagram. For example,

Theorem 1 (G. Creede [15]). *X is semi-metric iff (=if and only if) it is first-countable and semi-stratifiable.*

3. Characterizations

There are various characterizations obtained of generalized metric spaces, which help us to study their different aspects and often lead us to further results. As for M-space the following is fundamental.

Theorem 2 (Morita [40]). *X is an M-space iff there is a quasi-perfect map f from X onto a metric space Y.*

The following theorems show how deeply paracompact M-spaces, which are especially interesting among M-spaces, are related to metric spaces and compact spaces.

Theorem 3 (Morita [40], Arhangelskii [2]). *The following conditions are equivalent.*

(i) *X is paracompact M,*

(ii) *X is paracompact p,*

(iii) *there is a perfect map f from X onto a metric space Y.*

Theorem 4 (V. Kljušin [34], Morita [43], P. Zenor). *X is paracompact M iff it is the limit of an inverse system of metric spaces with perfect bonding maps.*

Theorem 5 (J. Nagata [50]). *X is paracompact M iff it is homeomorphic to a closed set of the product of a metric space and a compact Hausdorff space.*

The last theorem leads us to the following two questions:

1. *Find out a universal space Z such that X is a paracompact M-space with weight α iff it is homeomorphic to a closed set of Z.* (It is also worth while to find universal spaces for the other classes of generalized metric spaces including those of type 2.)

2. *Is every M-space homeomorphic to a closed set of the product of a metric space and a countably compact space?* (The converse is easily seen to be true.) A special case of the first problem is answered as follows.

Theorem 6 (Nagata [50]). *X is a paracompact, topologically complete* (in the sense of Čech) *space with weight α iff it is homeomorphic to a closed set of $H(A) \times P(A)$, where $H(A)$ and $P(A)$ are the Hilbert space with α coordinates and the product of α closed intervals, respectively.*

As well-known *p*-space and a little stronger concept, *strict p-space* are defined in terms of open families in Stone-Čech compactification, and this fact prevents us from handling them with more easiness or comparing them with *M*-space. D. Burke [12] gave internal characterizations of *p*- and strict *p*-spaces and studied their relations with $w\Delta$-spaces.

Theorem 7. *A Tychonoff space X is p iff there is a sequence $\{\mathcal{U}_i\}$ of open covers of X satisfying: If $x \in X$ and $x \in U_i \in \mathcal{U}_i$, $i = 1, 2, \ldots$, then*

(i) $\bigcap_{i=1}^{\infty} \overline{U}_i$ *is compact,*

(ii) *if $x_n \in \bigcap_{i=1}^{n} \overline{U}_i$, $n = 1, 2, \ldots$, then $\{x_n\}$ has a cluster point.*

Turning to spaces of type 2, the following characterization is basic for σ-spaces, which indicates that the natures of "base" and "*net* (or *network*)" are essentially different despite their similar definitions, and this fact contributes to the advantageous properties of σ-spaces in comparison with metric spaces.

Theorem 8 (F. Siwiec and J. Nagata [62]). *The following conditions for a regular space X are equivalent.*

(i) *X is σ,*

(ii) *X has a σ-closure preserving net,*

(iii) *X has a σ-discrete net.*

Since we have seen so many generalized metric spaces emerging up one after another, it is natural to try to characterize them in a unified manner. One of such attempts is done by R. Heath, R. Hodel and others by use of open nbds (= neighborhoods). Let $\{U(n, x) \mid n = 1, 2, \ldots\}$ be a sequence of open nbds of $x \in X$; then we consider e.g. the following conditions.

(1) $\{U(n, x) \mid n = 1, 2, \ldots\}$ is a nbd base for x.

(2) If $y \in U(n, x)$, then $U(n, y) \subset U(n, x)$.

(3) If $x \notin F$ for a closed set F, then $x \notin \bigcup\{U(n, y) \mid y \in F\}$ for some n. (This is equal to: If $x \in U(n, x_n)$, $n = 1, 2, \ldots$, then x is a cluster point of $\{x_n\}$.)

(4) If $x \notin F$ for a closed set F, then $x \notin [\bigcup\{U(n, y) \mid y \in F\}]^-$ for some n.

(5) If $\{x, x_n\} \subset U(n, y_n)$, $n = 1, 2, \ldots$, then x is a cluster point of $\{x_n\}$.

(6) If $\{x, x_n\} \subset U(n, y_n)$, $n = 1, 2, \ldots$, then $\{x_n\}$ has a cluster point.

(7) If $U(n, x) \cap U(n, x_n) \neq \emptyset$, $n = 1, 2, \ldots$, then x is a cluster point of $\{x_n\}$.

Theorem 9. *Generalized metric spaces are characterized in terms of existence of* $\{U(n, x)\}$ *satisfying the above conditions as follows: Semi-stratifiable* = (3) (Creede [16]), *semi-metric* = (1) *and* (3) (Heath [20]), σ = (2) *and* (3) (Heath-Hodel [26]), *stratifiable* = (4) (Heath [21]), *Nagata space* (= *stratifiable and first countable*) = (7) (Heath [21]), M_2 = (2) *and* (4) (Nagata [54]), *developable* = = (5) (Heath [20]), $w\Delta$ = (6) (Heath [20]). (Developable spaces are related to spaces of both type 1 and type 2, because every developable space is semi-metric and σ, and every Tychonoff developable space is strict p.)

A merit of this characterization is to help us to better understand relations between different spaces. In fact (4) was used by Heath to prove the implication stratifiable $\rightarrow \sigma$, and Hodel used (5) and (6) to study the relation between developable spaces and $w\Delta$-spaces. Ceder's question, $M_3 \rightarrow M_2$? is restated as follows: Given nbds satisfying (4) and nbds satisfying (2) and (3), is it then possible to construct nbds satisfying (4) and (2) at the same time?

4. Mappings

At the first Prague Symposium P. Alexandroff [1] asked the following question: *Which spaces can be represented as images (or inverse images) of nice spaces by nice maps?* Many interesting works have been done to answer this question and Arhangelskii [5] gave a good survey on the results obtained by 1966 and posed many interesting problems, some of which were answered since then. As for generalized metric spaces efforts along this idea coincide with those to characterize various generalized metric spaces in a unified manner, i.e., as the images (or inverse images) of metric spaces by adequate maps. Typical results in this respect are Theorems 2 and 3. Besides we can characterize many spaces as images of metric spaces as partially seen in the following theorem. (The left hand of equality shows the characterized space, and the right hand the used map. * denotes that the map is multi-valued, and † that the space is characterized as the image of 0-dimensional metric space and "0-dimensional" is essential. We need regularity for the characterizations of M_2 and 1st countable M_2 The reader is referred to the papers in parentheses for the definitions of used maps.)

Theorem 10. *Semi-stratifiable = semi-stratifiable* (Nagata [55]), *semi-metric = = a certain condition* (Heath [21]), *$\sigma = \sigma$ and one-to-one* (Nagami [76]), *$\sigma = \sigma$-locally finite* (Michael [39]), *σ = s. perfect** (Nagata [52]), *stratifiable = stratifiable* (Nagata [55]), *Nagata = a certain condition* (Heath [21]), *M_2 = q-open and q-closed*†* (Nagata [55]), *M_2 and first countable = almost open and q-closed†* (Nagata [55]), *developable = open π* (Arhangelskii [4], Heath [21]), *M^* = perfect and Y-countably compact** (Nagata [52]), *strict p = open π and bi-Y-compact** (Nagata [55]).

Dealing with the same problem from a little different point of view, many efforts also have been done to characterize the images of metric spaces by given (popular) maps though the image spaces are not necessarily nice as generalizations of metric spaces. As well-known, the images of metric spaces by open (S. Ponomarev [59], S. Hanai [71]), pseudoopen (Arhangelskii [3]), quotient (Franklin [19]), bi-quotient (Michael [39]), closed (N. Lašnev [26]), open compact (Arhangelskii [5]) and open *s*-maps (Ponomarev [59]) are characterized. In this respect T. Hoshina [28] recently answered Arhangelskii's problem [5] by *characterizing the images of metric spaces by quotient s-maps.* There are also works to *characterize images of M-spaces,* e.g. *by open* (Nagata [49]), *quotient and bi-quotient* (Nagata [47]), *pseudoopen* (T. Rishel [60]) *and almost open* (T. Chiba [14]) *maps* as well as *the images of paracompact M-spaces by open maps* (H. Wicke [67]). (According to recent news Morita and Rishel have characterized images of *M*-spaces by closed maps.) The perfect images of *M*-spaces were recently characterized by Nagata [55] as follows.

Theorem 11. *X is an M^*-space iff it is the image of an M-space by a perfect map.*

Theorem 11 was proved using the characterization of an M^*-space in Theorem 10. Unfortunately, the same method will not be applicable for characterization of perfect images of *p*-spaces. In this connection J. M. Worrell [77] answered Arhangelskii's question [5] in the negative by giving an example of a *p*-space whose perfect image is not *p*. It was also announced by Worrell and Wicke that the perfect image of a θ-refinable *p*-space is *p*. We can characterize σ-spaces by developing the characterization in Theorem 10 as follows:

Definition 3. Let (X, X') be a pair of a space X and its subspace X'. If X has a sequence $\{\mathcal{U}_i\}$ of locally finite open covers satisfying

(H) for every $x \in X'$ and every nbd V of x in X, there is $U \in \bigcup_{i=1}^{\infty} \mathcal{U}_i$ such that $x \in U \subset V$, then (X, X') is called a *partially metric space* (*or half-metric space*).

Theorem 12 (Nagata [55]). *A regular space Y is a σ-space iff there is a half-metric space (X, X') and a perfect map f from X onto Y such that $f(X') = Y$ iff there is a half-metric space (X, X') and a closed map f from X onto Y such that $f(X') = Y$.*

5. Metrization and other aspects

As for metrization of generalized metric spaces, a basic rule is that metrizability of a space is equal to a condition of type 1 plus a condition of type 2 (plus some additional condition). This rule was first observed by Borges [9], [10] and Okuyama [56]; the latter proved that metrizability $= M + \sigma +$ paracompact. Their theorems were improved by several mathematicians, especially by F. Slaughter, who observed that metrizability $= M + \sigma + T_2$. Probably the best result along this line of efforts is the following theorem due to T. Shiraki [61]. (See also [73].)

Definition 4 ([31]). X is a *wM-space* if it has a sequence $\{\mathscr{U}_i\}$ of open covers satisfying: If $x_i \in S^2(x, \mathscr{U}_i)$, $i = 1, 2, \ldots$, then $\{x_i\}$ clusters.

Definition 5 ([62]). A collection $\mathscr{U}$ of closed sets is a *ct-net* if for every $x \in X$, $\bigcap\{U \mid x \in U \in \mathscr{U}\} = \{x\}$. A space with a σ-closure preserving *ct*-net is $\sigma^\#$.

Every M^*-space is *wM*, every *wM*-space is $w\Delta$, and every semi-stratifiable space is $\sigma^\#$, which coincides with Hodel's *α-space.*

Theorem 13. *A T_2-space X is metrizable iff it is wM and $\sigma^\#$.*

Another important factor of metrizability is *point-countable base.* V. Filippov proved in his celebrated paper [18] that *every paracompact M-space with a point-countable base is metrizable.* This theorem was generalized by Nagata [48], Slaughter [63], Shiraki [61], J. Suzuki and others. Let us give a version of their theorem.

Definition 6. A collection $\mathscr{U}$ of open sets is called a *p-base* if for each $x \in X$, $\bigcap\{U \mid x \in U \in \mathscr{U}\} = \{x\}$.

This concept is given various names, e.g. pseudo-base, T_1-cover, separating open cover, etc. Since it is proving useful as a generalization of a base, a standard terminology should be decided.

Theorem 14. *A T_2-space X is metrizable iff it is M^* and has a point-countable p-base.*

The extension of Filippov's theorem in another direction was done by Michael [39]. The following is T. Shiraki's [61] improved version. (Essentially the same theorem is given also in [78].)

Definition 7. X is a Σ-*space* if it has a sequence $\{\mathcal{U}_i\}$ of locally finite closed covers satisfying: If $x_i \in C(x, \mathcal{U}_i) = \bigcap\{U \mid x \in U \in \mathcal{U}_i\}$, $i = 1, 2, \ldots$, then $\{x_i\}$ clusters.

This notion due to K. Nagami [44] is interesting as it generalizes both M^* and regular σ-spaces.

Theorem 15. *X is metrizable iff it is collectionwise normal Σ and has a point-countable base.*

This theorem is easily derived from the following theorem of Michael and Slaughter [78], which is interesting in its own right.

Theorem 16. *Every Σ-space X with a point-countable p-base is σ.*

Each of the above three metrization theorems implies none of the others; a more unified theory is desirable in this aspect.

"Developable space" is a classical example of generalized metric space. Some authors chose this space (instead of metric space) as their starting point to study generalized metric spaces in their papers. In this connection generalizations of paracompactness like *σ-paracompactness* ([5]) and *θ-refinability* ([69]) are proving useful, because general developable spaces satisfy only these conditions weaker than paracompactness.

Ishiwata [33], for example, has done an extensive investigation on inverse images of developable spaces by perfect and quasi-perfect maps. The following theorem may be compared with Theorem 13.

Theorem 17 (Burke [12]). *A regular space is developable iff it is $w\Delta$ and $\sigma^\#$.*

Čoban [75] and Hodel [27] also got interesting results about developability of spaces.

The importance of a generalized metric space largely depends on how beautifully theorems can be extended from metric spaces. Many of the previously mentioned theorems imply such extensions. Borges [8] extended the classical Dugundji's extension theorem on map of a metric space to a stratifiable space. Lašnev's theorem [35] (given a closed map f from a metric space X into Y, then $Y = \bigcup_{n=0}^{\infty} Y_n$, where each Y_n, $n > 0$ is discrete and $f^{-1}(y)$ is compact for each $y \in Y_0$) was extended by Filippov [18] to paracompact M-space, by R. S. Stoltenberg [64] to normal semi-stratifiable space and by Nagata [53] to a class of spaces including wM-spaces and semi-metric spaces. Nagami [44] generalized Morita's theorem on metric space as follows: *Let X be a paracompact Σ-space and Y a paracompact P-space in the sense of Morita* [40]; *then $X \times Y$ is paracompact.* Extension is not always an easy task. The author's conjecture [45] at the 2nd Prague Symposium: $\dim X \times Y \leqq \dim X + \dim Y$ *for paracompact M-spaces*? still remains open while the same is a well-known theorem for metric spaces and compact spaces. There is another interesting aspect. Morita [43] developed a theory on paracompactification of M-space. He defined $\mu(X)$ for every Tychonoff space X as the completion of X relative to its finest uniformity to show that *$\mu(X)$ is paracompact M if X is M* and μX has especially interesting properties.

There are many other generalizations (related to type 1 or 2) of metric spaces which have not been discussed here, M' ([74]), $M^{\#}$ ([62]), $\Sigma^{\#}$ ([39]), *strong* Σ ([44]), Σ^* ([58]), *k-semi-stratifiable* ([37], *submetrizable* (= *contractible onto a metric space* ([34])) to name just a few. The definitions of all generalizations mentioned by now contain some countability in the sense that they are defined in terms of existence of countably many covers, countably many collections of sets, etc. H. Tamano's [65] *elastic space* and J. Vaughan's [66] *linearly stratifiable space* are probably the first attempts to generalize stratifiable space by dropping countability from the definition. All of the discussed spaces were generalizations of *general* metric space. There are interesting generalizations of special metric spaces, e.g. Worell and Wicke [69] generalized complete metric space, and Michael [38] generalized separable metric space. However, for those spaces and others another survey should be given since we have already exceeded the expected length of lecture.

References

[1] *P. S. Alexandroff:* On some results concerning topological spaces and their continuous mappings. General Topology and its Relations to Modern Analysis and Algebra, (I) (Proc. (First) Prague Topological Sympos., 1961). Publishing House of the Czechoslovak Academy of Sciences, Prague, 1962, 41—54.

[2] *A. V. Arhangelskii:* On a class of spaces containing all metric and all locally bicompact spaces. Soviet Math. *4* (1963), 751—754.

[3] *A. V. Arhangelskii:* Some types of factor mappings and the relations between classes of topological spaces. Soviet Math. *4* (1963), 1335—1338.

[4] *A. V. Arhangelskii:* Factor mappings of metric spaces. Soviet Math. *5* (1964), 368—371.

[5] *A. V. Arhangelskii:* Mappings and spaces. Russian Math. Surveys *21* (1966), 115—162.

[6] *A. V. Arhangelskii:* The closed image of a metric space can be condensed to a metric space. Soviet Math. *7* (1966), 1109—1112.

[7] *E. Berney:* A completely regular semi-metric space with no σ-discrete net. Notices Amer. Math. Soc. *16* (1969), 852.

[8] *C. J. R. Borges:* On stratifiable spaces. Pacific J. Math. *17* (1966), 1—16.

[9] *C. J. R. Borges:* On metrizability of topological spaces. Canad. J. Math. *20* (1968), 795—804.

[10] *C. J. R. Borges:* A survey of M-spaces: open questions and partial results. General Topology and its Applications *1* (1971), 79—84.

[11] *D. K. Burke:* On subparacompact spaces. Proc. Amer. Math. Soc. *23* (1969), 655—663.

[12] *D. K. Burke:* On p-spaces and $w\Delta$-spaces. Pacific J. Math. *35* (1970), 285—296.

[13] *J. Ceder:* Some generalizations of metric spaces. Pacific J. Math. *11* (1961), 105—126.

[14] *T. Chiba:* On q-spaces. Proc. Japan Acad. *45* (1969), 453—456.

[15] *G. Creede:* Semi-stratifiable spaces. Proc. Topology Conference, Arizona State Univ., 1967, 318—323.

[16] *G. Greede:* Concerning semi-stratifiable spaces. Pacific J. Math. *32* (1970), 47—54.

[17] *V. Filippov:* On the perfect image of a paracompact *p*-space. Soviet Math. *8* (1967), 1151—1153.
[18] *V. Filippov:* On feathered paracompacta. Soviet Math. *9* (1968), 161—164.
[19] *S. Franklin:* Spaces in which sequences suffice. Fund. Math. *57* (1965), 107—115.
[20] *R. W. Heath:* Arcwise connectedness in semi-metric spaces. Pacific J. Math. *12* (1962), 1301—1319.
[21] *R. W. Heath:* On open mappings and certain spaces satisfying the first countability axiom. Fund. Math. *62* (1965), 91—96.
[22] *R. W. Heath:* A note on cosmic spaces. Proc. Amer. Math. Soc. *17* (1966), 868—870.
[23] *R. W. Heath:* Stratifiable spaces are σ-spaces. Notices Amer. Math. Soc. *16* (1969), 761.
[24] *R. W. Heath:* An easier proof that a certain countable space is not stratifiable. (To appear.)
[25] *R. W. Heath:* A survey of the theory of semi-metric spaces. General Topology and its Applications (to appear).
[26] *R. Heath and R. Hodel:* Characterizations of σ-spaces. (To appear.)
[27] *R. E. Hodel:* Moore spaces and $w\Delta$-spaces. Pacific J. Math. (to appear).
[28] *T. Hoshina:* On the quotient *s*-images of metric spaces. (To appear.)
[29] *T. Ishii:* On closed mappings and *M*-spaces I, II. Proc. Japan Acad. *43* (1967), 752—761.
[30] *T. Ishii:* On *M*- and M^*-spaces. Proc. Japan Acad. *44* (1968), 1028—1030.
[31] *T. Ishii:* On *wM*-spaces I, II. Proc. Japan Acad. *46* (1970), 5—15.
[32] *T. Ishiwata:* The product of *M*-spaces need not be an *M*-space. Proc. Japan Acad. *45* (1969), 154—156.
[33] *T. Ishiwata:* On inverse images of developable spaces. (Japanese.) Proc. Symposium on General Topology, Tokyo, 1970.
[34] *V. Kljušin:* Perfect mappings of paracompact spaces. Soviet Math. *5* (1964), 1583—1586.
[35] *N. S. Lašnev:* Continuous decompositions and closed mappings of metric spaces. Soviet Math. *6* (1965), 1504—1506.
[36] *N. S. Lašnev:* Closed images of metric spaces. Soviet Math. *7* (1966), 1219—1221.
[37] *D. J. Lutzer:* Semi-metrizable and stratifiable spaces. General Topology and its Applications *1* (1971), 43—48.
[38] *E. A. Michael:* $\aleph_0$-spaces. J. Math. Mech. *15* (1966), 983—1002.
[39] *E. A. Michael:* On Nagami's Σ-spaces and some related matters. Proc. Wash. State Univ. Conference on General Topology, 1970, 13—19.
[40] *K. Morita:* Products of normal spaces with metric spaces. Math. Ann. *154* (1964), 365—382.
[41] *K. Morita:* Some properties of *M*-spaces. Proc. Japan Acad. *43* (1967), 869—872.
[42] *K. Morita:* A survey of the theory of *M*-spaces. General Topology and its Applications *1* (1971), 49—55.
[43] *K. Morita:* Topological completions and *M*-spaces. Sci. Rep. Tokyo Kyoiku Daigaku Sect. A *10* (1970), 271—288.
[44] *K. Nagami:* Σ-spaces. Fund. Math. *65* (1969), 169—192.
[45] *J. Nagata:* A survey of dimension theory. General Topology and its Relations to Modern Analysis and Algebra, II (Proc. Second Prague Topological Sympos., 1966). Academia, Prague, 1967, 259—270.
[46] *J. Nagata:* Modern General Topology. Amsterdam-Groningen, 1968.
[47] *J. Nagata:* Quotient and bi-quotient spaces of *M*-spaces. Proc. Japan Acad. *45* (1969), 25—29.
[48] *J. Nagata:* A note on Filippov's theorem. Proc. Japan Acad. *45* (1969), 30—33.
[49] *J. Nagata:* Mappings and *M*-spaces. Proc. Japan Acad. *45* (1969), 140—144.
[50] *J. Nagata:* A note on *M*-space and topologically complete space. Proc. Japan Acad. *45* (1969), 541—543.
[51] *J. Nagata:* Lectures on generalized metric spaces. Lecture note, University of Florida, 1969.

[52] *J. Nagata:* On multi-valued mappings and generalized metric spaces. Proc. Japan Acad. *46* (1970), 936—940.
[53] *J. Nagata:* On closed mappings of generalized metric spaces. Proc. Japan Acad. *47* (1971), 181—184.
[54] *J. Nagata:* Characterizations of some generalized metric spaces. Notices Amer. Math. Soc. (to appear).
[55] *J. Nagata:* Some theorems on generalized metric spaces. Collected Papers in memory of F. Hausdorff (to appear).
[56] *A. Okuyama:* Some generalizations of metric spaces, their metrization theorems and product spaces. Sci. Rep. Tokyo Kyoiku Daigaku Sect. A *9* (1967), 236—254.
[57] *A. Okuyama:* A survey of the theory of σ-spaces. General Topology and its Applications *1* (1971), 57—63.
[58] *A. Okuyama:* On a generalization of Σ-spaces. (To appear.)
[59] *V. I. Ponomarev:* Axiom of countability and continuous mappings. Bull. Acad. Polon. Sci. Sér. Sci. Math. Astronom. Phys. *8* (1960), 127—134.
[60] *T. W. Rishel:* A characterization of pseudo-open images of M-spaces. Proc. Japan Acad. *45* (1969), 910—912.
[61] *T. Shiraki:* On some metrization theorems. (To appear.)
[62] *F. Siwiec and J. Nagata:* A note on nets and metrization. Proc. Japan Acad. *44* (1968), 623—627.
[63] *F. G. Slaughter:* Note on a theorem of J. Nagata. Proc. Japan Acad. (to appear).
[64] *R. A. Stoltenberg:* A note on stratifiable spaces. Proc. Amer. Math. Soc. *23* (1969), 294—297.
[65] *H. Tamano and J. E. Vaughan:* Paracompactness and elastic spaces. Proc. Amer. Math. Soc. *28* (1971), 299—303.
[66] *J. E. Vaughan:* Linearly stratifiable spaces. (To appear.)
[67] *H. H. Wicke:* On the Hausdorff open continuous images of Hausdorff paracompact p-spaces. Proc. Amer. Math. Soc. *22* (1969), 136—140.
[68] *H. H. Wicke:* Open continuous images of certain kinds of M-spaces and completeness of mappings and spaces. General Topology and its Applications *1* (1971), 85—100.
[69] *J. M. Worrell, Jr. and H. H. Wicke:* Characterizations of developable topological spaces. Canad. J. Math. *17* (1965), 820—830.
[70] *J. M. Worrell, Jr. and H. H. Wicke:* On the open continuous images of paracompact Čech complete spaces. Pacific J. Math. (to appear).
[71] *S. Hanai:* On open mappings II. Proc. Japan Acad. *37* (1961), 233—238.
[72] *J. Nagata:* Problems on generalized metric spaces II. Proc. Houston Conference on General Topology, 1971 (to appear).
[73] *T. Ishii and T. Shiraki:* Some properties of wM-spaces. Proc. Japan Acad. *47* (1971), 167—172.
[74] *T. Ishiwata:* Generalizations of M-spaces I, II. Proc. Japan Acad. *45* (1969), 359—367.
[75] *M. M. Čoban:* σ-paracompact spaces. Vestnik Moskov. Univ. Ser. I Mat. Meh. *24* (1969), 20—27.
[76] *K. Nagami:* σ-spaces and product spaces. Math. Ann. *181* (1969), 109—118.
[77] *J. M. Worrell, Jr.:* A perfect mapping not preserving the p-space property. (To appear.)
[78] *E. Michael and F. Slaughter:* A note on Σ-spaces with point-countable separating open covers. (To appear.)

UNIVERSITY OF PITTSBURGH, PITTSBURGH, PENNSYLVANIA

RAMIFICATION SYSTEMS AND SPACES OF ULTRAFILTERS

S. NEGREPONTIS

Montréal

Let α be an infinite cardinal. $U(\alpha)$ denotes the space of uniform ultrafilters on α. $\Omega(\alpha^+)$ denotes the space of subuniform ultrafilters on α^+, i.e., those ultrafilters on α^+ which are not uniform, but which have only elements of cardinality at least α. The Stone-Čech compactification of a space S is denoted by βS.

Theorem. *If* $\alpha = \alpha^{\underline{\alpha}}$, *then* $\beta(\Omega(\alpha^+)) \setminus \Omega(\alpha^+)$ (= *the growth of* $\Omega(\alpha^+)$) *has the following property*: *for every non-empty closed subset* F *of* $\beta(\Omega(\alpha^+)) \setminus \Omega(\alpha^+)$ *which is equal to the intersection of at most* α *open-and-closed sets and every family* $\{V_\eta, \eta < \alpha\}$ *of non-empty pairwise disjoint open-and-closed subsets of* $\beta(\Omega(\alpha^+)) \setminus \Omega(\alpha^+)$ *there are* p *in* F *and an open-and-closed set* N, *with* p *in* N *and such that* $|\{\eta < \alpha : V_\eta \cap N \neq \emptyset\}| < \alpha$.

The property of this Theorem was studied by the author in The Existence of certain uniform ultrafilters, Ann. of Math. *90* (1969), 23–32, where it was proved that the space $U(\alpha)$ has the property (for regular α).

Corollary. *If* $\alpha = \alpha^{\underline{\alpha}}$ *then* $U(\alpha^+)$ *is not homeomorphic to* $\beta(\Omega(\alpha^+)) \setminus \Omega(\alpha^+)$. *In particular* $\beta(\Omega(\alpha^+)) \setminus \Omega(\alpha^+)$ *is not* C^**-embedded in* $\beta(\alpha^+)$.

For $\alpha = \omega$ this last statement has been proved by Mrs. N. M. Warren (Doctoral Dissertation, University of Wisconsin, 1970) using the fact that ω^+ has property Q (in the sense of Erdös and Tarski, On some problems involving inaccessible cardinals, Essays on the Foundations of Mathematics, Magnes Press, 1966, pp. 50–82). By contrast we can derive the more general result of E. Specker (slightly improved by D. Monk):

Corollary. *If* $\alpha = \alpha^{\underline{\alpha}}$ *then* α^+ *has property* Q *in the sense of Erdös and Tarski.*

Finally the following result can be proved directly.

Theorem. *If* α *is weakly compact* (*i.e.*, α *is strongly inaccessible and does not have property* Q), *then the space* $N(\alpha)$ *of non-uniform, non-principal ultrafilters on* α *is* C^**-embedded in* $\beta(\alpha)$.

The details will appear in Trans. Amer. Math. Soc.

ON COMPLETIONS OF CONVERGENCE COMMUTATIVE GROUPS

J. NOVÁK

Praha

In this paper the notions of Cauchy sequences and completeness for convergence commutative groups are introduced. For any such group L a completion L_1 is constructed containing L as a dense subspace and a subgroup. Finally, examples are given which show that L can have more than one completion.

1.

Let L be a point set, $\mathfrak{L}$ a convergence and λ a convergence closure for L (λA is the set of all $\lim x_n \in L$ such that $\bigcup x_n \subset A$). We have a convergence space $(L, \mathfrak{L}, \lambda)$. Instead of $\lim x_n = x$ we sometimes write $(\{x_n\}, x) \in \mathfrak{L}$ or $\mathfrak{L}$-$\lim x_n = x$. Using the transfinite induction we define successive closures $\lambda^\xi A$ in the following manner

$$\lambda^0 A = A\,, \quad \lambda^1 A = \lambda A\,, \quad \lambda^\xi A = \bigcup_{\eta<\xi} \lambda\lambda^\eta A\,.$$

The closure operator λ^{ω_1} is a topology for L. A subset $A \subset L$ is λ-closed if $\lambda A = A$. A subset $B \subset L$ is λ^{ω_1}-dense in L provided that $\lambda^{\omega_1} B = L$.

Definition 1. Let $(L_1, \mathfrak{L}_1, \lambda_1)$ be a convergence space. Define $(\{x_n\}, x) \in \mathfrak{L}_1^*$ whenever for each subsequence $\{x_{n_i}\}$ of $\{x_n\}$ there is a subsequence $\{x_{n_{i_k}}\}$ of $\{x_{n_i}\}$ such that $(\{x_{n_{i_k}}\}, x) \in \mathfrak{L}_1$. We say that $\mathfrak{L}_1^*$ is a *star convergence*[1]) *of the convergence* $\mathfrak{L}$.

It is easy to prove that $\mathfrak{L}_1 \subset \mathfrak{L}_1^*$, $(\mathfrak{L}_1^*)^* = \mathfrak{L}_1^*$ and $\lambda_1^* = \lambda_1$.

Let $(L, \mathfrak{L}, \lambda)$ be a convergence space and assume that $\mathfrak{L}$ is a star convergence. Let $+$ be a commutative group operation on L. If $\lim x_n = x$ and $\lim y_n = y$ implies that $\lim (x_n - y_n) = x - y$ then we have [3] (see also [4]) a convergence commutative group (abbr. a cc group). It will be denoted $(L, \mathfrak{L}, \lambda, +)$.

Definition 2. Let $(L, \mathfrak{L}, \lambda, +)$ be a cc group. The collection of all pairs $(\{x_n\}, \{y_n\})$, where $\{x_n\}$, $\{y_n\}$ are sequences of points of L such that $\lim (x_{i_n} - y_{j_n}) = 0$ for all subsequences $\{i_n\}$ and $\{j_n\}$ of $\{n\}$ will be denoted ϱ. A sequence $\{x_n\}$ of points of L is called a *Cauchy sequence* (in L) if $(\{x_n\}, \{x_n\}) \in \varrho$.

---

[1]) P. Urysohn [5] calls $\mathfrak{L}_1^*$ the convergence a posteriori. $\mathfrak{L}_1^*$ is sometimes called the maximal or largest convergence [1], [2].

Lemma 1. *If* $(\{x_n\}, \{y_n\}) \in \varrho$ *then both* $\{x_n\}$ *and* $\{y_n\}$ *are Cauchy sequences.*

Lemma 2. *Each subsequence of a Cauchy sequence is a Cauchy sequence.*

Lemma 3. *If* $(\{x_n\}, x) \in \mathfrak{L}$ *then* $\{x_n\}$ *is a Cauchy sequence.*

Lemma 4. *Let* $\{x_n\}$ *be a sequence and* $\{y_n\}$ *a Cauchy sequence. Then* $(\{x_n\}, \{y_n\}) \in \in \varrho$ *iff* $\lim (x_n - y_n) = 0$.

Lemma 5. *If* $(\{x_n\}, \{x'_n\}) \in \varrho$ *and* $(\{y_n\}, \{y'_n\}) \in \varrho$ *then* $(\{x_n - y_n\}, \{x'_n - y'_n\}) \in \varrho$.

Lemma 6. *If* $\lim x_n = x$, *then* $\lim y_n = x$ *iff* $(\{x_n\}, \{y_n\}) \in \varrho$.

The proofs of Lemmas 1–6 are easy and only hints are given here:

Lemma 1. $0 = \lim (x_{i_n} - y_{j_n}) - \lim (x_{j_n} - y_{j_n}) = \lim (x_{i_n} - x_{j_n})$.
Lemma 2. Evident.
Lemma 3. $\lim x_{i_n} = x = \lim x_{j_n}$ implies $\lim (x_{i_n} - x_{j_n}) = 0$.
Lemma 4. $\lim (x_n - y_{j_n}) - \lim (y_n - y_{j_n}) = \lim (x_n - y_n)$.
Lemma 5. $\lim (x_{i_n} - x'_{j_n}) - \lim (y_{i_n} - y'_{j_n}) = \lim ((x_{i_n} - y_{i_n}) - (x'_{j_n} - y'_{j_n}))$.
Lemma 6. $\lim x_n = x = \lim y_n$ implies $0 = \lim (x_{i_n} - x) + \lim (x - y_{j_n}) = = \lim (x_{i_n} - y_{j_n})$. Now, if $\lim x_n = x$ and $(\{x_n\}, \{y_n\}) \in \varrho$ then $0 = \lim (x - x_n) + + \lim (x_n - y_n) = \lim (x - y_n) = x - \lim y_n$.

From Definition 2 and from Lemma 1 it easily follows that ϱ is an equivalence relation on the set of all Cauchy sequences. The class of all Cauchy sequences which are equivalent to a Cauchy sequence $\{x_n\}$ will be denoted $[\{x_n\}]$. Evidently, $\lim x_n = x$ iff $[\{x_n\}] = [\{x\}]$, $\{x\}$ being the constant sequence.

Definition 3. A subset A of a cc group $(L, \mathfrak{L}, \lambda, +)$ is called *complete* provided that each Cauchy sequence $\{x_n\}$, $x_n \in A$, converges to a point of A.

Lemma 7. *A subset A of a complete cc group* $(L, \mathfrak{L}, \lambda, +)$ *is complete iff it is* λ*-closed.*

The easy proof is omitted.

Let $(L_\iota, \mathfrak{L}_\iota, \lambda_\iota, +_\iota)$, $\iota \in I$ be cc groups. Denote $L = \mathsf{X}\{L_\iota; \iota \in I\}$ the Cartesian product of L_ι, $\mathfrak{L}$ the coordinatewise convergence on L, λ the convergence closure for L induced by $\mathfrak{L}$ and $+$ the coordinatewise group operation on L. Then $\mathfrak{L}$ is a star convergence [2] and we have a Cartesian convergence commutative group $(L, \mathfrak{L}, \lambda, +)$.

Lemma 8. *Let* $(L, \mathfrak{L}, \lambda, +)$ *be a Cartesian cc group defined by cc groups* $(L_\iota, \mathfrak{L}_\iota, \lambda_\iota, +_\iota)$, $\iota \in I$. *Then* $\{(x^n_\iota)\}_{n=1}^\infty$ *is a Cauchy sequence in* $(L, \mathfrak{L}, \lambda, +)$ *iff* $\{x^n_\iota\}_{n=1}^\infty$ *is a Cauchy sequence in* $(L_\iota, \mathfrak{L}_\iota, \lambda_\iota, +_\iota)$ *for each* $\iota \in I$.

The proof is evident.

Lemma 9. *Let $(L_\iota, \mathfrak{L}_\iota, \lambda_\iota, +_\iota)$, $\iota \in I$, be complete cc groups. Let $(L, \mathfrak{L}, \lambda, +)$ be their Cartesian cc group. Let G be a subgroup of the group $(L, +)$. Then $\lambda^{\omega_1}G$ is the smallest complete convergence group containing G as a subgroup.*

Proof. $\lambda^{\omega_1}G$ is the smallest λ-closed subgroup of L containing G as a subgroup. L is complete by Lemma 8. Hence the assertion instantly follows from Lemma 7.

2.

Definition 4. Let $(L, \mathfrak{L}, \lambda, +)$ be a cc group. A cc group $(L_1, \mathfrak{L}_1, \lambda_1, +)$ is called a *completion* of $(L, \mathfrak{L}, \lambda, +)$ if it is complete and such that L is a $\lambda_1^{\omega_1}$-dense subspace of $(L_1, \mathfrak{L}_1, \lambda_1)$ and a subgroup of $(L_1, +)$.

Theorem 1. *Each cc group $(L, \mathfrak{L}, \lambda, +)$ has a least one completion $(L_1, \mathfrak{L}_1, \lambda_1, +)$.*

The proof of Theorem 1 is divided into two parts **A** and **B**. In the first part a cc group $(L_1, \mathfrak{L}_1^*, \lambda_1, +)$ is constructed such that L is a dense subspace of L_1 and a subgroup of L_1. In the second part a definition and a lemma are given and it is proved that $(L_1, \mathfrak{L}_1^*, \lambda_1, +)$ is complete.

A. Let $(L, \mathfrak{L}, \lambda, +)$ be a cc group. Let X be a point set of power $|X| > 2^{|L|}$ containing L as a subset. Let g be a one-to-one map on the set of all classes $[\{x_n\}]$ of Cauchy sequences into X such that $g([\{x_n\}])$ is a point x of L iff $(\{x_n\}, x) \in \mathfrak{L}$. Denote L_1 the set of all points $g([\{x_n\}])$ in X. Then $L \subset L_1 \subset X$. Now define a binary operation $\dotplus$ on L_1:

Definition 5. Let $z = g([\{x_n\}])$ and $t = g([\{y_n\}])$ be points of L_1. By Lemma 5, $\{x_n + y_n\}$ is a Cauchy sequence in L. We put $z \dotplus t = g([\{x_n + y_n\}])$.

In view of Lemma 5 the operation $\dotplus$ does not depend on representatives of classes.

If $x \in L$, $y \in L$ then $x \dotplus y = g([\{x\}]) + g([\{y\}]) = g([\{x + y\}]) = x + y$. Consequently, we may write $+$ instead of $\dotplus$ on L_1. If $\{x_n\}$ is a Cauchy sequence in L, then $-g([\{x_n\}]) = g([\{-x_n\}])$.

Statement 1. *$(L_1, +)$ is a commutative group containing $(L, +)$ as a subgroup.*

The proof follows instantly from Definition 5.

Definition 6. Let $\mathfrak{L}_1$ be the set of all pairs $(\{z_n\}, z)$, $z_n \in L_1$, $z \in L_1$ such that there is a Cauchy sequence $\{x_m\}$, $x_m \in L$, with the property[2] $z - z_n = g([\{x_m\}]) - x_n$.

---

[2]) If $\{y_n\}$, $y_n \in L$, is another Cauchy sequence then $g([\{x_n\}]) - x_n = g([\{y_n\}]) - y_n$ iff $x_n - y_n$ is a constant point in L for each n. It follows that $(\{z_n\}, z) \in \mathfrak{L}_1$ can be defined by more than one Cauchy sequence in L.

Statement 2. *$\mathfrak{L}_1$ is a convergence on L_1.*

Proof. First prove that $(\{z_n\}, z') \in \mathfrak{L}_1, (\{z_n\}, z'') \in \mathfrak{L}_1$ implies $z' = z''$. As a matter of fact, let $\{x_m\}$ and $\{y_m\}$ be Cauchy sequences in L such that $z' - z_n = a - x_n$, $z'' - z_n = b - y_n$ where $a = g([\{x_m\}])$ and $b = g([\{y_m\}])$. Denote $c = a - b + (z'' - z')$. Then $y_n = x_n - c$. It follows $c \in L$ and so $b = g([\{x_m - c\}]) = a - c = b - (z'' - z')$. Hence $z'' - z' = 0$.

Evidently $(\{z\}, z) \in \mathfrak{L}_1$ for each $z \in L_1$ and $(\{z_n\}, z) \in \mathfrak{L}_1$ implies that $(\{z_{n_i}\}, z) \in \mathfrak{L}_1$ for each subsequence $\{n_i\}$ of $\{n\}$. It follows that $\mathfrak{L}_1$ is a convergence on L_1.

Statement 3. *$(\{z_n\}, z) \in \mathfrak{L}_1$ and $(\{t_n\}, t) \in \mathfrak{L}_1$ implies $(\{z_n - t_n\}, z - t) \in \mathfrak{L}_1$.*

Proof. Let $z - z_n = a - x_n$ and $t - t_n = b - y_n$ where $a = g([\{x_n\}])$ and $b = g([\{y_n\}])$. Then $z - t - (z_n - t_n) = a - b - (x_n - y_n) = g([\{x_n - y_n\}]) - (x_n - y_n)$. Hence $\mathfrak{L}_1$-lim $(z_n - t_n) = (z - t)$ by Definition 6.

Let us notice that $\mathfrak{L}_1$ need not be a star convergence on L_1. It is easy to see that $\mathfrak{L}_1 = \mathfrak{L}_1^*$ iff $\mathfrak{L}_1 = \mathfrak{L}$, i.e., iff $(L, \mathfrak{L}, \lambda, +)$ is a complete cc group.

Statement 4. *Let $\mathfrak{L}_1^*$ be a star convergence of the convergence $\mathfrak{L}_1$. Then $(L_1, \mathfrak{L}_1^*, \lambda_1, +)$ is a cc group.*

Proof. Let $(\{a_n\}, a) \in \mathfrak{L}_1^*$ and $(\{b_n\}, b) \in \mathfrak{L}_1^*$ and let $\{a_{n_i} - b_{n_i}\}$ be any subsequence of $\{a_n - b_n\}$. Then, by Definition 1, there is a subsequence $\{n_{i_k}\}_{k=1}^{\infty}$ of $\{n_i\}_{i=1}^{\infty}$ such that $(\{a_{n_{i_k}}\}, a) \in \mathfrak{L}_1$ and $(\{b_{n_{i_k}}\}, b) \in \mathfrak{L}_1$. Consequently, from Statement 3 it follows that $(\{a_{n_{i_k}} - b_{n_{i_k}}\}, a - b) \in \mathfrak{L}_1$. Hence $(\{a_n - b_n\}, a - b) \in \mathfrak{L}_1^*$ by Definition 1.

Now, we are going to prove

Lemma 10. *If $(\{x_n\}, 0) \in \mathfrak{L}_1^*$, $x_n \in L$, then $(\{x_n\}, 0) \in \mathfrak{L}$.*

Proof. Let $\{x_{n_i}\}$ be a subsequence of $\{x_n\}$. Since $\mathfrak{L}_1^*$ is a star convergence, there is a subsequence $\{x_{n_{i_k}}\}$ of $\{x_{n_i}\}$ such that $(\{x_{n_{i_k}}\}, 0) \in \mathfrak{L}_1$. By Definition 6 there is a Cauchy sequence $\{y_m\}_{m=1}^{\infty}$, $y_m \in L$, in L such that $-x_{n_{i_k}} = g([\{y_m\}]) - y_k$. It follows that $g([\{y_m\}]) \in L$ and consequently $\mathfrak{L}$-lim $(g([\{y_m\}]) - y_k) = 0$. Hence $\mathfrak{L}$-lim $(-x_{n_{i_k}}) = 0$. Since $\mathfrak{L} = \mathfrak{L}^*$ we have $\mathfrak{L}$-lim $x_n = 0$.

From Lemma 10 immediately follows

Statement 5. *L is a subspace of L_1.*

From Statements 1–5 it follows that $(L_1, \mathfrak{L}_1^*, \lambda_1, +)$ is a cc group containing $(L, \mathfrak{L}, \lambda. +)$ as a subgroup and a subspace such that $\lambda_1^{\omega} L = \lambda_1 L = L_1$.

B. Definition 7. Points z, t of L_1 are called *equivalent* provided that $z - t \in L$. The class of all points of L_1 which are equivalent to z will be denoted $[z]$.

Lemma 11. *If* $(\{z_n\}, 0) \in \mathfrak{L}_1$, $z_n \in L_1$, *then* $z_n \in [z_1]$ *for each natural n.*

Proof. By Definition 6 there is a Cauchy sequence of points $x_m \in L$ such that $-z_n = g([\{x_m\}]) - x_n$. Hence $z_1 - z_n = x_1 - x_n \in L$ for each n.

Statement 6. $(L_1, \mathfrak{L}_1, \lambda_1, +)$ *is a complete cc group.*

Proof. Let $\{a_n\}$ be a Cauchy sequence in L_1. Then, by Definition 2, $(a_{i_n} - a_{j_n}, 0) \in \in \mathfrak{L}_1^*$, $\{i_n\}$ and $\{j_n\}$ being any subsequences of $\{n\}$. Consider two cases:

(1) There is a subsequence $\{b_n\}$ of $\{a_n\}$ such that $i \neq j$ implies $b_i \notin [b_j]$. Then $\{b_n\}$ is one-to-one. Construct a subsequence $\{b_{i_n}\}$ of $\{b_n\}$ as follows: Put $i_1 = 2$. Suppose we have just chosen $k - 1$ naturals i_m such that $2^{m-1} < i_m \leqq 2^m$ and that no two distinct members of the sequence $\{b_m - b_{i_m}\}_{m=1}^{k-1}$ are equivalent. Consider the sequence $\{b_k - b_n\}_{n=2^{k-1}+1}^{2k}$. It contains 2^{k-1} points $b_k - b_n$. Because $2^{k-1} > > k - 1$ there is a natural i_k, $2^{k-1} < i_k \leqq 2^k$ such that the point $b_k - b_{i_k}$ fails to be equivalent to any of the points $b_m - b_{i_m}$ (otherwise there would be two indices n_1, n_2, $2^{k-1} < n_1 < n_2 \leqq 2^k$ such that $b_k - b_{n_1} \in [b_k - b_{n_2}]$, i.e., $b_{n_1} \in [b_{n_2}]$, which is impossible). In such a way we have a sequence $\{b_n - b_{i_n}\}_{n=1}^{\infty}$ no two distinct members of which are equivalent. By Lemma 11, no its subsequence $\mathfrak{L}_1$-converges to 0. On the other hand, $\{b_n\}$ is a Cauchy sequence, by Lemma 2. Hence $(\{b_n - b_{i_n}\}, 0) \in \in \mathfrak{L}_1^*$. In view of Definition 1 we have a contradiction.

It follows that the case (1) cannot occur.

(2) There exist a point $z \in L$ and a subsequence $\{c_n\}$ of $\{a_n\}$ such that $c_n \in [z]$ for each n. Hence $c_n = z + r_n$, r_n being suitable points of L. Since $\{c_n\}$ is a Cauchy sequence we have $(\{c_{i_n} - c_{j_n}\}, 0) \in \mathfrak{L}_1^*$ for any subsequences $\{i_n\}$ and $\{j_n\}$ of $\{n\}$. Consequently, $(\{r_{i_n} - r_{j_n}\}, 0) \in \mathfrak{L}_1^*$. According to Lemma 10 and Definition 2, $\{r_n\}$ is a Cauchy sequence in L. Denote $c = g([\{r_n\}])$. Since $c - r_n = g([\{r_n\}]) - r_n$ we have $(\{r_n\}, c) \in \mathfrak{L}_1 \subset \mathfrak{L}_1^*$, by Definition 5, and so $(\{c_n\}, c + z) \in \mathfrak{L}_1^*$, by Statement 4. Hence $\mathfrak{L}_1^*$-lim $a_n = c + z$, by Lemma 6.

3.

Examples. Let X be a non void point set. Let $\mathfrak{L}$ denote the usual set convergence on the system $\mathbf{X}$ of all subsets of X. Then $(\mathbf{X}, \mathfrak{L}, \lambda, \div)$ is a complete cc group. As a matter of fact, if $\{A_n\}$, $A_n \in \mathbf{X}$, is a Cauchy sequence then Lim inf A_n = Lim sup A_n. Otherwise, there would be two subsequences $\{i_n\}$ and $\{j_n\}$ of $\{n\}$ and a point $x \in \in A_{i_n} \div A_{j_n}$ for each n. This is a contradiction. From Lemma 7 it follows that each ring of sets $\mathbf{R} \subset \mathbf{X}$ considered as a cc group has a completion in $\mathbf{X}$, viz. the sigma ring $\mathbf{S}(\mathbf{R})$ over $\mathbf{R}$, because $\lambda^{\omega_1}\mathbf{R} = \mathbf{S}(\mathbf{R})$.

Let $\mathscr{F}$ be the class of all real valued functions on X. Let $\mathfrak{L}'$ be the convergence on $\mathscr{F}$ at each point. From Lemma 9 it follows that $(\mathscr{F}, \mathfrak{L}', \lambda', +)$ is a complete cc group. Now, if X is the real line R_1, $\mathscr{C}$ the class of all continuous functions on R_1 and $\mathscr{B}$ the class of all Baire functions then $\lambda^{\omega_1}\mathscr{C} = \mathscr{B}$. Hence $\mathscr{B}$ is a completion of $\mathscr{C}$, by Lemma 7.

A cc group can have several completions which are not homeomorphic. This will be illustrated by following examples.

Let R_1 be the set of all real numbers and R the set of all rational numbers. Let u_1 and u be the usual topologies for R_1 and R. Then the usual topological group $(R, \mathfrak{R}, u, +)$ of rationals is a cc group. It has two different completions. One of them is $(R_1, u_1, +)$ and the other, by Theorem 1, is the cc group $(R_1, \mathfrak{R}_1^*, \lambda_1, +)$ of real numbers the closure of which differs from the usual closure for reals. From Lemma 11 we deduce that no subsequence of the sequence $\{n^{-1}\sqrt{2}\}$ $\mathfrak{R}_1^*$-converges to 0. Hence $0 \in u_1 \bigcup_n n^{-1}\sqrt{2} - \lambda_1 \bigcup_n n^{-1}\sqrt{2}$.

Let $\mathbf{F}$ be the class of all finite subsets of an infinite set X. Then $(\mathbf{F}, \mathfrak{L}, \lambda, \div)$ is a cc group. There are two completions of $\mathbf{F}$, both consisting of all countable subsets of X. The convergence of the first completion is the usual set convergence whereas the convergence $\mathfrak{L}_1^*$ of the other completion from Theorem 1 is different from the usual set convergence. Notice that, by Lemma 11, the sequence of disjoint infinite sets has no subsequence $\mathfrak{L}_1^*$-converging to $\emptyset$.

Also the cc group $\mathscr{C}$ consisting of all continuous functions $f(x)$, $x \in R_1$, has two different completions, one being the cc group $\mathscr{B}$ of all Baire functions and the other is a subgroup of $\mathscr{B}$ with a special convergence $\mathfrak{L}_1^*$ at each point defined in Theorem 1 (Definition 6).

References

[1] *F. Hausdorff:* Gestufte Räume. Fund. Math. *25* (1935), 486—502.

[2] *J. Novák:* On convergence spaces and their sequential envelopes. Czechoslovak Math. J. *15* (*90*) (1965), 74—100.

[3] *J. Novák:* On convergence groups. Czechoslovak Math. J. *20* (*95*) (1970), 357—374.

[4] *O. Schreier:* *L*-Gruppen. Abhandlungen aus dem math. Seminar, 1926, 15.

[5] *P. Urysohn:* Sur les classes ($\mathscr{L}$) de M. Fréchet. Enseignement Math. *25* (1926), 77—83.

INSTITUTE OF MATHEMATICS OF THE CZECHOSLOVAK ACADEMY OF SCIENCES, PRAHA

STRONGLY ZERO-DIMENSIONAL SPACES

P. NYIKOS

Pittsburgh

In this paper I will review some old and not-so-old results on strongly zero-dimensional spaces, but my main purpose is to bring out two important unsolved problems and their many ramifications.

A strongly zero-dimensional space is a Tychonoff space whose Čech-Stone compactification βX is totally disconnected. There are many varied conditions which are equivalent to this one:

Theorem. *Let X be a Tychonoff space. The following conditions are equivalent:*

1. *βX is totally disconnected.*

2. *Given two disjoint zero-sets Z_1 and Z_2 of X, there exists a clopen (closed-and-open) subset C of X such that $Z_1 \subset C$, $Z_2 \cap C = \emptyset$.*

3. *Every zero-set of X is a countable intersection of clopen sets.*

4. *In the ring $C(X)$ of continuous real-valued functions, any two elements which generate the same principal ideal are associates. (One could also use the ring $C^*(X)$ of all bounded continuous real-valued functions here.)*

5. ad $C^*(X) = 0$. *In other words, every bounded continuous real-valued function on X is a uniform limit of continuous functions of finite range.*

There are many other equivalent conditions (cf. [5], [8]).

There is a kind of covering dimension which coincides with the analytic dimension of $C^*(X)$ and also with the Lebesgue covering dimension of βX [3, Chapter 16]. It is defined the same way as Lebesgue covering dimension except that cozero-sets are used in the place of open sets. A number of authors have taken to using "dim" for this kind of dimension on the grounds that it coincides with Lebesgue covering dimension for normal spaces and in some respects is more satisfactory than Lebesgue dimension for non-normal spaces. Under this system, dim $X = 0$ if, and only if, X is strongly zero-dimensional. In this paper, I will use the notation "dim" in this sense. I will also use the notation "D_S" for "strongly zero-dimensional".

One attractive feature of strong zero-dimensionality is the natural way it comes up, as illustrated by the following results. A space is D_S and metrizable if, and only if, it is T_0 and has a σ-locally finite base of clopen sets [4]. A space is D_S and paracompact if, and only if, it is T_1 and every open cover has a locally finite clopen refinement [1]. A space is D_S, normal, and countably paracompact if, and only if, it is Tychonoff and every countable open cover has a locally finite clopen refinement. In each case, we take either the definition of a class of spaces or a theorem which gives an equivalent condition (like the Nagata-Smirnov metrization theorem) and substitute the word "clopen" for "open" in the right place. (Sometimes, as above, we can even weaken the usual separation condition.) The same can be done for the definitions of perfect normality, collectionwise normality, and normality.

An even more attractive feature of strong zero-dimensionality is that conditions which seem much more special often turn out to be equivalent to it. For example, it is not hard to show that a space which admits a non-Archimedean metric[1]) is D_S; but who would suspect that the converse is also true (as shown by de Groot [4])? Similarly, on hearing the condition, "every open cover can be refined to a partition into clopen sets", one might think, what a pity this convenient property is so much more special than strong zero-dimensionality and paracompactness together! Yet the two latter conditions do imply the former.

But now let me get on to the two main unsolved problems. Briefly, they are:

(1) Is every product of D_S spaces itself D_S?

(2) Is every closed subspace of a realcompact D_S space itself D_S?

I became interested in the second problem through some work which led me last year to a negative solution to a long-standing problem: can every real-compact space with a base of clopen sets be embedded as a closed subspace of a product of countable discrete spaces? The problem arose from an article by Engelking and Mrówka [2] where spaces admitting such an embedding were called "N-compact" spaces, after the countable discrete space N of natural numbers.

At the time I started work on this problem, the following facts were known [5]: every D_S realcompact space is N-compact; every N-compact space has a base of clopen sets and is realcompact; and a metric space Δ described by Roy [9] is realcompact and has a base of clopen sets, but is not D_S : $\dim \Delta = 1$. In short, the class of N-compact spaces is sandwiched between two similar, but distinct, classes of spaces. I showed Δ is not N-compact, thus settling the nature of one containment but leaving unanswered the question of whether every N-compact space is D_S. That question remains open.

In fact, that question is equivalent to (2). To see this, one need only assemble the following information: every D_S realcompact space is N-compact; a closed subspace of an N-compact space is itself N-compact; and the product of arbitrarily many copies of the space N is D_S because every continuous real-valued function factors through a countable subproduct.

[1]) A metric d is *non-Archimedean* if it satisfies the strong triangle inequality: $d(x, z) \leqq \max \{d(x, y), d(y, z)\}$ for all x, y, z.

Since any product of N-compact spaces is N-compact, an affirmative solution to (2) would show also that any product of D_S realcompact spaces is itself D_S, a question which is itself open, even for finite products.

All results obtained thus far are trifling compared to what we would know if the answer to either (1) or (2) turned out to be yes. It is even unknown whether every metrizable N-compact space, or every separable normal N-compact space, is D_S. About the only general result on (2) is that every Lindelöf N-compact space is D_S.

K. Morita has shown that an arbitrary product of D_S Lindelöf Σ-spaces is D_S. Using the characterization of D_S paracompact spaces given above, and methods in Nagami's paper on Σ-spaces [7], one can show that a countable product of D_S paracompact Σ-spaces is itself a D_S paracompact Σ-space.

As for finite products, every general rule seems to be the zero-dimensional case of some theorem for arbitrary finite dimensions:

Theorem (Morita) [6]. *Let X be a paracompact space and let Y be a countable union of locally compact paracompact spaces. Then*

$$\dim (X \times Y) \leqq \dim X + \dim Y .$$

Theorem (Kodama). *Let X and Y be such that $X \times Y$ is countably paracompact and normal, with Y metrizable. Then*

$$\dim (X \times Y) \leqq \dim X + \dim Y .$$

Theorem (Morita). *Let X be an M-space and let Y be either metric or locally compact and paracompact. Then*

$$\dim (X \times Y) \leqq \dim X + \dim Y .$$

And, of course, there is the trivial result that if Y is discrete, then $\dim (X \times Y) = \dim X$ for any X at all. Otherwise the results consist of merely knocking down specific spaces. For example, P. Roy and I recently showed that the Sorgenfrey plane[2]) is a D_S space. But our proof does not even extend to Sorgenfrey 3-space, nor does a proof by R. Heath and D. Lutzer of the same fact[3]).

But unless one is looking for a counterexample, it seems almost a waste of time to work on finite products when there remains unanswered the sweeping question of whether every N-compact space is strongly zero-dimensional. There is a theorem by Herrlich that a Hausdorff space with a base of clopen sets is N-compact if, and only if, every clopen ultrafilter on X with the countable intersection property is fixed [5].

---

[2]) The Sorgenfrey line S is a D_S Lindelöf space, while the Sorgenfrey plane $S \times S$ is not even normal and does not come under any of the general results.

[3]) Added in proof: M. Mrówka has recently obtained the result that all powers of the Sorgenfrey line are strongly zero-dimensional.

Perhaps one could take any space with a base of clopen sets and, on the assumption that it is not D_S, exhibit a free clopen ultrafilter with the countable intersection property. That seems to be the most promising approach so far.

References

[1] *R. L. Ellis:* Extending continuous functions on zero-dimensional spaces. Math. Ann. *186* (1970), 114–122.

[2] *R. Engelking and S. Mrówka:* On *E*-compact spaces. Bull. Acad. Polon. Sci. Sér. Sci. Math. Astronom. Phys. *6* (1958), 429–436.

[3] *L. Gillman and M. Jerison:* Rings of Continuous Functions. Van Nostrand Co., Princeton, New Jersey, 1960.

[4] *J. de Groot:* Non-Archimedean metrics in topology. Proc. Amer. Math. Soc. *7* (1956), 945–953.

[5] *H. Herrlich:* $\mathfrak{E}$-kompakte Räume. Math. Z. *96* (1967), 228–255.

[6] *K. Morita:* On the product of paracompact spaces. Proc. Japan Acad. *39* (1963), 559–563.

[7] *K. Nagami:* Σ-spaces. Fund. Math. *65* (1969), 169–192.

[8] *P. Nyikos:* Not every zero-dimensional realcompact space is *N*-compact. Bull. Amer. Math. Soc. *77* (1971), 392–396.

[9] *P. Roy:* Nonequality of dimensions for metric spaces. Trans. Amer. Math. Soc. *134* (1968), 117–132.

IDEALS OF OPERATORS ON BANACH SPACES AND NUCLEAR LOCALLY CONVEX SPACES

A. PIETSCH

Jena

Many theorems about nuclear locally convex spaces can be proved without using special properties of nuclearity. We only need the fact that nuclear operators form an ideal. Since the same is true for Schwartz spaces in what follows we present a general theory of $\mathfrak{S}$-spaces which are related to an arbitrary ideal $\mathfrak{S}$ of operators.

1. Ideals of operators on Banach spaces

Let $\mathfrak{L}$ be the class of all bounded linear operators between arbitrary Banach spaces. The set of operators $S \in \mathfrak{L}$ which map from the Banach space E into the Banach space F is denoted by $\mathfrak{L}(E, F)$.

A subclass $\mathfrak{S}$ of $\mathfrak{L}$ is called an *ideal* if for the sets

$$\mathfrak{S}(E, F) := \mathfrak{S} \cap \mathfrak{L}(E, F)$$

the following axioms are satisfied:

(I_1) If $S \in \mathfrak{L}(E, F)$ and $\dim S(E) < \infty$ then $S \in \mathfrak{S}(E, F)$.
(I_2) If $S_1, S_2 \in \mathfrak{S}(E, F)$ then $S_1 + S_2 \in \mathfrak{S}(E, F)$.
(I_3) If $S \in \mathfrak{S}(E, F)$ and $R \in \mathfrak{L}(F, G)$ then $RS \in \mathfrak{S}(E, G)$.
(I_4) If $T \in \mathfrak{L}(E, F)$ and $S \in \mathfrak{S}(F, G)$ then $ST \in \mathfrak{S}(E, G)$.

The class $\mathfrak{F}$ of all bounded linear operators with finite dimensional range is the smallest ideal.

2. Locally convex spaces of type $\mathfrak{S}$

Let p be a seminorm on the linear space E. We denote by $E(p)$ the quotient space $E/N(p)$, $N(p) := \{x \in E : p(x) = 0\}$, with the elements $x(p) := x + N(p)$ and the norm $\|x(p)\| := p(x)$. The Banach space $\tilde{E}(p)$ will be the complete hull of $E(p)$.

If p and q are seminorms such that $q(x) \leqq c\, p(x)$ for all $x \in E$, where c is a constant, we write $q \prec p$. Then $N(p) \subset N(q)$ and a bounded linear operator

$E(p, q)$ from $E(p)$ onto $E(q)$ is defined by

$$E(p, q)\, x(p) := x(q)\,.$$

$\tilde{E}(p, q)$ will be the unique extension of $E(p, q)$ to a bounded linear operator from $\tilde{E}(p)$ into $\tilde{E}(q)$.

A system P of seminorms on a linear space E is called *saturated* if the following axioms are satisfied:

(P_1) If $p \in P$ and $q \prec p$ then $q \in P$.
(P_2) If $p_1, p_2 \in P$ then there exists $p \in P$ such that $p_1 \prec p$, $p_2 \prec p$.
(P_3) If $x \in E$ such that $p(x) = 0$ for all $p \in P$ then $x = o$.

A subsystem P_0 of a saturated system P of seminorms is a *basis* if for each $p \in P$ there is $p_0 \in P_0$ such that $p \prec p_0$.

A *locally convex space* $[E, P]$ is a linear space E with a saturated system P of seminorms. Let $\mathfrak{S}$ be an ideal of operators, then the locally convex space $[E, P]$ is called of *type* $\mathfrak{S}$, or $\mathfrak{S}$-*space*, if for some basis system P_0 of seminorms the following property holds:

(S) For every $q \in P_0$ there exists $p \in P_0$ such that $q \prec p$ and $\tilde{E}(p, q) \in \mathfrak{S}$.

It is easy to see that for a locally convex space of type $\mathfrak{S}$ every basis system of seminorms has property (S).

The class of locally convex $\mathfrak{S}$-spaces is denoted by $\boldsymbol{L}_{\mathfrak{S}}$.

3. Examples

3.1. The ideal of operators with finite dimensional range. A locally convex space $[E, P]$ is of type $\mathfrak{F}$ if and only if every Banach space $\tilde{E}(p)$, $p \in P$, has finite dimension, i.e., $[E, P]$ is a locally convex space with the weak topology.

3.2. The ideal of compact operators. An operator $S \in \mathfrak{L}(E, F)$ is called compact if $S(U_E)$, $U_E := \{x \in E : \|x\| \leqq 1\}$, is a precompact subset of F. The class $\mathfrak{K}$ of compact operators is the oldest known ideal. The Schwartz spaces are the locally convex spaces of type $\mathfrak{K}$ (cf. [4]).

3.3. The ideal of L_p-factorable operators. An operator $S \in \mathfrak{L}(E, F)$ is called L_p-factorable, $1 \leqq p \leqq \infty$, if there exists a measure space $[\Omega, B, \mu]$ such that

$$S : E \xrightarrow{A} L_p[\Omega, B, \mu] \xrightarrow{Y} F\,,$$

where $A \in \mathfrak{L}(E, L_p)$ and $Y \in \mathfrak{L}(L_p, F)$ (cf. [7], [13]). The class $\mathfrak{L}_p$ of L_p-factorable operators is an ideal. Banach spaces of type $\mathfrak{L}_p$ were considered by J. Lindenstrauss,

A. Pełczyński and H. Rosenthal (cf. [8], [9]). The case $p = 2$ is of special interest since a locally convex space $[E, P]$ is of type $\mathfrak{L}_2$ if and only if there exists a basis system P_0 of seminorms such that every $p \in P_0$ can be obtained from a semi-scalarproduct $(.,.)_p$ by $p(x) = (x, x)_p^{1/2}$.

3.4. The ideal of nuclear operators. An operator $S \in \mathfrak{L}(E, F)$ is called nuclear if there exist functionals $a_1, a_2, \ldots \in E'$ and elements $y_1, y_2, \ldots \in F$ such that

$$Sx = \sum_k \langle x, a_k \rangle y_k \quad \text{for all} \quad x \in E$$

and

$$\sum_k \|a_k\| \, \|y_k\| < \infty .$$

The class $\mathfrak{N}$ of nuclear operators is an ideal. The nuclear locally convex spaces are the locally convex spaces of type $\mathfrak{N}$ (cf. [6], [11]).

3.5. The ideal of absolutely p-summing operators. An operator $S \in \mathfrak{L}(E, F)$ is called absolutely p-summing, $0 < p < \infty$, if there exists a constant $c \geqq 0$ such that for every finite system of elements $x_1, \ldots, x_m \in E$ the inequality

$$\{\sum_k \|Sx_k\|^p\}^{1/p} \leqq c \sup_{\|a\| \leqq 1} \{\sum_k |\langle x_k, a \rangle|^p\}^{1/p}$$

holds. The class $\mathfrak{P}_p$ of absolutely p-summing operators is an ideal, and we obtain the nuclear locally convex spaces as the locally convex spaces of type $\mathfrak{P}_p$ (cf. [12]).

3.6. The ideal of $\mathfrak{S}_p^{\text{app}}$-operators. The approximation numbers of an operator $S \in \mathfrak{L}(E, F)$ are defined by

$$s_k(S) := \inf \{\|S - A\| : A \in \mathfrak{F}(E, F), \dim A(E) < k\}$$

for $k = 1, 2, \ldots$. The operators $S \in \mathfrak{L}$ with

$$\sum_k s_k(S)^p < \infty$$

form the ideal $\mathfrak{S}_p^{\text{app}}$, $0 < p < \infty$. The nuclear locally convex spaces are the locally convex spaces of type $\mathfrak{S}_p^{\text{app}}$ (cf. [11], [14]).

3.7. The ideal of $\mathfrak{S}_0^{\text{app}}$-operators. Let

$$\mathfrak{S}_0^{\text{app}} := \bigcap_{p>0} \mathfrak{S}_p^{\text{app}} ,$$

then the locally convex spaces of type $\mathfrak{S}_0^{\text{app}}$ are the so-called strictly nuclear locally convex spaces (cf. [1], [2], [10]).

4. Ideals of sequences

Let $\mathfrak{l}_\infty$ be the ring of all bounded sequences (σ_k). A subset $\mathfrak{s}$ of $\mathfrak{l}_\infty$ is called an *ideal* if the following axioms are satisfied:

(E_1) If $\{k : \sigma_k \neq 0\}$ is finite then $(\sigma_k) \in \mathfrak{s}$.
(E_2) If $(\sigma_k^{[1]}), (\sigma_k^{[2]}) \in \mathfrak{s}$ then $(\sigma_k^{[1]} + \sigma_k^{[2]}) \in \mathfrak{s}$.
(E_3) If $(\sigma_k) \in \mathfrak{s}$ and $(\varrho_k) \in \mathfrak{l}_\infty$ then $(\varrho_k \sigma_k) \in \mathfrak{s}$.
(E_4) If $(\sigma_k) \in \mathfrak{s}$ and π is a permutation of the natural numbers then $(\sigma_{\pi(k)}) \in \mathfrak{s}$.

The connection between ideals of operators and ideals of sequences is described in the following (cf. [3], [5])

Theorem. *Let $\mathfrak{S}$ be an ideal of operators. Then the set $\mathfrak{s}$ of sequences $(\sigma_k) \in \mathfrak{l}_\infty$ such that $D \in \mathfrak{S}(l_2, l_2)$, where $D(\xi_k) := (\sigma_k \xi_k)$, is an ideal. Moreover, $S \in \mathfrak{S}(l_2, l_2)$ if and only if $(s_k(S)) \in \mathfrak{s}$.*

An operator $S \in \mathfrak{L}(E, F)$ is called *l_2-factorable* if there are operators $A \in \mathfrak{L}(E, l_2)$ and $Y \in \mathfrak{L}(l_2, F)$ such that $S = YA$. The class $\mathfrak{H}$ of l_2-factorable operators is an ideal.

Theorem. *Let $\mathfrak{S}$ be an ideal of operators such that $\mathfrak{S} \subset \mathfrak{H}$. Then the class $\mathbf{L}_{\mathfrak{S}}$ is uniquely determined by the corresponding ideal $\mathfrak{s}$ of sequences.*

Now we state some lemmas.

Lemma 1. *Let $\mathfrak{s}$ be an ideal of sequences and let $m = 1, 2, \ldots$. If*

$$\sigma_k^{(m)} := \sigma_i \quad \text{for} \quad k = (i-1)\,m + j\,, \quad i = 1, 2, \ldots, j = 1, \ldots, m\,,$$

then $(\sigma_k) \in \mathfrak{s}$ implies $(\sigma_k^{(m)}) \in \mathfrak{s}$.

Proof. We obtain the result as follows:

$$(\sigma_1, \sigma_2, \sigma_3, \ldots) \in \mathfrak{s}\,,$$
$$(\sigma_1, 0, \sigma_3, \ldots) \in \mathfrak{s}\,,$$
$$(\underbrace{\sigma_1, \ldots, 0, 0}_{m}; \underbrace{0, 0, \ldots, 0, 0}_{m}; \underbrace{\sigma_3, \ldots, 0, 0}_{m}; \ldots) \in \mathfrak{s}\,,$$
$$(0, 0, \ldots, \sigma_1; 0, 0, \ldots, 0, 0; 0, 0, \ldots, \sigma_3; \ldots) \in \mathfrak{s}\,,$$
$$(\sigma_1, \ldots, \sigma_1; 0, 0, \ldots, 0, 0; \sigma_3, \ldots, \sigma_3; \ldots) \in \mathfrak{s}\,,$$
$$(0, 0, \ldots, 0, 0; \sigma_2, \ldots, \sigma_2; 0, 0, \ldots, 0, 0; \ldots) \in \mathfrak{s}\,,$$
$$(\sigma_1, \ldots, \sigma_1; \sigma_2, \ldots, \sigma_2; \sigma_3, \ldots, \sigma_3; \ldots) \in \mathfrak{s}\,.$$

Lemma 2. *Let $\mathfrak{S}$ and $\mathfrak{s}$, resp. $\mathfrak{T}$ and $\mathfrak{t}$, be corresponding ideals of operators or sequences. Then the following conditions are equivalent:*

(1) *If* $(\sigma_k^{[1]}), \ldots, (\sigma_k^{[m]}) \in \mathfrak{s}$ *then* $(\sigma_k^{[1]} \ldots \sigma_k^{[m]}) \in \mathfrak{t}$.

(2) *If* $S_1, \ldots, S_m \in \mathfrak{S}(l_2, l_2)$ *then* $S_1 \ldots S_m \in \mathfrak{T}(l_2, l_2)$.

Proof. (1) → (2): If $k = (i-1)m + j$, $i = 1, 2, \ldots, j = 1, \ldots, m$, then

$$s_k(S_1 \ldots S_m) \leqq s_{(i-1)m+1}(S_1 \ldots S_m) \leqq s_i(S_1) \ldots s_i(S_m) = s_k^{(m)}(S_1) \ldots s_k^{(m)}(S_m).$$

Since Lemma 1 implies $(s_k^{(m)}(S_j)) \in \mathfrak{s}$ for $j = 1, \ldots, m$ we obtain $(s_k(S_1 \ldots S_m)) \in \mathfrak{t}$. Consequently, $S_1 \ldots S_m \in \mathfrak{T}(l_2, l_2)$.

(2) → (1): The proof is left to the reader.

Lemma 3. *Let* $\mathfrak{S}$ *and* $\mathfrak{s}$ *be corresponding ideals of operators or sequences. Then the following conditions are equivalent:*

(M) *If* $S_1, S_2, \ldots \in \mathfrak{S}(l_2, l_2)$ *then there exist operators* $X_1, X_2, \ldots \in \mathfrak{L}(l_2, l_2)$, $B_1, B_2, \ldots \in \mathfrak{L}(l_2, l_2)$, *and* $S \in \mathfrak{S}(l_2, l_2)$ *such that*

$$S_h = B_h S X_h \quad \text{for} \quad h = 1, 2, \ldots.$$

(m) *If* $(\sigma_k^{[1]}), (\sigma_k^{[2]}), \ldots \in \mathfrak{s}$ *then there exist positive numbers* $\varrho_1, \varrho_2, \ldots$ *and* $(\sigma_k) \in \mathfrak{s}$ *such*

$$|\sigma_k^{[h]}| \leqq \varrho_h |\sigma_k| \quad \text{for} \quad k = 1, 2, \ldots.$$

Remark. Condition (M) is satisfied for every ideal $\mathfrak{S}$ of operators which is complete with respect to a quasinorm.

5. Equivalent ideals of operators

We have seen that the class of nuclear locally convex spaces can be obtained from different ideals, e.g. $\mathfrak{N}$, $\mathfrak{P}_p$, and $\mathfrak{S}_p^{\text{app}}$, $0 < p < \infty$. Consequently, if $\mathfrak{S}$ and $\mathfrak{T}$ are ideals of operators, it is useful to know necessary and sufficient conditions for the coincidence of locally convex spaces of type $\mathfrak{S}$ and $\mathfrak{T}$.

Theorem. *Let* $\mathfrak{S}$ *and* $\mathfrak{T}$ *be ideals of operators. If there exists a natural number n such that*

$$S_1 \in \mathfrak{S}(E_1, E_0), \ldots, S_n \in \mathfrak{S}(E_n, E_{n-1}) \quad \textit{implies} \quad S_1 \ldots S_n \in \mathfrak{T}(E_n, E_0)$$

then $\boldsymbol{L}_{\mathfrak{S}} \subset \boldsymbol{L}_{\mathfrak{T}}$.

Now we prove a partial converse.

Theorem. *Let* $\mathfrak{S}$ *and* $\mathfrak{T}$ *be ideals of operators such that* $\boldsymbol{L}_{\mathfrak{S}} \subset \boldsymbol{L}_{\mathfrak{T}}$. *If* $\mathfrak{S}$ *satisfies condition* (M) *and* $\mathfrak{S} \subset \mathfrak{H}$ *then there exists a natural number n such that*

$$S_1 \in \mathfrak{S}(E_1, E_0), \ldots, S_n \in \mathfrak{S}(E_n, E_{n-1}) \quad \textit{implies} \quad S_1 \ldots S_n \in \mathfrak{T}(E_n, E_0).$$

Proof. (1) In the first step we consider $(\sigma_k) \in \mathfrak{s}$ such that

$$\sigma_1 \geqq \sigma_2 \geqq \ldots > 0 .$$

Let

$$E := \{x = (\xi_k) : \sum_k \sigma_k^{-2l} |\xi_k|^2 < \infty, \, l = 1, 2, \ldots\}$$

and

$$p_l(x) := \{\sum_k \sigma_k^{-2l} |\xi_k|^2\}^{1/2}, \quad l = 1, 2, \ldots .$$

Moreover, we define by

$$D(\xi_k) := (\sigma_k \xi_k)$$

and

$$I_l(\xi_k) := (\sigma_k^{-l} \xi_k), \quad l = 1, 2, \ldots$$

an operator $D \in \mathfrak{S}(l_2, l_2)$ and isomorphisms $I_l \in \mathfrak{L}(\tilde{E}(p_l), l_2)$. Consequently, it follows from the commutative diagram

$$\begin{array}{ccc} \tilde{E}(p_{l+1}) & \xrightarrow{\tilde{E}(p_{l+1}, p_l)} & \tilde{E}(p_l) \\ \Big\downarrow{\scriptstyle I_{l+1}} & & \Big\uparrow{\scriptstyle I_l^{-1}} \\ l_2 & \xrightarrow[D]{} & l_2 \end{array}$$

that the locally convex space $[E, (p_l)]$ is of type $\mathfrak{S}$. Since $[E, (p_l)]$ is also of type $\mathfrak{T}$ there exists a natural number m such that $\tilde{E}(p_m, p_0) \in \mathfrak{T}$. Therefore, the commutative diagram

$$\begin{array}{ccc} \tilde{E}(p_m) & \xrightarrow{\tilde{E}(p_m, p_0)} & \tilde{E}(p_0) \\ \Big\uparrow{\scriptstyle I_m^{-1}} & & \Big\downarrow{\scriptstyle I_0} \\ l_2 & \xrightarrow[D^m]{} & l_2 \end{array}$$

implies that $D^m \in \mathfrak{T}(l_2, l_2)$. Hence $(\sigma_k^m) \in \mathfrak{t}$.

(2) Let us suppose that for every natural number $h = 1, 2, \ldots$ there exist $(\sigma_k^{[h,1]}), \ldots, (\sigma_k^{[h,h]}) \in \mathfrak{s}$ such that $(\sigma_k^{[h,1]} \ldots \sigma_k^{[h,h]}) \notin \mathfrak{t}$. Since $\mathfrak{s}$ satisfies condition (m) we find positive numbers $\varrho_{h,i}$, $h = 1, 2, \ldots, i = 1, \ldots, h$, and $(\sigma_k) \in \mathfrak{s}$ such that

$$|\sigma_k^{[h,i]}| \leqq \varrho_{h,i} |\sigma_k| .$$

Without loss of generality we may assume that

$$|\sigma_1| \geqq |\sigma_2| \geqq \ldots > 0 .$$

Consequently, $(\sigma_k^m) \in \mathfrak{t}$, where m is a natural number. Finally, we obtain $(\sigma_k^{[m,1]} \dots \sigma_k^{[m,m]}) \in \mathfrak{t}$. Contradiction.

(3) Since there is a natural number m such that

$$(\sigma_k^{[1]}), \dots, (\sigma_k^{[m]}) \in \mathfrak{s} \quad \text{implies} \quad (\sigma_k^{[1]} \dots \sigma_k^{[m]}) \in \mathfrak{t}$$

it follows from Lemma 2 that

$$S_1, \dots, S_m \in \mathfrak{S}(l_2, l_2) \quad \text{implies} \quad S_1 \dots S_m \in \mathfrak{T}(l_2, l_2)\,.$$

We put $n = 2m + 1$. If $S_i \in \mathfrak{S}(E_i, E_{i-1})$, $i = 1, \dots, n$, then we find factorizations $S_i = X_i A_i$, $A_i \in \mathfrak{L}(E_i, l_2)$ and $X_i \in \mathfrak{L}(l_2, E_{i-1})$. Consequently,

$$S_1 \dots S_n = X_1 \dots (A_{2j-1} S_{2j} X_{2j+1}) \dots A_{2m+1} \in \mathfrak{T}(E_n, E_0)\,.$$

6. Permanence properties

Without proofs we state some permanence properties.

Proposition. *The complete hull of a locally convex $\mathfrak{S}$-space is of type $\mathfrak{S}$.*

Proposition. *The product of an arbitrary set of locally convex $\mathfrak{S}$-spaces is of type $\mathfrak{S}$.*

An ideal $\mathfrak{S}$ of operators is called *injective* if the following axiom is satisfied (cf. [19]):

(J) Let $J \in \mathfrak{L}(F, F_0)$ be an injection (one-to-one operator with closed range) then $S \in \mathfrak{L}(E, F)$ and $JS \in \mathfrak{S}(E, F_0)$ imply $S \in \mathfrak{S}(E, F)$.

The ideals $\mathfrak{K}$, $\mathfrak{L}_2$, $\mathfrak{H}$, $\mathfrak{S}_0^{\text{app}}$, and $\mathfrak{P}_p$, $0 < p < \infty$, are injective.

Proposition. *Let $\mathfrak{S}$ be an injective ideal of operators or let $\mathfrak{S} \subset \mathfrak{L}_2$. Then every subspace of a locally convex $\mathfrak{S}$-space is of type $\mathfrak{S}$.*

An ideal $\mathfrak{S}$ of operators is called *surjective* if the following axiom is satisfied:

(Q) Let $Q \in \mathfrak{L}(E_0, E)$ be a surjection (operator onto E) then $S \in \mathfrak{L}(E, F)$ and $SQ \in \mathfrak{S}(E_0, F)$ imply $S \in \mathfrak{S}(E, F)$.

The ideals $\mathfrak{K}$, $\mathfrak{L}_2$, $\mathfrak{H}$, and $\mathfrak{S}_0^{\text{app}}$ are surjective.

Proposition. *Let $\mathfrak{S}$ be an surjective ideal of operators or let $\mathfrak{S} \subset \mathfrak{L}_2$. Then every quotient space of a locally convex $\mathfrak{S}$-space is of type $\mathfrak{S}$.*

7. Locally convex spaces of type $\mathfrak{L}_p$

It is easy to see that

$$L_{\mathfrak{N}} \subset L_{\mathfrak{L}_p} \quad \text{for all} \quad p \in [1, \infty].$$

On the other side, from the results of [15] follows that

$$L_{\mathfrak{N}} = L_{\mathfrak{L}_1} \cap L_{\mathfrak{L}_p} \quad \text{for all} \quad p \in (1, \infty]$$

and

$$L_{\mathfrak{N}} = L_{\mathfrak{L}_\infty} \cap L_{\mathfrak{L}_p} \quad \text{for all} \quad p \in [1, \infty).$$

References

[1] *B. S. Brudovskii:* Associated nuclear topology, type $\mathfrak{s}$ mappings and strongly nuclear spaces. (Russian.) Dokl. Akad. Nauk SSSR *178* (1968), 271—273.

[2] *B. S. Brudovskii:* Type $\mathfrak{s}$ mappings of locally convex spaces. (Russian.) Dokl. Akad. Nauk SSSR *180* (1968), 15—17.

[3] *D. J. H. Garling:* On ideals of operators in Hilbert spaces. Proc. London Math. Soc. *17* (3) (1967), 115—138.

[4] *H. G. Garnir, M. De Wilde and J. Schmets:* Analyse fonctionelle I. Basel/Stuttgart, 1968.

[5] *I. Z. Gochberg and M. G. Krein:* Introduction to theory of linear nonselfadjoint operators. (Russian, English translation.) 1965.

[6] *A. Grothendieck:* Produits tensoriels topologiques et espaces nucléaires. Mem. Amer. Math. Soc. *16*, 1955.

[7] *W. B. Johnson:* Factoring compact operators. Israel J. Math. *9* (1971), 337—345.

[8] *J. Lindenstrauss and A. Pełczyński:* Absolutely summing operators in L_p-spaces and their applications. Studia Math. *29* (1968), 275—326.

[9] *J. Lindenstrauss and H. Rosenthal:* The L_p-spaces. Israel J. Math. *7* (1969), 325—349.

[10] *A. Martinéau:* Sur une propriété universelle de l'espace des distribution de M. Schwartz. C. R. Acad. Sci. Paris Sér. A—B *259* (1964), 3162—3164.

[11] *A. Pietsch:* Nukleare lokalkonvexe Räume. Berlin, 1965.

[12] *A. Pietsch:* Absolut-p-summierende Abbildungen in normierten Räumen. Studia Math. *28* (1967), 333—353.

[13] *A. Pietsch:* l_p-faktorisierbare Operatoren in Banachräumen. Acta Sci. Math. (Szeged) *31* (1970), 117—123.

[14] *A. Pietsch:* Ideale von S_p-Operatoren in Banachräumen. Studia Math. *38* (1970), 59—69.

[15] *A. Pietsch:* Absolutely-p-summing operators in L_p-spaces. Séminaire L. Schwartz, Paris, 1970—71.

[16] *M. S. Ramanujan:* Power series spaces $\Omega(\alpha)$ and associated $\Omega(\alpha)$-nuclearity. Math. Ann. *189* (1970), 161—168.

[17] *B. Rosenberger:* φ-nukleare Räume. Dissertation, Bonn, 1970.

[18] *P. Spuhler:* Ω-nukleare Räume. Dissertation, Bonn, 1970.

[19] *I. Stephani:* Injektive Operatorenideale über der Gesamtheit aller Banachräume und ihre topologische Erzeugung. Studia Math. *38* (1970), 105—124.

A COMPACTNESS CRITERION FOR HAUSDORFF ADMISSIBLE (JOINTLY CONTINUOUS) CONVERGENCE STRUCTURES IN FUNCTION SPACES

H. POPPE

Greifswald

By an L-space $(X, \lim)$ we understand a set X and a mapping lim from the set of all filters of X into the set of all subsets of X which satisfies the following conditions:

(1) For each $x \in X$, $x \in \lim [x]$, where $[x]$ denotes the ultrafilter containing $\{x\}$.

(2) $x \in \lim \psi$ and $\psi \subset \varrho$ implies $x \in \lim \varrho$.

$(X, \lim)$ is called a convergence space and lim a convergence structure for X. $(X, \lim)$ is called an L^*-space, iff lim satisfies:

(3) If ψ is a filter on X and for each ultrafilter $\pi \supset \psi$, $x \in \lim \pi$ holds, then $x \in \lim \psi$.

For $x \in X$, the filter $\mathfrak{U}(x) = \bigcap\{\psi : x \in \lim \psi\}$ is called the neighborhood filter at x. If lim satisfies:

(4) For each $x \in X$, $x \in \lim \mathfrak{U}(x)$,

then $(X, \lim)$ is called a U-space ("Umgebungsraum") or a pretopological space.

$(X, \lim)$ is called a Hausdorff convergence space iff $x \in \lim \psi$ and $y \in \lim \psi$ implies $x = y$, that is for each converging filter ψ, $\lim \psi$ consists of a single element.

In the sequel let X and Y denote L-spaces. By Y^X we understand the set of all functions from X into Y and by $C(X, Y)$ the set of all continuous functions. ω denotes the evaluation map $\omega : Y^X \times X \to Y$, that is, $\omega(f, x) = f(x)$, and a convergence structure lim for Y^X or for $C(X, Y)$ is called *admissible (jointly continuous, conjoining)* iff ω: $(Y^X, \lim) \times X \to Y$ is continuous.

A very useful convergence structure for $C(X, Y)$ is that of continuous convergence.

Definition 1. Let $\mathfrak{F}$ be a filter on $C(X, Y)$; $\mathfrak{F}$ is said to *converge continuously to* $f \in C(X, Y)$, $\mathfrak{F} \xrightarrow{c} f$ or $f \in c\text{-}\lim \mathfrak{F}$, iff for each $x \in X$ and each filter ψ, $\psi \to x$ implies $\omega(\mathfrak{F} \times \psi) \to f(x)$, where $\mathfrak{F} \times \psi$ denotes the cartesian product of the filters.

J. L. Kelley and A. P. Morse [1] defined the notion of even continuity, which is a generalization of equicontinuity, for sets of functions from a topological space X into a topological space Y. This notion can be extended to the case of X and Y being only convergence spaces:

Definition 2. Let $H \subset C(X, Y)$, H is called *evenly continuous* iff for each $x \in X$, $y \in Y$, each filter $\mathfrak{F}$ on $C(X, Y)$ such that $H \in \mathfrak{F}$ and each filter ψ on X, $\omega(\mathfrak{F} \times [x]) \to y$ and $\psi \to x$ implies $\omega(\mathfrak{F} \times \psi) \to y$.

Remark. For information about convergence spaces, properties of the convergence structure of continuous convergence and of even continuity see [3], [4], [5] and especially [6].

Proposition 1. *Let $H \subset C(X, Y)$ and $\mathfrak{F}$ be a filter on $C(X, Y)$ such that $H \in \mathfrak{F}$ and $\mathfrak{F}$ converges pointwise to $f \in C(X, Y)$. If H is evenly continuous, then $\mathfrak{F}$ converges to f continuously.*

Proposition 2. *Let* lim *be a convergence structure for $C(X, Y)$. Then* lim *is admissible for $C(X, Y)$ iff* c-lim $\leq$ lim, *that is,* lim $\mathfrak{F} = f$ *implies* c-lim $\mathfrak{F} = f$ *for each filter $\mathfrak{F}$ on $C(X, Y)$.*

Remark. For proofs of Propositions 1 and 2 see [6].

Now we are able to formulate a compactness criterion for a Hausdorff admissible convergence structure lim for $C(X, Y)$. $H \subset C(X, Y)$ is called compact relative to lim iff each ultrafilter on $C(X, Y)$ containing H lim-converges to an element of H.

Theorem 1. *Let X be an L-space and Y a Hausdorff and regular L*-space and let* lim *be a Hausdorff admissible convergence structure for $C(X, Y)$ (that means, $(C(X, Y)$, lim$)$ is an L-space); let $H \subset C(X, Y)$. The following conditions are necessary and sufficient for the compactness of H relative to* lim:

(a) *H is closed in $C(X, Y)$ relative to* lim.

(b) *$H(x) = \{f(x) : f \in H\}$ is compact for each $x \in X$.*

(c) *H is evenly continuous.*

(d) *If $\mathfrak{F}$ is an ultrafilter on $C(X, Y)$ such that $H \in \mathfrak{F}$ and $\mathfrak{F} \xrightarrow{c} f \in C(X, Y)$, then $f \in$* lim $\mathfrak{F}$.

Remark. For the proof of Theorem 1 see [6]. The application of Theorem 1 to particular situations consists in finding conditions which imply condition (d) of Theorem 1. We will illustrate this by two examples.

A) The convergence structure of strictly continuous convergence.

Definition 3. Let $\mathfrak{F}$ be a filter on $C(X, Y)$ (or on Y^X); $\mathfrak{F}$ *converges strictly continuously* to f, $\mathfrak{F} \xrightarrow{str.c} f$ or $f \in$ *str.* c-lim $\mathfrak{F}$, iff for each filter ψ on X the convergence of $f\psi$ to $y \in Y$ implies $\omega(\mathfrak{F} \times \psi) \to y$.

Comparing it with the definition of continuous convergence, we see at once that c-lim $\leq$ *str.* c-lim in $C(X, Y)$ holds, that is, *str.* c-lim is admissible for $C(X, Y)$. Moreover, if Y is Hausdorff, $(C(X, Y)$, *str.* c-lim$)$ is Hausdorff, too.

Theorem 2. 1. *Let X be a pretopological space and Y a regular topological space; let $H \subset C(X, Y)$. The following conditions are sufficient for the compactness of H in $C(X, Y)$ relative to str. c-*lim:

(a) *H is closed relative to str. c-*lim.

(b) *$H(x)$ is compact for each $x \in X$.*

(c) *H is evenly continuous.*

(d) *If $\mathfrak{F}$ is an ultrafilter on $C(X, Y)$, $H \in \mathfrak{F}$, π is an ultrafilter on X and $y \in Y$, then $\omega(\mathfrak{F} \times \pi) \to y$ whenever there exists for every neighborhood V of y a set $B_V \in \pi$ with the following property: if $x \in B_V$, then there is a set $H_x \in \mathfrak{F}$, $H_x \subset H$, and a neighborhood U_x of x such that $\omega(H_x \times U_x) \subset V$.*

2. *Let X be a pretopological space and Y a Hausdorff and regular pretopological space. If $H \subset C(X, Y)$ is compact relative to str. c-*lim, *then the conditions* (a), ..., (d) *hold.*

Proof. 1. We show that condition (d) implies the corresponding condition (d) of Theorem 1. Let $\mathfrak{F}$ be an ultrafilter on $C(X, Y)$, containing H and converging continuously to $f \in C(X, Y)$. We must show that $\mathfrak{F}$ converges strictly continuously to f. For this it is sufficient to show that for each ultrafilter π on X, $f\pi \to y$ implies $\omega(\mathfrak{F} \times \pi) \to y$, since Y is an L^*-space. Now let V be an open neighborhood of y; since $f\pi \to y$, there exists $B_V \in \pi$ such that $f(B_V) \subset V$; therefore for $x \in B_V$, V is a neighborhood of $f(x)$; since $f \in c\text{-lim}\, \mathfrak{F}$ and $\mathfrak{U}(x) \to x$, we find $A_x \in \mathfrak{F}$ and $U_x \in \mathfrak{U}(x)$ such that $\omega(A_x \times U_x) \subset V$; we have $H_x = A_x \cap H \in \mathfrak{F}$ and $\omega(H_x \times U_x) \subset V$; hence the suppositions of (d) are satisfied and it follows $\omega(\mathfrak{F} \times \pi) \to y$ and hence $f \in str.\ c\text{-lim}\, \mathfrak{F}$. Then the compactness of H relative to *str.* c-lim follows from a c-lim-compactness criterion, which can be found in [3] or [6].

2. If H is compact relative to *str.* c-lim, then, as is easy to see, H is closed relative to *str.* c-lim. Moreover H is compact relative to c-lim, since *str.* c-lim is admissible for $C(X, Y)$. Then conditions (b) and (c) of Theorem 2 follow from the same c-lim-compactness criterion which was mentioned above. We now show that (d) holds. We assume that the suppositions of condition (d) are fulfilled. Let $\mathfrak{F}$ be an ultrafilter on $C(X, Y)$ such that $H \in \mathfrak{F}$, π an ultrafilter on X and $y \in Y$; since H is compact relative to *str.* c-lim, there exists $f \in H$ such that $f \in str.\ c\text{-lim}\, \mathfrak{F}$; we shall show that $f\pi \to y$. Let $V \in \mathfrak{U}(y)$; Y is a regular pretopological space and hence we have $\mathfrak{U}(y) \to y$, which implies $\mathfrak{U}^\lambda(y) \to y$, where the filter $\mathfrak{U}^\lambda(y)$ is generated by $\{U^\lambda : U \in \mathfrak{U}(y)\}$, $U^\lambda = \{y \in Y : \text{there exists a filter } \psi \text{ on } Y \text{ such that } U \in \psi \text{ and } \psi \to y\}$. We then find $V_1 \in \mathfrak{U}(y)$ such that $V_1^\lambda \subset V$; by the supposition of (d), for V_1 there exists $B_1 \in \pi$ such that for $x \in B_1$ there exist sets $H_x \in \mathfrak{F}$, $H_x \subset H$ and $U_x \in \mathfrak{U}(x)$ such that $\omega(H_x \times U_x) \subset V_1$; we have $f \in c\text{-lim}\, \mathfrak{F}$ and therefore $\omega(\mathfrak{F} \times \mathfrak{U}(x)) \to f(x)$, since X is a pretopological space; but $\omega(H_x \times U_x) \subset V_1$ implies $V_1 \in \omega(\mathfrak{F} \times \mathfrak{U}(x))$ and hence $f(x) \in V_1^\lambda \subset V$; thus we have $f(B_1) \subset V$, which implies $f\pi \to y$; since $f \in str.\ c\text{-lim}\, \mathfrak{F}$, it follows that $\omega(\mathfrak{F} \times \pi) \to y$ and hence (d) is shown.

B) A "graph topology" for $C(X, Y)$.

Definition 4. Let X and Y be topological spaces; for $f \in Y^X$ we denote by $\Gamma(f)$ the graph of f, that is, $\Gamma(f) = \{(x, f(x)) : x \in X\} \subset X \times Y$. Let G be an open set in $X \times Y$ and let $(G) = \{f \in C(X, Y) : \Gamma(f) \subset G\}$; then $\{(G) : G \text{ open in } X \times Y\}$ is a basis for a topology for $C(X, Y)$, which we denote by $\tau_{\mathfrak{S}_a}$.

Remark. The topology $\tau_{\mathfrak{S}_a}$ is obtained by the Tychonoff hyperspace topology, restricted to the set of all graphs of the functions from $C(X, Y)$ (see [7]). It was first considered by Naimpally [2]. For a proof of the following proposition see [7].

Proposition 3. *Let X, Y be topological spaces.*

1) *If X is a T_1-space and Y a Hausdorff space, then $(C(X, Y), \tau_{\mathfrak{S}_a})$ is Hausdorff.*

2) *If X is regular, then the $\tau_{\mathfrak{S}_a}$-convergence is finer than the continuous convergence, that is, by Proposition 2 $\tau_{\mathfrak{S}_a}$ is admissible for $C(X, Y)$.*

Theorem 3. *Let X be a regular T_1-space, Y a Hausdorff and regular space; let $H \subset C(X, Y)$.*

The following conditions are necessary and sufficient for the $\tau_{\mathfrak{S}_a}$-compactness of H:

(a) *H is closed in $(C(X, Y), \tau_{\mathfrak{S}_a})$.*

(b) *$H(x)$ is compact for each $x \in X$.*

(c) *H is evenly continuous.*

(d) *Let $\mathfrak{F}$ be an ultrafilter on $C(X, Y)$ such that $H \in \mathfrak{F}$. For each open set $G \subset X \times Y$ such that $pr_X G = X$ there exist systems of open sets in X and in Y, viz. $(U_i)_{i \in I}$, $(V_i)_{i \in I}$, respectively, with the following properties: $(U_i)_{i \in I}$ is a cover of X, $\bigcup_{i \in I}(U_i \times \bar{V}_i) \subset G$, for $i \in I$ there exists $A_i \in \mathfrak{F}$, $A_i \subset H$ such that $\omega(A_i \times U_i) \subset V_i$. Then there exists $B \in \mathfrak{F}$ such that $\Gamma(B) = \{\Gamma(f) : f \in B\} \subset G$.*

Proof. 1. First we show that conditions (a), …, (d) are sufficient for the $\tau_{\mathfrak{S}_a}$-compactness of H. As in the proof of Theorem 2, for this purpose we only prove that condition (d) implies the corresponding condition (d) of Theorem 1. Let $\mathfrak{F}$ be an ultrafilter on $C(X, Y)$ such that $H \in \mathfrak{F}$ and $f \in c\text{-lim } \mathfrak{F}$; let G be open in $X \times Y$ and $\Gamma(f) \subset G$; for $x \in X$ there exist open sets $\tilde{U}_x$ of X and $\tilde{V}_x$ of Y such that $(x, f(x)) \in \tilde{U}_x \times \tilde{V}_x \subset G$; since Y is regular, for each $x \in X$ there exists an open set V_x such that $f(x) \in V_x \subset \bar{V}_x \subset \tilde{V}_x$; since $f \in c\text{-lim } \mathfrak{F}$, we find an open set U_x and $\tilde{A}_x \in \mathfrak{F}$ such that $x \in U_x \subset \tilde{U}_x$ and $\omega(\tilde{A}_x \times U_x) \subset V_x$. If $A_x = \tilde{A}_x \cap H$, the families $(U_x)_{x \in X}$, $(V_x)_{x \in X}$ fulfill the suppositions of condition (d). Therefore there exists $B \in \mathfrak{F}$ such that $\Gamma(B) = \{\Gamma(f) : f \in B\} \subset G$. But since G is an arbitrary set, this means that $\mathfrak{F} \xrightarrow{\tau_{\mathfrak{S}_a}} f$.

2. We show that the compactness of H relative to $\tau_{\mathfrak{S}_a}$ implies condition (d). Let $\mathfrak{F}$ be an ultrafilter on $C(X, Y)$ such that $H \in \mathfrak{F}$ and let G be an open subset of $X \times Y$ such

that $pr_X G = X$ and G fulfils the suppositions of (d); by the compactness of H there exists $f \in H$ such that $\mathfrak{F} \xrightarrow{\tau_{\mathfrak{S}_a}} f$; we shall prove that $\Gamma(f) \subset G$; then $\mathfrak{F} \xrightarrow{\tau_{\mathfrak{S}_a}} f$ implies the existence of $B \in \mathfrak{F}$ such that $\Gamma(B) \subset G$. For G there exist families of open sets $(U_i)_{i\in I}$ in X and $(V_i)_{i\in I}$ in Y and a family $(A_i)_{i\in I}$ of subsets of H such that $\bigcup_{i\in I} U_i \times \overline{V}_i \subset G$, $\omega(A_i \times U_i) \subset V_i$ for each $i \in I$ and $(U_i)_{i\in I}$ is a covering of X; we assume now that there exists $x_0 \in X$ such that $(x_0, f(x_0)) \notin \bigcup U_i \times \overline{V}_i$; hence we have $x_0 \in U_{i_0}$ and $f(x_0) \notin \overline{V}_{i_0}$ for some index $i_0 \in I$; hence there exists an open set W such that $f(x_0) \in W$ and $W \cap V_{i_0} = \emptyset$; since X is a T_1-space, $G_1 = (X - \{x_0\}) \times Y \cup (U_{i_0} \times W)$ is an open subset of $X \times Y$, containing $\Gamma(f)$, hence there exists $F \in \mathfrak{F}$ such that $\Gamma(F) \subset G_1$; now let g be some element of $F \cap A_{i_0}$; $g \in F$ implies $\Gamma(g) \subset G_1$ and hence $(x_0, g(x_0)) \in U_{i_0} \times W$ and therefore we have $g(x_0) \in W$; but on the other hand $g \in A_{i_0}$ and we have $\omega(A_{i_0} \times U_{i_0}) \subset V_{i_0}$ and hence $g(x_0) \in V_{i_0}$, too, which yields a contradiction. Therefore we have $\Gamma(f) \subset \bigcup U_i \times \overline{V}_i \subset G$, as was desired.

References

[1] *J. L. Kelley:* General Topology. Princeton, New Jersey, 1957.

[2] *S. A. Naimpally:* Graph topologies for function spaces. Trans. Amer. Math. Soc. *123* (1966), 267—272.

[3] *H. Poppe:* Stetige Konvergenz und der Satz von Ascoli und Arzelà. Math. Nachr. *30* (1965), 87—122.

[4] *H. Poppe:* Stetige Konvergenz und der Satz von Ascoli und Arzelà II. Monatsb. Deutsch. Akad. Wiss. Berlin *8* (4) (1966), 259—264.

[5] *H. Poppe:* Stetige Konvergenz und der Satz von Ascoli und Arzelà III, IV, V, VI. Proc. Japan Acad. *44* (1968), 234—239, 240—242, 318—321, 322—324.

[6] *H. Poppe:* Compactness in general function spaces. Deutscher Verlag der Wissenschaften zu Berlin (to appear).

[7] *H. Poppe:* Über Graphentopologien für Abbildungsräume I. Bull. Acad. Polon. Sci. Sér. Sci. Math. Astronom. Phys. *15* (1967), 71—80.

METRIC SPACES IN WHICH PROHOROV'S THEOREM IS NOT VALID

D. PREISS

Praha

A well-known Prohorov's theorem says that for every topologically complete metric space X, for every compact set M of measures on X with mass 1 (with the weak topology) and for every $\varepsilon > 0$ there exists a compact set $A \subset X$ such that $\mu(X - A) < \varepsilon$ for each $\mu \in M$. It has been a problem whether this theorem holds in every separable metric space which is a Borel subset of its completion. This problem can be solved by the help of the following theorem.

Theorem 1. *A coanalytic separable metric space is topologically complete if and only if it contains no countable dense-in-itself G_δ subspace.*

If Prohorov's theorem holds in some metric space X then it is easy to prove that it holds also in every G_δ subspace of X. With respect to Theorem 1 and to Sierpiński's result according to which every two countable dense-in-itself metric spaces are homeomorphic it is clear that the negative answer for some countable dense-in-itself metric space implies the negative answer for every separable coanalytic metric space which is not topologically complete. Such countable metric space can be constructed (namely the space of rational numbers), therefore the following theorem holds.

Theorem 2. *If X is a separable coanalytic metric space then Prohorov's theorem holds in X if and only if X is topologically complete.*

CHARLES UNIVERSITY, PRAHA

A CATEGORICAL GENERALIZATION OF COMPLETELY HAUSDORFF SPACES

G. PREUSS

Berlin

A topological space X is well known to be called a completely Hausdorff space if and only if for any two distinct points $x, y \in X$ there exists a continuous function f from X to the space $\mathbf{R}$ of real numbers with $f(x) \neq f(y)$. If we substitute the space $\mathbf{R}$ of real numbers in the definition of a completely Hausdorff space by the discrete space D_2 consisting of exactly two points we obtain the class of all spaces whose quasi-components consist only of a single point. Taking the Sierpinski space S (that means the topological space consisting of exactly two points and three open sets) instead of $\mathbf{R}$ we obtain the class of all T_0-spaces. Now let us consider a whole class $\mathscr{E}$ of (non-void) topological spaces and let us define a class $Q\mathscr{E}$ of topological spaces in the following way:

$X \in Q\mathscr{E} \Leftrightarrow$ For any two distinct points $x, y \in X$ there exists a space $E \in \mathscr{E}$ and a continuous map $f : X \to E$ with $f(x) \neq f(y)$.

Let us choose for $\mathscr{E}$ the class of all spaces with the topology of finite complements. Then $Q\mathscr{E}$ consists of exactly all T_1-spaces. It is impossible to find a space E such that $Q\{E\} = \{T_1\text{-spaces}\}$. This is an immediate consequence of a result of Herrlich that for each T_1-space X there is a regular T_1-space Y consisting of at least two points such that each continuous function f from Y into X is constant.

Now let $\mathscr{C}$ be a category and $\mathscr{A}$ a (full) subcategory. Then we define a full subcategory $Q_{\mathscr{C}}\mathscr{A}$ of $\mathscr{C}$ by defining the object class $|Q_{\mathscr{C}}\mathscr{A}|$ of $Q_{\mathscr{C}}\mathscr{A}$ by

$|Q_{\mathscr{C}}\mathscr{A}| = \{X \mid$ For any two distinct morphisms $\alpha, \beta : Z \to X$ there exists an object $A \in |\mathscr{A}|$ and a morphism $f : X \to A$ with $f \circ \alpha \neq f \circ \beta\}$.

(If we take for $\mathscr{C}$ the category of topological spaces (and continuous maps) and define $\mathscr{A}$ by $|\mathscr{A}| = \mathscr{E}$, then we obtain

$$|Q_{\mathscr{C}}\mathscr{A}| = Q\mathscr{E}$$

in the sense defined above.)

It turns out that $Q_{\mathscr{C}}\mathscr{A}$ is isomorphically closed as well as closed under formation of products (in the categorical sense) and extremal subobjects (that means subspaces in the topological case). If $\mathscr{C}$ is a "nice" category we may conclude that $Q_{\mathscr{C}}\mathscr{A}$ is an epireflective subcategory of $\mathscr{C}$, but it does not generally coincide with the epireflective hull $R_{\mathscr{C}}\mathscr{A}$ of $\mathscr{A}$ in $\mathscr{C}$ (i.e., the smallest epireflective [full and isomorphically closed] subcategory of $\mathscr{C}$ containing $\mathscr{A}$). But *if in addition $\mathscr{C}$ is balanced (that means every $\mathscr{C}$-bimorphism is a $\mathscr{C}$-isomorphism), we get the result*

$$(*) \qquad Q_{\mathscr{C}}\mathscr{A} = R_{\mathscr{C}}\mathscr{A}\,.$$

Take for $\mathscr{C}$ the category of compact T_2-spaces (and continuous maps) and define $\mathscr{A}$ by $|\mathscr{A}| = \{D_2\}$; then *we obtain from (*) the well known fact that a compact T_2- space is zero-dimensional if and only if each of its quasi-components consists of a single point.*

For a "nice" category (that means complete, locally and colocally small) the following theorem is valid:

Theorem. *The following statements are equivalent*:

(1) $X \in |Q_{\mathscr{C}}\mathscr{A}|$.

(2) *The reflection map* $r_x : X \to X_{R_{\mathscr{C}}\mathscr{A}}$ *is a monomorphism.*

(3) *X is subobject of a product of $\mathscr{A}$-objects (in $\mathscr{C}$).*

Corollary. *$Q_{\mathscr{C}}\mathscr{A}$ is the smallest (full and isomorphically closed) subcategory of $\mathscr{C}$ containing $\mathscr{A}$ and being closed under formation of subobjects and products.*

Example. The category of completely Hausdorff spaces (and continuous maps) is the smallest subcategory of the category $\mathscr{T}$ of topological spaces (and continuous maps) containing the space **R** of real numbers and being closed under formation of subobjects and products in $\mathscr{T}$.

Problem. If $\mathscr{C}$ is not balanced, find a "nice" condition on X such that $X \in |Q_{\mathscr{C}}\mathscr{A}|$ together with this condition implies $X \in |R_{\mathscr{C}}\mathscr{A}|$.

References

[1] *H. Herrlich:* Topologische Reflexionen und Coreflexionen. Lecture Notes in Mathematics *78*, Springer-Verlag, Berlin, 1968.

[2] *G. Preuss:* Allgemeine Topologie I, II. Vorlesungsausarbeitung FU Berlin, 1970/71.

FREIE UNIVERSITÄT BERLIN, BERLIN

BANACH ALGEBRAS WITH INVOLUTION

V. PTÁK

Praha

In this lecture we intend to present a report about some recent results in the theory of Banach algebras with involution. They are all based on an inequality in hermitian algebras discovered about a year ago which has brought considerable improvements in the theory of hermitian algebras as well as in abstract characterizations of C^*-algebras.

Let us begin by recalling some definitions and known results. Given a Hilbert space H, we denote by $B(H)$ the set of all bounded linear operators on H. It becomes a Banach algebra if we equip it with the operator norm and the usual algebraic structure. Also, it carries a natural involution, the mapping which assigns to each $T \in B(H)$ its adjoint T^*. A C^*-algebra is defined as a closed selfadjoint subalgebra of $B(H)$ which contains the identity operator I. The starting point of the theory of C^*-algebras is the 1943 paper of I. M. Gelfand and M. A. Naimark [3] where the first abstract characterization of a C^*-algebra was given. Before stating their result and describing its further developments let us mention some important properties of C^*-algebras which admit an abstract formulation and which might be expected to be characteristic for C^*-algebras.

First of all, let us mention some obvious relations between the involution and the norm. First of all, the involution is isometric

$$|T^*| = |T| \quad \text{for all} \quad T \in A .$$

Also, it is easy to see that $|T^*T| = |T|^2$, or, using the isometry of the involution

$$|T^*T| = |T^*|\,|T| \quad \text{for all} \quad T \in A .$$

Let us pass now to some less superficial properties which do not involve explicitly the norm.

The fact that the elements of a C^*-algebra A are operators on a given concrete space enables us to formulate certain properties of the operators in terms of vectors of the underlying space. One example of such a formulation is the description of an operator T based on the properties of its quadratic form (Tx, x). It is a well known fact that an operator T is selfadjoint if and only if its quadratic form assumes only real values for every $x \in H$. Also, the spectrum of a selfadjoint element $T \in A$ is real. (The spectrum of T in A is the same as the spectrum of T taken as an element of

$B(H)$.) An element $T \in A$ is said to be nonnegative if its quadratic form assumes nonnegative values for every $x \in H$. It follows from what has been said above that a nonnegative T is necessarily selfadjoint. Also, the spectrum of T is nonnegative. Given any operator $S \in A$, we can construct a nonnegative operator in A by taking the product S^*S. Indeed, it is easy to see that its quadratic form is nonnegative as follows

$$(S^*Sx, x) = (Sx, Sx) = |Sx|^2 \geqq 0 .$$

Furthermore, every nonnegative operator in A may be obtained in this manner. This follows from the following theorem which is not difficult to prove.

Let A be a C^*-algebra and $T \in A$. Then the following conditions are equivalent:

(1) the quadratic form (Tx, x) is nonnegative,
(2) T is selfadjoint and its spectrum is nonnegative,
(3) $T = B^2$ for a suitable selfadjoint $B \in A$,
(4) $T = S^*S$ for a suitable $S \in A$.

In particular, for every $T \in A$ the spectrum in A of $I + T^*T$ does not contain zero and hence $I + T^*T$ has an inverse in A.

Now there are essentially two possibilities of stating abstract analogues of these properties. The first one (the one which we shall adopt) consists in imposing conditions describing spectral properties of selfadjoint elements or of elements of the form T^*T. The other (inaugurated by Vidav [15]) consists in replacing the quadratic form (which requires the vectors of the underlying space) by other notions defined intrinsically in terms of the algebra itself.

Let us state now the classical result of Gelfand and Naimark.

Theorem 1. *Let A be a (complex unital) Banach algebra with an involution which satisfies the following conditions*:

1° $(x + y)^* = x^* + y^*$ *for all* $x, y \in A$,
2° $(xy)^* = y^*x^*$ *for all* $x, y \in A$,
3° $(\alpha x)^* = \alpha^* x^*$ *for all* $x \in A$ *and all complex* α,
4° $|x^*x| = |x^*|\,|x|$ *for all* $x \in A$,
5° $|x^*| = |x|$ *for each* $x \in A$,
6° $(1 + x^*x)^{-1}$ *exists for each* $x \in A$.

*Then A is isometrically *-isomorphic to a closed selfadjoint subalgebra of $B(H)$ for a suitable Hilbert space H.*

We observe that the first three conditions describe the standard algebraic properties of the involution: the *-operation is to be a conjugate-linear anti-isomorphism of period two. The fourth condition, $|x^*x| = |x^*|\,|x|$, relates the involution to the metric structure. The last two conditions are of a different character; their introduction was necessitated by the method of proof and the authors themselves conjectured that these two conditions are consequences of the first four. This is indeed the case but much work was necessary to develop methods which yield the result without the last two assumptions.

The fifth condition, $|x^*| = |x|$, asserts that the involution is isometric, in particular continuous. The sixth asserts that elements of the form $1 + x^*x$ have an inverse, in other words, that the spectrum of an element of the form x^*x does not contain any negative numbers.

Let us examine now more closely what sort of problems the suppression of the last two conditions introduces. First of all, without condition five, we do not have continuity of the involution; it follows that delicate methods have to be developed which avoid using the continuity of the involution. Extreme care is indicated since we can no longer use many facts which are obvious in the case of operator algebras.

To mention one: since we do not have continuity of the *-operation, we cannot assert that the set of all selfadjoint elements (the kernel of the mapping $f(x) = x - x^*$) is closed.

It was only in 1960 that the fact that conditions five and six are consequences of the first four conditions or, in other words, that – apart from the standard conditions concerning the involution – the condition $|x^*x| = |x^*|\,|x|$ alone is sufficient for a Banach algebra to be isomorphic to an operator algebra was established by I. M. Glimm and R. V. Kadison [5].

It is customary to use the abbreviation B^*-algebra for Banach algebra which satisfies the first four conditions of the preceding theorem. The result of Glimm and Kadison can thus be formulated shortly as follows: every B^*-algebra is a C^*-algebra. (It should be mentioned at this point that there is a similar but much stronger condition which appears sometimes instead of condition 4°. This condition, $|xx^*| = |x|^2$, implies immediately the isometry of the involution – indeed, it is equivalent to conditions 4° and 5° taken together. The proof that a Banach algebra satisfying 1°, 2°, 3° and $|x^*x| = |x|^2$ is C^* can, accordingly, use the isometry of the involution and can be established essentially by the methods of Gelfand and Naimark.)

To return to condition six: we observe that it has a purely algebraic character. Let us state, at this point, two more algebraic conditions which will play an important rôle later:

7° *if* $x \in A$ *is selfadjoint then* $\sigma(x)$ *is real,*

8° *for every* $x \in A$ *the spectrum* $\sigma(x^*x)$ *is nonnegative.*

Also, some more terminology: a Banach algebra satisfying 1°, 2° and 3° will be called a Banach algebra with involution. Let us stress the fact that continuity of the involution is not required. A Banach algebra with involution is said to be hermitian if it satisfies condition 7°. We observe first that 8° is a stronger form of 6°; we shall see later that they are equivalent. Furthermore, it is not difficult to see that, in any Banach algebra with involution, condition 8° implies condition 7°. Indeed, assume 8° and consider a selfadjoint $h \in A$ and a $\lambda \in \sigma(h)$. It follows that $\lambda^2 \in (\sigma(h))^2 =$

$= \sigma(h^2) = \sigma(h^*h)$; the last set consists of nonnegative numbers only by our condition, so that λ has to be real. The question whether, conversely, hermitian algebras also satisfy condition 8°, remained open until quite recently.

Algebraic investigations, notably by I. Kaplansky [7], have put into evidence an intimate relation between condition six and eight and the fact that the sum of two selfadjoint elements with nonnegative spectra is again an element whose spectrum is nonnegative.

Let us try now to describe what we take to be the decisive steps in the further development.

First of all, it was necessary to investigate the possibility of extracting selfadjoint square roots from selfadjoint elements in algebras where continuity of the involution is not assumed. A lemma of J. Ford [2] asserts the existence of a selfadjoint u such that $u^2 = h$ provided h is a selfadjoint element whose spectrum lies in the open right half-plane. This lemma, together with methods due to Kaplansky, has enabled S. Shirali and J. Ford [14] to prove that, in a hermitian algebra, elements of the form x^*x have nonnegative spectra. (We have seen already that an algebra with this property must be hermitian.)

The second important contribution is the theorem of B. Russo and H. A. Dye [13] according to which the unit cell of a C^*-algebra coincides with the closed convex hull of its unitary elements. This result has been greatly improved and exploited by T. W. Palmer [8]. Using a transformation introduced by V. P. Potapov [10], L. A. Harris [6] has been able to develop a very simple non-spatial method of proof.

In a recent paper [11] the author showed that, in a Banach algebra with hermitian involution, the spectral radius of any element x and the spectral radius of x^*x are related by the following fundamental inequality

$$|x|_\sigma^2 \leqq |x^*x|_\sigma .$$

It was this result which made it possible to recognize fully the importance of the function $p(x) = |x^*x|_\sigma^{1/2}$, the square root of the spectral radius of the element x^*x; it will play a rôle not unlike the modulus of a complex number.

Once the fundamental inequality is established it is fairly easy to prove the following theorem.

Theorem 2. *Let A be a Banach algebra with hermitian involution. Then $|x|_\sigma \leqq p(x)$ for all $x \in A$. The function p is an algebra pseudonorm on A which coincides with $|x|_\sigma$ on normal elements. The kernel $p^{-1}(0)$ coincides with the radical of A.*

It should be mentioned that the fundamental inequality is an algebraic analogue of the B^*-condition $|x^*x| = |x^*|\,|x|$. To see that, let us note first that the B^*-condition is equivalent to the inequality $|x^*|\,|x| \leqq |x^*x|$. This is an immediate consequence of the submultiplicativity of the norm. If we replace norms by spectral radii in the preceding inequality, we obtain – using the obvious fact that $|x^*|_\sigma = |x|_\sigma$ for each $x \in A$ – the inequality $|x|_\sigma^2 \leqq |x^*x|_\sigma$, which is, indeed, the fundamental inequality. It is possible also to show that this inequality is actually equivalent to the assumption that the involution is hermitian. This analogy between hermitian algebras and B^*-algebras suggests that it may be possible to obtain most of the results known for B^*-algebras under the purely algebraic assumption that the algebra is hermitian. This is, indeed, the case. It is also possible to obtain considerable simplifications in the proof that B^* is C^* and in the theory of algebras which may be renormed to become B^*. In particular, the fact that p is an algebra pseudonorm in hermitian algebras plays an important rôle in the proofs. This fact has a deeper significance: indeed, p is an algebra pseudonorm if and only if the algebra is hermitian.

Also, methods using the fundamental inequality can be used to establish the equivalence between 7° and 8° in a simple manner. Before we state a theorem summarizing the recent state of our knowledge of hermitian algebras, let us define an algebra pseudonorm which will enable us to prove an analogy of the theorem of Russo and Dye for hermitian algebras. We denote by q the function defined on A as follows. Given $x \in A$, let $q(x)$ be the infimum of the sums $\sum \lambda_t$, $t \in K$, where K is a finite set, λ_t are positive numbers and $x = \sum \lambda_t u_t$, $t \in K$ for suitable unitary elements u_t. (An element u is said to be unitary if $u^*u = uu^* = 1$.)

Theorem 3. *Let A be a Banach algebra with involution. Then the following conditions are equivalent*:

1 *the involution is hermitian,*
2 $|x|_\sigma \leqq p(x)$ *for every* $x \in A$,
3 $|x|_\sigma \leqq p(x)$ *for every* $x \in N(A)$,
4 $|x|_\sigma = p(x)$ *for every* $x \in N(A)$,
5 $|\frac{1}{2}(x^* + x)|_\sigma \leqq p(x)$ *for every* $x \in A$,
6 p *is subadditive,*
7 $|u|_\sigma = 1$ *for every unitary* $u \in A$,
8 $|u|_\sigma \leqq 1$ *for every unitary* $u \in A$,
9 $|u|_\sigma \leqq \beta$ *for every unitary* $u \in A$ *and some* β,
10 x^*x *has nonnegative spectrum for every* $x \in A$,
11 *the real part of every* $\lambda \in \sigma(x^*x)$ *is nonnegative,*
12 $\sigma(x^*x)$ *does not contain negative numbers,*
13 $1 + x^*x$ *is invertible for every* $x \in A$,
14 $p(x) = q(x)$ *for every* $x \in A$.

The proof is based on the fundamental inequality and is contained in the paper [18] submitted for publication. Let us mention some of the implications which seem to be particularly interesting. The fundamental inequality is equivalent to the involution being hermitian. Also, unitary elements have bounded spectral radii if and only if the algebra is hermitian. We obtain also the equivalence of our former conditions 6°, 7° and 8°. The equivalence of 1 and 6 (of the present theorem) shows that hermitian algebras generate their own metric structure by means of p and that hermitian algebras form the natural boundary within which the function p has nice properties. If the algebra is not hermitian then p will not be subadditive. Also, the equivalence of 1 and 14 shows that hermitian algebras (and only these) share with C^*-algebras the important property that the unit ball of p is the closed convex hull of the unitary elements, another purely algebraic description of the natural pseudonorm p.

Let us mention another result [11] which illustrates the intimate relation in hermitian algebras between the algebraic structure and the topology generated by p.

Theorem 4. *Let A be a Banach algebra with hermitian involution. Let f be a linear form on A such that $f(1) = 1$. Then the following conditions are equivalent:*

1 $f(x^*x) \geqq 0$ *for every* $x \in A$,
2 $f(x) \leqq p(x)$ *for every* $x \in A$.

If we denote by $S(A)$ the set of all linear forms on A which possess these properties then $p(x) = \sup \{f(x^*x)^{1/2};\, f \in S(A)\}$.

This result shows that the algebraic condition of positivity may be replaced by the topological condition of continuity with respect to p.

Also, the fundamental inequality together with a simple closed graph argument show that, in a hermitian algebra, p is continuous. Together with the fact that $p^{-1}(0)$ is the radical and the above theorem this gives a sufficient number of continuous representations (modulo the radical) in Hilbert space.

To show at least one of the applications of the theory let us prove a sharpening of the theorem that $B^* = C^*$.

Theorem 5. *Let A be a Banach algebra with involution which satisfies $|x^*x| = |x^*|\,|x|$ for each normal element $x \in A$. Then A is isometrically *-isomorphic to a C^*-algebra.*

Proof. Let $x \in A$ be normal. For every natural number n

$$|(x^*x)^n| = |(x^*)^n x^n| = |x^{*n}|\,|x^n|\,.$$

Hence $|x^*x|_\sigma = |x^*|_\sigma\,|x|_\sigma$ and A is hermitian by Theorem 3. Since $|h^2| = |h|^2$ for selfadjoint $h \in A$, we have $|h| = |h|_\sigma$. Hence $p(x)^2 = |x^*x|_\sigma = |x^*x| = |x^*|\,|x|$. Since unitary elements are normal, it follows from our assumption that $|u| = 1$ for every unitary $u \in A$, whence $|x| \leqq q(x)$ for each $x \in A$. The algebra being hermitian, we have $q(x) = p(x)$ for every x. It follows that $|x|^2 \leqq q(x)^2 = p(x)^2 = |x^*|\,|x|$ whence $|x^*| = |x|$ and $p(x) = |x|$. In particular, A is semisimple (the radical is $p^{-1}(0)$) and has, accordingly, a *-representation π on Hilbert space such that $|\pi(a)| = p(a)$ for $a \in A$. Hence $|a| = |\pi(a)|$ and π is an isometric *-isomorphism.

References

[1] *E. Berkson:* Some characterizations of C^*-algebras. Illinois J. Math. *10* (1966), 1—8.
[2] *J. W. M. Ford:* A square root lemma for Banach star-algebras. J. London Math. Soc. *42* (1967), 521—522.
[3] *I. M. Gelfand and M. A. Naimark:* On the imbedding of normed rings into the ring of operators on Hilbert space. Mat. Sb. *12* (1943), 197—213.
[4] *B. W. Glickfeld:* A metric characterization of $C(X)$ and its generalization to C^*-algebras. Illinois J. Math. *10* (1966), 547—556.
[5] *J. G. Glimm and R. V. Kadison:* Unitary operators in C^*-algebras. Pacific J. Math. *10* (1960), 547—556.
[6] *L. A. Harris:* Schwarz's lemma and the maximum principle in infinite dimensional spaces. Doct. Disser., Cornell Univ., 1969.
[7] *I. Kaplansky:* Symmetry of Banach algebras. Proc. Amer. Math. Soc. *3* (1952), 396—399.
[8] *T. W. Palmer:* Characterizations of C^*-algebras. Bull. Amer. Math. Soc. *74* (1968), 538—540.
[9] *T. W. Palmer:* The Gelfand-Naimark pseudo-norm on Banach *-algebras. J. London. Math. Soc. (2) *3* (1971), 59—66.
[10] *V. P. Potapov:* La structure multiplicative des fonctions matricielles J-contractives. Trudy Moskov. Mat. Obšč. *4* (1955), 125—236.
[11] *V. Pták:* On the spectral radius in Banach algebras with involution. Bull. London Math. Soc. *2* (1970), 327—334.
[12] *C. E. Rickart:* General theory of Banach algebras. Van Nostrand, Princeton, N. J., 1960.
[13] *B. Russo and H. A. Dye:* A note on unitary operators in C^*-algebras. Duke Math. J. *33* (1966), 413—416.
[14] *S. Shirali and J. W. M. Ford:* Symmetry in complex involutory Banach algebras. Duke Math. J. *37* (1970), 275—280.
[15] *I. Vidav:* Eine metrische Kennzeichnung der selbstandjungierten Operatoren. Math. Z. *66* (1956), 121—128.
[16] *B. Yood:* On axioms for B^*-algebras. Bull. Amer. Math. Soc. *76* (1970), 80—82.
[17] *D. A. Raikov:* To the theory of normed rings with involution. Dokl. Akad. Nauk SSSR *54* (1946), 387—390.
[18] *V. Pták:* Banach algebras with involution. Manuscripta Math. *6* (1972), 245—290.

INSTITUTE OF MATHEMATICS OF THE CZECHOSLOVAK ACADEMY OF SCIENCES, PRAHA

ON TOPOLOGICAL ENTROPY

B. RIEČAN

Bratislava

In this communication we introduce an abstract scheme including the topological entropy (see [1]) as well as the Kolmogoroff-Sinaj's entropy (see [2], [3]) and also some other invariants.

Let P be a set with a reflexive and transitive relation $\leqq$. Assume that on the set P an associative binary operation $\vee$ is defined such that $A \vee B \geqq A$ and $A \vee B \geqq$ $\geqq B$ for every $A, B \in P$. Further let $T : P \to P$ and $H : P \to \langle 0, \infty)$ be any functions satisfying the following conditions:

1. $H(\bigvee_{i=0}^{k} T^i(A)) \leqq H(\bigvee_{i=0}^{j} T^i(A)) + H(\bigvee_{i=j+1}^{k} T^i(A))$.
2. $T(A \vee B) = T(A) \vee T(B)$.
3. $H(T(A)) \leqq H(A)$.

Lemma. *Under these assumptions* $\lim 1/n\, H(\bigvee_{i=0}^{n-1} T^i(A))$ *exists for any* $A \in P$.

Definition. For any given P, T, H and $A \in P$ let us put $h(A, T) =$ $= \lim 1/n\, H(\bigvee_{i=0}^{n-1} T^i(A))$, $h(T) = \sup \{h(A, T);\ A \in P\}$; $h(T)$ is called the *entropy* of the triple (P, T, H).

Examples.

1. Topological entropy. Let X be a topological space, $f : X \to X$ a continuous map, P the family of all finite open coverings of X ($R_1 \leqq R_2$ iff R_2 is a refinement of R_1), $H(A) = \log \operatorname{card} A$, $T(A) = f^{-1}(A)$.

2. Kolmogoroff-Sinaj's entropy. Let (X, S, m) be a probability measure space, $f : X \to X$ a measure preserving transformation, P the family of all finite measurable decompositions A of X such that $A, f^{-1}(A), \ldots, f^{-k}(A)$ are independent for all k, $T(A) = f^{-1}(A)$, $H(A) = -\sum\{m(E) \log m(E);\ E \in A\}$.

3. Entropy of an automorphism of a Boolean algebra. Let B be a Boolean algebra, f an automorphism of B. Let P be the set of all finite decompositions of the greatest element of B. For $A \in P$ put $H(A) = \log \operatorname{card} A$, $T(A) = f(A)$.

Usually, if "two systems are isomorphic" then their entropies are equal. In general, two triples (P, T, H) and (R, S, G) are equivalent, if there is a bijection $U : P \to R$ with the following properties:

1. $U(A \vee B) = U(A) \vee U(B)$.
2. $T \circ U = U \circ S$.
3. $G(U(A)) = H(A)$.

Theorem 1. *If (P, T, H) and (R, S, G) are equivalent then their entropies are equal.*

We shall illustrate the preceding fact by the following three examples; the first two examples are well-known, the third one leads to a new result.

Let X_n be the set of all sequences $x = \{x_i\}_{i=-\infty}^{\infty}$ of integers $0, 1, \ldots, n-1$. The shift is the map $f : X_n \to X_n$ defined by the formula $f(\{x_n\}_{n=-\infty}^{\infty}) = \{y_n\}_{n=-\infty}^{\infty}$, where $y_n = x_{n+1}$ for every n. There are at least three natural structures on X_n:

1. Topology T_n with the subbase consisting of all cylinders $\{x; x_i = j\}$ and the shift f. It was proved in [1] that the topological entropy $h(f) = \log n$. It follows that there is no homeomorphism $g : X_n \to X_m$ $(n \neq m)$ commuting with the shifts.

2. The (Bernoulli) dynamical system (X_n, S_n, μ, f); here S_n is the σ-algebra generated by the cylinders; $\mu = \mathop{\times}\limits_{i=-\infty}^{\infty} \mu_i$ is the Cartesian product of probability measures μ_i; for all i, $\mu_i = \mu_0$ and μ_0 is defined by means of n-tuple $(p_0, p_1, \ldots, p_{n-1})$, i.e., $\mu_0(i) = p_i$, f is the shift. It is well-known that the Kolmogoroff-Sinaj's entropy $h(f) = -\sum p_i \log p_i$. Hence two Bernoulli systems with different entropies cannot be isomorphic. (Recently D. Ornstein [4] has proved the converse theorem.)

3. σ-algebras S_n generated by the cylinders and the automorphism f induced by the shift. Problem: Is there an isomorphism $g : S_n \to S_m$ commuting with the shifts?

Theorem 2. *If S_n is the σ-algebra generated by the cylinders, f is the automorphism of S_n generated by the shift and $h(f)$ is the entropy introduced in the third example, then $h(f) = \log n$.*

Corollary. *Given $n \neq m$, there is no isomorphism $g : S_n \to S_m$ commuting with the shifts.*

The last corollary was proved also in [5], but in another way.

Of course, also some further theorems can be proved in the general case. So $h(T^k) = k\, h(T)$, $h(T_1 \times T_2) = h(T_1) + h(T_2)$ and if $A \in P$ is an element such that $\{\bigvee_{i=0}^{n-1} T^i(A)\}_{n=0}^{\infty}$ "generates" the set P, then $h(T) = h(T, R)$.

Finally we list further examples satisfying the assumptions of our scheme:

4. Another type of topological entropy. Let P be the family of all open coverings of X having refinements of finite orders, $H(A) = \log \min \{\text{order } B; B \text{ is a refinement of } A\}$. This invariant probably corresponds to the topological dimension. If X is a topological space of finite dimension, then $\dim X \geqq e^{h(T)} - 1$.

5. Group endomorphism entropy (see [1]). Let G be an Abelian group, P the family of all finite subgroups, $A \leqq B$ iff $A \subset B$, T an endomorphism and $H(A) = \log \text{order } A$.

6. Entropy of a measure preserving transformation. Let P be a ring of sets (ordered by the inclusion), H a measure on P, T a measure preserving transformation.

7. Entropy of an operator. P is the system of all integrable functions (ordered as usually), H is the integral, $T(f) = f + g$ where g is a fixed non-positive function.

References

[1] *R. L. Adler, A. G. Konheim and M. H. Mc Andrew:* Topological entropy. Trans. Amer. Math. Soc. *114* (1965), 309—319.

[2] *A. N. Kolmogoroff:* A new metric invariant of transitive dynamical systems and automorphisms of Lebesgue spaces. Dokl. Akad. Nauk SSSR *119* (1958), 861—864.

[3] *Ja. G. Sinaj:* On the entropy of a metric automorphism. Dokl. Akad. Nauk SSSR *124* (1959), 980—983.

[4] *D. Ornstein:* Bernoulli shifts with the same entropy are isomorphic. Advances in Math. *4* (1970), 337—352.

[5] *I. Kluvánek and B. Riečan:* Some properties of Bernoulli schemes. Mat.-Fyz. Časopis Sloven. Akad. Vied *14* (1964), 84—88.

SLOVAK TECHNICAL UNIVERSITY, BRATISLAVA

NICE SPACES; NICE MAPS

T. RISHEL

Halifax — Tokyo

1. Introduction

At the first Prague Symposium, P. S. Alexandroff [1] asked the question "*Which topological spaces can be characterized as 'nice' continuous images of 'nice' spaces?*"

Much work has been done on this topic in ten years; this paper reviews some of this work.

In this paper all maps are considered to be both continuous and onto; N denotes the natural numbers throughout.

2. Metric Spaces

Perhaps the first characterization theorem of this type was that done independently by Ponomarev and Hanai about 1961.

Theorem 2.1. *A T_0-topological space Y is first countable iff it is the image of a metric space X by an open map $f : X \to Y$.*

In 1963 A. Arhangelskii introduced the pseudo-open maps.

Definition 2.2. A map $f : X \to Y$ is *pseudo-open* iff for every $y \in Y$ and every neighborhood U of $f^{-1}(y)$, $y \in \operatorname{Int} f(U)$.

Pseudo-open maps are precisely those quotient maps which are hereditary. Arhangelskii then stated the next result.

Theorem 2.3. *Y is the pseudo-open image of a metric space X iff Y is a Fréchet space* (i.e., a set $F \subseteq Y$ is closed iff for every $p \in F$ there exists a point-sequence $\{p_n\} \subseteq F$ such that $\{p_n\} \to p$).

In 1965, S. P. Franklin tackled the problem of quotients of metric spaces, and in an interesting paper [6] showed the following.

Theorem 2.4. *Y is a sequential space iff Y is the image of a metric space under a quotient map.*

Sequential spaces are defined as follows.

Definition 2.5. Y is *sequential* iff: U is open in Y iff every sequence converging to a point in U is residually (eventually) in U.

In his paper Franklin also gives an explicit proof of Arhangelskii's assertion of Theorem 2.3.

As well as not being hereditary, quotient maps also do not preserve products. Thus it is logical to consider a map which is stronger than quotient but preserves products. This was done by E. Michael [13].

Definition 2.6. $f: X \to Y$ is a *bi-quotient map* iff: given $\mathscr{B}$ a filterbase in Y, if $y \in \operatorname{Cl} B$ for all $B \in \mathscr{B}$, then there exists $x \in f^{-1}(y)$ such that $x \in \operatorname{Cl} f^{-1}(B)$ for all $f^{-1}(B) \in f^{-1}(\mathscr{B})$.

Michael has asserted the next statement.

Theorem 2.7. *Y is the image of a metric space X by a bi-quotient map f iff every maximal filter $\mathscr{F} \to x$ has a countable family $\{F_1, F_2, \ldots\} \subseteq \mathscr{F}$ such that $\{F_i\} \to x$.* (Such a space is called bi-sequential.)

P. Vopěnka has defined almost-open maps.

Definition 2.8. $f: X \to Y$ is *almost-open* iff for any $y \in Y$ there exists an $x \in f^{-1}(y)$ having a basis of open sets such that the image of each member of the basis is open.

Almost-open maps form a class between open and bi-quotient maps. The next statement is a simple modification of Ponomarev's Theorem 2.1.

Corollary 2.9. *Y is first countable iff Y is the image of a metric space under an almost-open map.*

F. Siwiec and V. Mancuso [26] altered the definition of bi-quotient maps as follows.

Definition 2.10. A map $f: X \to Y$ is *countably bi-quotient* iff, when $\{A_n : n \in N\}$ is a decreasing family accumulating at some $y \in Y$, then there exists an $x \in f^{-1}(y)$ such that $f^{-1}(A_n)$ accumulates at x.

Definition 2.11. Y is *countably bi-sequential* iff, when $\{F_n\}$ is a decreasing sequence of sets accumulating at $y \in Y$, there exist $y_n \in F_n$ such that the point-sequence $\{y_n\} \to y$.

Theorem 2.12. *Y is countably bi-sequential iff Y is the countably bi-quotient image of a metric space X.*

Perfect maps have long been known to preserve metrizability (see Dugundji [5]), as have simultaneously closed and open maps (sometimes called clopen). For a proof of this last, see Morita-Hanai [16].

Closed images of metric spaces have been characterized by N. Lašnev [11]. Their characterization is somewhat different from those discussed previously. First, let us state a definition.

Definition 2.13. A sequence of closed coverings $\mathscr{U}_i$ of a space X is said to be an *almost refining sequence* if and only if for any point x in X, any arbitrary system of sets $\{B_i\}$ such that $B_i \in \mathscr{U}_i$ and $x \in B_i$, is either hereditarily closure preserving or forms a network of the space at the point x.

Lašnev's result follows.

Theorem 2.14. *A T_1-space Y is the closed image of a metric space X iff*

a) *Y is Fréchet,*

b) *there exists in Y an almost refining sequence of hereditarily closure preserving closed covers forming a network.*

Such a space has been called a Lašnev space.

A useful map between closed and perfect is quasi-perfect.

Definition 2.15. $f: X \to Y$ is *quasi-perfect* iff f is closed and $f^{-1}(y)$ is countably compact for every $y \in Y$.

Quasi-perfect images of metric spaces are well known to be metric.

Definition 2.16. A map $f: X \to Y$ is an *s-map* iff the preimage of each point has a countable dense subset.

T. Hoshina [8] has characterized quotient s-images of metric spaces.

Definition 2.17. A T_1-space Y is a *q-s space* iff it has a point countable family A of subsets of X satisfying the following condition:

A set $F \subseteq Y$ is closed iff for every $x \in Y - F$ and each q-sequence $\{A_n : n \in N\} \subseteq \mathsf{A}$ with $x = \bigcap\{A_n : n \in N\}$, $F \cap A_n = \emptyset$ for some $n \in N$.

Theorem 2.18. *A T_1-space Y is the quotient s-image of a metric space X iff Y is a q-s space.*

Ponomarev [22] has studied the problem of open s-images of metric spaces.

Theorem 2.19. *Y has a point countable base iff Y is the open s-image of a metric space X.*

The following tables collect our results thus far, assuming T_1.

open s-map → quotient s-map → quotient ←┐
open → almost-open → bi-quotient → countably bi-quotient → pseudo-open
perfect → quasi-perfect → closed ─┘ (↑ pseudo-open)

point-countable base → q-s → sequential ←┐
metric → first-countable → bi-sequential → countably bi-sequential → Fréchet
Lašnev ─┘ (↑ Fréchet)

3. *M*-Spaces.

In 1969, J. Nagata suggested extending the characterization problem to M-spaces, a class of spaces defined by K. Morita.

Definition 3.1. X is an M-space iff it has a normal sequence of open covers $\mathcal{U}_1, \mathcal{U}_2, \mathcal{U}_3, \ldots$ such that each point-sequence $\{x_n\}$, of the form $x_n \in \mathrm{St}(x, \mathcal{U}_n)$ for every $n \in N$ and for fixed $x \in X$, has a cluster point.

M-spaces simultaneously generalize all metric spaces and all countably compact spaces. These spaces have been characterized as follows by Morita [17].

Theorem 3.2. *X is M iff it is the quasi-perfect inverse image of a metric space.*

It is thus clear that quasi-perfect images of M-spaces are M.

M-spaces are not necessarily paracompact, so the problem of characterization may also be studied for paracompact M-spaces. In the next tables and definitions we see the results of this study. (For proofs, see [3], [18], [20], [30], [23] and [27].)

Table 3.3. ***M*-spaces**

f	Y	additional condition
(1) open	q	Y regular
(2) almost-open	q	Y regular
(3) bi-quotient	bi-quasi-k	
(4) countably bi-quotient	countably bi-quasi-k	
(5) pseudo-open	singly bi-quasi-k	
(6) quotient	quasi-k	X, Y regular
(7) perfect	M	Y normal
(8) closed and open	M	X normal

Definition 3.4. A countable decreasing family of sets $\{U_1, U_2, \ldots\}$ is called a *q-sequence at* $y \in Y$ iff $y \in U_n$ for every $n \in N$ and if $\{y_n\}$ is a point-sequence such that $y_n \in U_n$ for every $n \in N$, then $\{y_n\}$ clusters.

Definition 3.5. Y is a *q-space* iff every $y \in Y$ has a q-sequence of neighborhoods.

Definition 3.6. Y is *bi-quasi-k* iff every maximal filter $\mathscr{F} \to y \in Y$ contains a subfilter $\{F_1, F_2, \ldots\}$ which is a q-sequence at y.

Definition 3.7. Y is *countably bi-quasi-k* iff for every decreasing family $\{F_n\}$ clustering at $y \in Y$, there exists a q-sequence $\{A_n\}$ at y such that $y \in \mathrm{Cl}\,(F_n \cap A_n)$ for every $n \in N$.

Definition 3.8. Y is *singly bi-quasi-k* iff the following holds: given $B \subseteq Y$, $y \in \mathrm{Cl}\, B$ iff there exists a q-sequence $\{U_1, U_2, \ldots\}$ at y such that $y \in \mathrm{Cl}\,(B \cap U_n)$ for every $n \in N$.

Definition 3.9. Y is *quasi-k* iff the following holds: $F \subseteq Y$ is closed in Y iff $F \cap C$ is relatively closed in C for every countably compact $C \subseteq Y$.

Table 3.10. Paracompact M-spaces

f	Y	additional conditions
(1) open	pointwise countable type	Y Hausdorff
(2) almost-open	pointwise countable type	Y Hausdorff
(3) bi-quotient	bi-k	
(4) countably bi-quotient	countably bi-k	
(5) pseudo-open	singly bi-k	
(6) quotient	k	X, Y regular
(7) perfect	paracompact M	Y normal
(8) closed and open	paracompact M	X, Y normal

Definition 3.11. Given a point $y \in Y$, a countable decreasing family $\{U_1, U_2, \ldots\}$ is said to form a *k-sequence* at y iff $\{U_1, U_2, \ldots\}$ is a q-sequence at y and $\bigcap\{U_n : n \in N\}$ is compact.

Definitions of the spaces given in Table 3.10 will be omitted, except to say that they are the same as those in Table 3.3 with the exceptions that the words "k-sequence" and "compact" replace the words "q-sequence" and "countably compact", respectively.

Similarly, $\mathfrak{C}$-spaces were defined by Ishii, Tsuda and Kunugi [10]. Such spaces are defined as follows.

Definition 3.12. A topological space X is said to be a *$\mathfrak{C}$-space* iff X has a normal sequence of open covers $\mathscr{U}_1, \mathscr{U}_2, \ldots$ such that every point-sequence $\{x_n\}$, of the form $x_n \in \mathrm{St}(x, \mathscr{U}_n)$ for every $n \in N$ and for fixed $x \in X$, has a subsequence with compact closure.

These spaces include paracompact M-spaces, and form a class of spaces whose product with every M-space is M.

Definition 3.13. Given $y \in Y$, a countable decreasing family of sets $\{U_1, U_2, \ldots\}$ is said to form an *r_0-sequence at y* iff it is a q-sequence at y and every point-sequence $\{x_n\}$, $x_n \in U_n$ for every $n \in N$, has a subsequence with compact closure.

Definition 3.14. A set $P \subseteq Y$, a topological space, is said to be *proto-compact* iff every sequence of points of P has a subsequence with compact closure.

Relative to these last two definitions, we can now form another table. Again definitions of the various spaces are omitted for brevity. (For a full description of results, see [10] and [25].)

Table 3.15. $\mathfrak{C}$-spaces

f	Y	additional conditions
(1) almost-open	r_0	Y regular
(2) bi-quotient	bi-proto-k	
(3) pseudo-open	singly bi-proto-k	
(4) quotient	proto-k	X, Y regular
(5) perfect	$\mathfrak{C}$	Y normal

4. Locally compact and separable metric spaces

Locally compact and separable metric spaces have been characterized by means of mappings. The next result, due to K. Morita, is well-known.

Theorem 4.1. *Local compactness is invariant under open maps.*

Michael has also shown [13] that bi-quotient maps preserve local compactness. Another result, due to D. E. Cohen [4], is by now classical.

Theorem 4.2. *Y is a k-space iff it is the quotient of a locally compact space X.*

Bi-quotient and countably bi-quotient maps do not have the same effect on locally compact spaces. The next result is due to F. Siwiec [27] and Michael [14].

Theorem 4.3. *Y is strongly k' iff it is the image of a locally compact space under a countably bi-quotient map.* (A *strongly k'-space* Y is one in which, given a decreasing sequence $\{U_n\}$ accumulating at $y \in Y$, there exists a compact $K \subseteq Y$ such that $y \in \mathrm{Cl}(U_n \cap K)$ for all $n \in N$.)

Images of locally compact spaces by pseudo-open maps are precisely the k'-spaces. This result is due to Arhangelskii [2].

Definition 4.4. Given a set $A \subseteq Y$, topological, suppose a point $y \in \mathrm{Cl}\, A$ iff there exists a compact set $K \subseteq Y$ such that $y \in \mathrm{Cl}\,(K \cap A)$. Then Y is said to be a *k'-space.*

E. Michael [12] has discussed two classes of spaces which relate to separable metric spaces.

Definition 4.5. An *$\aleph_0$-space* Y is a regular space with a countable k-network (a *k-network* is a family $\mathscr{P}$ of subsets P of Y such that if C is compact, U open in Y and $C \subseteq U$, then some $P \in \mathscr{P}$ has the property $C \subseteq P \subseteq U$).

Definition 4.6. A *cosmic space* Y is a regular space with a countable network.

Theorem 4.7. *Y is a cosmic space iff it is the continuous image of a separable metric space X.*

Michael has also shown, in the same paper, the next result.

Theorem 4.8. *Y is $\aleph_0$ and k iff Y is the quotient of a separable metric space.*

P. Strong [28] has noted that Y need only have a countable k-network, be sequential and Hausdorff in the above result. He has also shown the next statement.

Theorem 4.9. *Y is a Fréchet, Hausdorff space with a countable k-network iff Y is the image of a separable metric space by a pseudo-open map.*

Separable metric spaces are preserved under open, bi-quotient and countably bi-quotient maps (whenever the image space is regular).

5. Other results.

Many maps and spaces have occurred but once in characterization theorems. Hence results of this section will not seem as fully developed as in previous discussions. Questions will naturally occur to the reader.

Definition 5.1. $f : X \to Y$ is *compact-covering* if every compact set $K \subseteq Y$ is the image of a compact $C \subseteq X$.

Michael defined the above map and proved the next result [12].

Theorem 5.2. *Y is $\aleph_0$ iff Y is the image by a compact-covering map of a separable metric space.*

Definition 5.3. Y is a *c-space* iff the closure of any set U in Y is obtained from closures of countable sets in U.

Theorem 5.4. *Y is a c-space iff Y is the quotient image of copies of a countable discrete space with single points of its Stone-Čech compactification.*

The above characterization is due to this author [24], but such spaces were originally mentioned by S. G. Mrówka and R. C. Moore [15] under the title of "spaces determined by countable sets".

Nagata has recently shown [21] that images of T_1 M-spaces under perfect maps are precisely the M^*-spaces of T. Ishii.

References

[1] *P. Alexandroff:* On some results concerning topological spaces and their continuous mappings. General Topology and its Relations to Modern Analysis and Algebra, (I) (Proc. (First) Prague Topological Sympos., 1961). Publishing House of the Czechoslovak Academy of Sciences, Prague, 1962, 41—54.
[2] *A. Arhangelskii:* Bicompact sets and the topology of spaces. Soviet Math. Dokl. *4* (1963), 561—564.
[3] *T. Chiba:* On q-spaces. Proc. Japan Acad. *45* (1969), 453—455.
[4] *D. E. Cohen:* Spaces with weak topology. Quart. J. Math. Oxford Ser. *5* (2) (1954), 77—80.
[5] *J. Dugundji:* Topology.
[6] *S. P. Franklin:* Spaces in which sequences suffice. Fund. Math. *57* (1965), 107—115.
[7] *S. Hanai:* On open mappings II. Proc. Japan Acad. *37* (1961), 233—238.
[8] *T. Hoshina:* On the quotient s-images of metric spaces. Sci. Rep. Tokyo Kyoiku Daigaku Sect. A *10* (1970), 265—268.
[9] *T. Ishii:* On closed mappings and M-spaces I, II. Proc. Japan Acad. *43* (1967), 752—761.
[10] *T. Ishii, M. Tsuda and S. Kunugi:* On the products of M-spaces I, II. Proc. Japan Acad. *44* (1968), 897—903.
[11] *N. Lašnev:* Closed images of metric spaces. Soviet Math. Dokl. *7* (1966), 1219—1221.
[12] *E. Michael:* $\aleph_0$-spaces. J. Math. Mech. *15* (1966), 983—1002.
[13] *E. Michael:* Bi-quotient maps and cartesian products of quotient maps. Ann. Inst. Fourier (Grenoble) *18* (2) (1968), 281—302.
[14] *E. Michael:* Images of certain quotient maps. (To appear.)
[15] *R. C. Moore and S. G. Mrówka:* Topologies determined by countable objects. Notices Amer. Math. Soc. *11* (1964), 554.
[16] *K. Morita and S. Hanai:* Closed mappings and metric spaces. Proc. Japan Acad. *32* (1956), 10—14.
[17] *K. Morita:* Products of normal spaces with metric spaces. Math. Ann. *154* (1964), 365—382.
[18] *J. Nagata:* Mappings and M-spaces. Proc. Japan Acad. *45* (1969), 140—144.
[19] *J. Nagata:* Modern General Topology.
[20] *J. Nagata:* Quotient and bi-quotient spaces of M-spaces. Proc. Japan Acad. *45* (1969), 25—29.
[21] *J. Nagata:* Some theorems on generalized metric spaces. (To appear.)
[22] *V. Ponomarev:* Axioms of countability and continuous mappings. Bull. Acad. Polon. Sci. Sér. Sci. Math. Astronom. Phys. *8* (1960), 127—134.

This research was partially funded by Canadian National Research Council Grant A-3999.

[23] *T. Rishel:* A characterization of pseudo-open images of *M*-spaces. Proc. Japan Acad. *45* (1969), 910—912.
[24] *T. Rishel:* A class of spaces determined by sequences with their cluster points. Portugal. Math. *31* (1972), 187—192.
[25] *T. Rishel:* Images of class-$\mathfrak{C}$. (To appear.)
[26] *F. Siwiec and V. J. Mancuso:* Relations among certain mappings and conditions for their equivalence. (To appear.)
[27] *F. Siwiec:* On sequence-covering and countably bi-quotient mappings. (To appear.)
[28] *P. Strong:* To appear in General Topology and its Applications.
[29] *Din' N'e T'ong:* Preclosed mappings and Taimanov's theorem. Soviet Math. Dokl. *4* (1963), 1335—1338.
[30] *H. Wicke:* On the Hausdorff open continuous images of Hausdorff paracompact *p*-spaces. Proc. Amer. Math. Soc. *22* (1969), 136—140.

DALHOUSIE UNIVERSITY, HALIFAX, NOVA SCOTIA
TOKYO UNIVERSITY OF EDUCATION, TOKYO

BOX PRODUCTS

M. E. RUDIN

Madison

Suppose $\{X_n\}_{n\in\omega_0}$ is a family of topological spaces. Then $\prod_{n\in\omega_0} U_n$ is called a *box* provided U_n is a non-empty open subset of X_n for each $n \in \omega_0$. The box product of $\{X_n\}_{n\in\omega_0}$ is $\prod_{n\in\omega_0} X_n$ topologized by using the set of all boxes as a basis. The Continuum Hypothesis implies the box product of countably many locally compact σ-compact metric spaces is paracompact. The Continuum Hypothesis implies the box product of countably many σ-compact ordinals is paracompact. And the Generalized Continuum Hypothesis implies the box product of countably many ordinals is normal (but not paracompact) if one of the ordinals has uncountable cofinality which is greater than the cardinality of the other ordinals and is not the successor of a singular cardinal. Of course, if one of the ordinals is of uncountable cofinality less than the cardinality of another, no product including both can be normal.

CARDINALITIES OF BASES

H. M. SCHAERF

Montreal

Most definitions of bases are specializations of the following one:

Given a class $\mathscr{V}$ of subsets of a set R and a transitive relation $<$ on R, call a subset of the union $\bigcup\mathscr{V}$ of $\mathscr{V}$ a *$\mathscr{V}$-base* iff it contains, for each V in $\mathscr{V}$ and v in V, some $b < v$ which belongs to $\overline{V} = \{r \in \bigcup\mathscr{V} : \exists v' < r,\ v' \in V\}$.

In this paper we relate the least power $w\mathscr{V}$ of $\mathscr{V}$-bases (assumed to be >0) to two other cardinalities depending on a relation σ defined on $\bigcup\mathscr{V}$, and state a few applications. One of these cardinalities, denoted by cel σ, is the sup of the powers of all $E \subset \bigcup\mathscr{V}$ such that $a, b \in E$ and $a\sigma b$, $b\sigma a$ imply $a = b$. To define the other, call a class $\mathscr{G}$ of subsets of $\bigcup\mathscr{V}$ a *σ-grading of $\mathscr{V}$* iff its union contains, for each V in $\mathscr{V}$ and v in V, some $b < v$, $b \in V$, and for each F in $\mathscr{G}$ there is G in $\mathscr{G}$ with the following property: for each V in $\mathscr{V}$ and f in $F \cap V$ there is g in $G \cap V$ such that $f \geqq g'$ for each g' in G with $g\sigma g'$. Let $\sigma\mathscr{V}$ be the least power of such σ-gradings.

Theorem 1. *Let σ be any relation on $\bigcup\mathscr{V}$ such that $a\sigma b$ holds whenever there is V in $\mathscr{V}$ and v in V with $v < a \in V$ and $v < b$, and let $\overline{\mathscr{V}} = \{\overline{V} : V \in \mathscr{V}\}$. Then $w\mathscr{V}$ is finite iff both* cel σ *and $\sigma\overline{\mathscr{V}}$ are finite (which is sure if both* cel σ *and $\sigma\mathscr{V}$ are finite); otherwise*

$$w\mathscr{V} = \max(\text{cel}\,\sigma, \sigma\overline{\mathscr{V}}) \leqq \max(\text{cel}\,\sigma, \sigma\mathscr{V}).$$

Corollary. *For infinite $w\mathscr{V}$ the Suslin Property $w\mathscr{V} =$ cel σ holds iff $\sigma\overline{\mathscr{V}} \leqq$ cel σ and is implied by $\sigma\mathscr{V} \leqq$ cel σ.*

Subsequently let X be a topological space and wX its weight. Some applications of Theorem 1 to the determination of wX follow.

Direct applications. Let R be the class of all nonvoid subsets of X, $V(x)$ the class of all open neighbourhoods of $x \in X$, $\mathscr{V}$ the family of all $V(x)$ and $<$ the inclusion on R. Then $\bigcup\mathscr{V}$ is the topology of X, $\overline{\mathscr{V}} = \mathscr{V}$, $w\mathscr{V} = wX$, and Theorem 1 yields information on the weight of X for each σ in an infinite family of relations which contains the Intersection Relation ϱ ($A\varrho B$ iff A intersects B). Since cel ϱ is the cellularity cel X of X, the Corollary sheds some light on the problem of Suslin. (This method can be extended to arbitrary coverings of a set.)

Indirect applications. For each x in X let $V(x)$ be a class of subsets of X each of which contains x. Call the family $\mathscr{V}$ of all these classes a *σ-system on* X iff σ is any relation on $\bigcup\mathscr{V}$ with the following properties:

(i) $A\sigma B$ holds whenever $\mathscr{V}$ contains some $V(x)$ such that x is in B and A is in $V(x)$;

(ii) there is a σ-grading $\mathscr{G}$ of $\mathscr{V}$ which is *full*, i.e., such that $\bigcup\mathscr{G} = \bigcup\mathscr{V}$ and every element of $\mathscr{G}$ contains members of every $V(x)$;

(iii) the class of all $A \subset X$ such that for every x in A there is B in $V(x)$ with $B \subset A$ is the topology of X.

Theorem 2. *For every full σ-grading* $\mathscr{G}$ *of any σ-system on* X,

$$wX \leqq \text{cel}\,\sigma\,.\,\text{card}\,\mathscr{G}\,.$$

This theorem yields the following specialization.

Theorem 3. *If* X *is a uniformizable space and* uX *the least weight of uniformities compatible with the topology of* X*, then* $wX \leqq \text{cel}\,X\,.\,uX$.

Proofs and other applications will appear elsewhere.

ON THE SHAPE CLASSIFICATION OF MANIFOLD-LIKE CONTINUA

J. SEGAL

Seattle

In [4] S. Mardešić and the author classified all (n-sphere)-like continua as to their shape. It was shown that such continua are of the shape of a point, the n-sphere or the $(n-1)$-fold suspension of a solenoid. Shape is a classification of compacta which is coarser than homotopy type but which coincides with it on ANR's. K. Borsuk also considered the relation between likeness and shape in [1] where he gave an example of two compacta which are like each other but are of different shape. The general question of the shape classification of manifold-like continua remains open. However, in this note some recent progress in the case of projective plane-like continua is presented.

A metric continuum X is said to be *like a compactum* Y or *Y-like* provided, for each $\varepsilon > 0$, there is a mapping $f_\varepsilon : X \to Y$ onto Y such that $\operatorname{diam} f_\varepsilon^{-1}(y) < \varepsilon$ for any $y \in Y$. The connection between likeness and shape can be made through the following result of Mardešić and the author [3].

Theorem 1. *If Y is a polyhedron and X is a Y-like continuum, then X is the inverse limit of an inverse sequence of copies of Y.*

It was shown by I. Lončar and S. Mardešić in [2] that Y need only be required to be an ANR in this theorem.

The reader is referred to [4] or to Mardešić's survey of the theory of shapes appearing in this volume for a description of the ANR-system approach to shape. Suffice it to say here, that we work in a category whose objects are inverse sequences of ANR's and whose morphisms are defined as in the case of mappings of inverse sequences except that one only requires commutativity up to homotopy. One defines a homotopy of morphisms and obtains a notion of homotopy type of ANR-sequences. Then one shows that for a compact metric space X this homotopy type is independent of the ANR-sequence associated with it in a particular way. So as far as shape is concerned one might as well use any inverse limit representation given by Theorem 1. Finally two compact metric spaces X and Y are said to be of the same *shape* if they have ANR-sequences associated with them of the same homotopy type.

We next describe a new class of projective plane-like spaces called solenoidal projective planes. These play the same role in the case of projective plane-like continua that solenoids did for (1-sphere)-like continua. Our description relies on P. Olum's analysis [5] of the twisted degree of mappings of the projective plane P into itself. A map $f: P \to P$ for which the induced homomorphism on the fundamental group of P into itself is an isomorphism, i.e.,

$$(1) \qquad f_{\#}: \Pi_1(P) \overset{\approx}{\to} \Pi_1(P)$$

(since $\Pi_1(P) = Z_2$ there is only one such isomorphism), is determined up to homotopy class by the absolute value of the twisted degree of f given by

$$(2) \qquad f^*: H^2(P, Z^t) \to H^2(P, Z^t).$$

Here Z^t denotes the group of twisted integers and both cohomology groups are infinite cyclic. We assume that for each of the infinite cyclic groups in (2) one of the two possible generators is chosen. Then the homomorphism f^* carries the generator of the first group into some integral multiple of the generator of the second. This integer, which defines the homomorphism uniquely, is called the *twisted degree* of the mapping f. The twisted degree agrees with the usual degree on orientable manifolds. In the special case of mappings of the projective plane into itself for which (1) holds, only odd integers occur as twisted degrees. Moreover, in this case two such maps are homotopic if and only if the absolute values of their twisted degrees are equal.

Let $Q = (q_1, q_2, \ldots)$ be a sequence of odd primes. Then we define the inverse sequence $\{X_n, p_{nn+1}\}$ where each X_n is a copy of P and the map $p_{nn+1}: X_{n+1} \to X_n$ is determined as follows. Let S^1 be the unit circle in the complex plane and represent S^2 as $S^1 \times [-1, 1]$ with the appropriate identifications. Let $g: S^2 \to S^2$ be the map which sends (x, t) to (x^{q_n}, t) and let $h: S^2 \to P$ be the identification map which identifies (x, t) and $(-x, -t)$. Since q_n is odd, p_{nn+1} is well defined by $p_{nn+1}(h(x, t)) = h(g(x, t))$. We define the *solenoidal projective plane* P_Q to be the inverse limit of the inverse sequence just described. Two sequences of odd primes $Q = (q_1, q_2, \ldots)$ and $Q' = (q'_1, q'_2, \ldots)$ are said to be *equivalent*, $Q \sim Q'$, provided it is possible to delete a finite number of terms from each so that every prime occurs the same number of times in each of the deleted sequences.

Theorem 2. *Two solenoidal projective planes P_Q and $P_{Q'}$, are of the same shape iff $Q \sim Q'$.*

In analogy to the case of the 2-sphere we have the following result.

Theorem 3. *Every projective plane-like continuum X has the shape of a point, the projective plane or a solenoidal projective plane.*

To prove this theorem one first uses Theorem 1 to get X as the inverse limit of an inverse sequence of projective planes. Then one considers (case 1) if infinitely many of the bonding maps induce the zero homomorphism or (case 2) if they all induce an isomorphism on the fundamental group of P into itself. Case 1 has two subcases depending on whether or not infinitely many of the maps are null-homotopic. One can then show that either subcase implies that X has the shape of a point. Case 2 also has two subcases depending on whether or not all bonding maps with sufficiently large indices have twisted degree one. If they do then X is of the shape of P. If not then X is of the shape of some solenoidal projective plane. (Details of the proofs can be found in [6].)

References

[1] *K. Borsuk:* A note on the shape of quasi-homeomorphic compacta. Comment. Math. Prace Mat. *14* (1970), 25—34.

[2] *I. Lončar and S. Mardešić:* A note on inverse sequences of ANR's. Glasnik Mat. Ser. III *3* (1968), 41—48.

[3] *S. Mardešić and J. Segal:* ε-mappings onto polyhedra. Trans. Amer. Math. Soc. *109* (1963), 146—164.

[4] *S. Mardešić and J. Segal:* Shapes of compacta and ANR-systems. Fund. Math. *72* (1971), 57—75.

[5] *P. Olum:* Mappings of manifolds and the notion of degree. Ann. of Math. *58* (1953), 458—480.

[6] *J. Segal:* Shape classification of projective plane-like continua. Glasnik Mat. Ser. III *6* (1971), 365—371.

EXTENDING POINT-FINITE COVERS

L. I. SENNOTT

Fairfax

A space X is said to be *strongly normal* if it is normal and given a locally finite family $(F_\alpha)_{\alpha \in I}$ of closed subsets of X, there exists a locally finite family $(G_\alpha)_{\alpha \in I}$ of open subsets of X such that $F_\alpha \subset G_\alpha$ for all $\alpha \in I$. Katětov [4] proved that a space is strongly normal if and only if it is collectionwise normal and countably paracompact. Considering spaces in which every point-finite family of closed sets expands to a point-finite family of open sets would not be fruitful. Note that $\{x\}_{x \in [0,1]}$ is a point-finite family of closed subsets of $[0, 1]$, and there does not exist a point-finite family $(G_x)_{x \in [0,1]}$ of open subsets of $[0, 1]$ such that $x \in G_x$ for all $x \in [0, 1]$.

A more meaningful direction for generalization is to consider a point-finite open cover $(H_\alpha)_{\alpha \in I}$ of a closed subspace F of a topological space X and to ask when there exists a point-finite family $(G_\alpha)_{\alpha \in I}$ of open subsets of X such that $H_\alpha \subset G_\alpha$ for all $\alpha \in I$. Spaces having this property are characterized and the class of such spaces is seen to contain all perfectly normal, collectionwise normal spaces. It is also shown that the class of perfectly normal, collectionwise normal spaces has the property that every point-finite cozero set cover of a closed subspace F extends to a point-finite cozero set cover of X that is locally finite on the complement of F.

These ideas should be useful in connection with the unsolved problem of finding a broad class of spaces for which a polytope with the weak topology will be an AE. Any continuous function from X into a polytope with either the weak topology or the metric topology gives rise to a point-finite partition of unity on X that is not necessarily locally finite.

We will first define three new concepts and then use these concepts to characterize those spaces in which every point-finite open cover on a closed subspace can be extended to a point-finite cover on the whole space.

Definition 1. A topological space X satisfies (α) if given a closed subspace F of X and a pairwise disjoint family $(H_\alpha)_{\alpha \in I}$ of open subsets of F, there exists a point-finite family $(G_\alpha)_{\alpha \in I}$ of open subsets of X such that $H_\alpha \subset G_\alpha$ for all $\alpha \in I$.

A topological space X satisfies (β) if given a discrete family $(S_\alpha)_{\alpha \in I}$ of subsets of X, there exists a point-finite family $(G_\alpha)_{\alpha \in I}$ of open subsets of X such that $S_\alpha \subset G_\alpha$ for all $\alpha \in I$.

A topological space X satisfies (γ) if given a closed subspace F of X and $(A_n)_{n\in N}$ an increasing countable closed cover of F, there exists $(B_n)_{n\in N}$ an increasing closed cover of X such that $B_n \cap F \subset A_n$ for all $n \in N$.

A family $\mathcal{S} = (S_\alpha)_{\alpha\in I}$ of subsets of X is of order n, written ord $\mathcal{S} = n$ if every element of X is in no more than n members of $\mathcal{S}$ and there is an element of X which is in exactly n members of $\mathcal{S}$. The family $\mathcal{S}$ is of finite order if ord $\mathcal{S} = n$ for some n.

Note that every collectionwise normal space satisfies (β) and every perfectly normal space satisfies (γ). We will show later that every hereditary collectionwise normal space satisfies (α). We will now use these concepts to characterize those spaces in which every point-finite open cover of finite order on a closed subspace extends to a point-finite family on the whole space.

Theorem 2. *Let X be a topological space. The following are equivalent:*

1. *The space X satisfies (α) and (β).*
2. *Given F a closed subspace of X and $\mathcal{H} = (H_\alpha)_{\alpha\in I}$ a point-finite open cover of F of finite order, there exists a point-finite family $(G_\alpha)_{\alpha\in I}$ of open subsets of X such that $H_\alpha \subset G_\alpha$ for all $\alpha \in I$.*

Proof. The proof of 1. ⇒ 2. will proceed by induction on ord $\mathcal{H}$. Suppose ord $\mathcal{H} = 1$. Then $\mathcal{H}$ is a discrete collection of subsets of X. Since X satisfies (β), there exists a point-finite family $(G_\alpha)_{\alpha\in I}$ of open subsets of X such that $H_\alpha \subset G_\alpha$ for all $\alpha \in I$.

Assume the theorem is true for all point-finite open covers with order $\leqq m$ of a closed subset of X, and let $(H_\alpha)_{\alpha\in I}$ be a point-finite open cover of F with order $m + 1$. Let $I^* = \{J \subset I \mid |J| = m + 1\}$. Let $H_J = \bigcap\{H_\alpha \mid \alpha \in J\}$ and consider the family $(H_J)_{J\in I^*}$. This is a pairwise disjoint family of open subsets of F, so by (α), there exists a point-finite family $(G_J)_{J\in I^*}$ of open subsets of X such that $H_J \subset G_J$ for all $J \in I^*$.

Consider $A = \{x \in F \mid |K(x)| \leq m\}$, where $K(x) = \{\alpha \in I \mid x \in H_\alpha\}$. The set A is closed in F, hence in X, since if x is a point of F not in A, then $\bigcap\{H_\alpha \mid \alpha \in K(x)\}$ is a neighborhood of x missing A. Now $(H_\alpha \cap A)_{\alpha\in I}$ is a point-finite open cover of A and has order $\leqq m$. By the induction hypothesis, there exists a point-finite family $(M_\alpha)_{\alpha\in I}$ of open subsets of X such that $H_\alpha \cap A \subset M_\alpha$ for all $\alpha \in I$.

Let $N_\alpha = M_\alpha \cup \bigcup\{G_J \mid \alpha \in J\}$. Clearly N_α is open in X and $H_\alpha \subset N_\alpha$ for all $\alpha \in I$. It remains to show that $(N_\alpha)_{\alpha\in I}$ is point-finite. If x is in X, there exists K^* a finite subset of I^* such that if $x \in G_J$, then $J \in K^*$. Similarly, there is a finite subset K of I such that if $x \in M_\alpha$, then $\alpha \in K$. Let $K' = K \cup \bigcup\{J \mid J \in K^*\}$. Then K' is a finite subset of I and if $x \in N_\alpha$, then $\alpha \in K'$. This completes the induction step.

To show that 2. ⇒ (α), let F be a closed subset of X and let $(H_\alpha)_{\alpha\in I}$ be a pairwise disjoint family of open subsets of F. Then $(H_\alpha)_{\alpha\in I} \cup \{F\}$ is a point-finite open cover of F, so by 2., there is a point-finite open family $(G_\alpha)_{\alpha\in I}$ of X such that $H_\alpha \subset G_\alpha$

for all $\alpha \in I$. To show that X satisfies (β) let $(S_\alpha)_{\alpha \in I}$ be a discrete family of subsets of X. Then $(\text{cl } S_\alpha)_{\alpha \in I}$ is a discrete family and hence a point-finite open cover of $F = \bigcup_\alpha \text{cl } S_\alpha$. By 2., there exists a point-finite family $(G_\alpha)_{\alpha \in I}$ of open subsets of X such that $S_\alpha \subset \subset \text{cl } S_\alpha \subset G_\alpha$ for all $\alpha \in I$. This completes the proof.

Corollary 3. *Let X be a topological space. The following are equivalent:*

1. *The space X satisfies (α) and (β).*
2. *Every point-finite open cover of finite order of a closed subspace of X extends to a point-finite open cover of X.*

Proof. To show 1. ⇒ 2., let $(H_\alpha)_{\alpha \in I}$ be a point-finite open cover of finite order of a closed subspace F of X. By Theorem 2, there exists a point-finite family $(G_\alpha)_{\alpha \in I}$ of open subsets of X such that $H_\alpha \subset G_\alpha$ for all $\alpha \in I$. For each $\alpha \in I$, let M_α be an open subset of X such that $M_\alpha \cap F = H_\alpha$. Fix $\alpha_0 \in I$ and let $N_{\alpha_0} = (M_{\alpha_0} \cap G_{\alpha_0}) \cup (X - F)$. For $\alpha \neq \alpha_0$, let $N_\alpha = M_\alpha \cap G_\alpha$. Note that $(N_\alpha)_{\alpha \in I}$ is a point-finite open cover of X and that $N_\alpha \cap F = H_\alpha$ for all $\alpha \in I$, hence it is the desired extension of $(H_\alpha)_{\alpha \in I}$.

The proof of 2. ⇒ 1. is clear.

By making a stronger assumption about the space X we can require the extended family to be locally finite on $X - F$.

Corollary 4. *Let X be a hereditary collectionwise normal space. Then every point-finite open cover of finite order of a closed subspace F of X extends to a point-finite open cover of X that is locally finite on $X - F$.*

Proof. The proof follows the proof of Theorem 2 with some modifications. If $\mathscr{H} = (H_\alpha)_{\alpha \in I}$ is a point-finite open cover of finite order of a closed subspace F of X, we will show that there exists a point-finite family $(G_\alpha)_{\alpha \in I}$ of open subsets of X, locally finite on the complement of F, such that $H_\alpha \subset G_\alpha$ for all $\alpha \in I$. By the same construction as in Corollary 3, the family $(G_\alpha)_{\alpha \in I}$ can be modified to give a point-finite open cover of X that is locally finite on $X - F$ and that extends $\mathscr{H}$.

We will indicate the modifications in the proof of Theorem 2 that are necessary. If ord $\mathscr{H} = 1$, then $\mathscr{H}$ is a discrete family of subsets of X. Since X is collectionwise normal, there exists a discrete family $(G_\alpha)_{\alpha \in I}$ of open subsets of X such that $H_\alpha \subset G_\alpha$ for all $\alpha \in I$.

In completing the induction step, note that $\mathscr{K} = (H_J)_{J \in I^*}$ is a pairwise disjoint family of open subsets of F. If $B = \{x \in F \mid \mathscr{K} \text{ is not discrete at } x\}$, then B is closed in X, is contained in F, and $\mathscr{K}$ is a discrete family of subsets of $X - B$. Since X is hereditary collectionwise normal, there exists a discrete (in $X - B$) family $(G_J)_{J \in I^*}$ of open subsets of $X - B$ such that $H_J \subset G_J$ for all $J \in I^*$. Note that each G_J is open in X and $(G_J)_{J \in I^*}$ is discrete on the complement of F. By the induction hypothesis the family $(M_\alpha)_{\alpha \in I}$ can be assumed to be locally finite on $X - F$. Then it is easy to show that $(N_\alpha)_{\alpha \in I}$ is also locally finite on $X - F$. This completes the proof.

From Corollary 4 and Theorem 2 it is clear that every hereditary collectionwise normal space satisfies (α). The next theorem will characterize the extendability of point-finite open covers that do not necessarily have finite order.

Theorem 5. *Let X be a topological space. The following are equivalent:*

1. *The space X satisfies (α), (β) and (γ).*
2. *Given F a closed subspace of X and $(H_\alpha)_{\alpha\in I}$ a point-finite open cover of F, there exists a point-finite family $(G_\alpha)_{\alpha\in I}$ of open subsets of X such that $H_\alpha \subset G_\alpha$ for all $\alpha \in I$.*

Proof. To show 1. $\Rightarrow$ 2., let $(H_\alpha)_{\alpha\in I}$ be a point-finite open cover of F, a closed subspace of X. For each $n \in N$, let $A_n = \{x \in F \mid |K(x)| \leqq n\}$. Then $(A_n)_{n\in N}$ is an increasing closed cover of F, so by (γ), there is an increasing closed cover $(B_n)_{n\in N}$ of X such that $B_n \cap F \subset A_n$ for all $n \in N$. The family $(H_\alpha \cap B_n \cap F)_{\alpha\in I}$ is a point-finite open cover of order n of $B_n \cap F$. By 1. $\Rightarrow$ 2. of Theorem 2, there exists $(G_\alpha^n)_{\alpha\in I}$, a point-finite family of open subsets of X such that $H_\alpha \cap B_n \cap F \subset G_\alpha^n$ for all $\alpha \in I$.

Let $G_\alpha = G_\alpha^1 \cup \bigcup\{G_\alpha^n \cap (X - B_{n-1}) \mid n \geqq 2\}$. Each G_α is clearly open. If x is an element of H_α, let n_0 be the first n such that $x \in B_n$. If $n_0 = 1$, then $x \in H_\alpha \cap B_1 \cap \cap F$, hence $x \in G_\alpha^1$. If $n_0 > 1$, then $x \notin B_{n_0-1}$. But $x \in H_\alpha \cap B_{n_0} \cap F$, so $x \in G_\alpha^{n_0}$, therefore $x \in G_\alpha$. Hence $H_\alpha \subset G_\alpha$ for all $\alpha \in I$.

It remains to show that $(G_\alpha)_{\alpha\in I}$ is point-finite. Let x be an element of X. There exists a finite subset N^* of N such that $x \in X - B_n$ if and only if $n \in N^*$. Given $n \in N$, there exists a finite subset K_n of I such that if $x \in G_\alpha^n$, then $\alpha \in K_n$. Let $K = K_1 \cup \cup \bigcup\{K_{n+1} \mid n \in N^*\}$. Then K is a finite subset of I. If $x \in G_\alpha^1$, then $\alpha \in K_1 \subset K$, and if $x \in G_\alpha^n \cap (X - B_{n-1})$ for some $n \geqq 2$, then $n - 1 \in N^*$, hence $\alpha \in K_n \subseteq K$.

By Theorem 2, we already have that 2. $\Rightarrow$ (α) and (β), hence it remains to show that X satisfies (γ). Given an increasing closed cover $(A_n)_{n\in N}$ of a closed subset F of X, note that $\{F\} \cup (F - A_n)_{n\in N}$ is a point-finite open cover of F. By 2., there exists $(G_n)_{n\in N}$, a point-finite family of open subsets of X such that $F - A_n \subset G_n$ for all $n \in N$. Let $H_n = \bigcup\{G_m \mid m \geqq n\}$, and let $B_n = X - H_n$. It is easy to show that $(B_n)_{n\in N}$ is an increasing closed cover of X such that $B_n \cap F \subset A_n$ for all $n \in N$. Hence X satisfies (α), (β) and (γ), and the proof is completed.

Theorem 6. *Let X be a collectionwise normal, perfectly normal space. Then every point-finite open cover of a closed subspace F of X extends to a point-finite open cover of X that is locally finite on $X - F$.*

Proof. Again, only the necessary modifications of the proof of Theorem 5 will be indicated. Let $(H_\alpha)_{\alpha\in I}$ be a point-finite open cover of F. Since X is perfectly normal, $X - F = CZ(f)$, where $0 \leqq f(x) \leqq 1$, for all $x \in X$. If we let $B_n = \{x \in X \mid f(x) \geqq \geqq 1/n\} \cup A_n$, then $(B_n)_{n\in N}$ is an increasing closed cover of X such that $B_n \cap F \subset A_n$ for all $n \in N$. This choice of $(B_n)_{n\in N}$ should be substituted throughout the proof of Theorem 5. By a result of Hodel [3], every perfectly normal, collectionwise normal space is hereditary collectionwise normal. By virtue of Corollary 4, for each n, the family $(G_\alpha^n)_{\alpha \in I}$ may be assumed to be locally finite on $X - F$. We will show that $(X - B_n)_{n \in N}$ is locally finite on $X - F$. From there it is easy to show that the family $(G_\alpha)_{\alpha \in I}$ constructed in the proof of Theorem 5 is locally finite on $X - F$. If $x \in X - F$, there is a natural number p such that $1/p < f(x)$. If $y \in \{x \in X \mid 1/p < f(x)\} \cap \cap (X - B_n)$, then $1/p < f(y) < 1/n$, hence $n < p$. Hence $(X - B_n)_{n \in N}$ is locally finite on $X - F$. This completes the proof.

Our final result concerns the extension of a point-finite cozero set cover of a closed subspace to a point-finite cozero set cover on the whole space. We need an idea gleaned from Hanner [2].

Lemma 7. *Let F be a closed subspace of a normal space X and let f be a continuous function on F into $[0, 1]$ such that $CZ(f)$ is contained in some open set G of X. Then there is a continuous extension f^* of f to X such that $CZ(f^*)$ is also contained in G.*

Proof. Let g be any continuous extension of f to X with values in $[0, 1]$. Define a function h on $F \cup (X - G)$ as follows: If $x \in F$, then $h(x) = f(x)$, and if $x \in X - G$, then $h(x) = 0$. It is easy to see that h is well-defined and continuous on $F \cup (X - G)$, so h has an extension h^* to X, with values in $[0, 1]$. Then $f^* = g \wedge h^*$ is the desired extension of f.

Theorem 8. *Let X be a collectionwise normal, perfectly normal space. Then every point-finite cozero set cover of a closed subspace F of X extends to a point-finite cozero set cover of X that is locally finite on $X - F$.*

Proof. Let $\mathscr{H} = (CZ(f_\alpha))_{\alpha \in I}$ be a point-finite cozero set cover of F. By Theorem 6, there exists a point-finite open cover $(G_\alpha)_{\alpha \in I}$ of X that extends $\mathscr{H}$ and is locally finite on $X - F$. By Lemma 7, for each $\alpha \in I$ there exists an extension g_α of f_α to X such that $CZ(g_\alpha) \subset G_\alpha$. Since X is perfectly normal, $X - F = CZ(g)$ where $0 \leq g(x) \leq 1$ for all $x \in X$. Fix $\alpha_0 \in I$ and let $f_{\alpha_0}^* = g \vee g_{\alpha_0}$. For $\alpha \neq \alpha_0$, let $f_\alpha^* = g_\alpha$. Then $(CZ(f_\alpha^*))_{\alpha \in I}$ is the desired extension of $\mathscr{H}$.

The author wishes to thank Professor Richard A. Alò, Carnegie-Mellon University, for his assistance.

Professor J. C. Smith has informed the author that condition (α) implies condition (β). Therefore, the statement of Theorem 2 can be improved by deleting condition (β).

References

[1] *R. A. Alò and L. I. Sennott:* Extending linear space-valued functions. Math. Ann. *191* (1971), 79–86.

[2] *O. Hanner:* Retraction and extension of mappings of metric and non-metric spaces. Ark. Mat. *2* (1952), 315–360.

[3] *R. E. Hodel:* Total normality and the hereditary property. Proc. Amer. Math. Soc. *17* (1966), 462–465.

[4] *M. Katětov:* Extension of locally finite covers. (Russian.) Colloq. Math. *6* (1958), 145–151.

GEORGE MASON UNIVERSITY, FAIRFAX, VIRGINIA

A NOTE ON RUDIN'S EXAMPLE OF DOWKER SPACE

P. SIMON

Praha

One assumption on a topological space occurs very frequently in mathematics – the property of being normal and countably paracompact. E.g. in such a space every Baire measure can be extended into a Borel measure [3]. In normal and countably paracompact space realcompactness can be described without (explicit or implicit) use of the notion of zero-set [1].

Last year Mrs. M. E. Rudin gave an example of a normal Hausdorff space Y, which is not countably paracompact ([4], [5], [6]). It seemed quite natural to study some other properties of the space Y in order to show the importance of the assumption of countable paracompactness.

Let us recall the following definitions from [1] and [2].

A topological space will be called *almost realcompact*, iff, whenever $\mathscr{A}$ is a maximal centered collection of open sets such that $\{\bar{A} \mid A \in \mathscr{A}\}$ has the countable intersection property (abbr. CIP), then $\bigcap\{\bar{A} \mid A \in \mathscr{A}\}$ is non-void. A topological space will be called *closed complete*, iff, whenever $\mathscr{A}$ is a maximal centered collection of closed sets with CIP, then $\bigcap\mathscr{A}$ is non-void. A topological space will be called *Baire-Borel complete*, if every maximal centered collection of zero sets $\mathscr{Z}$ with CIP has non-void intersection whenever there exists some maximal centered collection of Borel sets $\mathscr{B}$ with CIP such that $\mathscr{B} \supset \mathscr{Z}$.

Theorem 1. *The space Y is neither almost realcompact nor realcompact.*

Theorem 2. *The space Y is not Baire-Borel complete.*

Theorem 3. *There exists maximal centered collection $\mathscr{Z}$ with* CIP, *consisting of zero-sets in Y, which cannot be extended to maximal centered collection of closed sets with* CIP.

Theorem 4. *The space Y is closed complete.*

References

[1] *Z. Frolík:* On almost realcompact spaces. Bull. Acad. Polon. Sci. Sér. Sci. Math. Astronom. Phys. *9* (4) (1961), 247—250.
[2] *Z. Frolík:* Complete measurable spaces. (To appear.)
[3] *J. Mařík:* The Baire and Borel measures. Czechoslovak Math. J. *7* (*87*) (1957), 248—253.
[4] *M. E. Rudin:* On Dowker space. (Preprint.)
[5] *M. E. Rudin:* A normal space *X* for which $X \times I$ is not normal. Bull. Amer. Math. Soc. *77* (2) (1971), 246.
[6] *M. E. Rudin:* A normal space *X* for which $X \times I$ is not normal. Fund. Math. *63* (2) (1972), 179—186.

CHARLES UNIVERSITY, PRAHA

DIE FORTSETZUNG STETIGER HOMOMORPHISMEN VON δ-HALBGRUPPEN

L. SKULA

Brno

1. Einleitung. In der Zahlentheorie spielt die sog. Divisorentheorie eines Integritätsbereiches eine grosse Rolle. Dieser Begriff wurde von Borewicz und Šafarevič in [2] neuartig und übersichtlich eingeführt. Die äquivalente Formulierung dieses Begriffes ist, wie folgt (Skula [6]):

Unter einer *Divisorentheorie eines Integritätsbereiches* R versteht man einen Homomorphismus h der multiplikativen Halbgruppe R' von dem Integritätsbereich R in eine Halbgruppe $\mathfrak{D}$ mit eindeutiger Primfaktorzerlegung, wobei die folgenden Axiome gelten:

1° $r_1, r_2 \in R' \Rightarrow [r_1 \underset{R'}{/} r_2 \Leftrightarrow h(r_1) \underset{\mathfrak{D}}{/} h(r_2)]$,

2° $\mathfrak{a} \in \mathfrak{D} \Rightarrow \mathfrak{a}$ ist der grösste gemeinsame Teiler der Menge $\{h(r) : r \in R', \mathfrak{a} \underset{\mathfrak{D}}{/} h(r)\}$.

(Das Symbol / bezeichnet die Teilbarkeitsrelation.)

Den Begriff der Divisorentheorie können wir auf die kommutativen Halbgruppen mit Einselement und mit Kürzungsregel gleicherweise übertragen. Clifford ([3]) nennt diese Divisorentheorie „normal ideal arithmetics". Wir werden diese Halbgruppen mit einer Divisorentheorie studieren und wir beschränken uns auf die Halbgruppen, die ausser dem Einselement keine Einheit besitzen. Solche Halbgruppen nennen wir δ-Halbgruppen. Unter diesen Voraussetzungen ist der Homomorphismus h in der Definition der Divisorentheorie ein Isomorphismus, den wir als eine identische Einbettung betrachten und den wir mit e_G bezeichnen. Die Halbgruppe $\mathfrak{D}$ bezeichnen wir mit $\mathfrak{c}G$.

Für jede δ-Halbgruppe $\mathfrak{c}G$ gilt also:

1° $g_1, g_2 \in G \Rightarrow [g_1 \underset{G}{/} g_2 \Leftrightarrow e_G(g_1) \underset{\mathfrak{c}G}{/} e_G(g_2) \Leftrightarrow g_1 \underset{\mathfrak{c}G}{/} g_2]$,

2° $\mathfrak{a} \in \mathfrak{c}G \Rightarrow \mathfrak{a}$ ist der grösste gemeinsame Teiler der Menge $\{e_G(g) : g \in G, \mathfrak{a} \underset{\mathfrak{c}G}{/} e_G(g)\} = \{g \in G : \mathfrak{a} \underset{\mathfrak{c}G}{/} g\}$.

2. Die δ-Topologie. Eine nichtleere Untermenge I einer δ-Halbgruppe G heisst ein *δ-Ideal (der Halbgruppe G)*, wenn folgendes gilt:

$$I = \{g \in G : (g_1, g_2 \in G, g_1 \mid ig_2 \text{ für jedes } i \in I) \Rightarrow g_1 \mid gg_2\} .$$

(Vgl. Prüfer [5].)

Die Menge aller δ-Ideale von G bezeichnen wir mit $\mathfrak{J}(G)$. Diese Menge ist durchschnittsabgeschlossen und erzeugt gemeinsam mit der leeren Menge als System aller abgeschlossenen Mengen eine Topologie auf G, die wir mit δ_G bezeichnen. Diese Topologie ist eine U-Topologie in dem Sinne von Čech ([4]). Diese ist sogar eine KU-Topologie, aber sie ist allgemein weder eine A-Topologie (additive Topologie) noch eine B-Topologie ($\delta_G\{g\} \neq \{g\}$ für $g \in G$). Die Operation auf G ist bei dieser Topologie stetig.

Wenn wir jetzt die Kategorie $\mathscr{K}$ aller δ-Halbgruppen mit δ-stetigen Homomorphismen betrachten, dann erzeugen alle Halbgruppen mit eindeutiger Primfaktorzerlegung eine reflektive Unterkategorie $\mathscr{D}$ von $\mathscr{K}$, und $e_G : G \to \mathfrak{c}G$ ist die $\mathscr{D}$-Reflexion von G. Der Morphismus e_G ist ein Epimorphismus, also ist $\mathscr{D}$ eine epireflektive Unterkategorie.

Das heisst: Jeder δ-stetiger Homomorphismus f einer δ-Halbgruppe G in eine Halbgruppe $\mathfrak{D}$ mit eindeutiger Primfaktorzerlegung ist eindeutig auf einen δ-stetigen Homomorphismus $\bar{f}$ von $\mathfrak{c}G$ fortsetzbar:

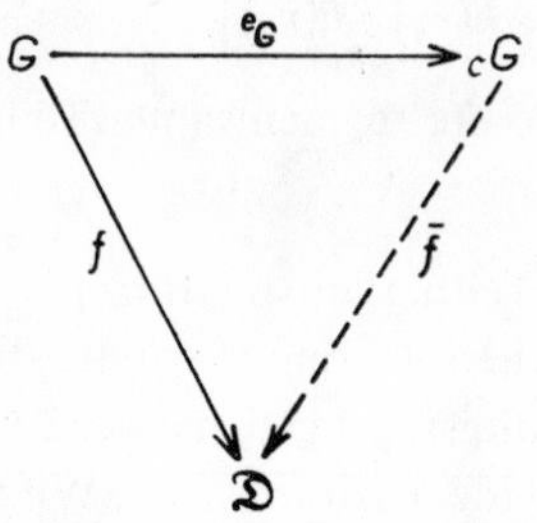

Wenn f ein Homomorphismus von $\mathfrak{D}_1$ in $\mathfrak{D}_2$ ist, wobei $\mathfrak{D}_1, \mathfrak{D}_2 \in \mathscr{D}$ sind, dann ist f genau dann eine δ-stetige Abbildung, wenn es für jede Untermenge $\emptyset \neq M \subseteq$ $\subseteq \mathfrak{D}_1$ folgendes gilt:

$$f(D(M)) = D(f(M)) .$$

(D bezeichnet den grössten gemeinsamen Teiler.)

3. Die δ^*-Topologie. Die A-Modifikation (d.h. die additive Modifikation) der Topologie δ_G auf einer δ-Halbgruppe G heisst δ_G^*-Topologie. Diese Topologie ist eine T_0-Topologie und die Operation auf der Halbgruppe G ist bei dieser Topologie stetig.

Analogische Resultate wie für die Kategorie der δ-Halbgruppen, gelten auch für die Kategorie der δ_1-Halbgruppen.

Eine δ_1-*Halbgruppe* G ist eine δ-Halbgruppe, wenn folgendes Axiom gilt:

$$\mathfrak{a}_1, \mathfrak{a}_2 \in \mathfrak{c}G \Rightarrow \text{es gibt } \mathfrak{i} \in \mathfrak{c}G \text{ mit } (\mathfrak{a}_2, \mathfrak{i}) = 1_{\mathfrak{c}G} \text{ und } \mathfrak{a}_1 \,.\, \mathfrak{i} \in G\,.$$

(Vgl. Arnold [1].)

Wir bezeichnen mit $\mathcal{K}_1$ die Kategorie aller δ_1-Halbgruppen mit δ^*-stetigen Homomorphismen. *Dann ist auch* $\mathcal{D}$, *die Kategorie aller Halbgruppen mit eindeutiger Primfaktorzerlegung mit* δ^*-*stetigen Homomorphismen, eine epireflexive Unterkategorie von* $\mathcal{K}_1$ *und* $e_G : G \to \mathfrak{c}G$ *is die* $\mathcal{D}$-*Reflexion von* G.

Für einen Homomorphismus f *von* $\mathfrak{D}_1$ *in* $\mathfrak{D}_2$, *wobei* $\mathfrak{D}_1, \mathfrak{D}_2 \in \mathcal{D}$ *ist, gilt:* f *ist genau dann eine* δ^*-*stetige Abbildung, wenn die Menge*

$$\{\mathfrak{p} \textit{ ist ein Primelement von } \mathfrak{D}_1 : \mathfrak{d} \mid f(\mathfrak{p})\}$$

stets endlich für jedes Element $\mathfrak{d} \in \mathfrak{D}_2$, $\mathfrak{d} \neq 1_{\mathfrak{D}_2}$, *ist.*

4. Anwendung an die Ringtheorie. Unter der multiplikativen Halbgruppe eines Integritätsbereiches R verstehen wir die übliche multiplikative Halbgruppe von R, in der wir die assoziierten Elemente identifizieren. Die multiplikative Halbgruppe $\bar{R}$ eines Integritätsbereiches R mit einer Divisorentheorie ist eine δ_1-Halbgruppe (Skula [6]).

Wenn k der Quotientenkörper von R ist und K/k eine endliche Erweiterung ist, so besitzt die ganzabgeschlossene Hülle S von R in K eine Divisorentheorie, also ist die multiplikative Halbgruppe $\bar{S}$ von S eine δ_1-Halbgruppe (Borewicz, Šafarevič [2]).

Aus den Resultaten in [2] folgt: *Die von der Norm von* K *in* k *erzeugte Abbildung* N *von* $\bar{S}$ *in* $\bar{R}$ *ist ein* δ^*-*stetiger Homomorphismus, der sich eindeutig auf einen* δ^*-*stetigen Homomorphismus* $\bar{N}$ *von* $\mathfrak{c}\bar{S}$ *auf* $\mathfrak{c}\bar{R}$ *fortsetzen lässt:*

$$\begin{array}{ccc} \bar{S} & \xrightarrow{\;N\;} & \bar{R} \\ {\scriptstyle e_{\bar{S}}}\downarrow & & \downarrow{\scriptstyle e_{\bar{R}}} \\ \mathfrak{c}\bar{S} & \underset{\bar{N}}{\dashrightarrow} & \mathfrak{c}\bar{R} \end{array}$$

Auch *die schlichten Homomorphismen von den Dedekindschen Ringen erzeugen* δ^*-*stetige Homomorphismen ihrer multiplikativen Halbgruppen.*

Literatur

[1] *I. Arnold:* Ideale in kommutativen Halbgruppen. Rec. Math. Moscow *36* (1929), 401—407.

[2] *S. I. Borewicz und I. R. Šafarevič:* Zahlentheorie. Deutsche Übersetzung. Basel und Stuttgart. Birkhäuser, 1966.

[3] *A. H. Clifford:* Arithmetic and ideal theory of commutative semigroups. Ann. of Math. *39* (1938), 594—610.

[4] *E. Čech:* a) Topological papers of Eduard Čech. Academia, Prague, 1968.
b) Topological spaces. Academia, Prague, 1966.

[5] *H. Prüfer:* Untersuchungen über die Teilbarkeitseigenschaften in Körpern. J. Reine Angew. Math. *168* (1932), 1—36.

[6] *L. Skula:* Divisorentheorie einer Halbgruppe. Math. Z. *14* (1970), 113—120.

J. E. PURKYNĚ UNIVERSITY, BRNO

PROPERTIES OF EXPANDABLE SPACES

J. C. SMITH

Blacksburg

Introduction. In [7] J. C. Smith and L. L. Krajewski investigated the properties of spaces in which locally finite collections can be expanded to point-finite or locally finite open collections. A number of results concerning these "expandable" spaces were obtained as well as the relationship of this class of spaces with the class of collectionwise normal spaces. In this paper we investigate the properties of the class of spaces in which locally finite collections can be expanded to point-finite or locally finite cozero collections. In § 1 results analogous to those in [7] are obtained for "cozero-expandable" spaces. In § 2 it is shown that every expandable screenable space is paracompact. In § 3 we observe that a class of spaces weaker than the class of collectionwise normal spaces has properties similar to the class of collectionwise normal spaces.

1. Definitions and preliminary results

Definition 1.1. A space X is called *cz-expandable* if for every locally finite collection $\{F_\alpha : \alpha \in A\}$ of subsets of X, there exists a locally finite collection $\{G_\alpha : \alpha \in A\}$ of cozero subsets of X, such that $F_\alpha \subseteq G_\alpha$ for all $\alpha \in A$. We will refer to collections with the above property as being "cozero expandable".

Remark. The above definition is analogous to the definitions of various "expandable" spaces found in [7]. By restricting the cardinality of the index set A, analogous cardinality-dependent definitions can be obtained. These definitions are omitted. The reader is referred to [7] for the definitions of expandable spaces, almost expandable spaces, discretely expandable spaces, etc.

Definition 1.2. (i) A space X is called (*almost*) *cz-expandable* if every locally finite collection of subsets of X can be expanded to a (point-finite) locally finite cozero collection.

(ii) A space X is called (*almost*) *discretely cz-expandable* if every discrete collection of subsets of X can be expanded to a (point-finite) locally finite cozero collection.

(iii) A space X is called *(almost) boundedly cz-expandable* if every bounded locally finite collection of subsets of X can be expanded to a (point-finite) locally finite cozero collection.

The following two theorems are obvious from the results in [7] and the fact that the locally finite union of cozero sets is a cozero set.

Theorem 1.3. *Let X be a normal space. Then*

(i) *X is (almost) expandable if and only if X is (almost) cz-expandable.*

(ii) *X is (almost) discretely expandable if and only if X is (almost) discretely cz-expandable.*

(iii) *X is (almost) boundedly expandable if and only if X is (almost) boundedly cz-expandable.*

Theorem 1.4. (i) *A space X is almost discretely expandable if and only if X is almost boundedly expandable.*

(ii) *A space X is (almost) discretely cz-expandable if and only if X is (almost) boundedly cz-expandable.*

Theorem 1.5. (Morita and Dowker). *Let $\{G_1, G_2, \ldots\}$ be a countable collection of cozero subsets of X and $G = \bigcup_{i=1}^{\infty} G_i$. Then there exists a collection of cozero subsets $\{H_1, H_2, \ldots\}$ such that $G = \bigcup_{i=1}^{\infty} H_i$ and,*

(i) *$H_i \subseteq G_i$ for each i,*

(ii) *$\{H_1, H_2, \ldots\}$ is locally finite with respect to points in G. That is, for each $x \in G$, there exists an open (in X) neighborhood of x which intersects only finitely many members of $\{H_1, H_2, \ldots\}$.*

Proof. The proof is found in Theorem 2.7 of [5].

We now obtain a result analogous to Theorem 2.8 of [7].

Theorem 1.6. *A space X is cz-expandable if and only if X is discretely cz-expandable and $\aleph_0$-cz-expandable.*

Proof. The sufficiency is clear.

Necessity. Let $\mathscr{F} = \{F_\alpha : \alpha \in A\}$ be a locally finite collection of closed subsets of X. For each integer $n \geqq 0$, define $S_n = \{x \in X : \operatorname{ord}(x, \mathscr{F}) \leqq n\}$ so that $\{S_0, S_1, \ldots\}$ is an open cover of X. Since X is countably paracompact and $S_n \subseteq S_{n+1}$ for each n, there exists by Theorem 5 of [2] a locally finite open cover $\{U_0, U_1, \ldots\}$ of X, such that $U_n \subseteq \overline{U}_n \subseteq S_n$. Now define $\mathscr{F}_n = \{\overline{U}_n \cap F_\alpha : \alpha \in A\} = \{F(n, \alpha) : : \alpha \in A\}$ for each n, so that $\mathscr{F}_n$ is a bounded locally finite collection in X. By Theorem

1.4 above there exists for each n, a locally finite cozero collection $\mathcal{G}_n = \{G(n, \alpha) : : \alpha \in A\}$ such that $F(n, \alpha) \subseteq G(n, \alpha)$ for each $\alpha \in A$. Furthermore, since X is $\aleph_0$-*cz*-expandable, there exists a locally finite cozero collection $\{V_0, V_1, \ldots\}$ such that $\bar{U}_n \subseteq V_n$. Now define $G^*(n, \alpha) = G(n, \alpha) \cap V_n$ and let $G_\alpha^* = \bigcup_{n=0}^{\infty} G^*(n, \alpha)$ for each $\alpha \in A$. Then $\{G_\alpha^* : \alpha \in A\}$ is a locally finite cozero collection such that $F_\alpha \subseteq G_\alpha^*$ for each $\alpha \in A$. Hence X is *cz*-expandable.

2. Subset theorems and screenable spaces

Theorem 2.1. *The Closed Subset Theorem holds for the following properties:*

(i) *cz-expandable,*
(ii) *almost cz-expandable,*
(iii) *discretely cz-expandable,*
(iv) *almost discretely cz-expandable.*

Proof. The proofs follow in the same manner as those in Theorem 7.2 of [7].

In [7] the authors left the following question open: Are expandable screenable spaces paracompact? We now answer this question in the affirmative.

Theorem 2.2. *Every expandable screenable space is paracompact.*

Proof. Let $\mathcal{G}$ be an open cover of X. Then $\mathcal{G}$ has a σ-disjoint open refinement $\mathcal{U} = \bigcup_{i=1}^{\infty} \mathcal{U}_i$, where $\mathcal{U}_i = \{U_\alpha : \alpha \in A_i\}$. Define $U_i = \bigcup_{\alpha \in A_i} U_\alpha$ so that $\{U_1, U_2, \ldots\}$ is a countable open cover of X. Since X is countably paracompact, there exists a locally finite open refinement $\{V_1, V_2, \ldots\}$. Now let $\mathcal{V}_i = \{V_i \cap U_\alpha : \alpha \in A_i\}$ and $\mathcal{V} = \bigcup_{i=1}^{\infty} \mathcal{V}_i$. Then $\mathcal{V}$ is a point-finite open refinement of $\mathcal{G}$ so that X is metacompact. As consequence of Theorem 4.2 of [7], X is paracompact.

Remark. It is well known that locally finite cozero covers are normal. In fact if X is normal, then every countable point-finite open cover is normal. The following result illustrates a somewhat unusual property enjoyed by discretely *cz*-expandable spaces.

Definition 2.3. A cover $\mathcal{G}$ of a space X is called a *cz-θ-refinable cover* if $\mathcal{G}$ has a refinement $\mathcal{U} = \bigcup_{i=1}^{\infty} \mathcal{U}_i$ satisfying:

(i) $\mathcal{U}_i$ is a cozero cover of X.

(ii) For $x \in X$ there exists an integer $n(x)$ such that x belongs to only finitely many members of $\mathcal{U}_{n(x)}$.

Theorem 2.4. *Let X be discretely cz-expandable. Then every cz-θ-refinable cover of X is normal.*

Proof. Let $\mathscr{G}$ be an open cover of X and $\mathscr{U} = \bigcup_{i=1}^{\infty} \mathscr{U}_i$ be a refinement of $\mathscr{G}$ satisfying the above properties. For each i we construct a sequence $\{\mathscr{G}(i, j) : j = 0, 1, \ldots\}$ of open collections such that

(1) $\mathscr{G}(i, j)$ is a locally finite cozero collection for each j.

(2) Each member of $\mathscr{G}(i, j)$ is a subset of some member of $\mathscr{U}_i$.

(3) If $x \in X$ and x belongs to at most j members of $\mathscr{U}_i$, then $x \in \bigcup\{G \in \mathscr{G}(i, k) : 0 \leq k \leq j\}$.

(4) If $x \in \bigcup\{G : G \in \mathscr{G}(i, j)\}$ then x belongs to at least j members of $\mathscr{U}_i$.

The proof is by induction on j. Define $\mathscr{G}(i, 0) = \emptyset$ and assume $\mathscr{G}(i, j)$ has been constructed satisfying (1)–(4) above for $0 \leq j \leq n$. We now construct $\mathscr{G}(i, n+1)$.

Let $\mathscr{U}_i = \{U_\alpha : \alpha \in A_i\}$, $\mathscr{B} = \{B \subseteq A_i : |B| = n+1\}$ and $G(i, j) = \bigcup\{G : G \in \mathscr{G}(i, j)\}$. Define $F(B) = [X - \bigcup_{j=0}^{n} G(i, j)] \cap [X - \bigcup\{U_\alpha : \alpha \in A_i - B\}]$ for each $B \in \mathscr{B}$. Then $\mathscr{F} = \{F(B) : B \in \mathscr{B}\}$ is a closed collection. We assert that $\mathscr{F}$ is discrete. Let $x \in X$.

Case 1: x belongs to more than $n+1$ members of $\mathscr{U}_i$, say $U(i, 1), U(i, 2), \ldots, U(i, k)$. Then $U(x) = \bigcap_{j=1}^{k} U(i, j)$ is an open neighborhood of x which intersects no member of $\mathscr{F}$.

Case 2: x belongs to less than $n+1$ members of $\mathscr{U}_i$. From (3) above $x \in \bigcup\{G \in \mathscr{G}(i, j) : 0 \leq j \leq n\}$ which intersects no member of $\mathscr{F}$.

Case 3: x belongs to exactly $n+1$ members of $\mathscr{U}_i$; say $\{U_{\alpha_k} : k = 1, 2, \ldots, n+1\}$ Then $U(x) = \bigcap_{k=1}^{n+1} U_{\alpha_k}$ intersects no member of $\mathscr{F}$ other than $F(B)$ where $B = \{\alpha_1, \alpha_2, \ldots, \alpha_{n+1}\}$.

Since X is discretely cz-expandable, there exists a locally finite cozero collection $\mathscr{K} = \{K(B) : B \in \mathscr{B}\}$ such that $F(B) \subseteq K(B)$ for each $B \in \mathscr{B}$. Now define $L(B) = K(B) \cap [\bigcap_{\alpha \in B} U_\alpha]$ and $\mathscr{G}(i, n+1) = \{L(B) : B \in \mathscr{B}\}$. It is clear that (1), (2), and (4) above are satisfied by $\mathscr{G}(i, n+1)$. To show (3), let $x \in X$ such that x belongs to at most $n+1$ members of $\mathscr{U}_i$. If $x \notin \bigcup_{j=0}^{\infty} G(i, j)$, then $x \in F(B)$ for some $B \in \mathscr{B}$ and hence belongs to some member of $\mathscr{G}(i, n+1)$. Now define $H(i, j) = \bigcup\{G : G \in \mathscr{G}(i, j)\}$ so that $\mathscr{H} = \{H(i, j) : i = 1, 2, \ldots; j = 0, 1, \ldots\}$ is a countable cozero cover of X. By Theorem 1.5 above $\mathscr{H}$ has a locally finite cozero refinement $\mathscr{H}^* = \{H^*(i, j) : i = 1, 2, \ldots; j = 0, 1, \ldots\}$ such that $H^*(i, j) \subseteq H(i, j)$. Let $\mathscr{G}^*(i, j) = \{G \cap H^*(i, j) : G \in \mathscr{G}(i, j)\}$

for each i and each j. Then $\mathscr{G}^* = \bigcup_{i=1}^{\infty} \bigcup_{j=0}^{\infty} \mathscr{G}^*(i,j)$ is a locally finite cozero refinement of $\mathscr{G}$. Since $\mathscr{G}^*$ is normal, $\mathscr{G}$ is normal.

Corollary 2.5. *Let X be a discretely cz-expandable space. Then every point-finite cozero cover of X is normal.*

3. *PF*-normal spaces

Definition 3.1. A space X is called *PF-normal* if every point-finite open cover is normal.

Theorem 3.2. *Collectionwise normal ⇒ PF-normal ⇒ normal.*

Proof. The proof of the first implication is Theorem 2 of [3] and the proof of the second is obvious.

Remark. The following examples show that neither of the reverse implications above is true.

Example 3.3. The space F of Bing [1] is *PF*-normal but not collectionwise normal. Michael [3] has shown that this space has the property that every point-finite open cover has a locally finite open refinement and hence is normal.

Example 3.4. Michael's modification of Bing's example G is normal but not *PF*-normal. Otherwise G would be fully normal and hence collectionwise normal.

The above examples show that closed subsets of *PF*-normal spaces need not have this property. It should be noted however that *PF*-normal spaces have many of the properties enjoyed by collectionwise normal spaces.

Theorem 3.5. *Let X be a PF-normal space. Then*

(i) *X is paracompact if and only if X is metacompact.*
(ii) *X is expandable if and only if X is almost expandable.*
(iii) *X is discretely expandable if and only if X is almost discretely expandable.*

Proof. Part (i) is obvious. Part (iii) follows from the same argument as does part (ii). We now prove part (ii). Let $\mathscr{F} = \{F_\alpha : \alpha \in A\}$ be a locally finite collection of closed subsets of X. Since X is almost expandable, there exists a point-finite collection $\mathscr{G} = \{G_\alpha : \alpha \in A\}$ of open subsets of X such that $F_\alpha \subseteq G_\alpha$ for all $\alpha \in A$. Since $\mathscr{G}^* = \mathscr{G} \cup \{X - \bigcup_{\alpha \in A} F_\alpha\}$ is a point-finite open cover of X, $\mathscr{G}^*$ is normal. Hence there exists a locally finite open cover $\mathscr{V} = \{V_\delta : \delta \in D\}$ such that $\{\mathrm{St}(V_\delta, \mathscr{V}) : \delta \in D\}$

refines $\mathscr{G}^*$. Define $H_\alpha = \mathrm{St}\,(F_\alpha, \mathscr{V})$ for all $\alpha \in A$. Then $\mathscr{H} = \{H_\alpha : \alpha \in A\}$ is a locally finite open collection such that $F_\alpha \subseteq H_\alpha$ for all $\alpha \in A$. Therefore X is expandable.

Corollary 3.6. *A space X is PF-normal and countably paracompact if and only if every σ-point-finite open cover is normal.*

References

[1] *R. H. Bing:* Metrization of topological spaces. Canad. J. Math. *3* (1951), 175—186.
[2] *J. Mack:* Directed covers and paracompact spaces. Canad. J. Math. *19* (1967), 649—654.
[3] *E. Michael:* Point-finite and locally finite coverings. Canad. J. Math. *7* (1955), 275—279.
[4] *K. Morita:* Star-finite coverings and the star-finite property. Math. Japon. *1* (1948), 60—68.
[5] *K. Nagami:* Dimension Theory. Academic Press, 1970.
[6] *H. L. Shapiro:* Extensions of pseudometrics. Canad. J. Math. *18* (1966), 981—998.
[7] *J. C. Smith and L. L. Krajewski:* Expandability and collectionwise normality. Trans. Amer. Math. Soc. *160* (1971), 437—451.

VIRGINIA POLYTECHNIC INSTITUTE AND STATE UNIVERSITY, BLACKSBURG, VIRGINIA

ON THE LATTICE OF TOPOLOGIES

A. K. STEINER

Edmonton

The family Σ of all topologies on a set X forms a lattice under the partial ordering of inclusion. The largest element, 1, is the discrete topology and the smallest element, 0, is the trivial topology. Since the intersection of any family of topologies on X is again a topology on X, Σ is a complete lattice.

The topologies of the form $\{X, E, \emptyset\}$ for $\emptyset \neq E \neq X$ are the atoms and every topology on X, different from 0, is the supremum of atoms.

The topologies $\tau(x, \mathscr{U}) = \{E \subset X : x \notin E \text{ or } E \in \mathscr{U}\}$, where $x \in X$ and $\mathscr{U}$ is an ultrafilter different from the fixed ultrafilter at x, are co-atoms of Σ and in 1964, O. Fröhlich [12] showed that co-atoms also generate Σ. The co-atoms, called *ultraspaces* by Fröhlich, fall into two classes: the *principal ultraspaces,* when $\mathscr{U}$ is a principal (or fixed) ultrafilter on X, and the *nonprincipal* (or T_1) *ultraspaces,* when $\mathscr{U}$ is a nonprincipal (or free) ultrafilter.

The nonprincipal ultraspaces generate a complete sublattice Λ of Σ which is the lattice of T_1-topologies. Λ has 1 as the finest element and the cofinite topology, $\mathscr{C}$ as the coarsest.

The principal ultraspaces generate a sublattice Π of Σ of *principal topologies.* The elements of Π are characterized by the property that arbitrary intersections of open sets are open. Π is a complete lattice sharing the largest and smallest elements with Σ . Π is a meet-complete sublattice of Σ but is not a complete sublattice since the atoms of Σ are in Π [27].

A topology on X which is neither T_1 nor principal is a *mixed topology.* A mixed topology can be represented as the infimum of a T_1-topology and a principal topology, but this representation need not be unique. The supremum of two mixed topologies can be either T_1, principal, or mixed; the infimum however, can never be T_1. Other than this, not much is known about mixed topologies.

In 1958, Hartmanis [17] proved that the lattice of topologies on a finite set is complemented. Since every topology on a finite set is principal, $\Sigma = \Pi$. From the fact that Π is generated by principal ultraspaces, a simple proof shows that Π is always complemented, regardless of the cardinality of the set X.

Theorem 1. Π *is a complemented lattice.*

Proof. For $\tau \in \Pi$, let $\mathscr{D}$ be the decomposition of X defined by: $x, y \in D \in \mathscr{D}$ if and only if either $\tau \leqq \tau(x, \mathscr{U}(y))$ or $\tau \leqq \tau(y, \mathscr{U}(x))$. Choose one element x_D from each $D \in \mathscr{D}$. If $\tau_1 = \bigwedge\{\tau(x_D, \mathscr{U}(x_E)) : D, E \in \mathscr{D}\}$ and τ_2 is the infimum of all ultraspaces $\tau(x, \mathscr{U}(y))$ for which $\tau \leqq \tau(y, \mathscr{U}(x))$ but $\tau \nleqq \tau(x, \mathscr{U}(y))$, then $\tau_1 \wedge \tau_2$ is a complement for τ in Π.

In 1961, Gaifman [13], [14], showed that Σ is complemented if X is a countable set. To obtain this result, he proved that if every T_1-topology on a set has a complement, then every topology has one.

In 1966, the author [27] generalized Gaifman's result, and with the following theorems proved that Σ is a complemented lattice.

Theorem 2. *If every T_1-topology on a set X has a principal complement (i.e., a complement lying in Π), then every topology on X has a principal complement.*

Theorem 3. *If every T_1-topology with no isolated points has a principal complement, then every T_1-topology has a principal complement.*

Theorem 4. *A T_1-topology with no isolated points has a principal complement.*

Theorem 5. *The lattice Σ is complemented. Moreover, each topology has a principal complement.*

In 1968, Van Rooij [31] independently proved that Σ is complemented, using ideas similar to those in the proof of Theorem 4. His work did not depend upon the ultraspaces, nor upon Gaifman's work. Briefly, Van Rooij first proved that if a topological space can be well-ordered so that initial segments are closed, then the topology has a complement. Then, if τ is any topology on X, X can be inductively decomposed into a well-ordered sequence of subsets $\{X_\alpha\}_{\alpha<\gamma}$, so that for each $\alpha < \gamma$, X_α is dense in $\bigcup\{X_\beta : \alpha \leqq \beta < \gamma\}$ and each X_α is itself a well-ordered set with initial segments closed in the induced topology $\tau|X_\alpha$. Each $\tau|X_\alpha$ has a complement and these are used to define a complement for τ. In going through the construction of the complements, it is easy to see that they are all principal topologies.

Complementation in Σ is by no means unique, as has been shown by P. Schnare [24], [25]. He proved

Theorem 6 (Schnare [25]). *Every proper topology on an infinite set X has at least $|X|$ complements and at most $2^{2^{|X|}}$ complements ($2^{|X|}$ principal complements). Moreover, these bounds are the best possible.*

The lattice Λ of T_1-topologies is not complemented as can be seen from a very simple counterexample [28].

Example. Let (X_1, τ_1) be an infinite set with the cofinite topology, (X_2, τ_2) an infinite set with the discrete topology and let (X, τ) be the topological sum of X_1 and X_2. Assume τ has a complement τ' in Λ. For each $x \in X$, $\{x\} \in \tau \vee \tau'$. If $\{x\} \in \tau'$ for all $x \in X_1$, then $X_1 \in \tau \wedge \tau'$, but X_1 is not cofinite. Thus there must be an $x \in X_1$ such that $\{x\} \notin \tau'$. But $\{x\} = U \cap V$ where $U \in \tau'$ and $V \in \tau$. Since $X_1 - V$ is finite, $U \cap X_1$ must be finite. But τ' is a T_1-topology, so there is a $U^* \in \tau'$ such that $\emptyset \neq U^* \subset \subset U \cap X_2$. Thus, $U^* \in \tau \wedge \tau'$ but is not cofinite. Thus, if $\tau \vee \tau' = 1$, $\tau \wedge \tau' \neq \mathscr{C}$.

However, many T_1-topologies do have complements in Λ: the hyperplanes of Bagley [4]; the nonprincipal ultraspaces and their finite intersections; the order topology on a well-ordered set, [29]; and the usual topology on the reals [30].

B. A. Anderson [1], [2], [3] has used the technique of the author and E. F. Steiner for providing a complement for the reals, and has found large classes of T_1-topologies with T_1-complements. He also has some bounds on the number of T_1-complements.

Theorem 7 (Anderson [1]). *If X is an infinite set, there is a family $L \subset \Lambda$ such that $|X| \leq L \leq 2^{|X|}$ and any two elements of L are complementary.*

The same result also holds for Σ and Π.

If τ and τ' are T_1-complements, knowledge about τ gives almost none about τ', as Anderson has shown.

Theorem 8 (Anderson [1]). *Every set of cardinal c has a T_1-topology τ such that for any T_1-topology σ on a set S of cardinality c, τ has a T_1-complement with a subspace homeomorphic to (S, σ). An analogous statement holds in Σ.*

No characterization has as yet been given for those topologies in Λ which do not have T_1-complements.

Ultraspaces may be studied easily because of their point-ultrafilter representation. Many topological properties of ultraspaces have been studied by the author [27] and J. Girhiny [15].

Theorem 9. *For an ultraspace $\tau(x, \mathscr{U})$ the following are equivalent:*

(a) *$\mathscr{U}$ is nonprincipal,*
(b) *$\tau(x, \mathscr{U})$ satisfies T_1 to T_5,*
(c) *$\tau(x, \mathscr{U})$ is totally disconnected,*
(d) *$\tau(x, \mathscr{U})$ is zero-dimensional,*
(e) *$\tau(x, \mathscr{U})$ is regular (completely regular).*

Theorem 10. *For any ultraspace $\tau(x, \mathscr{U})$, the following are equivalent:*

(a) *$\mathscr{U}$ is principal,*
(b) *$\tau(x, \mathscr{U})$ is locally compact,*
(c) *$\tau(x, \mathscr{U})$ is locally connected,*
(d) *$\tau(x, \mathscr{U})$ satisfies the first axiom of countability.*

Those topologies in Σ which are maximal P or minimal P, for various topological properties P, have frequently been studied [5], [7], [9], [10], [18], [20], [21], [22], [23], [26] and many have been characterized. Most of these characterizations involve the representation of the space as an infimum of ultraspaces.

Theorem 11 ([27]). *A space is maximal regular if and only if it is a nonprincipal ultraspace or is of the form* $\tau(x, \mathcal{U}(y)) \wedge \tau(y, \mathcal{U}(x))$ *for* $x \neq y$.

A comprehensive list has been given by J. Girhiny [15] for twenty-seven such properties. Two interesting results of his are:

Theorem 12 (Girhiny [15]). *There are no maximal second countable spaces.*

Theorem 13 (Girhiny [15]). *If* τ *is first countable and* $\tau < \tau(x, \mathcal{U})$ *with* $\mathcal{U}$ *nonprincipal, then there is a* τ' *which is first countable and* $\tau < \tau' < \tau(x, \mathcal{U})$.

The characterizations of maximal locally compact, maximal connected, and minimal totally disconnected topologies are not known.

Girhiny [16] has also investigated when a topological property is preserved by the lattice operations of coarser and finer, and by finite and infinite meets and joins.

Although not purely a lattice problem, the question arises as to when the property of a space being minimal P is preserved under products. Since minimal $T_{31/2}$ (T_4) spaces are compact Hausdorff [6], minimality is always preserved by products. That the product of minimal Hausdorff spaces is minimal Hausdorff has been proved in [11], [18], and [19]. The question as to whether the product of minimal T_3 spaces is minimal T_3 has not been answered.

References

[1] *B. A. Anderson:* Families of mutually complementary topologies. Proc. Amer. Math. Soc. *29* (1971), 362–368.

[2] *B. A. Anderson:* A class of topologies with T_1-complements. Fund. Math. *69* (1970), 267–277.

[3] *B. A. Anderson and D. G. Stewart:* T_1-complements of T_1 topologies. Proc. Amer. Math. Soc. *23* (1969), 77–81.

[4] *R. W. Bagley:* On the characterization of the lattice of topologies. J. London Math. Soc. *30* (1955), 247–249.

[5] *V. K. Balachandran:* Minimal bicompact spaces. J. Indian Math. Soc. *12* (1948), 47–48.

[6] *M. P. Berri:* Minimal topological spaces. Trans. Amer. Math. Soc. *108* (1963), 97–105.

[7] *M. P. Berri:* Categories of certain minimal topological spaces. J. Austral. Math. Soc. *4* (1964), 78–82.

[8] *M. P. Berri:* The complement of a topology for some topological groups. Fund. Math. *58* (1966), 159–162.

[9] *M. P. Berri and R. H. Sorgenfrey:* Minimal regular spaces. Proc. Amer. Math. Soc. *14* (1963), 454—458.

[10] *N. Bourbaki:* Espaces minimaux et espaces complément séparés. C. R. Acad. Sci. Paris Sér. A—B *205* (1941), 215—218.

[11] *D. Doitchinov:* Un théoréme sur les espaces minimaux. Bull. Sci. Math. *93* (1969), 33—36.

[12] *O. Frölich:* Das Halbordnungssystem der Topologischen Räume auf einer Menge. Math. Ann. *156* (1964), 79—95.

[13] *H. Gaifman:* The lattice of all topologies on a denumerable set. Notices Amer. Math. Soc. *8* (1961), 356.

[14] *H. Gaifman:* Remarks on complementation in the lattice of all topologies. Canad. J. Math. *18* (1966), 83—88.

[15] *J. Girhiny:* Minimal and maximal topologies. McMaster Math. Report No. 32, McMaster Univ., Hamilton, Ont., 1970.

[16] *J. Girhiny:* Preservation of topological properties under lattice operations and relations. McMaster Math. Report No. 33, McMaster Univ., Hamilton, Ont., 1970.

[17] *J. Hartmanis:* On the lattice of topologies. Canad. J. Math. *10* (1958), 547—553.

[18] *H. Herrlich:* T_v-Abgeschlossenheit und T_v-Minimalität. Math. Z. *88* (1965), 285—294.

[19] *S. Ikenaga:* Product of minimal topological spaces. Proc. Japan Acad. *40* (1964), 329—331.

[20] *R. E. Larson:* Minimal T_0-spaces and minimal T_D-spaces. Pacific. J. Math. *31* (1969), 451—458.

[21] *Teng-Sun Liu:* A note on maximal T_1-topologies. Portugal. Math. *18* (1959), 235—236.

[22] *A. Ramanathan:* Maximal Hausdorff spaces. Proc. Indian Acad. Sci. Sect A *26* (1947), 31—42.

[23] *C. T. Scarborough and R. M. Stephenson:* Minimal topologies. Colloq. Math. *19* (1968), 215—219.

[24] *P. S. Schnare:* Multiple complementation in the lattice of topologies. Fund. Math. *62* (1968), 53—59.

[25] *P. S. Schnare:* Infinite complementation in the lattice of topologies. Fund. Math. *64* (1969), 249—255.

[26] *N. Smythe and C. A. Wilkins:* Minimal Hausdorff and maximal compact spaces. J. Austral. Math. Soc. *3* (1963), 167—171.

[27] *A. K. Steiner:* The lattice of topologies: Structure and complementation. Trans. Amer. Math. Soc. *122* (1966), 379—398.

[28] *A. K. Steiner:* Complementation in the lattice of T_1-topologies. Proc. Amer. Math. Soc. *17* (1966), 884—886.

[29] *A. K. Steiner and E. F. Steiner:* Topologies with T_1-complements. Fund. Math. *61* (1967), 23—28.

[30] *A. K. Steiner and E. F. Steiner:* A T_1-complement for the reals. Proc. Amer. Math. Soc. *19* (1968), 177—179.

[31] *A. C. M. van Rooij:* The lattice of all topologies is complemented. Canad. J. Math. *20* (1968), 805—807.

ON A CLOSED RANGE THEOREM FOR NONLINEAR OPERATORS

S. SWAMINATHAN

Halifax

Let X and Y be Banach spaces, T a bounded linear operator from X to Y and T^* its conjugate from Y^* to X^*. It can be shown that the range of T is closed if and only if it is the set of all y in Y for which $\langle y, y^* \rangle = 0$ for y^* in ker T^*. The operator T is called normally solvable if, for y in Y, the equation $Tx = y$ has a solution if and only if $y \in (\ker T^*)^\perp$. Then the closed range theorem is equivalent to the statement that the operator T is normally solvable if and only if $T(X)$ is closed in Y.

When T is nonlinear and Fréchet differentiable it is possible to obtain closed range theorems by defining normal solvability of T for suitably restricted X and Y. In [3] S. I. Pohožaev considers a uniformly convex Y and defines T to be normally solvable when

(i) for any y in Y, there is a sequence $\{y_n\}$ such that $y_n \to y$ and for every y_n there exists $x_n \in X$ minimizing the functional $\|Tx - y_n\|$, and

(ii) for any such sequence $\{y_n\}$ if $T(x_n) - y_n \in [\ker T'(x_n)^*]^\perp$ then $y \in T(X)$.

His result can be stated in the following form: Let X be a Banach space and Y a Banach space which admits nearest points, i.e., for each closed set M in Y, the set of all x in Y, for which there is a y in M with $\|x - y\| = d(x, M)$, is dense in Y. Let T be a possibly nonlinear Fréchet differentiable operator from X to Y. The operator T is normally solvable if and only if the range $T(X)$ is closed in Y.

D. E. Wulbert [4] has shown that, besides uniformly convex Banach spaces, the following two classes of Banach spaces admit nearest points: (a) $2R$ Banach spaces or 2-fully convex Banach spaces of Ky Fan and I. Glicksberg [see 2, p. 113]. X is defined to be such a space when if $\{x_n\}$ is a sequence in X such that $\|x_n\| = 1$ for every n, and $\|x_m + x_n\| \to 2$ as $m, n \to \infty$, then $\{x_n\}$ is a Cauchy sequence. (b) Uniformly smooth [see 2, p. 113] Banach spaces satisfying the property that if a sequence $\{x_n\}$ converges weakly to x and if $\|x_n\| \to \|x\|$, then x_n converges strongly to x. Since there exist $2R$ Banach spaces which are not isomorphic to a uniformly convex space we have a positive answer to the question raised by S. I. Pohožaev in [3]. It should be pointed out that F. E. Browder [1] has formulated the underlying theory in a very elegant setting by considerably sharpening and generalizing the result to X a locally convex space and Y any Banach space.

References

[1] *F. E. Browder:* The Fredholm alternative for nonlinear operators. Bull. Amer. Math. Soc. *76* (1970), 993–998.
[2] *M. M. Day:* Normed Linear Spaces. Springer Verlag, 1962.
[3] *S. I. Pohožaev:* Normal solvability of nonlinear equations in uniformly convex spaces. Funkcional. Anal. i Priložen. *3* (2) (1969), 80–84.
[4] *D. E. Wulbert:* Differential theory for nonlinear approximation. (To appear.)

DALHOUSIE UNIVERSITY, HALIFAX, NOVA SCOTIA

SOME SET-THEORETIC CONSISTENCY RESULTS IN TOPOLOGY

F. D. TALL[1])

Toronto

1. Introduction

It has become evident in the past few years that many questions in general topology are independent of the usual set-theoretic axioms (e.g. Zermelo-Fraenkel Set Theory, including the Axiom of Choice). The ramifications of these developments for the study of topology are far from clear. I for one have no intuition as to whether Souslin spaces or separable normal non-metrizable Moore spaces, for example, "really" exist, and therefore cannot say that the undecidability of their existence merely indicates the need for stronger axioms which settle these questions the "right" way. In this note, however, we confine ourselves to mentioning several models of set theory which provide differing answers to various topological questions.

To avoid stating cumbersome relative consistency theorems, we assume the existence of a model of set theory. Also for simplicity, all spaces are assumed to be T_1.

2. Two models of set theory and their topological properties

The two models I know most about exhibit contrasting behavior with respect to problems connected with Souslin's Conjecture (see e.g. [24]) or the Normal Moore Space Conjecture (see e.g. [5]). Let $\mathfrak{A}$ be the result of adjoining to a model of the generalized continuum hypothesis, $\aleph_2$ Cohen subsets of ω_1 (for amplification, see Section 4). Let $\mathfrak{B}$ be any model of *Martin's Axiom* [21] plus $2^{\aleph_0} > \aleph_1$.

In $\mathfrak{A}$, I am kindly informed by T. Jech, Souslin's Conjecture fails, and so there is [20], [19] a space which satisfies the countable chain condition (i.e., every collection of disjoint open sets is countable), although its square does not. Other consequences include the existence of a compact perfectly normal space which is not separable [18], a perfectly normal Lindelöf space with a point-countable base which is not metrizable [4], [22], and a hereditarily separable, normal space which is not Lindelöf [25].

---

1) Some of this material appeared in the author's doctoral dissertation [28], which was supported by NSF grants GP-5913 and GP-8501. The author also acknowledges support from grant A-7354 of the National Research Council of Canada.

In $\mathfrak{B}$, Souslin's Conjecture holds [27], as well as the stronger propositions that the countable chain condition is preserved by arbitrary products (K. Kunen; for a proof see [17, Chapter 5]), and that compact perfectly normal spaces are separable [16]. The status of the other two problems in $\mathfrak{B}$ is open.

There is a separable normal non-metrizable Moore space in $\mathfrak{B}$. (The existence of Example E of [5] follows from a lemma of J. Silver in Section 2.5 of [21] or from Theorem 3.5 of [7] plus Lemmas 8 and 9 of [23].) Consequently [13], [31], there is a metacompact normal Moore space (otherwise known as a normal space with a uniform base [1], or a perfectly normal space with a σ-point-finite base, or a normal space which is the image of a metric space under a continuous open map with compact point inverses; for proofs of these equivalences, see [2], [3], [9], [12], [13]) which is not metrizable. I conjecture there is even a countable chain condition space with these properties in $\mathfrak{B}$. Other topological consequences of Martin's Axiom appear in Chapter 3 of [7], Chapter 5 of [17], [32], and [33].

A separable normal non-metrizable Moore space is, among other things, a normal first countable space containing a closed discrete subspace $Y = \{y_\alpha\}_{\alpha<\omega_1}$ which is not *separated*, i.e., there do not exist mutually disjoint open sets $\{U_\alpha\}_{\alpha<\omega_1}$, $y_\alpha \in U_\alpha$. *In* $\mathfrak{A}$, however, *every closed discrete subspace of cardinality* $\aleph_1$ *in every normal space of character less than* $2^{\aleph_1}$ *is separated.* (The character of a space is the supremum of the local weights – e.g. a first countable space has character $\aleph_0$, which of course is less than $2^{\aleph_1}$.) The proof is sketched in Section 4. Among many consequences (see also [28], [29], [30], [31]) are that countable chain condition normal Moore spaces are metrizable, countable chain condition normal spaces with point-countable bases are Lindelöf, and locally compact, perfectly normal, subparacompact spaces of cardinality $\leq\aleph_1$ are paracompact. The first, third, and probably the second of these assertions are false in $\mathfrak{B}$.

3. Other models

Many other models have been constructed by set-theorists in the past few years in order to prove various consistency results. To my knowledge, their implications for topology have not been investigated. A fair number of topological problems are of course equivalent to set-theoretic ones, so certain isolated results may be obtained. An example of such equivalence is the following theorem. (The normal case for $\kappa = \aleph_0$, $\lambda = \aleph_1$ is due half to Jones [15], and half to Heath [14].)

Theorem. *Let* $\kappa \leq \lambda$ *be infinite cardinals. Then* $2^{2^\kappa} \geq \lambda$ $(2^\kappa \geq \lambda)$ $(2^\kappa \geq 2^\lambda)$ *if and only if there is a Hausdorff* (*resp. regular*) (*resp. normal*) *space of density* κ (*the density is the least cardinal of a dense set*) *containing a closed discrete subspace of cardinality* λ.

Given any "reasonable" (in a well-defined sense) non-decreasing function F mapping the class of cardinal numbers into itself, Easton [10] constructs a model of a set theory in which for all cardinals κ, $2^\kappa = F(\kappa)$. Thus *there is a model of set theory, for example, in which no separable normal space contains an uncountable closed discrete subspace, but in which there is a normal space of density* $\aleph_1$ *containing a closed discrete subspace of power* $\aleph_{83}$.

Bukovský [8] considers a model $\mathfrak{D}$ in which $2^{\aleph_0} = 2^{\aleph_1}$ but every uncoutable separable metric space has a subset which is not Borel. In [28] we showed

Theorem. *There is a separable first countable normal space containing an uncountable closed discrete subspace if and only if there is an uncountable separable metric space in which every subset is* F_σ.

(Numerous other equivalents may be found in [28] or [29].) Translating, we have that *in* $\mathfrak{D}$, *no first countable separable normal space contains an uncountable closed discrete subspace, but there is a separable normal space containing such a subspace.*

4. Separating closed discrete subspaces in $\mathfrak{A}$

In [28] we proved a general theorem from which it follows that closed discrete subspaces of cardinality $\aleph_1$ in normal spaces of character less than $2^{\aleph_1}$ are separated in $\mathfrak{A}$. This particular case is sufficiently interesting that I think it worthwhile to present here a sketch of the proof which will hopefully be accessible to topologists unacquainted with consistency proofs. We take our basic definitions and theorems concerning forcing from Shoenfield [26].

Definitions. A *notion of forcing* is a partially ordered set C having a largest element. We write $\leq_C$ for the ordering and 1_C for the largest element, dropping the C when the context is clear. E'ements of C are generally designated by p, q and r. If $p \leq q$, we say p is an extension of q. A subset D of C is *dense* if every element of C has an extension in D.

Let C be a notion of forcing in a model $\mathfrak{M}$ of set theory. A subset G of C is *C-generic* over $\mathfrak{M}$ (or simply *generic*) if the following conditions hold.

(G 1) $1 \in G$.

(G 2) For all $p \in G$ and $q \geq p$, $q \in G$.

(G 3) For all $p, q \in G$, p and q have a common extension in G.

(G 4) For all dense sets D in $\mathfrak{M}$, $G \cap D \neq 0$.

For example, let C be the collection of countable partial functions from ω_1 (i.e., functions with domain a countable subset of ω_1) into $\{0, 1\}$ in $\mathfrak{M}$, ordered by $p \leqq q$ if $p \supset q$. Then if G is C-generic over $\mathfrak{M}$, it is easy to verify that

$$G_0 = \{\alpha : p(\alpha) = 0 \text{ for some } p \in G\},$$

$$G_1 = \{\alpha : p(\alpha) = 1 \text{ for some } p \in G\}$$

are disjoint sets whose union is ω_1.

The basic result about generic sets is

Theorem. *Let $\mathfrak{M}$ be a countable model of set theory, and C a notion of forcing in $\mathfrak{M}$. Then there is a set G which is C-generic over $\mathfrak{M}$, and a countable model of set theory $\mathfrak{M}[G]$ which is the smallest model including $\mathfrak{M}$ and containing G.*

Our main result is

Theorem. *Let $\mathfrak{A}_0$ be a model of set theory plus the* GCH *(generalized continuum hypothesis). Let C_{ω_2} be the collection of countable partial functions from $\omega_2 \times \omega_1$ into $\{0, 1\}$ in $\mathfrak{A}_0$, ordered by $p \leqq q$ if $p \supset q$. Let G be C_{ω_2}-generic over $\mathfrak{A}_0$. Then $\mathfrak{A} = \mathfrak{A}_0[G]$ is a model of set theory plus the* GCH *in which every closed discrete subspace of cardinality $\leqq \aleph_1$ in any normal space of character $< 2^{\aleph_1}$ is separated.*

By Gödel [11] and standard arguments, we may assume the existence of a countable model of set theory plus GCH. We also make the usual remark that if desired our theorem can be translated into a relative consistency theorem.

If G is C_{ω_2}-generic over $\mathfrak{A}_0$, $\alpha < \omega_2$, and $G_\alpha = \{\langle \eta, \varepsilon \rangle : \langle \alpha, \eta, \varepsilon \rangle \in \text{some } p \in G\}$, then G_α is a function from ω_1 into $\{0, 1\}$. If $\alpha \neq \beta$, then $G_\alpha \neq G_\beta$. The G_α's – or more precisely the subsets they determine – are known as *Cohen subsets* of ω_1. G can be "recovered" from the G_α's, so we may write $\mathfrak{A}_0[G_\alpha : \alpha < \omega_2] = \mathfrak{A}_0[G]$. We also consider the models $\mathfrak{A}_\beta = \mathfrak{A}_0[G_\alpha : \alpha < \beta]$, $\beta < \omega_2$ (i.e., the smallest model including $\mathfrak{A}_0$ and containing each G_α, $\alpha < \beta$). It can be shown that every member of $\mathfrak{A}_0[G]$ of cardinality $\leqq \aleph_1$ appears in some $\mathfrak{A}_\beta$. It can also be shown that if C_β is the collection of countable partial functions from ω_1 into $\{0, 1\}$ in $\mathfrak{A}_\beta$, then G_β is C_β-generic over $\mathfrak{A}_\beta$.

With these preliminaries, we can state the key lemma and sketch how it is used to get the main theorem.

Lemma. *Let $\mathfrak{M}$ be a model of set theory. Let C be the collection of countable partial functions from ω_1 into $\{0, 1\}$ in $\mathfrak{M}$. Let G be C-generic over $\mathfrak{M}$. In $\mathfrak{M}$, let $\langle X, \mathscr{T} \rangle$ be a topological space, and $Y = \{y_\alpha\}_{\alpha < \omega_1}$ a closed discrete unseparated subspace, such that all its countable subsets are separated. In $\mathfrak{M}[G]$, $\mathscr{T}$ is the basis for a topology $\mathscr{T}(G)$ on X. As noted earlier, G yields disjoint subsets G_0, G_1 of ω_1, and hence disjoint subsets $Y_0 = \{y_\alpha : \alpha \in G_0\}$, $Y_1 = \{y_\alpha : \alpha \in G_1\}$ of Y. Then in $\mathfrak{M}[G]$ there do not exist disjoint open sets about the disjoint closed sets Y_0, Y_1 in the space $\langle X, \mathscr{T}(G) \rangle$.*

Assume Lemma. In $\mathfrak{A}$, let $\langle X, \mathscr{T}\rangle$ be a normal space of character $<2^{\aleph_1}$ and Y a closed discrete subspace of cardinality $\leqq \aleph_1$. We wish to show Y is separated. Since countable closed discrete subspaces of normal – indeed regular – spaces are separated, we may assume Y has cardinality $\aleph_1$. Since for convenience we are assuming GCH in $\mathfrak{A}_0$ and hence in $\mathfrak{A}$, $2^{\aleph_1} = \aleph_2$ in $\mathfrak{A}$. Therefore the character of X is $\leqq \aleph_1$. X and $\mathscr{T}$ may be large, but it is not too difficult to construct another normal space $\langle X', \mathscr{T}'\rangle$ containing Y as a closed discrete subspace with both X' and a basis $\mathscr{B}'$ of cardinality $\aleph_1$, such that Y is separated in $\langle X', \mathscr{T}'\rangle$ if and only if it is separated in $\langle X, \mathscr{T}\rangle$.

There is a $\beta < \omega_2$ such that X', $\mathscr{B}'$, Y are all in $\mathfrak{A}_\beta$. $\mathscr{B}'$ generates a topology $\mathscr{T}_\beta$ on X in $\mathfrak{A}_\beta$. Since $\mathscr{B}'$ is a basis for $\mathscr{T}'$ in $\mathfrak{A}$ and countable subsets of Y are separated there, it follows that countable subsets of Y are separated in $\langle X', \mathscr{T}_\beta\rangle$ in $\mathfrak{A}_\beta$. If Y were unseparated in $\langle X, \mathscr{T}\rangle$ and hence in $\langle X', \mathscr{T}'\rangle$, Y would be unseparated in $\langle X', \mathscr{T}_\beta\rangle$ since $\mathscr{T}'$ and $\mathscr{T}_\beta$ have the same basis. But if Y is unseparated in $\langle X', \mathscr{T}_\beta\rangle$, then by Lemma, in $\mathfrak{A}_\beta[G_\beta] = \mathfrak{A}_{\beta+1}$ there do not exist disjoint open sets in the space $\langle X', \mathscr{T}_\beta(G_\beta)\rangle$ about $Y_{\beta 0}$, $Y_{\beta 1}$.

To conclude that $\langle X', \mathscr{T}'\rangle$ is not normal, a contradiction, it only remains to verify that the adjunction of $\{G_\alpha\}_{\beta<\alpha<\omega_2}$ does not undo the destruction of normality. The proof, due to K. Kunen, unfortunately must be omitted since it would take too long to provide the necessary background in forcing. The proof of the key lemma cannot be given here for the same reason, but what we can demonstrate – modulo some details – is the weaker fact that in $\mathfrak{M}[G]$ there do not exist disjoint members of $\mathscr{T}$ about Y_0, Y_1.

Let $\mathfrak{M}$, C, G, X, $\mathscr{T}$, Y be as in the hypothesis of Lemma. Let $U_0 \supset Y_0$, $U_1 \supset Y_1$, $U_0, U_1 \in \mathscr{T}$. Then for each $y_\alpha \in Y$, there is a $V_\alpha \in \mathscr{T}$ containing y_α and included in U_0 or U_1, according to whether $y_\alpha \in Y_0$ or Y_1.

For $p \in C$, let $p_0 = \{\alpha : p(\alpha) = 0\}$, $p_1 = \{\alpha : p(\alpha) = 1\}$. Claim

$$D = \{p : \bigcup\{V_\alpha : \alpha \in p_0\} \cap \bigcup\{V_\alpha : \alpha \in p_1\} \neq 0\}$$

is dense. Once this is proved, we are done. D is a dense subset of C in $\mathfrak{M}$ so there is a $p \in G \cap D$. By definition $G_0 \supset p_0$, $G_1 \supset p_1$, so

$$\bigcup\{V_\alpha : \alpha \in G_0\} \cap \bigcup\{V_\alpha : \alpha \in G_1\} \neq 0,$$

and hence $U_0 \cap U_1 \neq 0$.

To see that D is dense, let p be an arbitrary member of C. Domain p is countable, so by hypothesis there exist open mutually disjoint W_α, $\alpha \in$ domain p, $y_\alpha \in W_\alpha \subset V_\alpha$. Also by hypothesis the collection

$$\{W_\alpha\}_{\alpha \in \text{domain } p} \cup \{V_\alpha\}_{\alpha \notin \text{domain } p}$$

does not separate Y, so there is an $\alpha_1 \notin$ domain p, and an α_2, such that $V_{\alpha_1} \cap V_{\alpha_2} \neq 0$.

If $\alpha_2 \notin$ domain p, define $q \in C$ by

$$q = p \cup \{\langle \alpha_1, 1 \rangle\} \cup \{\langle \alpha_2, 0 \rangle\}.$$

If $\alpha_2 \in$ domain p, say $p(\alpha_2) = \varepsilon$, define

$$q = p \cup \{\langle \alpha_1, 1 - \varepsilon \rangle\}.$$

In either case $q \leqq p$ and $q \in D$.

A clever forcing argument due to J. Silver plus a fact about locally countable covers, reduce Lemma to the case we have just considered. All details can be found in [28], as can remarks about extending Theorem.

The character restriction entered into the proof in a natural fashion. It is therefore not surprising that it is best possible. Example G of [5] is a normal space of character $2^{2^{\aleph_0}}$ containing a closed discrete unseparated subspace of cardinality $\aleph_1$.

References

[1] *P. S. Alexandrov:* On the metrisation of topological spaces. Bull. Acad. Polon. Sci. Sér. Sci. Math. Astronom. Phys. *8* (1960), 135—140.

[2] *A. Arhangel'skiĭ:* On mappings of metric spaces. Soviet Math. Dokl. *3* (1962), 953—956.

[3] *C. E. Aull:* Topological spaces with a σ-point-finite base. Proc. Amer. Math. Soc. *29* (1971), 411—416.

[4] *H. R. Bennett:* Quasi-developable spaces. Topology Conference, Arizona State University, 1967, Tempe, Arizona, 1968, 314—317.

[5] *R. H. Bing:* Metrization of topological spaces. Canad. J. Math. *3* (1951), 175—186.

[6] *R. H. Bing:* Challenging conjectures. Amer. Math. Monthly *74* (1967), 56—64.

[7] *D. D. Booth:* Countably indexed ultrafilters. Thesis, University of Wisconsin, Madison, 1969.

[8] *L. Bukovský:* Borel subsets of metric separable spaces. General Topology and its Relations to Modern Analysis and Algebra, II (Proc. Second Prague Topological Sympos., 1966). Academia, Prague, and Academic Press, New York, 1967, 83—86.

[9] *M. M. Čoban:* On σ-paracompact spaces. Vestnik Moskov. Univ. Ser. I Mat. Meh. *24* (1) (1969), 20—27.

[10] *W. B. Easton:* Powers of regular cardinals. Ann. Math. Logic *1* (1970), 139—178.

[11] *K. Gödel:* Consistency-proof for the generalized continuum hypothesis. Proc. Nat. Acad. Sci. U.S.A. *25* (1939), 220—224.

[12] *S. Hanai:* On open mappings, II. Proc. Japan Acad. *37* (1961), 233—238.

[13] *R. W. Heath:* Screenability, pointwise paracompactness and metrization of Moore spaces. Canad. J. Math. *16* (1964), 763—770.

[14] *R. W. Heath:* Separability and $\aleph_1$-compactness. Colloq. Math. *12* (1964), 11—14.

[15] *F. B. Jones:* Concerning normal and completely normal spaces. Bull. Amer. Math. Soc. *43* (1937), 671—677.

[16] *I. Juhász:* Martin's Axiom solves Ponomarev's problem. Bull. Acad. Polon. Sci. Sér. Sci. Math. Astronom. Phys. *18* (1970), 71—74.

[17] *I. Juhász:* Cardinal functions in topology. Mathematical Center Tract *34*, Mathematisch Centrum, Amsterdam, 1971.

[18] *D. Kurepa:* Le problème de Suslin et les éspaces abstraits. C. R. Acad. Sci. Paris Sér. A—B *204* (1937), 325—327.

[19] *D. Kurepa:* Sur une propriété caractéristique du continu linéaire et le probléme de Souslin. Publ. Inst. Math. (Beograd) *4* (1952), 97—108.

[20] *G. Kurepa:* La condition de Souslin et une propriété caractéristique des nombers réels. C. R. Acad. Sci. Paris Sér. A—B *231* (1950), 1113—1114.

[21] *D. Martin and R. M. Solovay:* Internal Cohen extensions. Ann. Math. Logic *2* (1970), 143—178.

[22] *V. I. Ponomarev:* Metrizability of a finally compact *p*-space with a point countable base. Soviet Math. Dokl. *8* (1967), 765—768.

[23] *F. Rothberger:* On some problems of Hausdorff and of Sierpiński. Fund. Math. *35* (1948), 29—46.

[24] *M. E. Rudin:* Souslin's conjecture. Amer. Math. Monthly *76* (1969), 1113—1119.

[25] *M. E. Rudin:* A normal hereditarily separable non-Lindelöf space. Illinois J. Math. (to appear).

[26] *J. R. Shoenfield:* Unramified forcing. Proc. Sympos. Pure Math. 13, part I, American Mathematical Society, Providence, Rhode Island, 1971, 357—382.

[27] *R. M. Solovay and S. Tennenbaum:* Iterated Cohen extensions and Souslin's problem. Ann. of Math. *94* (1971), 201—245.

[28] *F. D. Tall:* Set-theoretic consistency results and topological theorems concerning the normal Moore space conjecture and related problems. Thesis, University of Wisconsin, Madison, 1969.

[29] *F. D. Tall:* New results on the normal Moore space problem. Proceedings of the Washington State University Conference on General Topology, March 1970, Pullman, Washington, 1970, 120—125.

[30] *F. D. Tall:* A set-theoretic proposition implying the metrizability of normal Moore spaces. Proc. Amer. Math. Soc. *33* (1972), 195—198.

[31] *F. D. Tall:* On the existence of normal metacompact Moore spaces (normal Hausdorff spaces with uniform bases) which are not metrizable. Canad. J. Math. (to appear).

[32] *F. D. Tall:* Souslin's conjecture revisited. Proc. Bolyai János Math. Soc. Colloquium on Topology, Keszthely, Hungary, June 19—23, 1972 (to appear).

[33] *F. D. Tall:* The countable chain condition vs. separability — applications of Martin's Axiom.

THE MARTIN COMPACTIFICATION IN AXIOMATIC POTENTIAL THEORY

J. C. TAYLOR

Montreal — Paris

1. Introduction. Let **B** denote an open ball in $\mathbf{R}^n$ ($n \geqq 2$) of radius R and let S denote its boundary. The Poisson kernel $K(x, y)$ for B is defined by the formula

$$K(x, y) = \frac{R^{n-2}}{\alpha_n}\left[\frac{R^2 - \|x\|^2}{\|x - y\|^n}\right],$$

where α_n is the $(n - 1)$-dimensional surface area of S with $x \in B$ and $y \in S$.

It is well known that if f is a continuous real-valued function on S the function H_f defined by

$$H_f(x) = \int_S K(x, y) f(y)\, dy,$$

with dy Lebesgue measure on S, solves the classical Dirichlet problem for B with boundary value f (cf. Helms [19]).

Since $x \to K(x, y)$ is harmonic, for all $y \in S$, it follows that for each positive Radon measure μ on S the function H_μ defined by

$$H_\mu(x) = \int_S K(x, y)\, \mu(dy)$$

is positive and harmonic. The converse is true (theorem of Herglotz) and even further the measure μ is uniquely defined by the corresponding harmonic function. Hence, there is an integral representation for the positive harmonic functions on B in terms of the functions $K(., y)$, with $y \in S$, and the set $\mathscr{M}^+(S)$ of positive Radon measures on S.

In 1941 R. S. Martin showed that a similar integral representation holds for any bounded domain D in R^n [25]. Specifically, he proved the following result.

Theorem (R. S. Martin). *Let $D \subset R^n$ ($n \geqq 2$) be a bounded domain. Then there exists a metrizable compactification $\tilde{D}$ of D with the following properties*:

(1) *to each point $y \in \Delta = \tilde{D} \setminus D$ there corresponds a non-zero positive harmonic function $K(., y)$ such that $(x, y) \to K(x, y)$ is continuous on $D \times \Delta$*;

(2) *if* H_μ *is defined by*

$$H_\mu(x) = \int K(x, y)\,\mu(dy)$$

with $\mu \in \mathscr{M}^+(\Delta)$, *the correspondence* $\mu \to H_\mu$ *maps* $\mathscr{M}^+(\Delta)$ *onto the cone* $\mathscr{H}^+(D)$ *of positive harmonic functions on* D; *and*

(3) *there is a* G_δ*-set* $\Delta_1 \subset \Delta$ *such that the map* $\mu \to H_\mu$ *is a bijection of* $\{\mu \in \mathscr{M}^+(\Delta) \mid \mu(\Delta \setminus \Delta_1) = 0\}$ *onto* $\mathscr{H}^+(D)$.

Definition. The compact metric space $\tilde{D}$ is called the *Martin compactification of* D and the ideal boundary $\Delta = \tilde{D} \setminus D$ is called the *Martin boundary* of D. The G_δ-set Δ_1 is called the set of *minimal points of* Δ.

The purpose of this paper is to discuss certain aspects of subsequent research dealing with a description of the Martin compactification and to mention some open problems.

2. Axiomatic potential theory. Martin's original work has been extended to three different contexts: axiomatic potential theory; and probabilistic potential theory both discrete and non-discrete. The following is a very brief outline of Brelot's axiomatic theory of potential (cf. [4] 1960 and also [1] 1966, [11] 1971 for the more general theories).

Let E be a connected, locally connected, locally compact, non-compact space with a countable base. Denote by $\mathscr{H}$ a sheaf on E.

Axiom 1. $\mathscr{H}$ is a sheaf of vector spaces of continuous real-valued functions on E.

Definition 2.1. A relatively compact open set $W \subset E$ will be said to be $\mathscr{H}$*-regular* if to each $f \in \mathscr{C}(\partial W)$ there corresponds a unique function $H_f \in \mathscr{H}(W)$ such that:

(1) $f \cup H_f \in \mathscr{C}(\overline{W})$; and
(2) $(f \geqq 0)$ implies $(H_f \geqq 0)$.

Axiom 2. E has a base of $\mathscr{H}$-regular sets.

Axiom 3. Let $W \subset E$ be open and connected. If $(h_n)_n \subset \mathscr{H}(W)$ is increasing then either $\sup_n h_n \in \mathscr{H}(W)$ or it is identically equal to $+\infty$.

Definition 2.2. A sheaf $\mathscr{H}$ on E that satisfies Axioms 1, 2 and 3 will be called a (Brelot) *harmonic sheaf.*

To each regular set W there correspond the *harmonic measures* μ_x^W, $x \in W$, defined by

$$\langle \mu_x^W, \varphi \rangle = H_f(x),$$

where $f = \varphi \mid \partial W$ and $\varphi \in \mathscr{C}_c(E)$. It can be seen that the mapping $x \to \langle \mu_x^W, f \rangle$ is Borel-measurable for any non-negative Borel function f. Consequently, the operator H_W is a kernel (cf. [27] for a definition) where

$$H_W f(x) \text{ equals } f(x) \text{ if } x \notin W \text{ and } \langle \mu_x^W, f \rangle$$

if $x \in W$, $f \geqq 0$ Borel-measurable.

Definition 2.3. A lower semi-continuous function $u : E \to (-\infty, +\infty]$ is said to be *($\mathscr{H}$)-hyperharmonic* (resp. *($\mathscr{H}$)-hyperharmonic on an open set U*) if $H_W u \leqq u$ for each regular set W (resp. for each regular set W with $\overline{W} \subset U$). A hyperharmonic function is said to be *superharmonic* if it is finite on a dense set. A continuous function h is said to be *harmonic on an open set U* if h and $-h$ are both hyperharmonic on U. A superharmonic function u is called a *potential* if $h \leqq u$ and h harmonic (on E) implies $h \leqq 0$, and $u \geqq 0$.

Each non-negative superharmonic function u has a unique Riesz decomposition $u = p + h$, where p is a potential and h is harmonic.

The *support* of a hyperharmonic function u is defined to be the complement of the largest open set on which it is harmonic.

Hypothesis I. There exists a positive potential.

Assuming this hypothesis one can prove that for each $y \in E$ there exists a potential with support $\{y\}$. It is natural to ask if such potentials are unique, up to a constant. While this is not true in general (see Constantinescu and Cornea [10] 1968 for a counter example) in a very large number of cases this is in fact so.

Hypothesis II. (The hypothesis of proportionality.) For each $y \in E$, if p_1 and p_2 are potentials with support $\{y\}$ there exists $\lambda > 0$ with $p_1 = \lambda p_2$.

Definition 2.4. A lower semi-continuous function $G : E \times E \to [0, +\infty]$ is called a *Green function for $\mathscr{H}$* if

(1) G is continuous off the diagonal; and
(2) for each $y \in E$, $x \to G(x, y)$ is a potential of support $\{y\}$.

In her thesis Mme Hervé proved that Hypotheses I and II imply the existence of Green functions for $\mathscr{H}$ ([17] 1962, Proposition 18.1).

3. Examples of harmonic sheaves. Let E be an open set in $\boldsymbol{R}^n$ $(n \geqq 2)$ and let $\mathscr{H}$ be the sheaf of $\mathscr{C}^2$-functions h for which $Lh = 0$ where L has one of the following forms:

(1) $L = \Delta$ (classical potential theory);

(2) $L = \sum_{i,j} a_{ij} \dfrac{\partial^2}{\partial x_i \, \partial x_j} + \sum_i b_i \dfrac{\partial}{\partial x_i} + c$

where i) b_i, c and a_{ij} are locally Lipschitz, and

ii) (a_{ij}) is a symmetric matrix whose associated quadratic form is positive definite [17] 1962;

(3) $L = \sum_i \frac{\partial}{\partial x_i}\left(\sum_j a_{ij} \frac{\partial}{\partial x_j}\right)$ with the a_{ij} Lebesgue measurable, (a_{ij}) symmetric and $\sum_{i,j} a_{ij}\xi_i\xi_j \geqq \varepsilon(\sum_i \xi_i^2)$ for some $\varepsilon > 0$, uniformly on E [18] 1964.

Then $\mathscr{H}$ is a harmonic sheaf that satisfies hypotheses I and II.

Information on further refinements of the example (2) is given in the survey article [7] 1970, where Bony's important work [2] 1967 on determination of an elliptic operator by a harmonic sheaf is also discussed.

If E is a $\mathscr{C}^2$-manifold then any elliptic operator L on E that satisfies (2) or (3) locally defines a harmonic sheaf. In particular if E is a Riemannian manifold, the Laplace-Beltrami operator defines a harmonic sheaf on E.

Finally, the harmonic functions on a Riemann surface E form a harmonic sheaf on E.

4. The Martin compactification. Let $(K_\alpha)_{\alpha\in I}$ be a family of continuous functions $K_\alpha : E \setminus D_\alpha \to \bar{R}$, $D_\alpha \subset E$ compact. Then, a very slight modification of arguments given by Constantinescu and Cornea in [9] 1963 shows that there exists a unique compactification $\tilde{E}$ of E with the following properties:

(1) for each α, K_α extends continuously to $\tilde{E} \setminus D_\alpha$; and

(2) the extended functions separate the points of $\tilde{E} \setminus E$ (see [31] 1970).

This compactification will be said to be *defined by* $(K_\alpha)_{\alpha\in I}$ (it is a Q-compactification in the terminology of [9]) and can be described in various equivalent ways (uniform structures, proximity spaces, etc.).

Let $x_0 \in E$ and let G be a Green function for $\mathscr{H}$. Define $K(x, y)$ to be 1 if $x = x_0 = y$ and to be $G(x, y)/G(x_0, y)$ otherwise. The *Martin compactification of E determined by* $\mathscr{H}$ is defined to be the compactification defined by the family $(K_x^*)_{x\in E}$, where $K_x^*(y) = K(x, y)$. It is clearly independent of the choice of Green function in view of Hypothesis II and can be shown to be independent of x_0 (with a suitable topology on the cone of positive superharmonic functions it can be seen to be the closure of the set of extreme points of a compact base for this cone). Further, it is metrizable as the argument given in [5] 1971 (p. 112) shows.

In the case of the harmonic sheaf $\mathscr{H}$ defined by a suitable elliptic operator L, the Martin compactification for L defined by Shur in [29] 1962 coincides with the compactification defined by the sheaf.

Let $y \in \Delta = \tilde{E} \setminus E$. Then there is a sequence $(y_n)_n \subset E$ with $y = \lim_n y_n$, i.e., for each $x \in E$

$$\lim_{n\to\infty} K(x, y_n) = K_y(x)$$

exists. If $x_1 \in E$ and W is a connected open set containing x_0 and x_1 the functions $K(., y_n)$ are harmonic on W for sufficiently large n. They all take value 1 at x_0 and since Harnack type inequalities hold in this setting, the function K_y is locally a uniform limit of harmonic functions and hence harmonic.

The function K is extended to $E \times \tilde{E}$ by using the functions K_y, $y \in \Delta$ and the desired integral representation follows by means of the arguments of Martin (c.f. [5] p. 113 for an exposition). Hence, Martin's theorem holds for any Brelot harmonic sheaf that satisfies Hypotheses I and II.

5. The relationship between $\tilde{W}$ and $\overline{W}$. Let $W \subset E$ be relatively compact and connected. The sheaf $\mathscr{H}$ on E induces a harmonic sheaf on W which satisfies Hypotheses I and II whenever $\mathscr{H}$ does. Hence W has a Martin compactification $\tilde{W}$. If W is an open ball in $\boldsymbol{R}^n$ and $\mathscr{H}$ is the sheaf defined by Laplace's operator it follows from the results stated in the introduction that $\tilde{W} = \overline{W}$ (the closure of W). It is therefore natural to ask in the general case what relationship holds between $\tilde{W}$ and $\overline{W}$?

In 1933 de la Vallée Poussin [24] obtained a Poisson type integral representation for a domain W in $\boldsymbol{R}^3$ of finite connection bounded by a finite number of "sufficiently regular" surfaces (at each point of the boundary a tangent plane exists and the angle $\theta(x_1, x_2)$ between the normals at x_1 and x_2 tends to zero as $x_1 - x_2$ tends to zero) and a finite number of closed sets of capacity zero. The boundary points are identified with harmonic functions and as in the case of an open ball there is a bijection between $\mathscr{M}^+(\partial W)$ and $\mathscr{H}^+(W)$. When no exceptional sets of capacity zero exist in the boundary then W is a C^1-domain and so $\tilde{W} = \overline{W}$ by [20] 1970. It would be of interest to have a good modern exposition of this elegant paper [24]. For a very readable account of it and of Martin's work see Deny [12] 1947.

In 1970 R. A. Hunt and R. L. Wheeden in [20] published a proof of the fact that $\tilde{W} = \overline{W}$ (relative once again to Laplace's operator) if W is a bounded Lipschitz domain in $\boldsymbol{R}^n$ $(n \geqq 2)$.

If W is a simply connected plane domain with at least two boundary points the Riemann mapping theorem states that there is a conformal map $\Phi : W \to (|z| < 1)$. The definition of the prime ends of W given by Carathéodory in [6] 1913 implies that they are the inverse images under Φ of the traces on $(|z| < 1)$ of the neighbourhood filters in $(|z| \leqq 1)$ of the points of the circle $(|z| = 1)$. Hence, the prime ends are the points of the Martin boundary for W. For W equal to the open square $(\max\{|x|, |y|\} < 1)$ minus the lines A_n, where $A_n = \{(x, y) \mid y \leqq 0,\ x = 1 - 2^{-n}\}$ it is well known from the theory of prime ends (cf. [6]) that $\tilde{W}$ and $\overline{W}$ are not comparable compactifications (i.e., there is no continuous map from one onto the other which is the identity map on W).

If $A \subset E$ is a closed set of capacity zero then $E \setminus A$ is connected and dense in E. Since the cone of positive harmonic functions on $E \setminus A$ can be identified with the cone of superharmonic functions on E whose support lies in A, it then follows that $\widetilde{E \setminus A} = \tilde{E}$ with Martin boundary equal to $\Delta \cup A$, where $\Delta = \tilde{E} \setminus E$. Further, each point of A is minimal. Consequently, if $\tilde{W} = \overline{W}$ and $A \subset W$ is closed and of capacity zero then $\widetilde{W \setminus A} = \overline{W \setminus A}$.

In the case of a general harmonic sheaf nothing is known about the relation between $\tilde{W}$ and $\overline{W}$. For example, it is not known if E has a base of relatively compact domains W for which $\tilde{W} = \overline{W}$. More specifically, if L is an elliptic operator on an open set $U \subset \boldsymbol{R}^n$ (of the type that defines a Brelot sheaf) and if W is even a very "regular" relatively compact domain with $\overline{W} \subset U$ nothing is known about the relation between $\tilde{W}$ and $\overline{W}$. Clearly, if everything is "sufficiently regular" one expects to have $\tilde{W} = \overline{W}$.

The same question can be considered for $\overline{W}$ a compact submanifold with boundary of a Riemannian manifold with interior W, the sheaf here being defined by the Laplace-Beltrami operator. Again one expects that $\tilde{W} = \overline{W}$. Hopefully, a solution of the local problem will solve this global one.

6. The existence of non-minimal points. In all the examples so far discussed $\Delta = \Delta_1$. Martin [25] constructed an example of a bounded domain W in $\boldsymbol{R}^3$ for which $\Delta \neq \Delta_1$ (it even satisfied the condition used by de la Vallée Poussin in [24]). Ikegami [21] 1967 proved that in the case of the Laplace operator (in fact for the somewhat more general case of a Green space) that $\Delta \neq \Delta_1$ implies $\Delta \setminus \Delta_1$ is infinite. Toda [32] 1967 then showed that it is even uncountable. These results have not been proved for a general harmonic sheaf. As a side comment it is noted that while in probabilistic potential theory the minimal points are the points to which the random particle almost surely converges as time goes to infinity, no probabilistic interpretation has been given of the non-minimal points.

In [25], Martin asked whether in general Δ_1 is dense in Δ. This question was resolved in striking fashion by Constantinescu and Cornea in [8] 1958. They showed that to each integer $n \geqq 1$ there corresponds a hyperbolic Riemann surface E_n whose Martin boundary is connected and contains exactly n minimal points. It would be interesting to know what part of the standard $(n-1)$-simplex corresponds to the Martin boundary of E_n. One can ask for what compact convex sets K do there exist Brelot sheaves such that the corresponding set Δ_1 can be identified with $\mathscr{E}(K)$. In probabilistic potential theory the following articles provide results in this direction, [28] 1966 and [34] 1960.

These examples of Constantinescu and Cornea raise the question: is there a topological property of E which implies $\Delta = \Delta_1$ or more generally $\overline{\Delta}_1 = \Delta$? Note that for simply connected plane domains $\Delta = \Delta_1$ and that in the examples E_n the homology groups of $E_n \setminus A$, for any compact $A \subset E_n$, are presumably infinitely generated. For example, if E is contractible to a point is $\overline{\Delta}_1 = \Delta$?

7. Entrance and exit boundaries. In [9] Constantinescu and Cornea proved that the Martin compactification of a hyperbolic Riemann surface E is the compactification defined by the family of continuous functions g of the form

$$g = H_U f / H_U 1,$$

where f is continuous and bounded on E, $\complement U$ is compact and outer regular and H_U is the kernel that solves the Dirichlet problem for U. As a consequence it was proved in [9] that $\tilde{E} \geqq \hat{E}$, where $\hat{E}$ is the Stoilow compactification of E (i.e., the one defined by those continuous bounded functions f that are constant on the components of the complement of some compact set $A = A(f) \subset E$) and "$\geqq$" means that there is a continuous map from $\tilde{E}$ onto $\hat{E}$ which is the identity on E. This result is equivalent to having a bijection between the connected components of Δ and the ends of E.

While for a general sheaf H the above description of the Martin compactification is no longer valid, it is still true that $\tilde{E} \geqq \hat{E}$ [31] 1970. In the general setting this is a consequence of the fact that if $A \subset E$ is compact then $\tilde{E} \setminus A$ can be canonically embedded in the topological sum of the Martin compactifications of the connected components of $E \setminus A$.

The family of functions defined at the beginning of this section defines a compactification, *the entrance compactification* of E [30] 1969 (by analogy with the probabilistic entrance compactification defined by Doob in [13] 1959). It is not known, for a Brelot sheaf satisfying Hypotheses I and II whether the entrance and the Martin (or exit) compactifications coincide.

When the Green function G defines an adjoint sheaf $\mathscr{H}^*$ which satisfies the hypothesis of proportionality (Hypothesis II) the entrance compactification of E is the Martin compactification of E determined by $\mathscr{H}^*$ [30] 1969. If for a sufficiently regular elliptic operator L defined on a neighbourhood of a sufficiently regular relatively compact domain $W \subset R^n$ $(n \geqq 2)$ it follows that $\tilde{W} = \overline{W}$, then in this case entrance and exit compactifications will coincide (the sheaf $\mathscr{H}^*$ will be defined by the formal adjoint L^* of L).

B. Walsh in [33] 1969 has defined the notion of a *normal structure* $\mathscr{L}$ associated with a Brelot sheaf $\mathscr{H}$. It is a family $(N_A)_A$ of kernels indexed by a family of compact sets A whose interiors cover E such that the following conditions are satisfied:

N_1) f continuous implies $N_A f$ continuous;
N_2) $A_1 \subset A_2$ implies $N_{A_2} \circ N_{A_1} = N_{A_1}$; and
N_3) f bounded Borel implies $N_A f$ is harmonic on $\complement A$.

A normal structure $\mathscr{L}$ defines a subsheaf $\mathscr{H}^{\mathscr{L}}$ of $\mathscr{H}$ by requiring $N_A h = h$, $\forall A \subset \complement U$ if $h \in \mathscr{H}^{\mathscr{L}}(U)$ and so regulates the "behaviour at infinity" of the harmonic functions.

To each normal structure one can associate an entrance (and exit or Martin) compactification. Presumably, there are probabilistic interpretations of these compactifications.

Dynkin in [16] 1965 considered an abstract type of "boundary condition" which may have some relation to the notion of a *normal structure.*

8. Stability of the compactification. The Martin compactification of a locally compact space E with countable base depends a priori on the Brelot sheaf $\mathscr{H}$. In [31] 1970 the question of stability was considered and the following result proved.

Let $\mathscr{H}_1$ and $\mathscr{H}_2$ be two Brelot sheaves on a locally compact space such that both of them satisfy Hypotheses I and II. If there exists a compact set $A \subset E$ such that the sheaves agree on $E \setminus A$ then the corresponding Martin compactifications coincide and the corresponding sets of minimal points coincide.

For an elliptic operator L it would be of interest to know what perturbation of the coefficients, other than an "arbitrary" one on a compact set, leaves the Martin compactification invariant.

The stability question is not trivial as the following example shows. Consider E an open ball B of radius 1 in $\boldsymbol{R}^3$ minus the ray A equal to $(x = y = 0,\ z \geqq 0)$. Since A has capacity zero the Martin compactification of E corresponding to Laplace's operator is $\bar{B}$ and the Martin boundary is $\partial B \cup A$. By analogy with the prime ends of a slit disk it is natural to try to construct prime ends for E that ramify A, replacing each point except $(0, 0, 0)$ by a circle. This can be done by using a suitable elliptic operator. Specifically, let L be the operator E obtained by transporting Laplace's operator on X to E by means of a diffeomorphism, where X is the open ball B minus the cone $(x^2 + y^2 = z^2,\ z \geqq 0)$. The Martin boundary of X is homeomorphic to a 2-sphere (by [20]) and so the Martin compactification of E determined by L ramifies A as desired. Hence this compactification differs from the one defined by Laplace's operator.

This example serves also to emphasize a point made by other authors. Namely, the theory of prime ends is really a study of the Martin boundary. Furthermore, the work done by Kaufmann (cf. [23] 1930) and later Mazurkiewicz [26] 1945 generalizing the theory of prime ends to higher dimensions defined objects which seem to be less maniable than the Martin compactification. Brelot in [3] 1946 showed (for the Laplacian) that in general Martin's and Mazurkiewicz' compactifications are not comparable. It could be that Mazurkiewicz' compactification can always be obtained by using a suitable sheaf.

9. The description of the boundary associated with an elliptic operator L. According to Dynkin in [15] 1963 "the set of minimal nonnegative solutions of the elliptic differential equation $Lf = 0$ has received little attention and provides many interesting unsolved problems. We will state two such problems.

The first problem is: under what restrictions on the differential operator L can be the set of minimal functions be provided with the structure of a smooth manifold such that $K_y(x)$ is a smooth function of y and x $(y \in \Delta)$?

The second problem concerns the connections between the geometry of a complete Riemannian manifold and the Martin boundary of this manifold", in particular the relation of the dimension of the boundary to the dimension of the manifold.

Dynkin himself in [14] 1961 and Karpelevich in [22] 1963 have shown that for certain homogeneous spaces of non-positive curvature the set of minimal functions can be explicitly described and has dimension one less than that of the manifold.

If an open Riemannian manifold E is isometrically embedded onto the interior of a compact Riemannian manifold with boundary then it is to be expected that, under certain conditions, this boundary is the Martin boundary of E. In other words, the Martin boundary should be the "natural" boundary to add to such a manifold.

References

[1] *H. Bauer:* Harmonische Raüme und ihre Potentialtheorie. Lecture Notes in Mathematics *22*, Springer-Verlag, Berlin, 1966.

[2] *J. M. Bony:* Détermination des axiomatiques de théorie du potentiel dont les fonctions harmoniques sont différentiables. Ann. Inst. Fourier (Grenoble) *17* (1) (1967), 353–382.

[3] *M. Brelot:* Le problème de Dirichlet "ramifié". Ann. de l'Univ. de Grenoble (n.s.) *22* (1946), 167–200.

[4] *M. Brelot:* Lectures on Potential Theory. Tata Institute of Fundamental Research, Bombay, 1960.

[5] *M. Brelot:* On Topologies and Boundaries in Potential Theory. Lecture Notes in Mathematics *175*, Springer-Verlag, Berlin, 1971.

[6] *C. Carathéodory:* Über die Begrenzung einfach zusammenhängender Gebiete. Math. Ann. *73* (1913), 323–370.

[7] *C. Constantinescu:* Harmonic spaces and their connections with the semi-elliptic differential equations and with the Markov processes. Elliptische Differentialgleichungen, Band I. Herausgegeben von G. Anger, Akademie-Verlag, Berlin, 1970, 19–30.

[8] *C. Constantinescu und A. Cornea:* Über den idealen Rand und einige seiner Anwendungen bei der Klassifikation der Riemmanschen Flächen. Nagoya Math. J. *13* (1958), 169–233.

[9] *C. Constantinescu und A. Cornea:* Ideale Ränder Riemannscher Flächen. Ergebnisse der Mathematik und ihrer Grenzgebiete, Springer-Verlag, Berlin, 1963.

[10] *C. Constantinescu and A. Cornea:* Examples in the theory of harmonic spaces. Seminar über Potentialtheorie. Herausgegeben von H. Bauer, Lecture Notes in Mathematics *69*, Springer-Verlag, Berlin, 1968, 161–171.

[11] *C. Constantinescu and A. Cornea:* Potential Theory on Harmonic Spaces. Springer-Verlag, Berlin (to appear).

[12] *J. Deny:* Le principe des singularités positives de G. Bouligand et la représentation des fonctions harmoniques positives dans un domaine. Revue Sci. *85* (1947), 866–872.

[13] *J. L. Doob:* Discrete potential theory and boundaries. J. Math. Mech. *8* (1959), 433–458.

[14] *E. B. Dynkin:* Nonnegative eigenfunctions of the Laplace-Beltrami operator and Brownian motion in certain symmetric spaces. Soviet Math. *2* (1961), 1433–1436.

[15] *E. B. Dynkin:* Markov processes and problems in analysis. Amer. Math. Soc. Transl. Series 2 *31* (1963), 1–24.

[16] *E. B. Dynkin:* Martin boundary and positive solutions of some boundary value problems. Ann. Inst. Fourier (Grenoble) *15* (1) (1965), 275–282.

[17] *R. M. Hervé:* Recherches axiomatiques sur la théorie des fonctions surharmoniques et du potentiel. Ann. Inst. Fourier (Grenoble) *12* (1962), 415–571.

[18] *R. M. Hervé:* Un principe de maximum pour les sous-solutions locales d'une équation uniformément elliptique de la forme $Lu = -\sum_i (\partial/\partial x_i)(\sum_j a_{ij}(\partial u/\partial x_j)) = 0$. Ann. Inst. Fourier (Grenoble) *14* (1964), 493–507.

[19] *L. L. Helms:* Introduction to Potential Theory. Wiley-Interscience, New York, 1961.

[20] *R. A. Hunt and R. L. Wheeden:* Positive harmonic functions on Lipschitz domains. Trans. Amer. Math. Soc. *147* (1970), 507–527.

[21] *T. Ikegami:* On the non-minimal Martin boundary points. Nagoya Math. J. *29* (1967), 287–290.

[22] *F. I. Karpelevich:* Nonnegative eigenfunctions of the Laplace-Beltrami operator on symmetric spaces of nonpositive curvature. Soviet Math. *4* (1963), 1180.
[23] *B. Kaufmann:* Über die Berandung ebener und raümlicher Gebiete (Primendentheorie). Math. Ann. *103* (1930), 70—144.
[24] *C. de la Vallée Poussin:* Propriétés des fonctions harmoniques dans un domaine ouvert limité par des surfaces à courbure bornée. Ann. Scuola Norm. Sup. Pisa *2* (1933), 167—197.
[25] *R. S. Martin:* Minimal positive harmonic functions. Trans. Amer. Math. Soc. *49* (1941), 137—172.
[26] *S. Mazurkiewicz:* Recherches sur la théorie des bouts premiers. Fund. Math. *33* (1945), 177—228.
[27] *P. A. Meyer:* Probability and Potentials. Blaisdell, Waltham, Mass., 1966.
[28] *P. Ney and F. Spitzer:* The Martin boundary for random walk. Trans. Amer. Math. Soc. *121* (1966), 116—132.
[29] *M. G. Shur:* The Martin boundary for linear elliptic operators. Soviet Math. *3* (1962), 730—732.
[30] *J. C. Taylor:* The Martin boundary and adjoint harmonic functions. Contributions to Extension Theory of Topological Structures (Proc. Sympos., Berlin, 1967). Deutsch. Verlag Wissensch. Berlin, 1969, 221—223.
[31] *J. C. Taylor:* The Martin boundaries of equivalent sheaves. Ann. Inst. Fourier (Grenoble) *20* (1) (1970), 433—456.
[32] *N. Toda:* A remark on Ikegami's paper "On the nonminimal Martin boundary points". Proc. Japan Acad. *43* (1967), 308—309.
[33] *B. Walsh:* Flux in axiomatic potential theory I. Invent. Math. *8* (1969), 175—221.
[34] *T. Watanabe:* A probabilistic method in Hausdorff moment problem and Laplace-Stieltjes transform. J. Math. Soc. Japan *12* (1960), 192—206.

MCGILL UNIVERSITY, MONTREAL, QUEBEC
UNIVERSITÉ DE PARIS VI, PARIS

COVERING PROPERTIES AND PRODUCT SPACES

R. TELGÁRSKY

Bratislava

During the last five years General Topology has been noted for a rapid development of concepts and methods involving covering properties. The time is coming to make a coordination and a regulation in the investigations of so many covering axioms (see [12], Vocabulary on p. 74). A central role among them play Compactness and Paracompactness. The behavior of topological properties with respect to various kinds of mappings was investigated before and especially after the lecture of P. S. Aleksandrov at the first Prague Symposium. The Cartesian product $X \times Y$ of spaces X and Y does not inherit, in general, covering properties of X and Y. There are singular examples in this area (see [12], pp. 42–54). The following problem given by H. Tamano [7] in his lecture at the second Prague Symposium, is not yet solved:

"Which space X satisfies the condition that $X \times Y$ is normal for any paracompact space Y".

He proved there the statement: Let X be a completely regular space. Then $X \times Y$ is normal for any paracompact space Y iff $X \times Y$ is paracompact for any paracompact space Y.

We say that a paracompact space X is *productible*, if the product space $X \times Y$ is paracompact for any paracompact space Y.

X is productible, if any of the following conditions holds:

(1) X is compact (J. Dieudonné [1]).

(2) X is paracompact and σ-locally compact (K. Morita [4]).

(3) X is paracompact and it has a linearly locally finite covering by compact sets (H. Tamano [6]).

(4) X is a closed image of a paracompact, locally compact space (T. Ishii [2]).

(5) X is a closed image of a paracompact, perfectly normal, σ-locally compact space (M. Tsuda [14]).

(6) X is a closed image of a paracompact, σ-locally compact space (R. Telgársky [8], [9] and [10]).

(7) X has an order locally finite open cover $\{U_i : i \in I\}$ and a closed cover $\{F_i : i \in I\}$ by compact sets such that $F_i \subseteq U_i$ for each $i \in I$ (Y. Katuta [3]).

(8) X is paracompact and it has an order locally finite open cover $\{U_i : i \in I\}$

and a closed cover $\{F_i : i \in I\}$ by C-scattered sets such that $F_i \subseteq U_i$ for each $i \in I$ (R. Telgársky [8], [10]).

C-scattered space means: each its nonvoid closed subspace F has a point $p \in F$ with a compact neighborhood in F ([8], [9], [10]).

Among the previous conditions the following implications hold:

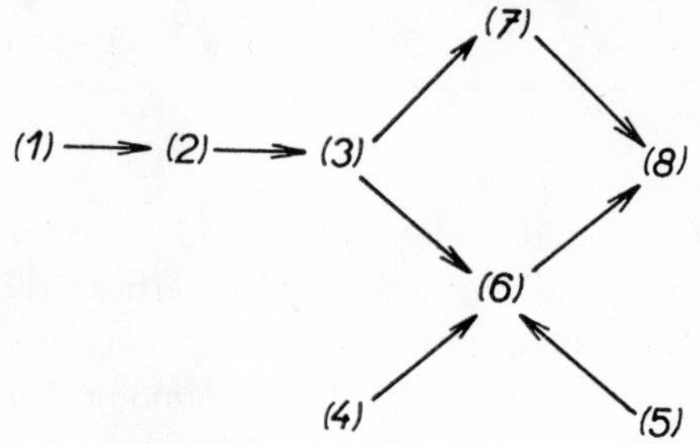

(9) X has a locally finite closed cover by productible spaces ([8], [10]).

(10) X is an F_σ-subset in a productible space ([8], [10]).

(11) X is the perfect image or the perfect preimage of a productible space ([8], [9], [10]).

(12) $X = Y \times Z$, where Y and Z are productible ([8], [10]).

(13) X is paracompact and it has a closed, locally compact subspace Y such that every closed subspace of X contained in $X \setminus Y$ is productible (J. Suzuki [5], R. Telgársky [8] and [10]).

(14) X is paracompact and it has a closed, σ-locally compact, G_δ-subset Y such that each closed subspace of X contained in $X \setminus Y$ is productible (R. Telgársky [8], [10]).

We are convinced that an appropriate generalization of "C-scattered" and "closure-preserving covering by compact sets" shall lead to the solution of the problem of H. Tamano.

For to see how the assumption "C-scattered" works for some stronger covering properties, we present the following results:

(a) If X is a paracompact scattered space and Y is an absolutely paracompact space, then the product space $X \times Y$ is absolutely paracompact (R. Telgársky [11]).

(b) Let $X = \{0\} \cup \{1/m + 1/n : m \in N \text{ and } n \in N\}$ equipped in the relative topology and Y be Cantor's set. Then X is totally hypocompact and scattered, Y is compact, $X \times Y$ is C-scattered, hypocompact, and absolutely paracompact, but $X \times Y$ is not totally hypocompact (R. Telgársky [11]).

(c) If X is a Lindelöf C-scattered space and Y is a Hurewicz space (Lindelöf space), then $X \times Y$ is a Hurewicz space (Lindelöf space, resp.) (R. Telgársky [11]).

(d) If X is a σ-compact regular space and Y is a Hurewicz space (Lindelöf space), then $X \times Y$ is a Hurewicz space (Lindelöf space, resp.) (R. Telgársky [11]).

In [13] we introduced the property: X has a closure-preserving covering by compact sets. We believe that a regular space with a closure-preserving covering by compact sets must be paracompact.[1]) If we consider completely regular spaces with a closure-preserving covering by compact sets, then the matter looks a bit simpler, because it suffices to prove only the normality of the spaces. It follows from the following statements:

(I) If X has the closure-preserving covering $\{C_i : i \in I\}$ by compact sets and Y is compact, then $X \times Y$ has the closure-preserving covering $\{C_i \times Y : i \in I\}$ by compact sets.

(II) If $X \times \beta X$ is normal, then X is paracompact – so reads the famous theorem of H. Tamano.

References

[1] *J. Dieudonné:* Une généralisation des espaces compacts. J. Math. Pures Appl. *23* (1944), 65–76.

[2] *T. Ishii:* On the product spaces and product mappings. J. Math. Soc. Japan *18* (1966), 166–181.

[3] *Y. Katuta:* A theorem on paracompactness of product spaces. Proc. Japan Acad. *43* (1967), 615–618.

[4] *K. Morita:* On the product of paracompact spaces. Proc. Japan Acad. *39* (1963), 559–563.

[5] *J. Suzuki:* Paracompactness of product spaces. Proc. Japan Acad. *45* (1969), 457–460.

[6] *H. Tamano:* Note on paracompactness. J. Math. Kyoto Univ. *3* (1963), 137–143.

[7] *H. Tamano:* Normality and product spaces. General Topology and its Relations to Modern Analysis and Algebra, II (Proc. Second Prague Topological Sympos., 1966). Academia, Prague, 1967, 349–352.

[8] *R. Telgársky:* C-dispersed spaces and topological product of paracompact spaces. Doctoral Dissertation, Institute of Mathematics, Polish Academy of Sciences, Wrocław, 1969.

[9] *R. Telgársky:* On the product of paracompact spaces. Bull. Acad. Polon. Sci. Sér. Sci. Math. Astronom. Phys. *17* (1969), 533–536.

[10] *R. Telgársky:* C-scattered and paracompact spaces. Fund. Math. *73* (1971), 59–74.

[11] *R. Telgársky:* Concerning product of paracompact spaces. Fund. Math. *74* (1972), 153–159.

[12] *R. Telgársky:* A survey of results concerning coverings and bases of topological spaces. Publ. č. 1, Mathematical Institute of the Slovak Academy of Sciences, Bratislava, 1971.

[13] *R. Telgársky:* Concerning two covering properties. Submitted to Proc. Japan Acad.

[14] *M. Tsuda:* On the normality of certain product spaces. Proc. Japan Acad. *40* (1964), 465–467.

INSTITUTE OF MATHEMATICS OF THE SLOVAK ACADEMY OF SCIENCES, BRATISLAVA

---

[1]) Added in proof. Professor E. Michael has kindly communicated me that our conjecture is disproved by H. Potoczny in his paper "A non-paracompact space which admits a closure-preserving cover of compact sets", which will be published in Proc. Amer. Math. Soc.

О ПОПЕРЕЧНИКАХ УРЫСОНА n-МЕРНОЙ ЭВКЛИДОВОЙ СФЕРЫ

В. М. ТИХОМИРОВ и Л. А. ТУМАРКИН

Москва

1. Пусть F — компактное метрическое пространство (с фиксированной метрикой). Поперечником Урысона $d_k^U(F)$, $k = 0, 1, 2, \ldots$, называется нижняя грань чисел $\varepsilon > 0$ таких, что имеется замкнутое ε-покрытие F кратности $k + 1$ [1].

Для компактов, лежащих в банаховом пространстве, имеются и другие характеристики их „массивности", именуемые поперечниками: поперечники по Александрову, Колмогорову, Бернштейну, Гельфанду [2] и ряд других. Поперечники во всех смыслах образуют монотонную последовательность: $d_0(F) \geqq \geqq d_1(F) \geqq \ldots$

Представляют интерес фигуры, у которых все ненулевые поперечники одинаковы. Для поперечников во всех перечисленных смыслах, кроме урысоновского, за такую фигуру, расположенную в гильбертовом пространстве, можно принять n-мерный шар (то есть центральное сечение шара в гильбертовом пространстве n-мерной плоскостью). К. А. Ситников [3] показал, что если эвклидово пространство само n-мерно, то шары в этом пространстве суть единственные фигуры, для которых $d_0 = d_1 = \ldots = d_{n-1}$, где d_k — поперечники по Александрову.

В тридцатых годах одним из авторов настоящей заметки была высказана гипотеза, что для n-мерной эквлидовой сферы (то есть границы $(n + 1)$-мерного шара) все ненулевые поперечники по Урысону одинаковы. В случае $n = 2$ так оно и есть [2].

2. Целью настоящей заметки является доказательство следующей теоремы, в которой E^{n+1} есть $(n + 1)$-мерное эвклидово пространство, где норма вектора $x = (x_1, \ldots, x_{n+1})$ определяется равенством $\|x\| = \sqrt{(x_1^2 + \ldots + x_{n+1}^2)}$, а

$$S^n = \left\{x \in E^{n+1} : \sum_{k=1}^{n+1} x_k^2 = 1\right\}.$$

Теорема. $d_K^U(S^n) = 2$ при $0 \leqq k \leqq (n - 1)/2$.

Доказательство. Пусть $[C_1, \ldots, C_M]$ есть некоторое замкнутое покрытие S^n кратности $k + 1$, где $k \leqq (n - 1)/2$. Обозначим через C_i^δ δ-раздутие множеств

$C_i \subset S^n$:

$$C_i^\delta = \{y \in S^n : \inf_{x \in C_i} \|x - y\| < \delta\}, \quad i = 1, \dots, M.$$

При достаточно малом $\delta > 0$ $[C_1^\delta, \dots, C_M^\delta]$ есть октрытое покрытие S^n той же кратности $k + 1$.

Обозначим через $\mathcal{N}$ нерв покрытия $[C_1^\delta, \dots, C_M^\delta]$ [4]. Реализуем $\mathcal{N}$ гомеоморфно в $E^{2k+1} \subset E^n$, что возможно в силу неравенства между k и n [4]. Пусть F — каноническое отображение S^n в $\mathcal{N} \subset E^n$ [4].

В силу теоремы Борсука об антиподах [5] найдутся две диаметрально-противоположные точки x_0 и $-x_0$, принадлежащие S^n, отображающиеся в одну точку $x \in \mathcal{N}$. В силу построения канонического отображения x_0 и $-x_0$ принадлежат одному и тому же множеству $C_{i_0}^\delta$. Значит, диаметр $C_{i_0}^\delta$ равен двум. В силу произвольности δ максимальный диаметр элементов покрытия $[C_1, \dots, C_M]$ таже равен двум. В силу же произвольности покрытия $[C_1, \dots, C_M]$ мы получаем, что $d_k^U(S^n) \geqq 2$. Откуда в силу того, что $d_0^U(S^n) = 2$ и из монотонности поперечников мы получаем, что $d_k^U(S^n) = 2$, $0 \leqq k \leqq (n-1)/2$, что и требовалось.

Литература

[1] *П. С. Урысон:* О канторовых многообразиях I. Труды по топологии и другим областям математики, т. 1, ГИТТЛ, Москва, 1951, 483.

[2] *В. М. Тихомиров и Л. А. Тумаркин:* О поперечниках компактов. Труды Международного Симпозия по топологии и ее применениях, Херцег-Нови, 25—31. 8. 1968, Югославия. Београд, 1969, 310—316.

[3] *К. А. Ситников:* Über die Rundheit der Kugel. Nachr. Akad. Wiss. Göttingen Math.-Phys. Kl. II *9* (1958), 213—215.

[4] *Л. С. Понтрягин:* Основы комбинаторной топологии. Москва—Ленинград, 1947, 30; 32; 35.

[5] *К. Борсук:* Drei Sätze über die *n*-dimensionale euklidische Sphäre. Fund. Math. *20* (1933), 177—190.

ON SOME PROBLEMS OF LOCAL APPROXIMABILITY IN COMPACT SPACES

G. TIRONI and R. ISLER[1])

Trieste

1. Introduction.

In this paper we study some problems concerning T_2 compact spaces. The main problem, not yet solved, is the following:

Problem 1. Is a compact separable and accessible Hausdorff space E first countable?

We give three examples showing that the conjecture is not true if the compact separable and accessible space is not Hausdorff or if the separable and accessible Hausdorff space fails to be compact or finally, if the compact Hausdorff space is neither separable nor accessible.

Another question arises if we require such a space to be Fréchet or sequential. In this paper it is proved that a T_2 compact, sequentially compact space is sequential provided that the topology induced by the sequential closure is Lindelöf.

Another interesting result deduced from a proposition of Mrówka is given, concerning the points of countable character in such spaces.

2. Let (E, τ) be a topological space where E is a set and τ a topological structure. Let us denote by (E, λ) a space where λ denotes the convergence structure on E (in the sense of Dolcher [5], or Novák [9]). Let (E, μ) be a closure space, i.e., a space E where a closure $\hat{A}$ is defined such that it does not necessarily satisfy the axiom of the closed closure $\hat{\hat{A}} = \hat{A}$. Let us denote by T, L, M, N the following functors T: $: (E, \lambda) \to (E, \tau)$; $L: (E, \tau) \to (E, \lambda)$; $M: (E, \lambda) \to (E, \mu)$; $N: (E, \mu) \to (E, \tau)$ (see [5], [7]).

Definition. A topological space (E, τ) is said to be *sequential* iff $TL(\tau) = \tau$; it is said to be *Fréchet* iff $ML(\tau) = \tau$.

Remark. These definitions are equivalent to the well-known ones; they are expressed in terms of categories and functors.

It is well known that $NM = T$.

1) This work was supported by the G.N.A.F.A. group of the Italian C.N.R.

Definition. Let (E, τ) be a separable space with a countable subset D dense in E; we shall say that E is *accessible from* D with the topology τ iff for every point x of E there is a one-to-one sequence of points of D converging to x.

Definition. A space E is said to be *sequentially compact* iff every sequence of points of E contains a convergent subsequence.

3. The main problem is to find sufficient conditions for a T_2 compact space to be first countable. The simple hypothesis that (E, τ) is a T_2 compact separable space is not sufficient. This is shown by the example of I^I with the product topology where $I = [0, 1]$.

Using a result of Mrówka [8]: "Under continuum hypothesis a T_2 compact space of cardinality $\mathfrak{c}$ has a subset of the same cardinality in which the first countability axiom holds", we can immediately prove the following propositions (continuum hypothesis is needed for Propositions 2, 3, 4):

Proposition 1. *A separable and accessible topological space* (E, τ) *has cardinality at most* $\mathfrak{c}$.

Proposition 2. *A* T_2 *compact separable accessible and homogeneous topological space* (E, τ) *is first countable.*

The proof follows immediately from Proposition 1 and from the result of Mrówka.

Proposition 3. *A* T_2 *compact separable and accessible topological group is metrizable.*

The proof follows from Proposition 2 and from a well-known criterion of the metrizability of topological groups with countable local basis (see [3]).

Proposition 4. *A* T_2 *compact topological space* (E, τ) *with cardinality at most* $\mathfrak{c}$ *(in particular, separable and accessible) has a dense subset* D *every point of which satisfies the first axiom of countability.*

Proof. Let x be a point of E and U an open neighbourhood of x. E is completely regular and therefore there exists a continuous function f on the set E into $[0, 1]$ such that $f(x) = 0$ and $f(\mathscr{C}U) = 1$ where $\mathscr{C}U$ is the complement of U in E. Let $C = f^{-1}(0)$. C is a subset of U and it is G_δ. Since C is closed in E, it is T_2 compact. By the result of Mrówka there is $y \in C$ which has a countable basis in the topology of C. But C is a G_δ and because of the transitivity of the character of a point (see [1], [2], [4]) y has a countable basis also in E.

The space E in Problem 1 is (1) compact; (2) Hausdorff; (3) separable; (4) accessible. We give now an example of a space satisfying (2), (3), (4) but not (1) which is not first countable.

Example 1. Let E be a square: $[0, 1] \times [0, 1]$. The neighbourhoods of points $(a, b) \neq (1, 0)$ are the same as in the usual topology of R^2 restricted on E. The basis of the neighbourhoods of $(1, 0)$ is the following family of sets:

$$V_{\bar{y},g}((1, 0)) = \{(x, y) : g(y) < x \leqq 1, y < \bar{y}\},$$

where g is a continuous function $g : [0, 1] \to [0, 1]$ such that $g(0) = 0$, $g(x) \neq 0$, $\forall x \neq 0$, and $\bar{y}$ a point such that $0 < \bar{y} \leqq 1$.

Now we construct a space satisfying (1), (3), (4) but not (2) which is not first countable.

Example 2. Let E be the set given by the union of the closed interval $I = [0, 1]$ and a point $x_0 \notin I$. The topology τ is the following: the neighbourhoods of the points of I are the same as in the usual topology for the interval. A basis of neighbourhoods of x_0 is given in the following way:

$$V_{x_1,\ldots,x_n}(x_0) = [I - \{x_1, x_2, \ldots, x_n\}] \cup \{x_0\}, \; x_i \in I \quad (i = 1, 2, \ldots, n).$$

We finally give an example of a topological space satisfying (1), (2) but neither (3) nor (4) which is not first countable.

Example 3. Let E be the set of all transfinite ordinals less than or equal to ω_1 where ω_1 is the first uncountable ordinal. Obviously, the space E is not first countable.

4. Problem 1 can be decomposed in a number of problems, every one of which can be interesting. For example, to find sufficient conditions for a compact topological space to be sequential or to be Fréchet. In fact, it is well-known that every first countable space is Fréchet and every Fréchet space is sequential ([6]). We have obtained the following result:

Lemma 1. *In a Hausdorff topological space (E, τ) the convergent sequences are the same as in the space $(E, TL(\tau))$. In particular, if (E, τ) is sequentially compact, so is $(E, TL(\tau))$.*

The proof is obvious.

Proposition 5. *Let (E, τ) be a T_2 compact, sequentially compact space. Let $(E, TL(\tau))$ be Lindelöf. Then (E, τ) is a sequential space.*

Proof. Let us consider the topological space (E, τ) and suppose that it is not sequential. This means $TL(\tau) \neq \tau$ and consequently $TL(\tau)$ is strictly finer than τ. This means that in the $TL(\tau)$ topology there are open sets not open in τ. Therefore

$(E, TL(\tau))$ cannot be a compact space. But we know that (E, τ) is sequentially compact and therefore every sequence in $(E, TL(\tau))$ has a convergent subsequence (by Lemma 1). Now $(E, TL(\tau))$ is not compact, therefore there exists an open covering of E that has no finite subcovering. But this covering must have a countable subcovering, because $(E, TL(\tau))$ is Lindelöf. Hence there exists a sequence in $(E, TL(\tau))$ without convergent subsequences and this contradicts the hypothesis that (E, τ) is sequentially compact. Therefore (E, τ) is sequential.

Remark. A compact and sequentially compact space need not be sequential. This is shown by Example 3.

References

[1] *A. V. Arhangel'skii:* On the cardinality of bicompacta satisfying the first axiom of countability. Soviet Math. Dokl. *10* (4) (1969), 951.

[2] *A. V. Arhangel'skii:* Suslin number and power. Characters of points in sequential bicompacta. (Russian.) Dokl. Akad. Nauk SSSR *192* (1970), 255—258.

[3] *N. Bourbaki:* Topologie Générale. 1968, Chap. 9, 55.

[4] *M. M. Čoban:* Perfect mappings and spaces of countable type. (Russian.) Vestnik Moskov. Univ. Ser. I Mat. Meh. *22* (6) (1967), 87—93.

[5] *M. Dolcher:* Topologie e strutture di convergenza. Ann. Scuola Norm. Sup. Pisa *14* (1960), 63—92.

[6] *S. P. Franklin:* Spaces in which sequences suffice. Fund. Math. *57* (1965), 107—115.

[7] *R. Isler:* Una generalizzazione degli spazi di Fréchet. Rend. Sem. Mat. Univ. Padova *41* (1968), 164—176.

[8] *S. Mrówka:* On the potency of compact spaces and the first axiom of countability. Bull. Acad. Polon. Sci. Sér. Sci. Math. Astronom. Phys. *6* (1958), 7—9.

[9] *J. Novák:* On some problems concerning multivalued convergences. Czechoslovak Math. J. *14* (*89*) (1964), 548—561.

WHEN CATEGORIES OF PRESHEAVES ARE BINDING[1])

V. TRNKOVÁ and J. REITERMAN

Praha

Following [8], a category is called *binding* if every category of universal algebras and homomorphisms can be fully embedded into it. If K is binding then every small category – and, under the assumption of non-existence of a proper class of measurable cardinals, every concrete category – admits a full embedding into K ([6], [9]). Particularly, every semigroup with unity can be represented as the semigroup of all endomorphisms of an object of K. Categories with the last property will be called *semibinding*.

While a lot of algebraic categories are binding ([4], [5], [11], [12], [13]), topological categories are often not even semibinding. Every semigroup of all continuous mappings from a topological space into itself contains idempotents (constant mappings) and therefore a non-trivial semigroup without non-identical idempotents cannot be represented in a topological category with continuous mappings as morphisms. Also the category of Hausdorff spaces and local homeomorphisms is not semibinding ([10]).

In the present note we give a certain criterion how far a category is from being binding. This is described by presenting the class P_K of all partially ordered sets k such that the category K^k of presheaves in K over k is binding. If K has an initial or a terminal object (all categories considered here have them both, namely, the empty space and the one-point space) then K is binding if and only if P_K is the class of all non-void partially ordered sets. Thus, roughly speaking , the bigger P_K is, the nearer K is to being binding. Analogously, denote S_K the class of all non-void partially ordered sets k such that K^k is semibinding.

The aim of this note is to describe the classes P_K (or S_K) for some categories familiar in topology. In particular the categories mentioned in Theorem 1 are binding.

The full text with proofs will be sent to Czechoslovak Math. J. The proofs of all theorems except Theorem 4 use the space M_1 from [2].

Definitions and conventions. The symbol $\dot{\subset}$ will be used for full embedding. As usual, every partially ordered set is considered as a thin category ($a \leqq b$ if and only if there exists a morphism from a to b).

---

[1]) Preliminary communication.

If K is a category, k a partially ordered set, denote by K^k the category of presheaves in K over k (= covariant functors from the thin category k to K) and their transformations.

Denote by **P** the class of all non-void partially ordered sets. Denote P_K (or S_K) the class of all $k \in \mathbf{P}$ such that K^k is binding (or semibinding, respectively).

The following three propositions are evident:

Proposition 1. *Let a category K have an initial or a terminal object. Then* $P_K = \mathbf{P}$ (or $S_K = \mathbf{P}$) *if and only if K is binding (or semibinding, respectively).*

Proposition 2. *If* $K \mathrel{\dot{\subset}} H$ *then* $P_K \subset P_H$ *and* $S_K \subset S_H$.

Proposition 3. *Let* $K \mathrel{\dot{\subset}} L \mathrel{\dot{\subset}} H$. *If* $P_K = P_H$ *(or* $S_K = S_H$*) then* $P_K = P_L = P_H$ *(or* $S_K = S_L = S_H$*, respectively).*

Theorem 1. $P_K = S_K = \mathbf{P}$ *for the following types of categories:*

1) *All subcategories of the category of metric spaces and open proximally continuous mappings containing the category of metric spaces and open Lipschitz mappings with bound* 1.

2) *All subcategories of the category of topological spaces and open continuous mappings containing the category of* T_1*-spaces and open local homeomorphisms.*

3) *All subcategories of the category of* T_1*-spaces and continuous locally one-to-one mappings containing all local homeomorphisms.*

Convention. If $G \subset \mathbf{P}$, put

$$\text{gen } G = \{k \in \mathbf{P}; (\exists h \in G)(h \mathrel{\dot{\subset}} k)\} .$$

Put $A = \text{gen }\{h_1\}$, $B = \text{gen }\{l_1\}$, $C = \text{gen }\{l_1, l_2\}$, $D = \text{gen }\{k_1, \ldots, k_{35}\}$ (see the figure on page 449).

Theorem 2. $P_K = S_K = A$ *for*

1) *all subcategories of the category of Hausdorff spaces and locally one-to-one continuous mappings, containing either the category of compact Hausdorff spaces and local homeomorphisms or the category of metrizable spaces and local homeomorphisms,*

2) *all subcategories of the category of Hausdorff spaces and open continuous mappings, containing either the category of compact Hausdorff spaces and open local homeomorphisms or the category of metrizable spaces and open local homeomorphisms.*

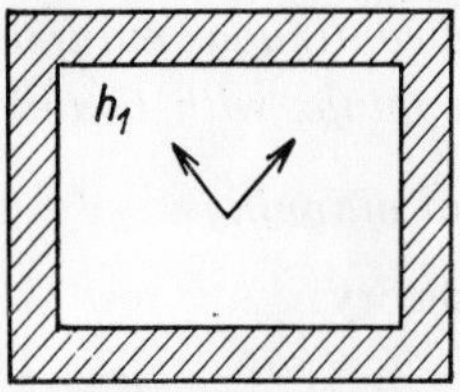

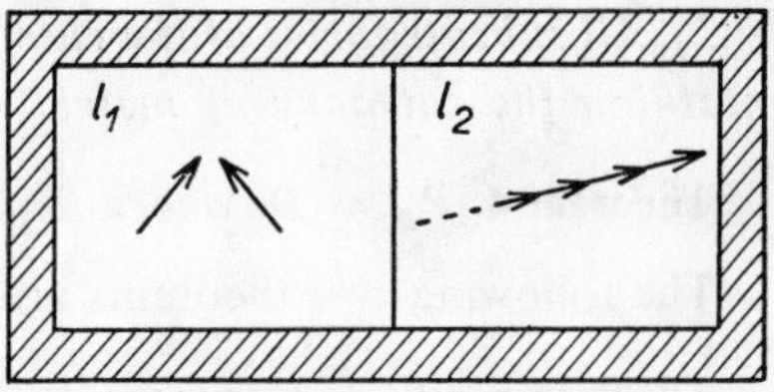

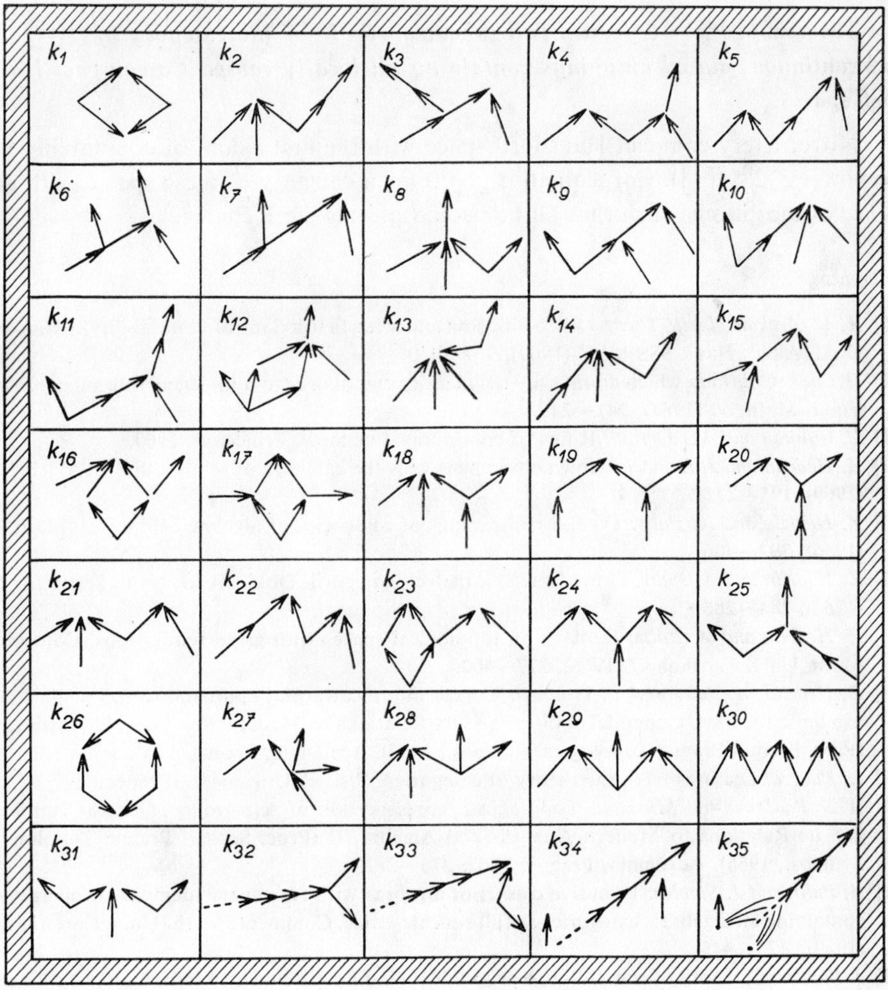

Theorem 3. $P_K = S_K = C$ *for the following types of categories:*

1) *All full subcategories of the category of topological (or proximity or uniform) spaces and continuous (or proximally continuous or uniformly continuous, respectively) mappings, containing all metrizable spaces.*

2) *All subcategories of the category of metric spaces and continuous mappings containing the category of metric spaces and Lipschitz mappings with bound* 1.

Theorem 4. $P_{\mathscr{S}} = D$, *where* $\mathscr{S}$ *is the category of sets and mappings.*

The following two theorems are consistent with the set-theory:

Theorem 5. $P_K = B$, *where* K *is the category of compact Hausdorff spaces and continuous mappings.*

Theorem 6. $P_K = C$ *for all full subcategories* K *of the category of* T_1*-spaces and continuous closed mappings containing all locally compact* σ*-compact Hausdorff spaces.*

Note. Every compact Hausdorff space with the first axiom of countability has the power $\leq 2^{\aleph_0}$ ([1]). Consequetly $P_K = \emptyset$ for a category of these spaces with any choice of morphisms such that all homeomorphisms are included.

References

[1] *A. V. Archangelskij:* The power of bicompacta with first axiom of countability. (Russian.) Dokl. Akad. Nauk SSSR *187* (1969), 967—970.

[2] *H. Cook:* Continua which admit only the identity mapping onto non-degenerate subcontinua. Fund. Math. *60* (1967), 241—249.

[3] *L. Gillman and M. Jerison:* Rings of continuous functions. Princeton, 1960.

[4] *Z. Hedrlín and J. Lambek:* How comprehensive is the category of semigroups? J. Algebra *11* (1969), 195—212.

[5] *Z. Hedrlín and A. Pultr:* On full embeddings of categories of algebras. Illinois J. Math. *10* (1966), 392—406.

[6] *Z. Hedrlín and A. Pultr:* O predstavlenii malych kategorij. Dokl. Akad. Nauk SSSR *160* (2) (1965), 284—286.

[7] *Z. Hedrlín and A. Pultr:* Remark on topological spaces with given semigroups. Comment. Math. Univ. Carolinae *7* (1966), 377—400.

[8] *Z. Hedrlín, A. Pultr and V. Trnková:* Concerning a categorial approach to topological and algebraic theories. General Topology and its Relations to Modern Analysis and Algebra, II (Proc. Second Prague Topological Sympos., 1966). Academia, Prague, 1967, 176—181.

[9] *L. Kučera:* Lectures from the theory of categories. Charles University. (Preprint.)

[10] *A. B. Paalman - de Miranda:* Topological representation of semigroups. General Topology and its Relations to Modern Analysis and Algebra, II (Proc. Second Prague Topological Sympos., 1966). Academia, Prague, 1967, 276—282.

[11] *A. Pultr and J. Sichler:* Primitive classes of algebras with two unary idempotent operations containing all algebraic categories as full subcategories. Comment. Math. Univ. Carolinae *10* (1969), 425—445.

[12] *J. Sichler:* Category of commutative grupoids is binding. Comment. Math. Univ. Carolinae *8* (1967), 753—755.

[13] *J. Sichler:* Concerning minimal primitive classes of algebras containing any category of algebras as a full subcategory. Comment. Math. Univ. Carolinae *9* (1968), 627—635.

[14] *V. Trnková and J. Reiterman:* On full embeddings of categories of algebras into categories of functors with thin domain. Comment. Math. Univ. Carolinae *12* (1971), 419—422.

CHARLES UNIVERSITY, PRAHA

A GENERAL FIXED POINT THEOREM

E. WATTEL

Amsterdam

In general topology there exist some fixed point theorems for contracting mappings. Their common characteristic is that they guarantee uniqueness of the fixed point by means of a principle of contraction relative to the metric of the space. In this paper we shall describe yet another fixed point theorem of the same sort, originating from a problem in differential equations, and we shall give a generalization to uniform spaces. First we shall state two well-known contraction theorems and we shall then prove the main theorem in a metric space and a uniform space separately. Eventually we shall show that Theorem 1 is a special case of Theorem 3, but we shall not include a similar proof for Theorem 2.

Theorem 1. (Banach). *Let (X, ϱ) be a complete metric space and let ϕ be a mapping from X into X such that there exists a positive real number α less than 1 with the property that $\varrho(\phi(x), \phi(y)) \leqq \alpha\, \varrho(x, y)$ for all x and y in X, then X contains one and only one point x_ϕ for which $\phi(x_\phi) = x_\phi$ holds.*

Theorem 2. *Let (X, ϱ) be a metric space and let ϕ be a mapping from X into itself such that $\overline{\phi(X)}$ is compact and $\varrho(\phi(x), \phi(y)) < \varrho(x, y)$ for all $x, y \in X$, then X contains one and only one fixed point relative to the mapping ϕ.*

Convention. If X is a space and ϕ is a mapping from X into X then $\phi^0(x)$ is the identity on X and $\phi^n(x) = \phi(\phi^{n-1}(x))$ for every natural number n and every $x \in X$. Clearly ϕ^n can be considered as a mapping from X into X.

Example. The following example is to show that Theorems 1 and 2 are independent. It meets the requirements of Theorem 2 but not of Theorem 1.

Let X be the collection of all real seqences $\{x_i\}_{i=1}^{\infty}$ with $|x_i| \leqq 2^{-i}$ (this is the Hilbert cube). We consider the usual metric. We define a contraction $\phi : X \to X$ by

$$\phi : \{x_i\}_{i=1}^{\infty} \mapsto \left\{\frac{i}{i+1}\, x_i\right\}_{i=1}^{\infty} .$$

ϕ is clearly a contraction on a compact space but there is no contraction constant $\alpha < 1$ such that $\varrho(\phi(x), \phi(y)) \leqq \alpha\, \varrho(x, y)$, for every x and y in X.

Theorem 3. *Let (X, ϱ) be a metric space, and let ϕ be a continuous function from X into X which satisfies the following properties:*

(i) $\exists x_0 \in X$ *such that* $\{\phi^n(x_0)\}_{n=1}^{\infty}$ *contains a convergent subsequence in* X.
(ii) $\forall x \in X$; $\forall y \in X$ *we have* $\lim_{n\to\infty} \varrho(\phi^n(x), \phi^n(y)) = 0$.

Then the space X contains exactly one fixed point relative to the transformation ϕ.

Proof. Since $\{\phi^n(x_0)\}_{n=1}^{\infty}$ contains a convergent subsequence in (X, ϱ) there exists an infinite subset M of the natural numbers such that $\{\phi^m(x_0) \mid m \in M\}$ is convergent. Let $\hat{x}_0$ be its limit. From the continuity of ϕ it follows that $\{\phi^{m+1}(x_0) \mid m \in M\}$ is convergent with limit $\phi(\hat{x}_0)$. Choose an $\varepsilon > 0$. From condition (ii) it follows that $\exists N_0$ such that for every natural number $n > N_0$ we have

$$\varrho(\phi^n(x_0), \phi^{n+1}(x_0)) < \tfrac{1}{3}\varepsilon .$$

Furthermore there exists an N_1 such that

$$\forall m \in M ; \quad m > N_1 \quad \text{we have} \quad \varrho(\phi^m(x_0), \hat{x}_0) < \tfrac{1}{3}\varepsilon$$

and

$$\forall m \in M ; \quad m > N_1 \quad \text{we have} \quad \varrho(\phi^{m+1}(x_0), \phi(\hat{x}_0)) < \tfrac{1}{3}\varepsilon .$$

Since M is infinite we conclude that $\varrho(\hat{x}_0, \phi(\hat{x}_0)) < \varepsilon$ for every positive number ε and therefore $\hat{x}_0$ has to be a fixed point of ϕ. Suppose that $\hat{y}_0$ is another fixed point, then $\lim_{n\to\infty} \varrho(\phi^n(\hat{x}_0), \phi^n(\hat{y}_0)) = 0$. Since $\hat{x}_0 = \phi(\hat{x}_0) = \phi^n(\hat{x}_0)$ and $\hat{y}_0 = \phi(\hat{y}_0) = \phi^n(\hat{y}_0)$ for all $n \in \mathbf{N}$ we have $\hat{x}_0 = \hat{y}_0$. Therefore $\hat{x}_0$ is the unique fixed point of the function ϕ.

Theorem 4. *Let X be a Tychonoff space and let ϕ be a continuous mapping from X into X. If there exists a compatible uniform structure $\mathscr{H}$ on X such that*

$\forall x \in X$, $\forall y \in X$, $\forall H \in \mathscr{H}$, $\exists N_0 \in \mathbf{N}$ *such that* $\forall n \in \mathbf{N}$ *with* $n > N_0$ *we have* $(\phi^n(x), \phi^n(y)) \in H$,

then the following conditions are equivalent:

(i) $\exists x_0 \in X$ *and an infinite subset M of $\mathbf{N}$ such that* $\{\phi^m(x_0) \mid m \in M\}$ *is a convergent sequence.*
(ii) *The space X contains exactly one fixed point $\hat{x}_\phi$ relative to ϕ.*
(iii) *For every $x \in X$ the sequence* $\{\phi^n(x) \mid n \in \mathbf{N}\}$ *converges.*

Proof. (i) $\Rightarrow$ (ii). Let $\hat{x}$ be the limit of $\{\phi^m(x_0) \mid m \in M\}$. From the continuity it follows that $\{\phi^{m+1}(x_0) \mid m \in M\}$ converges to $\phi(\hat{x})$. Let H be an arbitrary diagonal neighbourhood in $\mathscr{H}$. Then there exists a $K \in \mathscr{H}$ such that $K = K^{-1}$ and $K \circ K \circ K \subset H$. There exists an $N_0 \in \mathbf{N}$ such that

(a) $(\phi^m(x_0), \phi^{m+1}(x_0)) \in K$ for every $m \in \mathbf{N}$; $m \geqq N_0$.
(b) $(\phi^m(x_0), \hat{x}) \in K$ for every $m \in M$; $m \geqq N_0$.
(c) $(\phi^{m+1}(x_0), \phi(\hat{x})) \in K$ for every $m \in M$; $m \geqq N_0$.

Since M is infinite we can choose m sufficiently large in M and we conclude that

$$(\hat{x}, \phi(\hat{x})) \in K \circ K \circ K \subset H .$$

Since X is a Tychonoff space it follows that $\hat{x} = \phi(\hat{x})$. Suppose that $\hat{y}$ is another fixed point of ϕ, then

$$\forall H \in \mathscr{H} ; \quad \exists N \in \mathbf{N} ; \quad \forall n > N \quad \text{we have} \quad (\phi^n(\hat{y}), \phi^n(\hat{x})) = (\hat{y}, \hat{x}) \in H .$$

This implies that $\hat{y} = \hat{x}$. Therefore $\hat{x}$ is the unique fixed point of ϕ in X.

(ii) $\Rightarrow$ (iii). Let x_0 be the fixed point of ϕ in X and let x be an arbitrary point of X. Let U be a neighbourhood of x_0. Then U contains a neighbourhood of x_0 of the form: $\{y \mid (y, x_0) \in H\}$ for some $H \in \mathscr{H}$. By definition there exists an $N_0 \in \mathbf{N}$ such that for every $n > N_0$ we have $(\phi^n(x), \phi^n(x_0)) \in H$; hence $(\phi^n(x), x_0) \in H$. Therefore $\phi^n(x)$ is eventually in every neighbourhood of x_0, i.e., $\phi^n(x)$ converges to x_0.

(iii) $\Rightarrow$ (i). Obvious.

Proof of Theorem 1. Let x and y be two arbitrary points of X. Then $\varrho(\phi^{n+1}(x), \phi^{n+1}(y)) \leqq \alpha \cdot \varrho(\phi^n(x), \phi^n(y)) \leqq \alpha^{n+1} \cdot \varrho(x, y)$. Since $\alpha < 1$ we have $\lim_{n\to\infty} \varrho(\phi^n(x), \phi^n(y)) = 0$.

Moreover, for every $x \in X$ we have

$$\varrho(\phi^n(x), x) \leqq \sum_{i=1}^{n} \varrho(\phi^i(x), \phi^{i-1}(x)) \leqq \sum_{i=1}^{n} \alpha^{i-1} \cdot \varrho(\phi(x), x) \leqq \frac{1}{1-\alpha} \varrho(\phi(x), x) .$$

Therefore, for every k and $l \in \mathbf{N}$, $k \geqq l$ we have

$$\varrho(\phi^k(x), \phi^l(x)) \leqq \alpha^l \cdot \varrho(\phi^{k-l}(x), x) \leqq \frac{\alpha^l}{1-\alpha} \cdot \varrho(\phi(x), x) .$$

This implies that $\{\phi^n(x) \mid n \in \mathbf{N}\}$ is a Cauchy sequence and from the completeness of X it follows that its limit exists. We conclude that $\{\phi^n(x) \mid n \in \mathbf{N}\}$ satisfies the condition (i) in Theorem 3 and Theorem 1 follows from Theorem 3.

The paper has been prepared in cooperation with P. P. N. de Groen.

References

[1] *P. van Emde Boas, J. van de Lune and E. Wattel:* On the continuity of fixed points of contractions. Rapport ZW 1968—008, Mathematisch Centrum, Amsterdam.

IDENTIFYING HILBERT CUBES: GENERAL METHODS AND THEIR APPLICATION TO HYPERSPACES BY SCHORI AND WEST

J. E. WEST

Ithaca

In the past several years new methods of identifying Hilbert cubes have been discovered, and they have been applied by R. M. Schori and the author to hyperspaces of some one-dimensional Peano continua. In particular, we have solved affirmatively the conjecture of M. Wojdyslawski [11] that the hyperspace 2^I of all non-void, closed subsets of the interval is a Hilbert cube, when topologized by the Hausdorff metric. I am informed by Professor Kuratowski that the conjecture was originally posed in the 1920's and was well-known to topologists in Warsaw and other places at that time. This report outlines some of these methods and the Schori-West proof of this conjecture.

Definitions and notation

Hilbert cubes will be denoted generically by Q and viewed primarily as infinite products of closed intervals. A space X will be called a *Hilbert cube factor* (Q-factor) if $X \times Q = Q$. The term "map" implies continuity, and homeomorphisms are all surjections. A map is a *near-homeomorphism* if it is the uniform limit of homeomorphisms. Additionally, a map $f : X \to Y$ *stabilizes to a near-homeomorphism* if $f \times id : X \times Q \to Y \times Q$ is a near-homeomorphism, "*id*" denoting the identity map. The *mapping cylinder* $M(f)$ of $f : X \to Y$ is the quotient space of $X \times I \cup Y$ under the relation identifying each $(x, 0)$ with $f(x)$, and if A is a closed subset of X, then $M(f, A)$ is the *reduced mapping cylinder of f relative to A* and is obtained from $X \times I \cup Y$ by identifying each point (a, t) of $A \times I$ with $f(a)$. The collapsing map $c : M(f) \to Y$ or $c : M(f, A) \to Y$ is that retraction induced on $M(f)$ or $M(f, A)$ by identifying each (x, t) of $X \times I$ with $f(x)$.

A closed subset A of a separable, metric absolute neighborhood retract X is said to have *property Z in X* if for each open subset U of X, the inclusion $U \setminus A \to U$ is a homotopy equivalence. (This is almost R. D. Anderson's definition [2] and is equivalent for Hilbert cube manifolds and Hilbert manifolds.)

Methods of identifying Hilbert cubes known prior to the second Symposium

Before 1966 there were four ways known to tell that a space is a Hilbert cube.

(1) *Infinite-dimensional, convex compacta of Hilbert spaces are Hilbert cubes.* (O.-H. Keller [5].)

(2) *Inverse limits of Hilbert cubes and near-homeomorphisms are Hilbert cubes.* (A specialization of a theorem of M. Brown [3] having the following as an immediate corollary.)

(2′) *An inverse limit of Hilbert cube factors and maps which stabilize to near-homeomorphisms is a Hilbert cube factor.*

(3) *Dendra are Hilbert cube factors.* (R. D. Anderson [1]. This is a solution of a problem in the *Scottish Book* posed by K. Borsuk. The proof was not published, but one appears in [8]. A. Szankowski [7] gave another proof that an infinite product of triods, i.e., spaces homeomorphic to the letter "T", is a Hilbert cube, which also answers Borsuk's problem.)

(4) *An infinite product* $\prod_{i=1}^{\infty} X_i$ *of non-degenerate Q-factors is a Hilbert cube if for each i and every positive number* ε *there are a homeomorphism* $g : X_i \times Q \to \to I \times Q$ *and a map* $f : X \to I$ *such that the I-coordinate of* $g(x, q)$ *is within* ε *of* $f(x)$ *for each point* (x, q) *of* $X_i \times Q$. (R. D. Anderson [1]. This was never published, either. A proof occurs in [8].)

New methods of identifying Hilbert cubes

Shortly after the second Symposium, Anderson established

(5) *A union of two Hilbert cubes intersecting in a third with property Z in each of the others is a Hilbert cube.* (Not explicitly stated, but a corollary to Theorem 10.1 of [2].)

This may be reformulated in a slightly weaker way as

(5′) *The union of two Hilbert cube factors intersecting in a third with property Z in each of the others is a Hilbert cube factor.*

I was able to generalize (5′) to read

(6) *The union of two Hilbert cube factors intersecting in a third which has property Z in one of the others is a Hilbert cube factor.* (Not explicit but equivalent to Theorem 5.1 of [8].) A corollary of (6) is

(7) *A compact contractible polyhedron is a Hilbert cube factor; a locally compact polyhedron is a Hilbert cube manifold factor and a Hilbert manifold factor.*

Also, I was able in [8] to lift completely the last hypothesis of (4), obtaining

(8) *An infinite product of non-degenerate Hilbert cube factors is a Hilbert cube.*

Two results have appeared on near-homeomorphisms of Hilbert cubes, necessary to the application of (2) and (2′). Both begin with (7) but point in different directions. The next, due to D. Curtis [4] is the definitive result in the simplicial category.

(9) *A simplicial surjection between locally finite simplicial complexes stabilizes to a near-homeomorphism if and only if each point-inverse is compact and contractible.*

My own result deals with mapping cylinders [9]:

(10) *The mapping cylinder $M(f)$ of any map between Hilbert cube factors is itself a Hilbert cube factor; moreover, the collapse $c : M(f) \to Y$ stabilizes to a near-homeomorphism.*

This allows the extension of (7) to cell-complexes.

(7′) *A compact, contractible cell-complex is a Hilbert cube factor; a locally compact cell-complex is a Hilbert cube manifold factor and a Hilbert manifold factor.*

Another consequence of (10) is a reduced version.

(11) *If $f : K \to Y$ is a map of a finite cell-complex to a Hilbert cube factor and A is any closed subset of K, then $M(f, A)$ is a Hilbert cube factor and its collapse $c : M(f, A) \to Y$ stabilizes to a near-homeomorphism.*

The next result is of a different sort and has been useful in work on hyperspaces [10].

(12) *If X is a Hilbert cube factor resulting from the compactification of a Hilbert cube manifold M by the addition of a second Hilbert cube factor $A = X \setminus M$ having property Z in X, then X is a Hilbert cube.*

Using (2′), (7), (8), (9) or (10), and (12), I was able to obtain a first result on hyperspaces [10]:

(13) *The hyperspace $C(D)$ of all subcontinua of a dendron D is a Hilbert cube factor which is a Hilbert cube if and only if the branch points of D are dense in D.* (The branch points of D are those separating it into more than two components.)

The current state of affairs is expressed in the next three results on hyperspaces, which are joint work of mine with R. M. Schori.

(14) *The hyperspace 2^I of all non-void closed subsets of the unit interval is a Hilbert cube.*

(15) *If Γ is any non-degenerate, connected, finite graph, then 2^Γ is a Hilbert cube.*

(16) *For any non-degenerate dendron D, 2^D is a Hilbert cube.*

2^I is a Hilbert cube, I: The strategy of the proof

I now show how Schori and I use the results discussed above to prove (14). We use inverse limits with (2), (2′), (8) and (10) as our starting point. We use the symbol 2^I_X to denote those members of 2^I containing X. If X is a finite set, we simply list its members as subscripts and begin by reducing the problem to that of showing that 2^I_{01}, those members of 2^I containing both 0 and 1, is a Hilbert cube. This is easy because Schori has shown in [6] that 2^I is the cone over the cone over 2^I_{01} and because (1) shows that the cone over a Hilbert cube is a Hilbert cube. We next derive (11) from (10) for technical convenience. After this the proof falls into three conceptual divisions:

(i) 2^I_{01} *may be expressed as an inverse limit of spaces each homeomorphic to the infinite power of* 2^I_{01}.

(ii) 2^I_{01} *is a Hilbert cube factor, so by* (8) *each space in the inverse limit of* (i) *is a Hilbert cube.*

(iii) *Each bonding map of the inverse limit of* (i) *is a near-homeomorphism, so by* (2) 2^I_{01} *is a Hilbert cube.*

The inverse system of (i) is as follows: For each positive integer n let $\sigma(n) = \{1, 1/n, 1/(n+1), \ldots\}$ and let $r_n : 2^I_{\sigma(n)} \to 2^I_{\sigma(n-1)}$ be defined by

$$r_n(A) = A \cup [m_n(A), m_n(A) + \delta_n(A)] \cup [M_n(A) - \delta_n(A), M_n(A)],$$

where

$$m_n(A) = \max(A \cap [0, 1/(n-1)]), \quad M_n(A) = \min(A \cap [1/(n-1), 1]),$$

and

$$\delta_n(A) = \min\{|1/(n-1) - a| \mid a \in A\}.$$

(17) $2^I_{01} = \operatorname{inv\,lim}\{2^I_{\sigma(n)}, r_n\}$.

(18) *The natural function* $\psi: 2^I_{\sigma(n)} \to 2^{[1/n,1]}_{1/n1} \times 2^{[1/(n+1),1/n]}_{1/(n+1)1/n} \times \ldots$ *is a homeomorphism.*

These establish (i).

2^I is a Hilbert cube, II: 2^I_{01} is a Hilbert cube factor

We use another inverse limit for 2^I_{01} in order to establish that it is a Q-factor: For each real number $t \geqq 1$, let $F_t : 2^I_{01} \to 2^I_{01}$ be the map sending each set A to its closed $1/t$-neighborhood in I and let $B_n = F_n(2^I_{01})$. Let $f_n = F_{n(n-1)} \mid B_n$. Then $F_{n-1} = f_n F_n$ and

(19) $2^I_{01} = \text{inv lim}\,\{B_n, f_n\}$.

Observe that B_n is composed of all members A of 2^I_{01} such that for some m, $A = \bigcup_{k=1}^{m} [a_{2k-1}, a_{2k}]$ where $a_1 = 0$, $a_{2m} = 1$, $a_{2k+1} > a_{2k}$, and $a_{2k} - a_{2k-1} \geqq 2/n$ unless $k = 1$ or m, in which case $a_{2k} - a_{2k-1} \geqq 1/n$.

(20) *Each B_n is a Hilbert cube factor.*

This may be proved in two ways. First, B_n is contractible because the maps $\{F_t \mid B_n\}_{t \geqq 1}$ form a homotopy joining the constant map F_1 and the identity $(= \lim_{t \to \infty} F_t)$. Because B_n is a polyhedron, (7) applies. The second way is of use in proving (23) below. Let

$$B^m_n = \{A \in B_n \mid A \text{ has no more than } m \text{ components}\},$$
$$C^m_n = \{A \in B_n \mid A \text{ has exactly } m \text{ components}\}.$$

If φ is the map of C^m_n to R^{2m} sending each set A to the sequence $(a_1, \ldots, a_{2m})$ of its end-points, then φ embeds C^m_n in a $(2m-2)$-simplex Δ of R^{2m} and its image misses exactly the union X of every second face of Δ. Moreover, φ^{-1} extends naturally to a map of Δ carrying X into B^{m-1}_n. Because X is homeomorphic to I^{2m-3} and C^m_n is open in B^m_n, we have

(21) *B^m_n is homeomorphic to the mapping cylinder of $\varphi^{-1} \mid X : X \to B^{m-1}_n$.*

Because $B^1_n = \{I\}$, we have $B_n \times Q = Q$ by induction using (10). We prove each f_n stabilizes to a near-homeomorphism in (22) and (23): Let, for $0 \leqq t \leqq 1$, $\lambda_t : B_n \to B_n$ be given by

$$\lambda_t\Big(\bigcup_{k=1}^{m} [a_{2k-1}, a_{2k}]\Big) = \bigcup_{k=1}^{m} [a^t_{2k-1}, a^t_{2k}],$$

where

(i) $a^t_i = t a^1_i + (1-t)\, a_i$,

(ii) $a^1_{2k} = a_{2k} + 1/n$ unless $a_{2k+1} - a_{2k} \leqq 2/n$, in which case we set $a^1_{2k} = 1/2(a_{2k+1} + a_{2k})$, and

(iii) $a^1_{2k+1} = a_{2k+1} - 1/n$ unless $a_{2k+1} - a_{2k} \leqq 2/n$, in which case we set $a^1_{2k+1} = 1/2(a_{2k+1} + a_{2k})$.

(22) *For $0 < t < 1$, λ_t is an embedding of B_n in itself with image containing B_{n-1}. Moreover, $B_{n-1} \cup \lambda_t(B^m_n)$ is a reduced mapping cylinder of a map from a $(2m-3)$-cell to $B_{n-1} \cup \lambda_t(B^{m-1}_n)$.*

This is proved using the map $\varphi : C_n^m \to R^{2m}$ to obtain a geometric representation. If $c_{mt} : B_{n-1} \cup \lambda_t(B_n^m) \to B_{n-1} \cup \lambda_t(B_n^{m-1})$ is the collapse of this mapping cylinder and $g_{nt} = c_{2t}c_{3t} \ldots c_{nt}\lambda_t : B_n \to B_{n-1}$, then

(23) *The mapping cylinder structures of* (22) *may be chosen so that as* t *approaches* 1, g_{nt} *approaches* f_n *uniformly. As each* g_{nt} *stabilizes to a near-homeomorphism, so does* f_n.

Applying (2′) to (21) and (23), we have II.

2^I is a Hilbert cube, III: r_n is a near-homeomorphism

In the representation ψ of $2^I_{\sigma(n)}$ as an infinite power of 2^I_{01}, r_n appears as the stabilization of the map $r^{1/n} : 2^{[1/n,1]}_{1/n1} \to 2^{[1/n,1]}_{1/n1/(n-1)1}$ defined by the same formula. Therefore, by using (2) it is sufficient to establish

(24) *For any* t *in* (0, 1), *the retraction* $r^t : 2^I_{01} \to 2^I_{0t1}$ *defined in strict analogy to* r_n *stabilizes to a near-homeomorphism.*

We achieve (24) by using $\{B_n, f_n\}$ and

(25) *Let* $X = \operatorname{inv\,lim} \{X_n, f_n\}$ *and* $Y = \operatorname{inv\,lim} \{Y_n, g_n\}$ *be compact metric spaces, and for each* n *let* $h_n : X_n \to Y_n$ *be a map such that* $g_n h_n = h_{n-1} f_n$. *The induced map* $h = \operatorname{inv\,lim} \{h_n\} : X \to Y$ *is a near-homeomorphism if each* f_n *and each* h_n *are, so* h *stabilizes to a near-homeomorphism if each* f_n *and each* h_n *do.*

We express r^t as a limit map in the inverse sequence $\{B_n, f_n\}$ as follows: Let $H_n = B_n \cap 2^I_{t-1/n}$, $K_n = H_n \cap 2^I_{t+1/n}$, and $L_n = F_n(2^I_{0t1}) = B_n \cap 2^I_{[t-1/n,t+1/n]}$. Also, let $s_n : K_n \to L_n$ be the map defined by the formula $A \to A \cup [t - 1/n, t + 1/n]$, and let $h_n = s_n r^{t+1/n} r^{t-1/n} \mid B_n$.

(26) $2^I_{0t1} = \operatorname{inv\,lim} \{L_n, f_n \mid L_n\}$, $f_n h_n = h_{n-1} f_n$ *and* $r^t = \operatorname{inv\,lim} \{h_n\}$.

After this, all that remains is to prove

(27) *Each of* $r^{t-1/n} \mid B_n$, $r^{t+1/n} \mid H_n$, *and* s_n *stabilizes to a near-homeomorphism.*

We prove (27) by using (11) inductively on a reduced mapping cylinder construction of B_n, H_n, and K_n from H_n, K_n, and L_n, respectively. The sequence (25)–(27) establishes (26), hence III, so the proof is completed.

References

[1] *R. D. Anderson:* The Hilbert cube as a product of dendrons. Notices Amer. Math. Soc. *11* (1964), 572.

[2] *R. D. Anderson:* On topological infinite deficiency. Michigan Math. J. *14* (1967), 365—383.

[3] *M. Brown:* Some applications of an approximation theorem for inverse limits. Proc. Amer. Math. Soc. *11* (1960), 478—483.

[4] *D. Curtis:* Near homeomorphisms and fine homotopy equivalences. Compositio Math. (to appear).

[5] *O.-H. Keller:* Die Homoiomorphie der kompakten konvexen Mengen im Hilbertschen Raum. Math. Ann. *105* (1931), 748—758.

[6] *R. M. Schori:* Hyperspaces and symmetric products of topological spaces. Fund. Math. *63* (1966), 77—88.

[7] *A. Szankowski:* On factors of the Hilbert cube. Bull. Acad. Polon. Sci. Sér. Sci. Math. Astronom. Phys. *17* (1969), 703—709.

[8] *J. E. West:* Infinite products which are Hilbert cubes. Trans. Amer. Math. Soc. *150* (1970), 1—25.

[9] *J. E. West:* Mapping cylinders of Hilbert cube factors. General Topology and its Applications *1* (1971), 111—125.

[10] *J. E. West:* The subcontinua of a dendron form a Hilbert cube factor. Proc. Amer. Math. Soc. (to appear).

[11] *M. Wojdysławski:* Sur la contractilité les hyperespaces de continus localement connexes. Fund. Math. *30* (1938), 247—252.

CORNELL UNIVERSITY, ITHACA, NEW YORK

RELATIONS BETWEEN $\mathfrak{B}$-COMPLETENESS AND m-PARACOMPACTNESS

K. WICHTERLE

Praha

This communication studies the relations between m-paracompactness and $\mathfrak{B}$-completeness for some classes $\mathfrak{B}$ of directed sets. Most of the results are similar to [4], but the proofs are new and more simple. We shall show that m-paracompactness implies $\mathfrak{N}_m$-completeness in completely regular spaces. Equivalence between these two notions does not hold in general, but it takes place when the space is supposed to be a generalized order closure space (Theorem 2). The rest of the paper is devoted to the closed relations in $\mathfrak{B}$-spaces.

$\mathfrak{B}$ denotes any class of directed sets, $\mathfrak{N}$ the class of monotone ordered sets, $\mathfrak{N}_m = \{\langle D, \prec\rangle \in \mathfrak{N} \mid \text{card } D \leqq m\}$; a net is called $\mathfrak{B}$-net iff its domain belongs to $\mathfrak{B}$. A net N is called *remarkable* in $\mathscr{P} = \langle P, u\rangle$ iff $f \circ N$ converges in $I = [\![0, 1]\!]$ for each $f \in C = \mathscr{C}(\mathscr{P}, I)$, equivalently, if the range of N is in $\mathscr{P}$ and N converges in $\beta\langle P, \tilde{u}\rangle$ ($\tilde{u}$ being the completely regular modification of u). A closure space $\mathscr{P}$ is called $\mathfrak{B}$-*complete* iff every $\mathfrak{B}$-net remarkable in $\mathscr{P}$ converges in $\mathscr{P}$ (contrary to [3], we do not suppose in this definition that $\mathscr{P}$ is a $\mathfrak{B}$-space).

Theorem 1. *For every cardinal number* m, *every* m-*paracompact completely regular space is* $\mathfrak{N}_m$-*complete.*

Proof. Let $\mathscr{P} = \langle P, u\rangle$ be an m-paracompact completely regular space. It is sufficient to prove that each non-convergent $\mathfrak{N}_m$-net is not remarkable. Let N be such a net. Without loss of generality we may assume that N is one-to-one and such that $\mathbf{D}N = \langle \alpha, \in\rangle$ where $\alpha \leqq m$ is a regular ordinal.

Let us denote $U_\eta = P - uN[\![\eta, \rightarrow]\!]$ for each $\eta < \alpha$. Then $\mathscr{U} = \{U_\eta \mid \eta < \alpha\}$ is an increasing open cover of $\mathscr{P}$ (this follows easily from the fact that N does not converge) and card $\mathscr{U} \leqq m$. Therefore, by [2], there exists a locally finite open cover $\mathscr{Z}$ such that $\{uZ \mid Z \in \mathscr{Z}\}$ refines $\mathscr{U}$.

We shall construct an increasing map $d : \alpha \rightarrow \alpha$ and a disjoint and locally finite family $\{V_\xi \mid \xi < \alpha\}$ of open neighbourhoods of points $Nd\xi$.

Transfinite induction. $\eta = 0 : d0 = 0$, $V_0 = U_0 \cap Z_0$, where $Z_0 \in \mathscr{Z}$, $Nd0 \in Z_0$ and U_0 is an open neighbourhood of $Nd0$, which intersects only finitely many members of $\mathscr{Z}$.

Let $0 < \eta < \alpha$ and suppose that $d\xi$, V_ξ and Z_ξ have been defined for all $\xi < \eta$. Since α is regular and since $u[\mathscr{Z}]$ refines $\mathscr{U}$, then exists a $\lambda < \alpha$ with $N\lambda \notin$

$\notin \mathrm{u} \bigcup\{Z_\xi \mid \xi < \eta\}$. There exists a $Z_\eta \in \mathscr{Z}$ with $N\lambda \in Z_\eta$ and an open neighbourhood U_η of the point $N\lambda$, which interesects only finitely many members of $\mathscr{Z}$. For the induction step it remains to define:

$$\lambda = d\eta\,, \quad V_\eta = U_\eta \cap Z_\eta - \mathrm{u} \bigcup\{Z_\xi \mid \xi < \eta\}\,.$$

For each $\xi < \alpha$ we can choose a $g_\xi \in \mathsf{C}$ such that $g_\xi[P - V_\xi] = \{0\}$, $g_\xi N d_\xi = 1$. If S and $\alpha - S$ are cofinal subsets of α, the function $g = \sum\{g_\xi \mid \xi \in S\}$ is correctly defined, continuous and belongs to C. Moreover, gN is equal to 1 and 0 respectively on cofinal subsets $d[S]$ and $d[\alpha - S]$ of α, hence gN does not converge in I. Thus N is not remarkable, which completes the proof.

Proposition 1. *If some proper maximal filter $\langle j, \supset\rangle$ of open sets of a completely regular space $\mathscr{P}$ belongs to $\mathfrak{B}$ (or is a quotient of some element of $\mathfrak{B}$), then $\mathscr{P}$ is not $\mathfrak{B}$-complete.*

The proof is obvious: A net $\{N_U \mid U \in j\}$, where $N_U \in U$, converges to j in $\beta\mathscr{P}$.

Therefore, for separated spaces and sufficiently large $\mathfrak{M}$, *the $\mathfrak{M}$-completeness coincides with compactness (and hence with $\mathfrak{N}$-compactness)*. On the other hand, any infinite discrete space is $\mathfrak{N}$-complete.

Proposition 2. *Every product of completely regular paracompact spaces is $\mathfrak{N}$-complete.*

To prove Proposition 2 notice that any $\mathfrak{B}$-completeness is a productive property, and apply Theorem 1.

Proposition 2 enables us to show that the only if part in Theorem 1 cannot be true in general: Any non-paracompact product of paracompact spaces (e.g. Sorgenfrey's square) serves as an example of a non-paracompact $\mathfrak{N}$-complete space.

Theorem 2. *A generalized order closure space is $\mathfrak{N}_{\mathrm{m}}$-complete if and only if it is* m-*paracompact.*

Proof. Assume $\mathscr{P} = \langle P, \mathrm{u}\rangle$ is not m-paracompact. Without loss of generality we may assume that there exist an open-closed interval-like subspace $\mathscr{Q}'$ of $\mathscr{P}$, a point $z \in Q'$, a regular ordinal $\gamma \leqq m$ and an increasing net $N = \{N\xi \mid \xi \in \gamma\}$ such the open cover $\mathscr{W} = \{]\!]z, N\xi[\![\mid \xi \in \gamma\}$ of $Q = Q' \cap]\!]z, \rightarrow[\![$ is not uniformizable. N does not converge. The existence of such $\mathscr{W}$ follows by [2]; for the details, see [4].

Suppose $f \circ N$ does not converge in I for some $f : P \to [\![0, 1]\!]$. Then $f \circ N$ is frequently in two sets A and B separated in I and we can choose an increasing map $h : \gamma \to \gamma$ such that $N \circ h$ lies alternately in $f^{-1}[A]$ and $f^{-1}[B]$. For each $t \in Q$ we can define the minimal $m_t \in \alpha$ such that $t = Nhm_t$; then $]\!]z, Nh(m_t + 1)[\![$ is a neighbourhood of t. Since $\mathscr{W}$ is not uniformizable, there exists (see [1], p. 435) $R \subset Q$ and $y \in \mathrm{u}R - \bigcup\{]\!]z, Nh(m_t + 1)[\![\mid t \in R\}$. We can prove that $y \in \mathrm{u}f^{-1}[B] \cap$

$\cap\, \mathsf{u} f^{-1}[A]$, hence f is not continuous. Therefore, N is remarkable and $\mathscr{P}$ is not $\mathfrak{N}_{\mathsf{m}}$-complete.

Let $\mathscr{X}$ be a $\mathfrak{V}$-compact space (which means that every $\mathfrak{V}$-net ranging in X has an accumulation point in $\mathscr{X}$), *let the Cartesian product $\mathscr{X} \times \mathscr{X}$ be a $\mathfrak{V}$-space* (i.e., its closure u is determined by a convergence of $\mathfrak{V}$-nets). *Then the composition of any two (or finitely many) closed relations is a closed relation* (i.e., $(\mathsf{u}R = R \subset X \times X \,\&\, \mathsf{u}S = S \subset X \times X) \Rightarrow \mathsf{u}(R \circ S) = R \circ S$).

The problem arises whether the product of two closed relations is closed, provided that $\mathscr{X}$ is a $\mathfrak{V}$-compact $\mathfrak{V}$-space for some $\mathfrak{V}$.

Let $\mathscr{X}$ be separated, let D be discrete in $\mathscr{X}$, let M be a net converging to x in $\mathscr{X}$, let N be a net converging to $y \neq x$ in $\mathscr{X}$ such that $\aleph_0 \leqq \operatorname{card} \mathbf{E}M = \operatorname{card} \mathbf{E}N \leqq \operatorname{card} D$. Then there exist closed equivalences R and S on $\mathscr{X}$ such that neither $R \circ S$ nor its transitive envelope is closed.

References

[1] *E. Čech:* Topological spaces. Praha, 1965.
[2] *J. Mack:* Directed covers of paracompact spaces. Canad. J. Math. *19* (1967), 649–654.
[3] *K. Wichterle:* On $\mathfrak{V}$-convergence spaces. Czechoslovak Math. J. *18* (*93*) (1968), 569–588.
[4] *K. Wichterle:* Relations between the $\mathfrak{V}$-completeness and the paracompactness of closure spaces. Comment. Math. Univ. Carolinae 9 (4) (1968), 583–593.

INSTITUTE OF MATHEMATICS OF THE CZECHOSLOVAK ACADEMY OF SCIENCES, PRAHA

ON HOMEOMORPHISMS OF ∞-DIMENSIONAL BUNDLES

R. Y. T. WONG

Santa Barbara

We announce here several generalizations of results in [On homeomorphisms of infinite-dimensional bundles I, II, and III] to not necessarily separable locally trivial fibre bundles $\xi = (E, p, B)$ over polyhedron base space B with fibre a paracompact manifold M modeled on some Fréchet space F homeomorphic to F^∞, the countable infinite product of F by copies of itself. (Starting with Theorem 2 we will let M and (E, p, B) denote respectively such manifolds and bundles.) The following is our main lemma.

Theorem 1 [1]. *Let $\xi = (E, p, B)$ be a fibre bundle over (Hausdorff) space B with fibre F a metric absolute retract. Let $A \subset E$ be a closed set such that for each $b \in B$, the inclusion $j : p^{-1}(b) \setminus (A \cap p^{-1}(b)) \to p^{-1}(b)$ is a homotopical equivalence. Suppose (K, L) is a locally finite simplicial pair and f a map of $|K|$ into B, then each lifting f_1 of $f \mid |L|$ into $E \setminus A$ (that is, $f_1 : |L| \to E \setminus A$ such that $pf_1 = f \mid |L|$) can be extended to a lifting f^* of f into $E \setminus A$.*

A closed subset A of a space X is a *Z-set* if Interior $(A) = \emptyset$ and for each nonempty homotopically trivial open subset U of X, $U \setminus A$ remains homotopically trivial. By virtue of Theorem 1 we prove

Theorem 2. *Let K be a closed set in the total space $M \times B$ of the product bundle $(M \times B, p, B)$ over polyhedron B satisfying that for each $b \in B$, $K \cap p^{-1}(b)$ is a Z-set in $p^{-1}(b)$. Let $\mathcal{U}$ denote any open cover of $M \times B$. Then there is a fibre-preserving (that is, each $p^{-1}(b)$ being mapped into itself by f_t) homotopy $F = \{f_t\}$ of $M \times B$ into itself such that $f_0 =$ identity, $\mathrm{cl}\,(f_1(M \times B)) \cap K = \emptyset$ and F is limited by $\mathcal{U}$ (that is, each $F(\{x\} \times [0, 1]) \subset U$ for some $U \in \mathcal{U}$).*

The case where $B = \{\text{point}\}$ was announced earlier by D. Henderson. (Incidentally, our proof may be different from his.)

Hereafter all maps f of any $A \subset E$ into E will be fibre-preserving maps, that is, $pf(x) = p(x)$ for any $x \in E$. We also let $K_1, K_2, \ldots$ denote closed subsets of E such that for any $b \in B$, $K_i \cap p^{-1}(b)$ is a Z-set in $p^{-1}(b)$.

Theorem 3. *Let f be a homeomorphism of K_1 onto K_2. Then f can be extended to a homeomorphism $\tilde{f}$ of E provided that f is homotopic to the identity on K_1.*

Furthermore, if the homotopy is limited by some open cover $\mathscr{U}$ of E, we may choose $\tilde{f}$ to be isotopic to the identity and the isotopy be limited by $\mathrm{St}^{(4)}(\mathscr{U})$.

(We define $\mathrm{St}(\mathscr{U})$ to be the open cover of E consisting of all sets V such that for some $U \in \mathscr{U}$, $V = \bigcup\{W \in \mathscr{U} : W \cap U \neq \emptyset\}$.)

Theorem 4. *E is homeomorphic to* $E \setminus \bigcup_{i \geq 1} K_i$. *Furthermore, if we let φ denote the collection of all such homeomorphisms and if $\mathscr{U}$ is any open cover of E, we may choose $f \in \varphi$ to be isotopic to the identity and the isotopy be limited by $\mathscr{U}$.*

Using the same technique as in [4] we prove

Theorem 5. *Let $(M \times \Delta_n, p, \Delta_n)$ be a product bundle over n-simplex Δ_n and $f : M \times \Delta_n \to M \times \Delta_n$ be a map such that $f \mid M \times \partial\Delta_n$ is a homeomorphism of $M \times \partial\Delta_n$. Then $f \mid M \times \partial\Delta_n$ can be extended to a homeomorphism F of $M \times \Delta_n$. Furthermore, if $n = 1$ and the homotopy $\{f_t = f \mid M \times \{t\}\}$* is *limited by some open cover $\mathscr{U}$ of M, we may choose F to be limited by* $\mathrm{St}^{(10)}(\mathscr{U})$.

Corollary. *Any two homeomorphisms of M are isotopic if and only if they are homotopic.*

(This result was also announced by T. A. Chapman.)

References

[1] *R. Y. T. Wong:* On homeomorphisms of infinite-dimensional bundles, I.
[2] *T. A. Chapman and R. Y. T. Wong:* On homeomorphisms of infinite-dimensional bundles, II.
[3] *T. A. Chapman and R. Y. T. Wong:* On homeomorphisms of infinite-dimensional bundles, III.
[4] *R. Y. T. Wong:* Parametric extensions of homeomorphisms for Hilbert manifolds.

ON PSEUDO-OPEN MAPPINGS

D. ZAREMBA

Wrocław

Let us call a mapping[1]) $f : X \to f(X)$ of a topological space X to be *pseudo-open* if

(1) *the images $f(U)$ of open sets $U \subset X$ are open whenever U contains at least one component of the counter-image $f^{-1}(y)$, for which $f^{-1}(y) \cap U \neq \emptyset$.*

Now suppose that the topological spaces in further considerations are metric. It is not too hard to show that for a compact space X a mapping f is pseudo-open if and only if

(2) *the inequality $f^{-1}(y) \cap \mathrm{Ls}\, f^{-1}(y_n) \neq \emptyset$ implies the inequality $C \cap \mathrm{Ls}\, f^{-1}(y_n) \neq \emptyset$ for all components C of $f^{-1}(y)$*[2]).

In general, (1) is stronger than (2).

It follows from the definitions that for compact spaces the class of pseudo-open mappings contains the class of open ones as well as the class of monotone mappings, but it is strictly larger than the union of both of them. Some theorems which are true for monotone as well as for open mappings can be generalized to pseudo-open mappings. For example one can show that

a non-degenerate pseudo-open image of an arc is an arc.

The concept of pseudo-open mappings concerns the problem to give some sufficient conditions, for metric spaces $\{X, \varrho\}$ and for mappings $f : X \to f(X)$, such that the set $\{\varrho(x, y) : f(x) = f(y)\}$ is connected. It appears that for pseudo-open mappings of chainable continua the set $\{\varrho(x, y) : f(x) = f(y)\}$ is connected, and moreover, it is connected also for pseudo-open mappings of such continua X, for which

(3) *any continuum $Q \subset X \times X$ intersects the diagonal of $X \times X$ whenever $Q \cap \{x\} \times X \neq \emptyset$ and $Q \cap X \times \{x\} \neq \emptyset$*[3]).

---

[1]) A mapping means in this paper a continuous function.

[2]) $\mathrm{Ls}\, f^{-1}(y_n)$ denotes here superior limit of the sequence $\{f^{-1}(y_n)\}$.

[3]) Property (3) holds for all continua X of span 0 (for definition of the span of a space see A. Lelek: Disjoint mappings and the span of spaces, Fundamenta Mathematicae *55* (1964), 199—214), thus for all chainable continua (ibidem).

Therefore the following theorem holds:

If a continuum X has the property (3) *and a mapping* $f: X \to f(X)$ *is pseudo-open, then the set* $\{\varrho(x, y): f(x) = f(y)\}$ *is connected.*

While proving this theorem a stronger result is obtained, namely it appears that the set $\{(x, y): f(x) = f(y)\}$ is also connected.

The following lemma is useful to prove the theorem:

If $f: X \to f(X)$ *is a pseudo-open mapping of a compact space X onto a connected space* $f(X)$, *then* $f(C) = f(X)$ *for any component C of X.*

Detailed proofs will be published in Colloquium Mathematicum.

SPACES WITH REGULAR G_δ-DIAGONALS

P. ZENOR

Auburn

The proofs that are omitted in this note will appear in [6]. Recall that a subset H of the space X is a regular G_δ-set if there is a sequence $\{U_n\}$ of open sets containing H such that $H = \bigcap_{i=1}^{\infty} U_i = \bigcap_{i=1}^{\infty} U_i^-$. A space X has a (regular) G_δ-diagonal if $\{(x, x) : x \in X\}$ is a (regular) G_δ-set in $X \times X$. In [2], Ceder obtains the following characterization of spaces with G_δ-diagonals:

Theorem 1. *X has a G_δ-diagonal if and only if there is a sequence $\{\mathcal{G}_n\}$ of open covers of X such that if $x \in X$, then $x = \bigcap_{i=1}^{\infty} \mathrm{st}(x, \mathcal{G}_i)$.*

We have a comparable characterization of spaces with regular G_δ-diagonals:

Theorem 2. *X has a regular G_δ-diagonal if and only if there is a sequence $\{\mathcal{G}_n\}$ of open covers of X such that if x and y are distinct points of X, then there are an integer n and open sets U and V containing x and y respectively such that no member of $\mathcal{G}_n$ intersects both U and V.*

From Theorems 1 and 2, it is quite easy to see that any paracompact T_2-space with a G_δ-diagonal has a regular G_δ-diagonal. Also, it is a corollary to Theorem 2 that any space with a regular G_δ-diagonal is Hausdorff.

A development $\{\mathcal{G}_n\}$ for the space X is said to satisfy the 3-link property if it is true that if p and q are distinct points, then there is an integer n such that no member of $\mathcal{G}_n$ intersects both $\mathrm{st}(p, \mathcal{G}_n)$ and $\mathrm{st}(q, \mathcal{G}_n)$ (Heath [3]). According to Borges [1], a space X is a $w\Delta$-space if there is a sequence $\{\mathcal{G}_n\}$ of open covers of X such that if x is a point and, for each n, x_n is a point of $\mathrm{st}(x, \mathcal{G}_n)$, then the sequence $\{x_n\}$ has a cluster point.

Theorem 3. *Let X be a topological space. Then the following conditions are equivalent:*

(a) *X admits a development satisfying the 3-link property.*

(b) *X is a $w\Delta$-space with a regular G_δ-diagonal.*

(c) *There is a semi-metric d on X such that:*

(i) *If* $\{x_n\}$ *and* $\{y_n\}$ *are sequences both converging to* x, *then* $\lim_{n\to\infty} d(x_n, y_n) = 0$.

(ii) *If* x *and* y *are distinct points of* X *and* $\{x_n\}$ *and* $\{y_n\}$ *are sequences converging to* x *and* y *respectively, then there are integers* N *and* M *such that if* $n > N$, *then* $d(x_n, y_n) > 1/M$.

According to Morita [5], a space X is an M-space if there is a normal sequence $\{\mathscr{G}_n\}$ of open covers of X such that if x is a point and, for each n, x_n is a point of $\mathrm{st}(x, \mathscr{G}_n)$, then the sequence $\{x_n\}$ has a cluster point.

Theorem 4. *If* X *is a topological space, then the following conditions are equivalent*:

(a) X *is metrizable.*

(b) X *is a* T_1-M-*space such that* X^2 *is perfectly normal.*

(c) X *is an* M-*space with a regular* G_δ-*diagonal.*

(d) X *is a* T_1-M-*space such that* X^3 *is hereditarily normal.*

(e) X *is a* T_1-M-*space such that* X^3 *is hereditarily countably paracompact.*

(f) X *is an* M-*space that admits a one-to-one continuous function onto a metric space.*

Finally, in [1], Borges shows that if X is paracompact, locally connected and locally peripherally compact, then X is metrizable if and only if X has a G_δ-diagonal. Borges' result follows as a corollary to the following theorem:

Theorem 5. *If* X *is locally connected and locally peripherally compact, then* X *is metrizable if and only if* X *has a regular* G_δ-*diagonal.*

Proof. Let $\{\mathscr{U}_n\}$ be a sequence of open covers of X such that each member of $\mathscr{U}_n$ is connected and such that if p and q are distinct points, then there are open sets U and V containing p and q respectively and an integer n such that no member of $\mathscr{U}_n$ intersects both $\mathrm{st}(p, \mathscr{U}_n)$ and $\mathrm{st}(q, \mathscr{U}_n)$. We will first show that $\{\mathscr{U}_n\}$ is a development for X. To this end, let $x \in X$ and let U be an open set containing x. There is an open set V with a compact boundary such that $x \in V \subset U$. Suppose that, for each n, there is a member, say g_n, of $\mathscr{U}_n$ that contains x and intersects $X - V$. Then, since each g_n is connected, there is a point x_n of the boundary of V that is in g_n. Since the boundary of V is compact, the sequence $\{x_n\}$ has a cluster point, say x_0. It follows that $x_0 \in \bigcap_{n=1}^{\infty} \mathrm{cl}(\mathrm{st}(x, \mathscr{U}_n))$ which is a contradiction. It follows that X is developable. By Theorem 3, there is a development $\{\mathscr{G}_n\}$ for X that satisfies the 3-link property. Since X is locally connected, we may assume that, for each n, the members of $\mathscr{G}_n$ are connected. Let x denote a point of X and let U be an open set containing x. We will show that there is an integer n such that if $g \in \mathscr{G}_n$ and $g \cap \mathrm{st}(x, \mathscr{G}_n) \neq \emptyset$, then $g \subset U$. It will then follow that X is metrizable by Moore's Metrization Theorem

[4]. To this end, let V be an open subset of U containing x with compact boundary. Suppose that, for each n, there are members U_n and V_n of $\mathcal{G}_n$ such that $x \in U_n$, $U_n \cap \cap V_n \neq \emptyset$, and $(U_n \cap V_n) \cap (X - V) \neq \emptyset$. Since, for each n, $U_n \cup V_n$ is connected, there is a point x_n of $U_n \cup V_n$ in the boundary of V. Since the boundary of V is compact, there is a cluster point, x_0, of $\{x_n\}$. But it follows that, for each n, there is a member of $\mathcal{G}_n$ that intersects both of st $(x, \mathcal{G}_n)$ and st $(x_0, \mathcal{G}_n)$ which is a contradiction, from which the theorem follows.

References

[1] *C. J. R. Borges:* On metrizability of topological spaces. Canad. J. Math. *20* (1968), 795—803.
[2] *J. G. Ceder:* Some generalizations of metric spaces. Pacific J. Math. *11* (1961), 105—125.
[3] *R. W. Heath:* Metrizability, compactness and paracompactness in Moore spaces. Notices Amer. Math. Soc. *10* (1963), 105.
[4] *R. L. Moore:* A set of axioms for plane analysis situs. Fund. Math. *25* (1935), 13—28.
[5] *K. Morita:* Products of normal spaces with metric spaces. Math. Ann. *154* (1964), 365—382.
[6] *P. Zenor:* On spaces with regular G_δ-diagonals. Pacific J. Math. (to appear).

AUBURN UNIVERSITY, AUBURN, ALABAMA

LATTICES AND TOPOLOGY

S. P. ZERVOS

Athens

Abbreviations. Iff = if and only if; $\blacksquare$ denotes the end of a proof.

Notation. $P(E)$ = the set of all subsets of the set E.

1. The inverse of a function between two complete lattices

Motivation. Let E and H be nonvoid sets, f the extension $P(E) \to P(H)$ of a single(multiple)-valued function $\sigma : E \to H$ and f^{-1} (f^{+}) the inverse (superior inverse, in the sense given in Berge [3], p. 26) of $f : P(H) \to P(E)$ (f^{+} reduces to f^{-1} for single-valued σ). The general properties of f^{-1}, in relation to those of f, being in the background of many facts in General Topology (and elsewhere), it seems interesting to search for an analogously useful definition of the "inverse" f^{-1} of a single-valued function f between two complete lattices P and Q.

Notation. X, X', X'' and X_i (Y, Y', Y'' and Y_j) denote elements of $P(Q)$; let f be a single-valued function $P \to Q$, $X_0(Y_0)$ the minimum and $X_u(Y_u)$ the maximum of $P(Q)$. Let (X_i) $[(Y_j)]$ be the family $(X_i)_{i \in I}[(Y_j)_{j \in J}]$. $\{X \mid \ldots\}$ = the set of all X such that .... Notice: In Bourbaki and in our previous papers, the above-mentioned $f^{-1} : P(H) \to P(E)$ was written with -1 above f.

Permanent hypothesis. $\forall Y$, $\{X \mid f(X) \leqq Y\} \neq \emptyset$. In particular, this is fulfilled if $f(X_0) = Y_0$.

Definition 1. $\forall Y$, $f^{-1}(Y) = \bigvee\{X \mid f(X) \leqq Y\}$.

Special case. For a binary relation R on $E \times H$, $X (\subseteq E) \mapsto R(X)$ defines a single-valued function $f : P(E) \to P(H)$. The f^{-1} just defined gives then what can be called "the superior inverse R^{+}" of R (distinct from what is usually denoted by R^{-1} and in Bourbaki by -1 above R); it reduces to f^{+} when R is a function.

Abbreviations. $f(\bigvee) \leqq \bigvee f$ stays for: $\forall (X_i)$, $f(\bigvee X_i) \leqq \bigvee f(X_i)$; $f^{-1}(\bigvee) \geqq \bigvee f^{-1}$ stays for: $\forall (Y_j)$, $f^{-1}(\bigvee Y_j) \geqq \bigvee f^{-1}(Y_j)$; and all analogous abbreviations.

An immediate consequence of Definition 1 is

Proposition 1. a) $\forall X, f^{-1} \circ f(X) \geqq X$;
b) *f^{-1} is isotone; hence also $f^{-1}(\bigwedge) \leqq \bigwedge f^{-1}$ and $\bigvee f^{-1} \leqq f^{-1}(\bigvee)$;*
c) $f^{-1}(Y_u) = X_u$.

Obvious remarks: $[\forall Y, f \circ f^{-1}(Y) = Y] \Rightarrow [f$ is surjective$]$; $[\forall X, f^{-1} \circ f(X) = X] \Rightarrow [f^{-1}$ is surjective and f is injective$]$; $[f$ is isotone$] \Rightarrow [f(X_0) = Y_0]$.

Proposition 2. *If $f(\bigvee) \leqq \bigvee f$, then $\forall Y, f \circ f^{-1}(Y) \leqq Y$.*

Proof. $f^{-1}(Y) = \bigvee X_i$, with $f(X_i) \leqq Y$. Hence

$$f \circ f^{-1}(Y) = f(\bigvee X_i) \leqq \bigvee f(X_i) \leqq Y. \blacksquare$$

Proposition 3. *If $\forall Y, f \circ f^{-1}(Y) \leqq Y$, then* a) $f^{-1} \circ f \circ f^{-1} = f^{-1}$,
b) *$[f^{-1}$ is injective$] \Rightarrow [f$ is surjective$]$.*

Proof. a) $\forall Y$, by Definition 1, $[f(f^{-1}(Y)) = Y_1] \Rightarrow [f^{-1}(Y) \leqq f^{-1}(Y_1)]$ while, by the hypothesis, $Y_1 \leqq Y$ and, by Definition 1, $[Y_1 \leqq Y] \Rightarrow [f^{-1}(Y_1) \leqq f^{-1}(Y)]$ so that $f^{-1} \circ f \circ f^{-1}(Y) = f^{-1}(Y)$.

b) Keeping the notation from a), $f^{-1}(Y) = f^{-1}(Y_1)$ and the hypothesis that f^{-1} is injective implies that $Y = Y_1$, hence f is surjective. $\blacksquare$

Proposition 4. *Hypotheses*: $\forall Y, f \circ f^{-1}(Y) \leqq Y$; *$f$ is isotone.*

Conclusions: a) $f^{-1}(\bigwedge) = \bigwedge f^{-1}$, *i.e., f^{-1} is a complete $\bigwedge$-morphism*;
b) $f \circ f^{-1} \circ f = f$;
c) $f(\bigvee) \leqq \bigvee f$.

Hence f is a complete $\bigvee$-morphism.

Proof. a) It suffices to prove $\geqq$; but

$$[X \leqq \bigwedge f^{-1}(Y_i)] \Rightarrow [\forall i, X \leqq f^{-1}(Y_i)] \Rightarrow [\forall i, f(X) \leqq f \circ f^{-1}(Y_i) \leqq Y_i] \Rightarrow$$
$$\Rightarrow [f(X) \leqq \bigwedge Y_i] \Rightarrow [X \leqq f^{-1}(\bigwedge Y_i)].$$

b) By the hypothesis, $\forall X$, $f \circ f^{-1}(f(X)) \leqq f(X)$, while, by Proposition 1, $f^{-1} \circ f(X) \geqq X$. Hence $f \circ f^{-1} \circ f(X) \geqq f(X)$ which proves the equality.

c) $[Y \geqq \bigvee f(X_i)] \Rightarrow [\forall i, Y \geqq f(X_i)] \Rightarrow [\forall i, f^{-1}(Y) \geqq f^{-1} \circ f(X_i) \geqq X_i] \Rightarrow$
$\Rightarrow [f^{-1}(Y) \geqq \bigvee X_i] \Rightarrow [Y \geqq f \circ f^{-1}(Y) \geqq f(\bigvee)]. \blacksquare$

Corollary. *The hypotheses of Proposition 4 are equivalent to the supposition that f is a complete $\bigvee$-morphism.*

Proposition 5. *Under the hypotheses of Proposition 4,*

a) $[f(X) = Y] \Rightarrow [f \circ f^{-1}(Y) = Y]$;

b) $[f \textit{ is surjective}] \Rightarrow [\forall Y, f \circ f^{-1}(Y) = Y \textit{ and } f^{-1} \textit{ is injective}]$;

c) $[f^{-1} \textit{ is surjective}] \Rightarrow [\forall X, f^{-1} \circ f(X) = X]$;

d) $[f \textit{ is injective}] \Rightarrow [\forall X, f^{-1} \circ f(X) = X]$.

Proof. a) $[f(X) = Y] \Rightarrow [X \leqq f^{-1}(Y)]$; hence $Y = f(X) \leqq f \circ f^{-1}(Y) \leqq Y$.

b) $[f^{-1}(Y_1) = f^{-1}(Y_2)] \Rightarrow [f \circ f^{-1}(Y_1) = f \circ f^{-1}(Y_2)] \Rightarrow [Y_1 = Y_2]$.

c) $[X = f^{-1}(Y)] \Rightarrow [X \leqq f^{-1} \circ f(X) = f^{-1}(f \circ f^{-1}(Y)) \leqq f^{-1}(Y) = X]$.

d) $[f(X) = Y] \Rightarrow [f(f^{-1} \circ f(X)) = f \circ f^{-1}(Y) = Y = f(X)]$; the injectivity of f implies then the assertion.$\blacksquare$

Proposition 6. *If f is a complete $\bigvee$-morphism, then*

$$\forall Y \in f(P), \quad f^{-1}(Y) = \bigvee\{X \mid f(X) = Y\}.$$

Proof. Let $Y \in f(P)$ and set $E' = \{X \mid f(X) < Y\}$, $E'' = \{X \mid f(X) = Y\}$. Then obviously $E'' \neq \emptyset$, while, if $E' = \emptyset$, the assertion also is obvious. Suppose $E' \neq \emptyset$ and denote by X_i the elements of E' and by X_j those of E''; set $X' = \bigvee X_i$ and $X'' = \bigvee X_j$. Then $f(X' \vee X'') = f(\bigvee X_i) \vee f(\bigvee X_j) = (\bigvee f(X_i)) \vee (\bigvee f(X_j)) = Y$; hence $X' \vee X''$ is some X_j; therefore, $\bigvee X_j = X'' \leqq X' \vee X'' \leqq \bigvee X_j = X''$; hence $X' \vee X'' = X''$ and $X'' = f^{-1}(Y)$.$\blacksquare$

The following combined corollary of the preceding propositions seems to be useful.

Theorem 1. *Hypothesis*: *f is a complete $\bigvee$-morphism.*

Conclusions: a) *f^{-1} is a complete $\bigwedge$-morphism; hence, in particular, f^{-1} is isotone and $\bigvee f^{-1} \leqq f^{-1}(\bigvee)$;*

b) $\forall X, f^{-1} \circ f(X) \geqq X$ *and*, $\forall Y, f \circ f^{-1}(Y) \leqq Y$;

c) $f^{-1} \circ f \circ f^{-1} = f^{-1}$ *and* $f \circ f^{-1} \circ f = f$; *hence also* $(f \circ f^{-1})^2 = f \circ f^{-1}$ *and* $(f^{-1} \circ f)^2 = f^{-1} \circ f$;

d) $\forall Y \in f(P), f^{-1}(Y) = \bigvee\{X \mid f(X) = Y\}$;

e) *f is surjective iff f^{-1} is injective iff*, $\forall Y, f \circ f^{-1}(Y) = Y$;

f) *f is injective iff f^{-1} is surjective iff*, $\forall X, f^{-1} \circ f(X) = X$;

g) $f^{-1}(Y_u) = X_u$;

i) $f(X_0) = Y_0$.

Corollary. *$f^{-1} \circ f$ is a Kuratowski's closure operator.*

Theorem 2 (Characterization of f^{-1}). *Hypotheses*: *f is isotone and g is a single-valued isotone function $Q \to P$, such that*: $\forall X, g \circ f(X) \geqq X$ *and* $\forall Y, f \circ g(Y) \leqq Y$.

Conclusions: a) $g = f^{-1}$;

b) *f is a complete $\bigvee$-morphism.*

Proof. a) $\forall Y$, set $E_Y = \{X \mid f(X) \leqq Y\}$; by our permanent hypothesis, $E_Y \neq \emptyset$. $[X \leqq g(Y)] \Rightarrow [f(X) \leqq f \circ g(Y) \leqq Y] \Rightarrow [X \in E_Y]$; hence $\sup E_Y \geqq g(Y)$. $[f(X) \leqq \leqq Y] \Rightarrow [X \leqq g \circ f(X) \leqq g(Y)] \Rightarrow [X \leqq g(Y)]$; hence $\sup E_Y \leqq g(Y)$. Therefore, $g(Y) = \sup E_Y$; but $\sup E_Y = f^{-1}(Y)$.

b) f isotone and $\forall Y$, $f \circ f^{-1}(Y) \leqq Y$ was the hypothesis of Proposition 4; its conclusion c) is the above b). ■

Proposition 7. *When f is a complete surjective $\bigvee$-morphism, then $(f^{-1})^{-1} = f$ on $f^{-1}(Q)$; and if, in addition, $f^{-1}(Y_0) = X_0$, then $(f^{-1})^{-1} \leqq f$ on $P - f^{-1}(Q)$, with $<$ being actually possible.*

Proof. Set $g = f^{-1}$. Then $\forall Y' \in Q$, $g^{-1}(f^{-1}(Y')) = \bigvee\{Y \mid g(Y) \leqq f^{-1}(Y')\}$; $[f^{-1}(Y) \leqq f^{-1}(Y')] \Rightarrow [f \circ f^{-1}(Y) \leqq f \circ f^{-1}(Y')] \Rightarrow [Y \leqq Y']$; and since $g(Y') = = f^{-1}(Y')$, $g^{-1}(f^{-1}(Y')) = Y'$; so much for $(f^{-1})^{-1}$ on $f^{-1}(Q)$. Now, when $f^{-1}(Y_0) = = X_0$, $\bigvee\{Y \mid g(Y) \leqq X\}$ is $\neq \emptyset$ for all X, hence $(f^{-1})^{-1}$ is defined on P; then for all X, $g^{-1}(X) = \bigvee\{Y_i \mid f^{-1}(Y_i) \leqq X \leqq f^{-1}(f(X))\} \leqq f(X)$. That the case $<$ is actually possible is shown by the following example: $P = \{X_0 < X_1 < X_u\}$, $Q = \{Y_0 < Y_u\}$ and $f(X_0) = Y_0$, $f(X_1) = f(X_u) = Y_u$; then $f^{-1}(Y_0) = X_0$, and $(f^{-1})^{-1}(X_1) = = \bigvee\{Y_i \mid f^{-1}(Y_i) \leqq X_1\} = Y_0$. ■

Note 1. Proposition 7 is the only result where the hypothesis $f^{-1}(Y_0) = X_0$ is explicitly made.

Note 2. The hypothesis of Theorem 1 is not sufficient for $f^{-1}(\bigvee) \leqq \bigvee f^{-1}$.

Example. Let $P = \{X_0 < X_1 < X_u\}$, $Q = \{Y_0 < Y_i < Y_u \ (i = 1, 2);\ Y_0 = = Y_1 \wedge Y_2, Y_1 \vee Y_2 = Y_u\}$ and $f(X_0) = Y_0, f(X_1) = Y_1, f(X_u) = Y_u$. Then $f^{-1}(Y_0) = = f^{-1}(Y_2) = X_0, f^{-1}(Y_1) = X_1$ and $f^{-1}(Y_u) = X_u$. Hence $f^{-1}(Y_1 \vee Y_2) = f^{-1}(Y_u) = = X_u > X_1 = X_1 \vee X_0 = f^{-1}(Y_1) \vee f^{-1}(Y_0)$.

In order to assure that $f^{-1}(\bigvee) \leqq \bigvee f^{-1}$, we shall have to introduce a new notion, that of the "f-finer covering".

Definition 2. a) Given (X, Y) with $f(X) \leqq Y$ and coverings (X_i), (Y_j) of X, Y, respectively (i.e., $\bigvee X_i \geqq X$, $\bigvee Y_j \geqq Y$), (X_i) will be called "*f-finer*" than (Y_j) iff $\forall i$, $\exists j$ such that $f(X_i) \leqq Y_j$ (for $Q = P$ and for the identical mapping $f : P \to P$, "f-finer" reduces to the classical notion of "finer").

b) (P, Q) will be said to have the property of "*f-fineness*" iff $\forall (X, Y)$ with $f(X) \leqq Y$ and $\forall$ covering (Y_j) of Y, $\exists$ a covering (X_i) of X f-finer than (Y_j).

Proposition 8. *When f is a complete $\bigvee$-morphism and (Y_j) is a covering of Y, then $[\forall X$ with $f(X) \leqq Y$, $\exists$ a covering (X_i) of X f-finer than $(Y_j)] \Rightarrow [f^{-1}(\bigvee Y_j) = = \bigvee f^{-1}(Y_j)]$.*

Proof. It suffices to prove $f^{-1}(\bigvee) \leqq \bigvee f^{-1}$. $[X \leqq f^{-1}(\bigvee Y_j)] \Rightarrow [f(X) \leqq \leqq f \circ f^{-1}(\bigvee Y_j) \leqq \bigvee Y_j]$. If (X_i) is a covering of X f-finer than (Y_j), then $\forall i$, $\exists j_i$ with

$f(X_i) \leqq Y_{j_i}$. $[\forall i, f(X_i) \leqq Y_{j_i}] \Rightarrow [\forall i, X_i \leqq f^{-1} \circ f(X_i) \leqq f^{-1}(Y_{j_i})] \Rightarrow [X \leqq \bigvee X_i \leqq \leqq \bigvee f^{-1}(Y_{j_i}) \leqq \bigvee f^{-1}(Y_j)]$.$\blacksquare$

Theorem 3. *When f is a complete $\bigvee$-morphism, then (P, Q) has the property of f-fineness iff $f^{-1}(\bigvee) \Rightarrow \bigvee f^{-1}$, i.e., iff f^{-1} is a complete lattice morphism.*

Proof. 1) Suppose (P, Q) has the property of f-fineness and apply Proposition 8 with $Y = \bigvee Y_j$. 2) Suppose $f^{-1}(\bigvee) \leqq \bigvee f^{-1}$. Then, if $f(X) \leqq Y$ and $\bigvee Y_j \geqq Y$, $X \leqq \leqq f^{-1}(Y) \leqq f^{-1}(\bigvee Y_j) \leqq \bigvee f^{-1}(Y_j)$, hence $(f^{-1}(Y_j))$ is a covering of X with $f \circ \circ f^{-1}(Y_j) \leqq Y_j$.$\blacksquare$

Remark 1. Consequently, in the special case of point-mappings $f : P(E) \rightarrow \rightarrow P(H)$, not points themselves but the f-fineness of $(P(E), P(H))$ implied by them influenced the above generalized interconnection between f and f^{-1}. Points contributed rather to the properties of the extension $f : P(E) \rightarrow P(H)$ of $\sigma : E \rightarrow H$; so, for instance, the property of f, in that special case, to be injective iff $f(\bigwedge) = \bigwedge f$, is not necessarily shared by a complete $\bigvee$-morphism $f : P \rightarrow Q$, even when (P, Q) has the property of f-fineness. Both "if" and "only if" assertions fail, as it is shown by the following examples: 1) $P = \{X_0 < X_u\}$, $Q = \{Y_0 = Y_u\}$ and $f(X_0) = f(X_u) = = Y_0$; then $f(\bigwedge) = \bigwedge f$, but f is not injective. 2) $P = \{X_0 < X_i < X_u \ (i = 1, 2)$; $X_0 = X_1 \wedge X_2, X_1 \vee X_2 = X_u\}$, $Q = \{Y_0 < Y' < Y_i < Y_u \ (i = 1, 2); Y' = Y_1 \wedge Y_2$, $Y_1 \vee Y_2 = Y_u\}$ and $f(X_0) = Y_0$, $f(X_i) = Y_i$, $f(X_u) = Y_u$; then f is injective, but $f(X_1 \wedge X_2) = f(X_0) = Y_0 < Y' = Y_1 \wedge Y_2 = f(X_1) \wedge f(X_2)$.

Remark 2 (a "by-product" of Remark 1). The last example shows: If P is the four lattice and if for any injective complete $\bigvee$-morphism $f : P \rightarrow Q$, $f(\bigwedge) = \bigwedge f$, then Q cannot contain any sublattice of the form

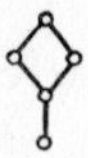

More generally, choosing a certain P and imposing conditions on all functions $f : P \rightarrow Q$ of a certain species in order to obtain, from the "outside", information for the "inside" of Q, seems to be a possibly useful "external" method for studying Q.

We close now our general treatment of f and f^{-1} and, for the first time in this paper, make the supposition that P and Q are complemented. The following hypothesis suggests itself: In P, $[X \wedge X' = X_0] \Rightarrow [\forall$ complement $\complement X$ of X, $X' \leqq \complement X]$; similarly in Q. However, Huntington's theorem ([4], p. 46; [7], p. 130) asserts that P and Q are then simply Boolean algebras. So, all that remains is to search for additional properties of f and f^{-1} in that special case.

Proposition 9. *Suppose that P and Q are Boolean algebras and f a $\bigvee$-morphism $P \rightarrow Q$. Then*

a) $\forall(X_1, X_2), f(X_1) - f(X_2) \leqq f(X_1 - X_2)$;

b) *if* $\{X \mid f(X) = Y_0\} = X_0$, *then* $[\forall(X_1, X_2), f(X_1) - f(X_2) = f(X_1 - X_2)] \Rightarrow$ $\Rightarrow$ [*f is injective*].

Proof. a) A direct consequence of the Boolean identity $(X_1 \wedge X_2) \vee$ $\vee (X_1 \wedge \complement X_2) = X_1$ and the Boolean implication $[Y \vee Y' = Y''] \Rightarrow [Y'' - Y' \leqq Y]$.

b) Obvious.∎

Note 3. Conjecture: In Proposition 9, b), $\Leftarrow$ holds as well.

Proposition 10. *Suppose that P and Q are complete Boolean algebras and f a complete $\bigvee$-morphism $P \to Q$. Then, if (P, Q) has the property of f-fineness, f^{-1} transforms Q to a complete subalgebra P' of P and complements in P', Q have the following properties*:

a) $\forall Y, f^{-1}(\complement Y) = \complement f^{-1}(Y)$;

b) *more generally*, $\forall(Y_1, Y_2), f^{-1}(Y_1 - Y_2) = f^{-1}(Y_1) - f^{-1}(Y_2)$.

Proof. a) By the isotoneity of f^{-1}, $f^{-1}(Y_0)$ and $f^{-1}(Y_u)$ is respectively the minimum and the maximum element of P'. $[Y \vee \complement Y = Y_u] \Rightarrow [f^{-1}(Y) \vee f^{-1}(\complement Y) =$ $= f^{-1}(Y_u)]$ and $[Y \wedge \complement Y = Y_0] \Rightarrow [f^{-1}(Y) \wedge f(\complement Y) = f^{-1}(Y_0)]$, which proves the assertion.

b) $f^{-1}(Y_1 - Y_2) = f^{-1}(Y_1 \wedge \complement Y_2) = f^{-1}(Y_1) \wedge \complement f^{-1}(Y_2)$.∎

One could think of applying the above facts to [12].

After this short digression to Boolean algebras, we come back to our initial P and Q (general complete lattices).

2. Topological considerations for the $\bigvee$-complete morphisms between complete lattices

Notation. $P_1(Q_1)$ is a $\bigvee$-complete sublattice of $P(Q)$ with X_0, $X_u(Y_0, Y_u)$; it will be its lattice of "open" elements.

Note 4. It is well-known that parts of General Topology have been generalized for (P, P_1) under supplementary hypotheses, especially for P_1; see, for instance, [1], [4], [6], [7] and, especially, [9] (among the initiators, M. H. Stone, A. Tarski and H. Wallman, in the years 1934–1938); however, even in the algebraically minded [5], P_1 is more special than here; we are concerned rather with $((P, P_1), (Q, Q_1))$ than with (P, P_1).

Definition 3. $\bigvee$-complete morphisms $f: P \to Q$ [with $f(P_1) \subseteq Q_1$ or with $f^{-1}(Q_1) \subseteq P_1$] will be called "*mappings*" ("*o-mappings*" or "*c-mappings*").

Definition 3 generalizes for arbitrary complete lattices the extension to power-sets of a single-valued function (in particular cases open or globally continuous single-valued function).

The composition of two mappings (o-mappings, c-mappings) gives again a mapping (o-mapping, c-mapping).

Note 5. In the case of the extension $f : P(E) \to P(H)$ of a multiple-valued function $\sigma : E \to H$ between topological spaces, the notion of a c-mapping is more general than that of an upper semicontinuous function in the sense of Berge ([3], p. 32, [4], p. 114–115), where $\forall X \in E, f(\{X\})$ has to be compact. According to [4] (p. 114), the notions of lower and upper semicontinuity of a multiple-valued function were introduced, in the thirties, independently by Bouligand and Kuratowski (see also: Kuratowski's "Topologie", II, 1961, p. 32, and Bouligand's "Titres et travaux scientifiques", 1961, p. 29).

For a generalization of topological facts concerning semicontinuous multiple-valued functions, independently of c-mappings, we refer the reader to Rosie Voreadou ([13], [14]).

Abbreviation. $\bigwedge$-complete distributivity = $\bigvee$-complete $\bigwedge$-distributivity.

Obviously, it holds the following

Metatheorem 1. *All results concerning globally continuous or open functions between topological spaces, which, together with their proofs, can be phrased exclusively in terms of open elements, without using distributivity (or with the use of distributivity or $\bigwedge$-complete distributivity) and with the use of only the above established results concerning the interconnection of f and f^{-1}, are a priori valid for $((P, P_1), (Q, Q_1))$ (where, moreover, P and Q are supposed, respectively, to be distributive or $\bigwedge$-completely distributive).*

Terminology (Bourbaki): Quasicompact = satisfying Heine-Borel-Lebesgue's axiom and not necessarily Hausdorff.

No special supposition on P or Q, P_1 or Q_1 is made in the following

Example. Let p and m be cardinals, m being sufficiently large; let q be an element of the Kurepa completion of the totally ordered set of cardinals $\leqq m$. Connexity and quasicompactness and, more generally, p-connexity and q-quasicompactness (see [16], [17] and [18]) are, substantially, notions for P_1 (or Q_1) (q-quasicompactness is even a notion for a q-complete semilattice). Then any surjective c-mapping $f : P \to Q$ such that (P, Q) has the property of f-fineness, preserves q-quasicompactness; if, in addition, $f^{-1}(Y_0) = X_0$, it also preserves p-connexity.

3. Adherent elements in not necessarily complemented lattices

Motivation. The introduction of "closed" elements in (P, P_1), when P is not necessarily complemented.

Additional notation. B, A and W denote respectively elements of P, P_1, Q_1; also, $C \in P$.

Definition 4. a) X will be called "*adherent*" to B, iff $\forall A$, $A \wedge X > X_0$ only if $A \wedge B > X_0$.

b) The join of all elements of P adherent to B will be called the "*adherence*" $\bar{B}$ of B.

Note 6. Even in the case of ordinary topology, no explicit use of adherent elements other than points or adherences (= closures) seems to have been made in the literature; consequently this simple notion seems new even in this case.

Abbreviations. X adh B = X is adherent to B; non adh = non adherent; etc.

Proposition 11. a) $\forall B$, X_0 adh B;

b) $[X > X_0] \Rightarrow [X \text{ non adh } X_0]$; *hence* $\bar{X}_0 = X_0$;

c) $\forall X$, X adh X; *hence* $X \leqq \bar{X}$ (*extensivity*);

d) $[X \geqq X'$ *and* X adh $B] \Rightarrow [X'$ adh $B]$;

e) $[C \geqq B$ *and* X adh $B] \Rightarrow [X$ adh $C]$; *hence* $[C \geqq B] \Rightarrow [\bar{C} \geqq \bar{B}]$ (*isotoneity*); *hence* $\bar{X}_1 \vee \bar{X}_2 \leqq \overline{X_1 \vee X_2}$;

f) $\forall A$, $[A \wedge B = X_0$ *and* $X \leqq A] \Rightarrow [X$ non adh $B]$.

Proof. Obvious; one uses the fact that Definition 4, a) is equivalent to: X adh B iff $[A \wedge B = X_0] \Rightarrow [A \wedge X = X_0]$. ∎

Definition 5. A single-valued function $f : P \to Q$ will be called "*surjective from below*" iff $\forall X$ and $\forall Y \leqq f(X)$, $\exists X_1 \leqq X$ with $f(X_1) = Y$; then, in particular, together with every $Y \in f(P)$, all $\leqq Y$ elements of Q belong to $f(P)$. (An analogous definition of "surjective from above" is obviously possible.)

Example. The extension of any function $E \to H$ to the power sets $P(E)$ and $P(H)$ is surjective from below.

Proposition 12. *Hypothesis*: *f is a surjective from below single-valued function $P \to Q$, such that f^{-1} is a $\bigwedge$-morphism, with $f^{-1}(Y_0) = X_0$ and $f^{-1}(Q_1) \subseteq P_1$.*

Conclusion: $\forall B$, $[X$ adh $B] \Rightarrow [f(X)$ adh $f(B)]$.

Proof. $[f(X)$ non adh $f(B)] \Rightarrow [\exists W$ with $W \wedge f(X) > Y_0$ and $W \wedge f(B) = Y_0]$. Since $Y_0 < W \wedge f(X) \leqq f(X)$, $\exists X_1 < X_0$ and $\leqq X$ such that $f(X_1) = W \wedge f(X)$;

then, $f(X_1) \leqq W$, hence $X_0 < X_1 \leqq f^{-1} \circ f(X_1) \leqq f^{-1}(W)$ and also $X_0 < X_1 \wedge \wedge f^{-1}(W)$; hence, a fortiori, $f^{-1}(W) \wedge X > X_0$. On the other hand, $[W \wedge f(B) = Y_0] \Rightarrow [f^{-1}(W) \wedge f^{-1} \circ f(B) = X_0] \Rightarrow [f^{-1}(W) \wedge B = X_0]$. Hence, X non adh B, contrary to our hypothesis. $\blacksquare$

An immediate corollary of Proposition 12 is

Theorem 4a. *When f is a surjective from below c-mapping with $f^{-1}(Y_0) = X_0$, then f transforms adherent elements to adherent elements.*

Definitions 3 – 5 make sense also when P_1 and Q_1 are any nonvoid subsets of P and Q respectively, not necessarily lattices; and Propositions 11 and 12 together with Theorem 4a hold as well. This and the definitions of the lower and upper semi-continuity at a point for multiple-valued functions between topological spaces given in ([4], p. 114) suggest the following combined "generalization".

Definition 6. A $\bigvee$-complete morphism $f : P \to Q$ will be called "*locally* **c**-*mapping at X*" (briefly "**c** *at X*"), iff $[W \wedge f(X) > Y_0] \Rightarrow [\exists A$ with $A \wedge X > X_0$ and $f(A) \leqq W]$.

Special case. Let P and Q be topological spaces, with $f : P \to Q$ a single-valued function and X a singleton $\{X^*\}$; then "**c** at X" means simply "continuous at the point X^*".

An obvious consequence of Proposition 12 and Definition 6 is

Theorem 4b. *When f is a surjective from below* **c** *at X mapping with $f^{-1}(Y_0) = X_0$, then* $[X \text{ adh } B] \Rightarrow [f(X) \text{ adh } f(B)]$.

Note 7. By Proposition 11, the notion of adherence given in Definition 4, b) is weaker than the classical one in that $\overline{\overline{X}} \leqq \overline{X}$ and $\overline{X}_1 \vee \overline{X}_2 \leqq \overline{X_1 \vee X_2}$ (consequences of isotoneity and extensivity). In the special case of power-sets, this operation $X \mapsto \overline{X}$ (by its definition compatible with P_1) gives a "generalized topological" [11] or "hypotopological" ([15], p. 356, [8]) space which was already considered by E. Čech and B. Pospíšil (see also the references in [15] to O. Ore).

Theorem 5. *When P is $\bigwedge$-completely distributive, then* a) *any join of elements adherent to B is also adherent to B;*

b) $B \mapsto \overline{B}$ *defines a Kuratowski's closure operator* (also, such an operator in the sense of McKinsey and Tarski [10], extended to lattices).

Proof. Since b) is an immediate consequence of Proposition 11 and a), it will suffice to prove a). Let (X_i) be a family of elements adh B. Since $\bigvee(A \wedge X_i) = A \wedge (\bigvee X_i)$, $[A \wedge (\bigvee X_i) > X_0] \Rightarrow [\bigvee(A \wedge X_i) > X_0]$; but $[\forall i,\ A \wedge X_i = X_0] \Rightarrow [\bigvee(A \wedge X_i) = X_0]$; hence, $[A \wedge (\bigvee X_i) > X_0] \Rightarrow [\exists i$ with $A \wedge X_i > X_0] \Rightarrow [A \wedge B > X_0]$. $\blacksquare$

Hence, it obviously holds the following

Metatheorem 2. *When P and Q are $\bigwedge$-completely distributive, all topological facts, which together with their proofs can be phrased exclusively in terms of open sets and closures (without complementation) and with the use of only the above established results concerning the interconnection of f and f^{-1}, are a priori valid for (P, P_1) and $((P, P_1), (Q, Q_1))$.*

Note 8. The fact that a complete lattice is $\bigwedge$-completely distributive iff it is Brouwerian shows that the above interdiction concerning the use of complements may, sometimes, become less strict; this is a point, where, independently of our present work, special results appear in previous literature.

References

[1] *J. Benabou:* Treillis locaux. Séminaire Dubreil/Pisot, Paris, 1957/58.
[2] *C. Berge:* Théorie générale des jeux à *n* personnes. Mémor. Sci. Math. *138*, Paris, 1957.
[3] *C. Berge:* Espaces topologiques. Fonctions multivoques. Paris, 1959 et 1965.
[4] *G. Birkhoff:* Lattice Theory. 3rd ed., Amer. Math. Soc., 1967.
[5] *J. Lévy-Bruhl:* Introduction aux structures algébriques. Paris, 1968.
[6] *C. H. Dowker and D. Papert:* On Urysohn's Lemma. General Topology and its Relations to Modern Analysis and Algebra, II (Proc. Second Prague Topological Sympos., 1966). Academia, Prague, 1967, 111—114.
[7] *R. Faure et E. Heurgon:* Structures ordonnées et algèbres de Boole. Paris, 1971.
[8] *P. B. Krikelis:* Certain relations between generalized topology and universal algebra. Proc. Internat. Sympos. on Topology and its Applications (Herceg-Novi, 1968). Savez Društava Mat. Fiz. i Astronom., Belgrade, 1969, 236—237.
[9] *L. Lésieur:* Les treillis en topologie, I et II. Séminaire A. Chatelet-Dubreil, Paris, 1953/54.
[10] *J. McKinsey and A. Tarski:* The algebra of topology. Ann. of Math. *45* (1944), 142—191.
[11] *Z. P. Mamuzić:* Introduction to General Topology. 1963.
[12] *R. Sikorski:* Boolean Algebras. 1960.
[13] *R. Voreadou:* Comm. Indiv. Congrès Intern. Math. Nice, 1970, 72.
[14] *R. Voreadou:* Sur certaines propriétés des fonctions quelconques entre espaces topologiques. C. R. Acad. Athènes, 1969.
[15] *S. P. Zervos:* Aspects modernes de la localisation des zéros des polynomes d'une variable. (Thèse.) Ann. Sci. École Norm. Sup. (*3*), *77* (4) (1960), 303—410.
[16] *S. P. Zervos:* Une généralisation du théorème de Bolzano pour la connexité. C. R. Acad. Sci. Paris Sér. A—B *260* (1965), 5979—5982.
[17] *S. P. Zervos:* Une généralisation abstraite du théorème topologique de Weierstrass pour la préservation de la quasi-compacité; une notion de dimension de quasi-compacité. C. R. Acad. Sci. Paris Sér. A—B *260* (1965), 6781—6784.
[18] *S. P. Zervos:* Une définition générale de la dimension. Séminaire Delange-Pisot-Poiton, Paris, 1965/66.

CONTENTS

Preface . . . 5
List of participants . . . 6
List of invited addresses . . . 8
List of communications . . . 9
Contributed papers . . . 15
AARTS, J. M.: Complementary Inductive Invariants and Dimension . . . 17
ALAS, O. T.: Uniform Continuity in Paracompact Spaces . . . 19
ALÒ, R. A.: Some Tietze Type Extension Theorems . . . 23
ANDERSON, R. D.: Some Open Questions in Infinite-Dimensional Topology . . . 29
ARCHANGELSKIJ, A. V.: On Cardinal Invariants . . . 37
AULL, C. E.: Point-Countable Bases and Quasi-Developments . . . 47
BANASCHEWSKI, B.: On Profinite Universal Algebras . . . 51
BARIT, W.: Contraction of some Spaces of Homeomorphisms . . . 63
BELLEN, A. and VOLČIČ, A.: Non-Cyclic Transformations and Uniform Convergence of Picard Sequences . . . 65
BINZ, E.: Recent Results in the Functional Analytic Investigations of Convergence Spaces 67
BORGES, C. R.: Four Generalizations of Stratifiable Spaces . . . 73
BORSUK, K.: Some Remarks Concerning the Theory of Shape in Arbitrary Metrizable Spaces . . . 77
CHARATONIK, J. J.: On the Fixed Point Property for Set-Valued Mappings of Hereditarily Decomposable Continua . . . 83
CHVALINA, J. and SEKANINA, M.: Realizations of Closure Spaces by Set Systems . . . 85
COMFORT, W. W. and NEGREPONTIS, S.: Continuous Functions on Products with Strong Topologies . . . 89
CSÁSZÁR, K.: *H*-Closed Extensions of Topological Spaces . . . 93
DANEŠ, J.: On Norms and Subsets of Linear Spaces . . . 97
DUDA, R.: One Result on Inverse Limits and Hyperspaces . . . 99
EFIMOV, B. A.: On the Imbedding of Extremally Disconnected Spaces into Bicompacta . . . 103
FLACHSMEYER, J.: Normal and Category Measures on Topological Spaces . . . 109
FLETCHER, P. and LINDGREN, W. F.: Topological Spaces which Admit a Compatible Complete Quasi-Uniformity . . . 117
FRIČ, R.: Sequential Envelope and Subspaces of the Čech-Stone Compactification . . . 123
FROLÍK, Z.: Topological Methods in Measure Theory and the Theory of Measurable Spaces 127
GAGRAT, M. and NAIMPALLY, S. A.: Proximity Approach to Topological Problems . . . 141
GÄHLER, S.: Über 2-Banach-Räume . . . 143
GARG, K. M. and NAIMPALLY, S. A.: On some Pretopologies Associated with a Topology 145
GERLITS, J.: On $\mathfrak{m}$-adic Spaces . . . 147
GORDH, G. R., JR.: On Monotone Decompositions of Smooth Continua . . . 149

GRIMEISEN, G.: On the Saturation of a Topological Partial Algebra with Respect to a Congruence Relation 151
DE GROOT, J.: On the Topological Characterization of Manifolds 155
HAGER, A. W.: Three Classes of Uniform Spaces 159
HAJEK, D. W. and STRECKER, G. E.: Direct Limits of Hausdorff Spaces 165
HAMBURGER, P.: On Internal Characterizations of Complete Regularity and Wallman-Type Compactifications 171
HANÁK, J.: Game-Theoretical Approach to some Modifications of Generalized Topologies 173
HEJCMAN, J.: Remarks on Dimensions of Mappings 181
HENRIKSEN, M.: On Difficulties in Embedding Lattice-Ordered Integral Domains in Lattice-Ordered Fields 183
HERRLICH, H.: A Generalization of Perfect Maps 187
HEWITT, E.: Harmonic Analysis and Topology 193
HURSCH, J. L. and VERBEEK, A.: A Class of Connected Spaces with many Ramifications 201
HUŠEK, M.: Simple Categories of Topological Spaces 203
IVANOVA, V. M. and IVANOV, A. A.: Continuous Mappings of Extensions of a Topological Space 209
IVANŠIĆ, I.: Disconnected Bounded *PL* Manifolds in Euclidean Spaces 215
JAYNE, J. E.: Topological Representations of Measurable Spaces 217
JONES, F. B.: The Utility of Empty Inverse Limits 223
JUHÁSZ, I.: Cardinal Functions on Products 229
KANNAN, V. and RAJAGOPALAN, M.: On Rigidity and Groups of Homeomorphisms . . . 231
KATĚTOV, M.: On Descriptive Classification of Functions 235
KIRKOR, A.: On Mild and Wicked Embeddings 243
KOLOMÝ, J.: Some Mapping and Fixed Point Theorems 245
KOUTNÍK, V.: On some Convergence Closures Generated by Functions 249
KRATOCHVÍL, P.: On a Convergence Property of Set Algebras 253
KRIKELIS, P. B.: On Condensation Numbers 257
KRONHEIMER, E. H.: Very Unlatticelike Ordered Spaces 259
KUČERA, L. and PULTR, A.: On a Mechanism of Choosing Morphisms in Concrete Categories 263
KUČERA, L. and PULTR, A.: The Category of Compact Hausdorff Spaces Is not Algebraic if there Are too many Measurable Cardinals 267
KURATOWSKI, K.: A General Approach to the Theory of Set-Valued Mappings 271
KUREPA, D.: Factorials and the General Continuum Hypothesis 281
MAGILL, K. D., JR.: Semigroups and Near-Rings of Continuous Functions 283
MAKAI, E., JR.: The Space of Bounded Maps into a Banach Space 289
MARDEŠIĆ, S.: A Survey of the Shape Theory of Compacta 291
MEYER, P. R.: On Total Orderings in Topology 301
MICHAEL, E.: On Two Theorems of V. V. Filippov 307
MIODUSZEWSKI, J.: On a Method which Leads to Extremally Disconnected Covers . . . 309
MITJAGIN, B.: Homotopical Structure of Linear Groups of Banach Spaces 313
MUSIELAK, J. and WASZAK, A.: A Contribution to the Theory of Modular Spaces 315
NAGATA, J.: A Survey of the Theory of Generalized Metric Spaces 321
NEGREPONTIS, S.: Ramification Systems and Spaces of Ultrafilters 333
NOVÁK, J.: On Completions of Convergence Commutative Groups 335
NYIKOS, P.: Strongly Zero-Dimensional Spaces 341
PIETSCH, A.: Ideals of Operators on Banach Spaces and Nuclear Locally Convex Spaces 345
POPPE, H.: A Compactness Criterion for Hausdorff Admissible (Jointly Continuous) Convergence Structures in Function Spaces 353

Preiss, D.: Metric Spaces in which Prohorov's Theorem Is not Valid 359
Preuss, G.: A Categorical Generalization of Completely Hausdorff Spaces 361
Pták, V.: Banach Algebras with Involution 363
Riečan, B.: On Topological Entropy 371
Rishel, T.: Nice Spaces; Nice Maps 375
Rudin, M. E.: Box Products 385
Schaerf, H. M.: Cardinalities of Bases 387
Segal, J.: On the Shape Classification of Manifold-Like Continua 389
Sennott, L. I.: Extending Point-Finite Covers 393
Simon, P.: A Note on Rudin's Example of Dowker Space 399
Skula, L.: Die Fortsetzung stetiger Homomorphismen von δ-Halbgruppen 401
Smith, J. C.: Properties of Expandable Spaces 405
Steiner, A. K.: On the Lattice of Topologies 411
Swaminathan, S.: On a Closed Range Theorem for Nonlinear Operators 417
Tall, F. D.: Some Set-Theoretic Consistency Results in Topology 419
Taylor, J. C.: The Martin Compactification in Axiomatic Potential Theory 427
Telgársky, R.: Covering Properties and Product Spaces 437
Тихомиров, В. М. и Тумаркин, Л. А.: О поперечниках Урысона n-мерной эвклидовой сферы 441
Tironi, G. and Isler, R.: On some Problems of Local Approximability in Compact Spaces 443
Trnková, V. and Reiterman, J.: When Categories of Presheaves Are Binding 447
Wattel, E.: A General Fixed Point Theorem 451
West, J. E.: Identifying Hilbert Cubes: General Methods and their Application to Hyperspaces by Schori and West 455
Wichterle, K.: Relations between $\mathfrak{B}$-Completeness and $\mathfrak{m}$-Paracompactness 463
Wong, R. Y. T.: On Homeomorphisms of ∞-Dimensional Bundles 467
Zaremba, D.: On Pseudo-Open Mappings 469
Zenor, P.: Spaces with Regular G_δ-Diagonals 471
Zervos, S. P.: Lattices and Topology 475